Proceedings of the

First Asian Workshop on Physical Modelling in Geotechnics

(ASIAFUGE 2012)

Proceedings of the

First Asian Workshop on Physical Modelling in Geotechnics

(ASIAFUGE 2012)

Indian Institute of Technology Bombay, Mumbai

November 14–16, 2012

Editors

Dr.-Ing. B.V.S. Viswanadham

Professor
Department of Civil Engineering
Indian Institute of Technology Bombay
Powai, Mumbai – 400 076
India

&

Dr. C. Gaudin

Professor
Centre for Offshore Foundation Systems
M053
The University of Western Australia
35 Stirling Highway
Crawley WA 6009
Australia

BLOOMSBURY

LONDON • NEW DELHI • NEW YORK • SYDNEY

First published, 2012

BLOOMSBURY PUBLISHING INDIA PVT. LTD.
London New Delhi New York Sydney

ISBN: 978-93-82563-38-9

Published by Bloomsbury Publishing India Pvt. Ltd.
VISHRUT Building, DDA Complex, Building No. 3, Pocket C-6 & 7
Vasant Kunj, New Delhi 110070

Printed at Anvi Composers
1 DDA Market, Block A-1B, Pachim Vihar, New Delhi 110063

Disclaimer

All the data and technical information furnished in this volume is based on the actual experimental investigations/research work carried-out by the respective investigators. The editors are not responsible for any discrepancies in the content and interpretations of results.

B.V.S. Viswanadham
C. Gaudin

Preface

With an aim of disseminating knowledge and expertise about the centrifuge based physical modelling techniques, the Technical committee TC 104 on Physical Modelling in Geotechnics of the *International Society of Soil Mechanics and Geotechnical Engineering* (ISSMGE) is organizing regional workshops first in Europe and now also in Asia. The first Asian workshop on Physical Modelling in Geotechnics (Asiafuge2012) was held in Mumbai, India in November 14-16, 2012 and was organised in association with Indian Institute of Technology Bombay, Mumbai, India, and the Indian Geotechnical Society Delhi with an emphasis on the application of centrifuge-based physical modelling for infrastructure development. Selected themes included *soft ground problems, foundations, deep excavations, slopes and embankments, earthquakes, climate change, ground improvement techniques, tunnels, offshore foundation systems, environmental geotechnics, geosynthetics and novel construction techniques in infrastructure geotechnics*.

In the last 30 years, centrifuge modelling has become a recognised powerful modelling technique to investigate a large range of geotechnical problems. Centrifuge modelling concerns the study of geotechnical processes using small-scale models subjected to acceleration fields of magnitude many times Earth's gravity. These tests have proven to be particularly valuable in revealing mechanisms of deformation and collapse and in providing data for validation of numerical analyses. A 4.5 m radius large beam geotechnical centrifuge, with a payload capacity of 2.5 t has been established at IIT Bombay in 2002. It is being used for carrying out research studies on a number of challenging problems, particularly those which are unique to India. India is making rapid strides towards becoming a global economic power. This has demanded massive investments in the infrastructure development such as metros, harbours, airports flyovers, bridges, etc. Most of the problems in geotechnical engineering are region specific and some times unique. Geotechnical centrifuge studies can contribute significantly in providing solutions to these complex problems and have lot of societal relevance.

This workshop is an opportunity for physical modelling experts from Asia, but also from Europe, the USA and Australia to showcase their research and share their expertise with the larger geotechnical community. These proceedings offer a unique compilation of high quality articles related to state of the art research about a wide range of geotechnical issues and notably about issues associated with infrastructure developments.

B.V.S. Viswanadham
C. Gaudin
Editors, Asiafuge 2012

Acknowledgements

The workshop organisers would like to take this opportunity to thank the authors of the manuscripts included in this and for their invaluable contributions. We would also like to express our gratitude to members of advisory committee, who were instrumental in making this workshop successful by their support and guidance.

It is our desirc that this compilation of articles provide an opportunity for the exchange of views among the academics, professional engineers and policy makers. It is our belief that all sectors will benefit through the sharing of knowledge and experience related to the understanding of behaviour of geotechnical structures through physical model testing, especially using centrifuge-based physical modelling. It is hoped that the information contained in these proceedings will contribute to a better understanding of physical modelling in geotechnics and the role it can play in assisting notably the infrastructure development in Asia and worldwide..

The editors would also like to acknowledge the support provided by Prof. Devang V. Khakhar, Director, Indian Institute of Technology Bombay, India and Prof. K.V. Krishna Rao, Head, Department of Civil Engineering, Indian Institute of Technology Bombay, India. Thanks are also due to Prof. D. Choudhury, Prof. A. Juneja and Prof. D.S. Murthy, who contributed immensely in achieving endeavours of this event.

Finally, the organisers would like to express their gratitude to students and staff, who have helped and put in number of hours in bringing this event to happen.

B.V.S. Viswanadham
C. Gaudin

Workshop Sponsors

DIAMOND SPONSOR

L&T GeoStructure

http://www.lntecc.com/home.htm/

GOLD SPONSORS

Board of Research in Nuclear Sciences (DAE) http://www.barc.gov.in/brns/

ओएनजीसी ONGC

http://www.ongcindia.com/ieot.asp/

एनटीपीसी NTPC Power of India

http://www.ntpc.co.in/

SILVER SPONSOR

DASSAULT SYSTEMES

http://www.3ds.com/SIMULIA/

BRONZE SPONSORS

V.T. Ganpule Ground Engineering Consultants

URGE — http://www.urground.com/

AFCONS INFRASTRUCTURE LIMITED — http://www.afcons.com/

http://www.lagasys.com/

NATIONAL INSTITUTE OF OCEAN TECHNOLOGY NIOT CHENNAI — http://www.niot.res.in/

MACCAFERRI India — http://www.maccaferri-india.com/

INDIAN GEOTECHNICAL SOCIETY — http://www.igs.org.in/

Contents

Examples of the Influence of Centrifuge Modelling on Offshore Geotechnical Engineering

C. Gaudin

Centre for Offshore Foundation Systems, The University of Western Australia, Australia
E-mail: christophe.gaudin@uwa.edu.au

ABSTRACT

Offshore geotechnical engineering is characterized by unusual soil and loading conditions, a continuously moving frontier resulting in the requirement for innovative geotechnical solutions, and the very high cost associated with in-situ soil characterization and field testing in remote and deep oceans. As a consequence, offshore engineering practices has probably benefitted more from centrifuge modelling than other domains of geotechnical engineering, since the first offshore projects performed in 1973. After a brief overview of the role, contribution and advantages of centrifuge modelling for offshore geotechnics, the paper illustrates the influence centrifuge modelling has had in recent years in improving the understanding of soil structure interaction and in the development of guidelines and recommendations in two particular areas via five examples; spudcan foundation and pipe-soil interaction.

1. INTRODUCTION

1.1 A Brief History of the Impact of Centrifuge Modelling on Offshore Geotechnics

Centrifuge modelling was applied for the first time for offshore geotechnics problems in 1973 at Manchester University. The behaviour and performance of gravity platforms to be established in the Gulf of Mexico were investigated in a large beam centrifuge (Rowe & Craig, 1981). The work encompassed a wide range of soil and loading conditions and provided pivotal insights into the failure mechanism taking place (Craig & Al-Saoudi, 1981). In the early days of the centrifuge, it was understood and acknowledged that centrifuge modelling could significantly contribute to design when novel conditions, or those not fully understood, prevailed (Craig, 1984).

Since the pioneering work in 1973 centrifuge research has expanded worldwide, initially under the impulsion of Prof. Schofield in Europe and Prof. Kimura in Japan. Initially, research focused first on phenomenological and site specific studies before progressively developing towards more general studies, including the observation of failure mechanisms and the understanding of soil-structure interaction. Eventually it aided the development of predictive design methods.

As geotechnical centrifuge techniques developed technically and scientifically, and with an increased need for performance data and understanding of offshore soil-structure interaction, the acceptance of the offshore community to the benefits of the centrifuge grew significantly. An important step in this process was the keynote address given by Professor Murff to the wider offshore community, at the Offshore Technology Conference (Murff, 1996). It advocated the benefits of centrifuge methods by providing key examples, notably related to suction caissons, drag anchors and jack-up foundations. The latter example mainly discussed the research on jack-up foundations sponsored by Exxon and performed in the Cambridge University centrifuge, and also at Oxford University (final report: Noble Denton &

Associates, 1987). The outcomes of this research were integrated in the original SNAME guidelines and are still influential on the draft ISO (2009) code of practice. Since then, the geotechnical offshore industry has continued to benefit from centrifuge methods, and both academic and industry users have developed a strong expertise in analysing centrifuge method outcomes, incorporating them into the development of new design tools, and developing deeper understanding in offshore structure-soil interaction.

1.2 Role and Contribution of Centrifuge Modelling

Offshore geotechnical engineering is characterized by unusual soil and loading conditions, a continuously moving frontier resulting in the requirement for innovative geotechnical solutions, and the very high cost associated with in-situ soil characterization and field testing in remote and deep oceans. As a consequence, offshore engineering practice has probably benefitted more from centrifuge modelling than other domains of geotechnical engineering. The role and contribution of centrifuge modelling to the design of offshore structures and the understanding of offshore structures-soil interaction has been addressed successively by Murff (1996), and more recently by Gaudin et al. (2010). A short summary of the roles of centrifuge modelling in offshore geotechnics practice is presented here, with a particular emphasis on spudcan installation and extraction and pipe soil interaction:

1. Provide performance data to calibrate analytical or numerical models. Centrifuge modelling provides homogeneous and well characterized soil conditions, known boundary conditions, accurate measurements of parameters and repeatable testing conditions, offering reliable performance data for a given idealized problem which can be used to calibrate analytical and numerical models. Pipeline lateral buckling models have been established for a fairly narrow range of soil condition. The use of the models for different soil conditions requires calibration data that can be easily and economically provided by centrifuge modelling.
2. Providing qualitative insights into soil-structure interaction and mechanisms. This aspect is particularly important when novel concepts or unusual conditions are encountered. Understanding the structure behaviour, observing the failure mechanism taking place, is a pivotal step into the development of sound design methodology. This is particularly relevant for spudcan installation and pipe-soil interaction were large deformations and specific soil flow dominates the soil-structure interaction.
3. Validate the structure design for a particular site or a specific design approach. By allowing the use of natural soils, sampled in-situ, by applying complex loading sequences directly relevant to design and by using a models replicating precisely prototypes, centrifuge modelling offers a valuable tool to validate or justify specific offshore structures. Pipeline design is still at an early development, where normalised recommendations have not been formulated yet. The totality of the pipeline development in the North West Shelf of Australia is supported by site-specific centrifuge data.
4. Investigating the feasibility of developing new foundation concept. The very good control of the testing conditions, the possibility of measuring a significant number of parameters and the relatively low cost of centrifuge testing compared to field testing make the centrifuge an ideal tool to develop new concepts and investigate the feasibility of particular foundations. New spudcan design, featuring skirts, jetting system and preloading systems are currently investigated using centrifuge modelling.
5. Characterising in-situ soils. The improvement of soils reconstitution methods allows the replication of complex soil stratigraphy using natural soils exhibiting the same properties as found in the field. The centrifuge offers the possibility of performing a large number of tests at a reduced cost, using characterization tools identical to the ones used in-situ.

1.3 Advantages of Centrifuge Methods to Investigate Spudcan and Pipeline Performance

The contributions listed in the previous section are the direct results of the advantages centrifuge methods provides when investigating offshore geotechnical issues. These advantages are intrinsic and are not limited to offshore geotechnics, but they are augmented in offshore geotechnical engineering, where guidelines and design rules play a lesser role than onshore, and where engineers' judgement is paramount in establishing the most technically sound and economically efficient design. Basing judgment on reliable and quality data from a range of modelling methods may make a significant difference between a satisfactory and an optimal design. While there are several examples to illustrate the advantages of centrifuge methods, including notably suction caissons and other anchoring systems, the paper focuses on two of the most recent examples.

Some particular aspects of these advantages, when associated to spudcan installation and extraction and pipe-soil interaction are:

- Accelerated time frame. Centrifuge modelling and testing require a limited volume of soil, accelerating sample preparation. For soft soil, the process may be further accelerated by in-flight self-weight consolidation. Similarly, testing sequences are considerably shortened still ensuring the correct drainage conditions. This allows data to be collected in a short time frame and at cost orders of magnitude lower than those associated with field testing. More particularly, for pipe testing, this enables the investigation of the pipe-soil interaction over a significant number of cycles, throughout the whole operating life of the field. For spudcan testing, series of six to ten tests in the centrifuge can be undertaken within 2 to 3 weeks, including soil sample preparation. With an ability to cover a wide variety of loading and/or soil conditions, the centrifuge also provides the ability to conduct parametric studies
- Accurate loading sequences. New motion control techniques permit users to replicate accurately complex loading sequences, including cyclic vertical and/or horizontal motion under either load or displacement control. This is particularly relevant to replicate the complex motion of the pipe at the touchdown zone during laying, the large deformations resulting from lateral buckling, the combined loading conditions generated by environmental loading during jack-up operation, or during re-installation events near existing footprints.
- Use of natural soils. By requiring a limited volume of soil, it is possible to use in-situ soil with reasonable supply cost. The use of natural soils for model tests is maybe more important for pipelines than for other geotechnical structures, because of the heavy remoulding and the large deformations experienced by the soil. This may trigger specific behaviour which would not be captured by common artificial laboratory soils.
- Accurate seabed characterization. The good control of the soil reconstitution process results in a homogeneous sample with known boundary conditions and stress history. By using standard soil characterisation tools in controlled conditions (such as cone penetrometer tests, shear vane tests and more recently T-bar penetrometer tests for soft soils (see Randolph, 2004)), and post-testing soil classification methods (such as particle size distribution or Atterberg limits performed on core samples taken from centrifuge samples), it is possible to determine the soil characteristics accurately. Since boundary conditions, material properties and soil stress history are well controlled in centrifuge models, quantitative results of the penetration resistance and patterns of soil flow may be used to validate numerical or analytical models rigorously.
- Enhanced instrumentation. New insights into spudcan behaviour have been derived by improved instrumentation developed for centrifuge modelling. For instance, by using visual image acquisition systems (such

as digital camera) and associated image processing systems (such as Particle Image Velocimetry (PIV), and photogrammetry, described for application to geotechnics by White et al., 2003), soil flow mechanisms evolving with spudcan penetration have provided the basis for a number of new predictive models. Accurate and detailed miniature pore pressure and total pressure measurements have also provided insights into mechanisms generating excess or negative pore pressures at the spudcan surface. By using similar image acquisition systems and pore pressure measurements, in addition to the measurements of load and displacements, one gains access to particular features of the pipe-soil interaction, such as the development of lateral berms during buckling, the creation of a trench during dynamic laying and the drainage conditions around the pipe during the pipe motion.

- Case-specific study. Centrifuge testing can be used to provide insight in some particular issues related to spudcan and pipeline design. This includes particular storm loading conditions or geometric conditions, such as sloped seabed for spudcan or scarp crossing for pipeline.

1.4 Centrifuge Modelling for Spudcan Performance and Pipe-Soil Interaction

Buckling and axial walking (due to changes in internal pressure and temperature in the pipeline) have only emerged recently as a major research topic. Significant advances have been made over the last 5 years, using data from 1g model tests (performed notably at Cambridge University, Cheuk et al., 2007, and at the Norwegian Geotechnical Institute (NGI), Dendani & Jaeck, 2007), in-situ tests (notably using the SMARTPIPE device developed by Fugro and BP, Jacob & Looijen, 2008, White et al. 2010) and also centrifuge tests. While each method has its own advantages and disadvantages (see Hill & Jacob, 2008 for a comparison of each method for pipelines), centrifuge tests have certainly boosted knowledge of pipeline-soil interaction and generated significant breakthroughs.

In contrast, all aspects of spudcan development and design has greatly benefited from centrifuge modelling. This is notably evident for the research establishing spudcan yield surface approaches on the Cambridge centrifuge (see Dean et al., 1993, Wong et al., 1993 amongst others), the bearing capacity of spudcans in silica and calcareous sands (e.g. Finnie & Randolph, 1994; Dean et al., 1993; Teh et al., 2006), the interaction between an installing spudcan and the nearby piles of a fixed jacket platform (e.g. Siciliano et al., 1990; Leung et al., 2006; 2008; Xie et al., 2006; 2010), the contributions of Ng & Lee (2002) to predicting spudcan settlements under cyclic loading, the problematic of spducan installation near existing footprint (Gaudin et al., 2012) and the combined VHM capacity of spudcan via the development of force-resultant models calibrated with centrifuge data (see Bienen et al., 2007 among many others).

The paper aims at illustrating some of these advances and breakthrough via five typical examples presented in Table 1 and developed in details in the following sections.

2. SPUDCAN INSTALLATION AND EXTRACTION

In the offshore oil and gas industry most drilling operations in water depths up to around 120 m are performed from self-elevating mobile jack-up units. These offshore platforms typically have a buoyant triangular hull, three independent truss-work legs, and foundations, commonly known as 'spudcans', that approximate large inverted cones. Roughly circular in plan, spudcans typically have a shallow conical underside, some with a sharp protruding spigot and can be in excess of 20 m in diameter, with shapes varying with manufacturer and rig.

Jack-up rigs are self-installing. They are towed to site with their legs elevated out of the water. On location, their legs are lowered to rest on the seabed. Once the jack-up has been positioned the spudcans are jacked until adequate bearing capacity exists for the hull to be lifted clear of

Table 1: Examples of Contribution of Centrifuge Modelling in Offshore Geotechnics

Example	*Problem Addressed*	*Centrifuge Technology Used*	*Insights / Outcomes*	*Application in Industry*	*Future Requirements*
1	Installation in clays: • Deep penetration • Back flow	• PIV and photogrammetry • Accurate load-displacement measurements	• Definition of transition from shallow to deep mechanisms • Method for predicting the depth at which flow around mechanism occurs • New bearing capacity formulae that account for deep penetration, as well as rate and sensitivity	• New design charts and formula for predicting preferential flow mechanism, as implemented in draft ISO (2009)	• Tests with varying soil sensitivity • Application into soils with intermediate drainage conditions • Assessing capacity increase and settlement due to dissipation of pore-pressures during and after preloading
2	Spudcan punch-through: Stiff-over-soft clay and-over-clay	• PIV and photogrammetry • Accurate load-displacement measurements • Miniature ball penetrometers	• Definition of punch-through mechanisms • Underpinning of analytical prediction methods	• New methods being trialled in industry, such as in InSafeJIP (Osborne et al., 2009)	• Testing in more complex multi-layered soils • Direct predictive methods based on in-situ penetrometer data • Application in unconventional soils, such as cemented hard layers
3	Spudcan extraction	• PIV and photogrammetry • Pore-pressure and total stress transducers • Water jetting using a syringe pump	• Mechanisms governing extraction, incl. additional capacity due to consolidation during long operations • New methodology for predicting extraction resistance whilst using water jetting	• InSafeJIP (2009, 2010)	• Extend current methodology for deep embedment • Investigating performance of top jetting • Investigating performance of cyclic extraction
4	Pipe dynamic laying	• Accurate load-displacement measurements	• Increase embedment compared to static laying due to soil remoulding and water entrainment • Dominance of lateral sweeping amplitude over cyclic vertical loading in embedment development		• Integrate water entrainment in embedment prediction • Investigate SCR soil interaction at the touchdown zone
5	Pipe thermal buckling	• PIV and photogrammetry • Accurate load-displacement measurements • Pore pressure measurements	• Mechanism governing lateral resistance including berm formation • Development of friction factors including pipe soil friction and berm resistance as a function of the number of cycles	• Friction factor formulations and embedment prediction (SAFEBUCK JIP)	• Extend friction factor formulation to drained and partially drained conditions • Develop VH yield envelope to predict breakout friction factors

the water. The spudcan foundations are then preloaded by pumping sea-water into ballast tanks in the hull. This 'proof tests' the foundations by exposing them to a larger vertical load than would be expected during service (usually by a factor of 1.3 to 2). The ballast tanks are emptied before drilling operations begin.

During the preloading process, challenges faced by the geotechnical engineer include accurate prediction of the penetration depth and ensuring the stability of the jack-up during penetration. Instabilities can notably occur due a rapid leg penetration during a 'punch-through' failure. In the latter the spudcan temporarily loses vertical capacity as it punches through a layer of stronger soil into underlaying softer conditions.

After the jack-up has been installed, it typically operates at the site for as little as days or as long as a number of years. When the jack-up is to be finally moved from the site the spudcan footings must be removed from the ground. Deep penetrations can make this operation difficult, with the time to pull the spudcans clear being reported to exceed one month in extreme circumstances. There is an industry need for better understanding of the extraction mechanisms and the development of a more efficient extraction procedure.

The following three examples present insights and outcomes provided by centrifuge modelling of spudcan installation and extraction on soft clay. On a modern jack-up the maximum vertical leg load during installation can exceed 140 MN and produce average vertical bearing pressures on the spudcan in excess of 400 kPa. In soft soils, this fully embeds the spudcan and can even cause penetration of over 30 m resulting in large deformation and significant soil flow. During extraction, undrained behaviour is expected resulting in suction developed at the spudcan invert increasing significantly the extraction resistance.

2.1 Example 1: Prediction of Back-Flow and Bearing Capacity in Homogenous Clay

In model tests in clay at 1g, the ratio of shear strength, su, to the effective overburden stress, σ'_{v0} is higher than for an offshore spudcan. However, producing correct stresses due to soil self-weight in experimental testing is particularly important for the continuous penetration of spudcans, where the soil flow mechanisms evolve with penetration depth and are affected by the strength ratio, s_u/σ'_{v0}. This is because the ratio directly controls the triggering of soil to flow around the footing from underneath to above (defined as backflow). This ratio can be correctly simulated in model tests in the centrifuge.

Spudcan penetration mechanisms were initially investigated in the centrifuge by installing dry spaghetti markers vertically across the centreline of the foundation in the undisturbed clay bed (Craig & Chua, 1990, 1991), or by inserting lead threads through a soil sample prior to testing (Murff, 1996). After completion of the test, the soil sample was dissected along the centreline, permitting the final deformation pattern to be observed. This only allowed the final mechanism to be considered and did not provide information on the ongoing mechanism changes, including when soil backflow occurred. Predictions of this were attempted by Springman & Schofield (1998). In the centrifuge, they used a mini video camera fixed to the model jack-up platform leg to capture the clay infill into the lattice leg. Although this provided good images of the surface soil deformation, mechanisms occurring within the soil could not be revealed.

By utilising new visualisation techniques for capturing soil flow mechanisms in the centrifuge, Hossain et al. (2005) provided a breakthrough in understanding the deep penetration mechanisms of a spudcan into soft clay. The system, which combines digital still photography, PIV and close range photogrammetry (GeoPIV, White et al., 2003), allows accurate resolution of the flow pattern around a 'half-object' penetrated adjacent to a transparent window (necessary due to the opacity of natural soils). The soil was confined within a purpose designed strongbox with a plexiglass window to allow observation of the soil deformations, with the box mounted within the drum centrifuge channel. Half-spudcan penetration tests were conducted at elevated gravity (50~200g) tight against the window of the strongbox. Images were captured continuously by a high resolution digital still camera sitting at right angles on a cradle within the channel. The experimental arrangement is shown in Figure 1.

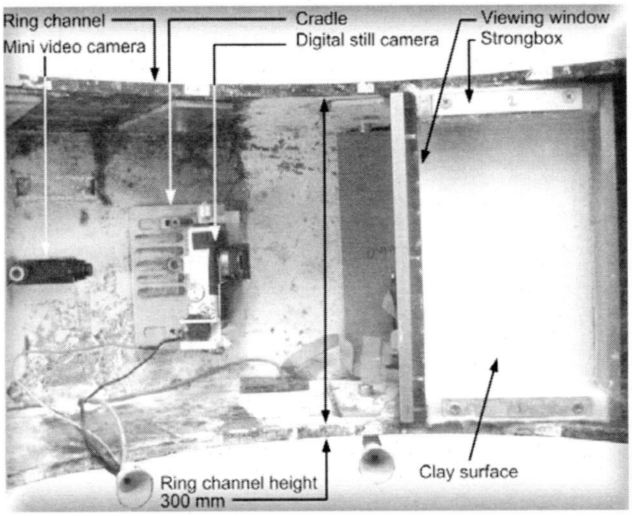

FIGURE 1: Set-up within the drum centrifuge channel for a half-spudcan test (after Hossain and Randolph, 2010)

In single layer clay, the soil flow patterns observed from centrifuge model tests and continuous penetration finite element analyses revealed three distinct mechanisms of soil flow around the advancing spudcan, as presented in Figure 2 (Hossain et al., 2005, 2006). At a certain stage of penetration, soil backflow is initiated and, in contrast to the recommendation in the current offshore design guidelines (SNAME, 2008), Figure 2 shows that this occurs not because of instability of the open cavity, but because of a preferential flow mechanism of soil from below the spudcan to above it. A new design chart was proposed (see Figure 3) along with a robust formula to estimate the limiting cavity depth, H, above the penetrating spudcan as:

$$\frac{H}{D} = \left(\frac{s_{uH}}{\gamma' D}\right)^{0.55} - \frac{1}{4}\left(\frac{s_{uH}}{\gamma' D}\right) \qquad ...(1)$$

where D is the spudcan diameter, γ' is the submerged unit weight of the clay and suH is the undrained shear strength at the backflow depth, H. Equation 1 has already replaced expressions that were based on hole collapse in offshore design guidelines (such as in the draft ISO, 2009).

It is significant as it modifies the bearing capacity calculation of a penetrating spudcan. Any soil backflow into the cavity created by spudcan penetration affects the bearing response in two ways: (a) by (partially) negating the surcharge contribution, γ'_d, and (b) by increasing the shear resistance (and hence N_c) as the failure mechanism now must pass through the backfilled soil. Since backflow provides a seal over the top of the spudcan, the above relationship also provides guidance on conditions where transient suctions may be sustainable beneath the spudcan, with a consequential increase in uplift resistance and

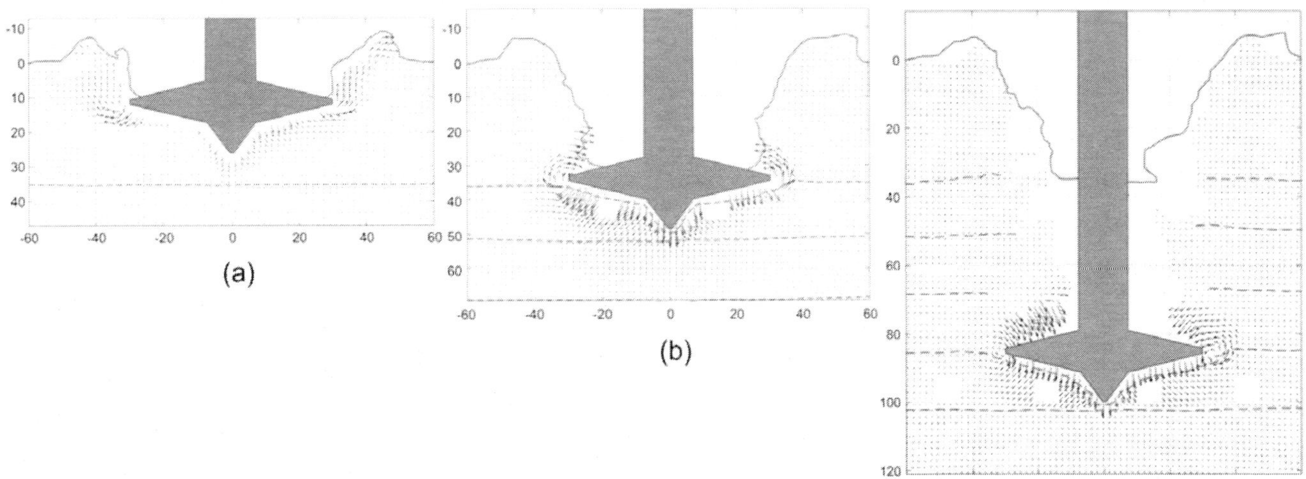

FIGURE 2: Soil flow mechanisms of spudcan penetration in single layer clay: (a) heave mechanism, (b) onset of backflow mechanism, and (c) deep flow mechanism (after Hossain et al., 2006)

moment capacity at low vertical loads. These were also shown to exist in centrifuge experiments of deeply embedded spudcans by Cassidy et al. (2004a), amongst others.

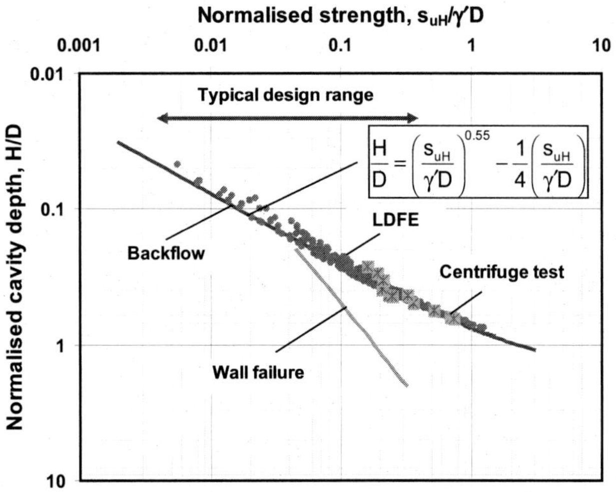

FIGURE 3: New design chart for estimating cavity depth after spudcan installation in clay (after Hossain et al., 2006)

2.2 Example 2: Installation in Strong-Over-Soft Clays

Jack-up installation and preloading in stratified deposits, where a strong layer overlays weaker soil, has always been a challenge to offshore engineers due to the potential for catastrophic 'punch-through' failure. Centrifuge testing has helped understand the problem, as the higher stress due to enhanced soil self-weight at elevated gravity has allowed researchers to more easily reconstitute stratified soil deposits. In 1g tests on the laboratory floor, it is arduous to achieve proper bonding (similar to field sediments) at the interface between two layers. However, this can be established more easily by simply testing under elevated gravity (Hossain & Randolph, 2010a), or by allowing for significant consolidation prior to testing (Teh et al., 2008; 2010; Lee, 2009). Homogeneous sand layers can be laid through sand raining techniques, including those developed for spraying sand in-flight (see Lee, 2009, for an example technique in a drum centrifuge). The development of such capabilities has focused research onto understanding spudcan punch-through mechanisms (Hossain & Randolph, 2010a; Teh et al., 2008; 2010; Lee, 2009; Lee et al., 2009).

In stiff-over-soft clays, the failure mechanisms reported by Kim (1978) from 1g model tests on a surface circular footing are inconsistent compared to the recent centrifuge visualisation on penetrating spudcan. Once again, half-spudcan penetration tests (at 100~200g) permitted the soil flow mechanisms to be captured with a digital camera (Hossain & Randolph, 2010a). The experimental evidence provided failure patterns at different spudcan penetration depths. The quantified soil flow vectors showed that punch-through was associated with the formation of distinct shear planes in the top layer (see Figure 4), and consequently a soil plug with the shape of an inverted truncated cone was forced down into the underlying soft layer.

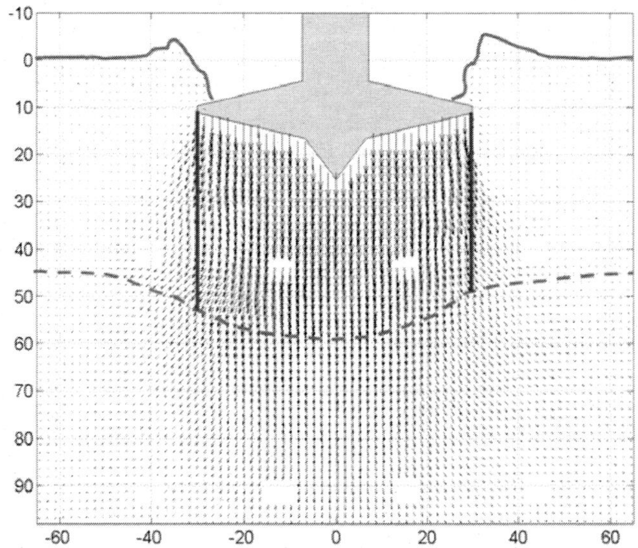

FIGURE 4: Spudcan punch-through on stiff-over-soft clay (from PIV analysis: axes in mm and in model scale) (after Hossain & Randolph, 2010a)

For a spudcan penetrating sand-over-clay the situation is even more complex. Spudcan behaviour in the sand layer is governed by confining stress, q_{clay}/q_{sand} ratio and relative thickness of the sand layer (t/D; where t is the thickness of the strong layer) and they are a function of stress level and operative friction and dilation angles. These vary with stress level and footing diameter, even on sand layers with similar relative density (Lee et al., 2009). Revealing this has taken centrifuge developments over a couple of decades. Initially, Craig & Chua (1990; 1991) depicted post-test snapshots of sand-over-clay punch-through behaviour through use of spaghetti in the sample. Following this Okamura et al. (1997) reported mechanisms for surface flat footings by employing a radiography technique. More recently and again employing the image-based analysis technique of half-spudcan penetration tests against a transparent window, the detailed progressive failure mechanisms at a sand-over-clay site were illustrated by Teh et al. (2008). The mechanism at the time of punch-through is shown in Figure 5. The key finding is the dilatancy characteristics play a key role in the form of the projected area beneath the advancing spudcan. They are suppressed with the increase of t/D and the strength su of the lower clay layer.

In order to develop a calculation method, the image results discussed above require augmentation with full load displacement profiles and numerical finite-element analysis. In both stiff-over-soft clay and sand-over-clay sites, the load-penetration response was also measured experimentally through full-spudcan penetration tests. In stiff-over-soft clay, centrifuge test data were also used to validate large deformation finite element (LDFE) analyses, before undertaking parametric analyses (Hossain & Randolph, 2010b).

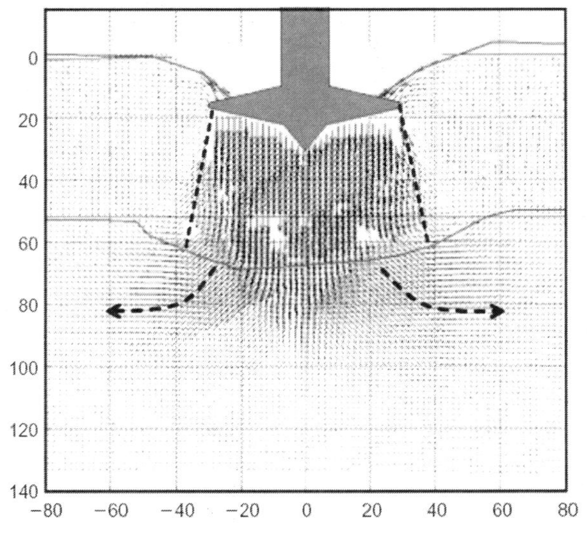

FIGURE 5: Observed mechanism of spudcan punch-through in sand-over-clay soils (from PIV analysis: axes in mm and in model scale) (after Teh et al., 2008)

The combination of half-spudcan visual evidence, full spudcan load-displacement profiles and LDFE results provided the basis for the development of a more rational mechanism-based design approach reported by Hossain & Randolph (2009). In sand-over-clay, Lee et al. (2009) and Teh et al. (2009) have reported new analytical design approaches accounting for the failure patterns, the stress level and dilatant response of the sand. All these approaches have seen evaluation against field case studies in the recently completed InSafeJIP (Osborne et al., 2008; 2009; InSafeJIP, 2009; 2010).

2.3 Example 3: Extraction with Water Jetting

Spudcan extraction often proves difficult and time-consuming in the field, especially in soft soils where deep penetrations are experienced. As no guidelines are currently available for spudcan extraction, various strategies to free the leg are usually employed through a trial and error process (InSafeJIP, 2009). Therefore, the key contributing factors to a successful extraction have remained uncertain to the jack-up industry. Aspects that increase the level of complexity in understanding spudcan extraction in the field include highly stratified soil conditions, difficulties in quantifying the changes in soil strength due to disturbance and reconsolidation, sometimes unquantifiable wave excitations, as well as difficulties in measuring the loading exerted during the extraction process on the jack-up. In order to develop sound predictive methods for the expected resistance and to devise successful extraction procedures, the governing soil failure mechanisms must first be established.

As detailed in the following example, centrifuge modelling has recently provided evidence of these extraction mechanisms and analytical models have been developed based on these results.

As was shown in examples 1 and 2, soil failure mechanisms can be visualised by testing half models against a transparent window in combination with PIV and photogrammetry. Purwana et al. (2006; 2008; 2009) employed this technique to gain insight into soil failure mechanisms during spudcan extraction, an example of which is shown in Figure 6.

In the left half of the figure the photo taken during the extraction test is presented, showing the spudcan and the clay to which coloured flock was added to allow for better visualisation of the soil movement. The right half of the figure depicts the corresponding vectors of soil displacement as obtained from PIV analysis. The results have proven valuable in determining the soil displacement pattern and understanding the mobilisation of soil resistance during spudcan extraction. Importantly, the effect on extraction of soil remoulding and consolidation during installation, preloading and holding of self-weight during operation could all be modelled in a timely manner in the centrifuge.

The centrifuge data can also be used as a basis for the development of predictive methods. Such a model for spudcan extraction resistance was proposed by Purwana et al. (2009). It was developed based on centrifuge extraction data from spudcan embedments of up to 1.5 diameters, and utilised experimental recordings of resistance at both the spudcan top and base, as well as pore pressures. A similar formulation for calculating extraction resistance was adopted in the InSafeJIP (2010) guidelines, also calibrated against centrifuge data.

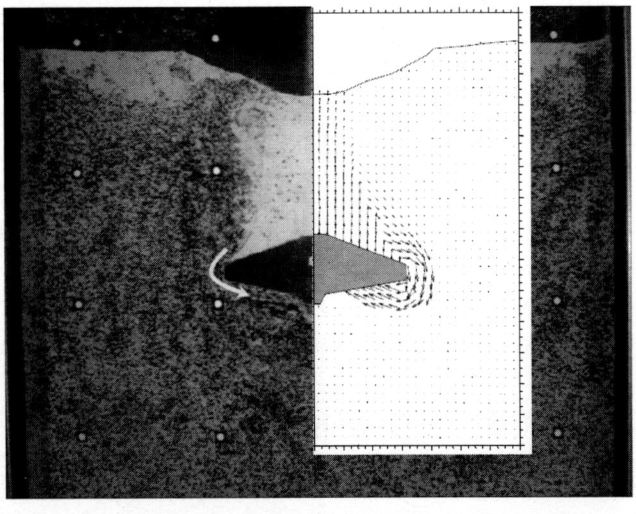

Figure 6: Visualisation of soil failure mechanism during spudcan extraction (after Purwana et al., 2008)

When insufficient uplift load is available through jack-up hull buoyancy, water jetting may ease spudcan extraction. Jetting with nozzles located at the spudcan top face aims at reducing resistance to extraction through fracturing and remoulding the soil. Jetting with nozzles at the spudcan invert (i.e. the base of the spudcan), on the other hand, aims at decreasing the negative pore pressures mobilised by the uplifting spudcan. This can extend to creating positive excess pore pressures at the spudcan invert, applying additional active uplift force.

Insight into this has been provided by a series of centrifuge tests, investigating various parameters including the extraction rate, the jetting flow rate and the jetting pressure (Bienen et al., 2009; Gaudin et al., 2011). A model spudcan with the ability to jet water at different flow rates during a centrifuge test was used, as shown in Figure 7. The model represents a 17.11 m diameter spudcan currently being used offshore. The spudcan invert featured three concentric circles of twelve jetting nozzles each. Only one nozzle ring was used at a time, the other two were blocked. The prototype spudcan jetting nozzles are 38 mm in diameter. As direct scaling of the nozzle diameter for testing at 200g was technically not feasible, the model jetting nozzles were 2 mm in diameter initially. This was reduced to 0.5 mm after the first testing series. Dimensional analysis, as developed in Gaudin et al. (2011), was used to determine the prototype flow rate according to the model pump flow rate, accounting notably for the nozzle diameter. A semi-circular outlet guard above each nozzle redirected the jetting flow along the bottom face of the spudcan, which is consistent with offshore practice.

Results demonstrated that the jetting efficiency relates to the ratio of the jetting volume flow to the volume of the void theoretically created at the invert due to uplift of the spudcan (defined as the filling ratio). The experimental data suggest optimum jetting performance not to coincide with vented extraction (i.e. a filling ratio of 1), but to a mechanism where localised flow-around the spudcan edge still occurs in addition to the uplift mechanism of the soil above the spudcan (Bienen et al. 2009; Gaudin et al. 2011). Jetting flow appears to be a dominant factor over jetting pressure, in contrast to general belief.

From the understanding of mechanism gained from the centrifuge tests, a conceptual framework for spudcan extraction with water jetting was established based on (i) the pullout force resulting from the buoyancy of the jack-up hull, and (ii) the filling ratio f. A relationship between the ratio of applied extraction force (Qdirect) to the expected extraction resistance without jetting (Qult) and the filling ratio f was developed (Bienen et al. 2009; Gaudin et al. 2011), which is shown in Figure 8 (centrifuge test results are represented by the open circles). The filling ratio can be related to the required parameter in the field, the jetting flow rate.

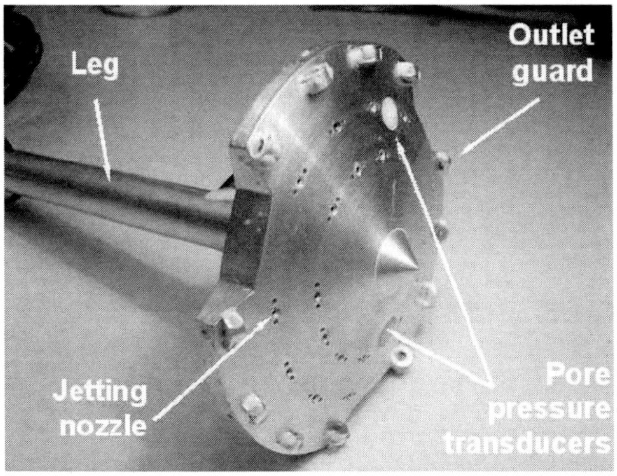

(a) Spudcan base with jetting nozzles and deflectors

(b) Water jetting in model spudcan

FIGURE 7: Model spudcan for extraction tests with water jetting (after Gaudin et al., 2011)

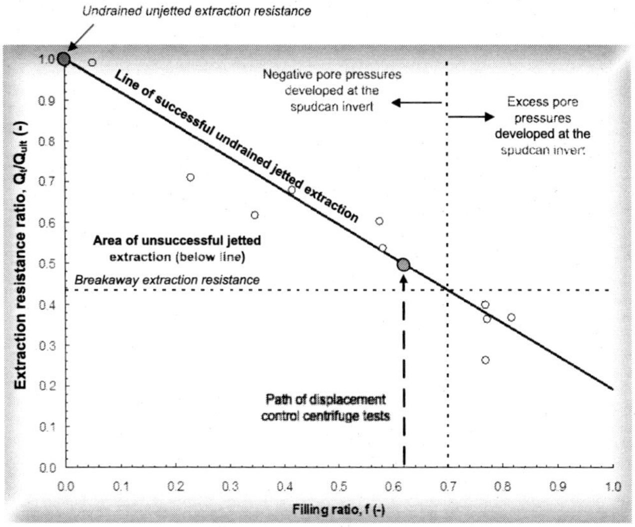

FIGURE 8: Conceptual chart of jetting extraction efficiency. The chart indicates, for a given extraction load, the required flow rate for a successful jetted extraction (after Gaudin et al., 2011)

3. PIPE-SOIL INTERACTION

Offshore pipelines are laid on the seabed and used as flow lines or trunk lines tied back to shore. They are becoming more and more common as the oil and gas extraction is taking place in deeper and more remote areas. Geotechnical design procedures for offshore pipeline and risers have not yet reached the maturity exhibited by design procedures for piles, foundations and anchoring systems. Reasons are multiple and relate mainly to the unknown geometry of the problem (the final embedment of the pipe is uncertain), the uncertainties of the soil conditions (both difficult to assess at very shallow depth and significantly affected by the installation process) and the very large deformations (and associated post-failure behaviour) the pipeline may experience (notably during lateral buckling).

The following two examples aims to illustrate some of the insight and breakthrough, obtained from both centrifuge testing performed to gain insight into specific aspects of pipeline interaction, and from centrifuge modelling performed to assist in the design of specific pipeline projects. It is not a state-of-the-art report of pipeline behaviour and design. Such reports have been presented by Cathie et al. (2005) and White & Cathie (2010).

3.1 Example 4: Pipe Dynamic Laying

The knowledge of the pipe embedment is pivotal for subsequent on-bottom stability and lateral buckling design. The effect of dynamic laying on pipe embedment has been first observed by Lund (2000) who concluded on the necessity to account for pipeline laying history in subsequent on-bottom stability design, as it results in excessive embedment compared to a static laying process.

During the lay process, an element of pipe moves through the touchdown zone, from an initial contact with the seabed to a stationary position, where the pipe weight is supported by an equal up-wards seabed reaction force. The dynamic behaviour of the pipe through this process is complex and difficult to replicate. It is a function of the following parameters:

- The profile of stress concentration in the touch-down zone due to the catenary shape, resulting in a vertical load on the soil varying along the pipe and higher, at some specific location, than the as-laid pipeline weight. It can be estimated, based on a structural analysis of the hanging pipeline using the planned lay tension and hang-off angle.
- The horizontal and vertical oscillation due to the motion of the laying vessel, resulting in a sweeping and damping of the soil at the touchdown zone. It can be estimated from the analysis of the motion of the vessel under for the wave motion considered.
- The number of oscillations during the entire lay process. It can be assessed, based on the estimated pipeline laying rate, the touchdown zone length and the wave-induced oscillation frequency.

Figure 9 presents a typical pipeline setup used to model dynamic laying and lateral buckling. The pipe is modelled as a short section (with an aspect ratio of at least 6 to prevent any end effect) and may be sandblasted to achieve a particular roughness. It is connected to a VHM loading arm, use to measure the horizontal loads, via shear load cell, insensitive to bending moment. The purpose of this load cell is to provide a very accurate reading of the vertical load applied (permitting a resolution of 1 N model) which is essential for pipelines, which have a self-weight orders of magnitude lower than typical geotechnical structures. Additional instrumentation may include pore pressure transducers at the pipe invert, as shown in Figure 9.

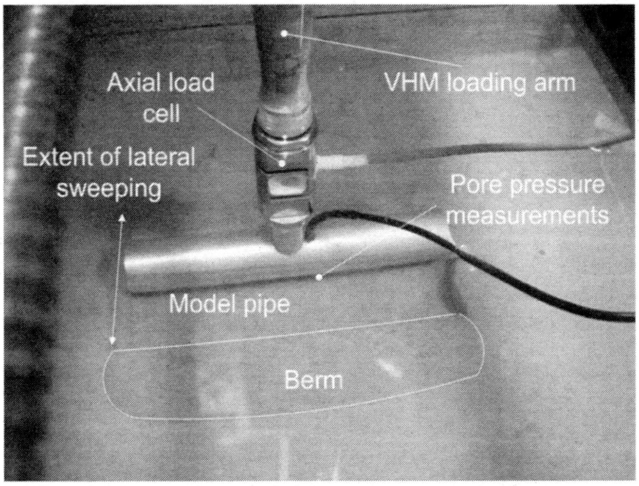

FIGURE 9: Typical pipeline model setup

A purposely developed motion control system (De Catania et al., 2010) is used to (i) apply a targeted vertical load depending on the stress concentration, via a feedback loop on the axial load cell and (ii) apply lateral oscillation of varying amplitude, linked via a second feedback loop to the pipe embedment. The following example presents centrifuge modelling performed recently to determine the final embedment of a flowline pipeline offshore Australia (Gaudin & White, 2009). In-situ soil was used and the dynamic motion applied replicated faithfully the one determined from the wave and vessel motion analysis.

The pipeline was divided into six sections, from the initial contact point to a far away position when the pipe was considered to be unaffected by the laying motion. The sic sections featured six different loading sequence. Each sequence included a cyclic vertical loading of a specified amplitude (from 0 to 1.55 times the as-laid pipeline weight Vlay) concurrent with horizontal cyclic motion, also of specified amplitude, as presented in Figure 10. The first sequence, modelling the first segment of the pipe at the touchdown zone, features a specific cyclic motion in which the pipe was pushed into the soil to a specified load of 0.7Vlay before being lifted up until separation between the soil and the pipe occurred.

The resulting accumulation of pipe embedment is presented in Figure 11 for each sequence as a function of the number of imposed cycles. The key observation from Figure 10 and Figure 11 is the much larger pipe embedment (about 0.24 m) compared to a static embedment (about 44 mm using formulation presented by Randolph & White, 2008a), even accounting for the stress concentration.

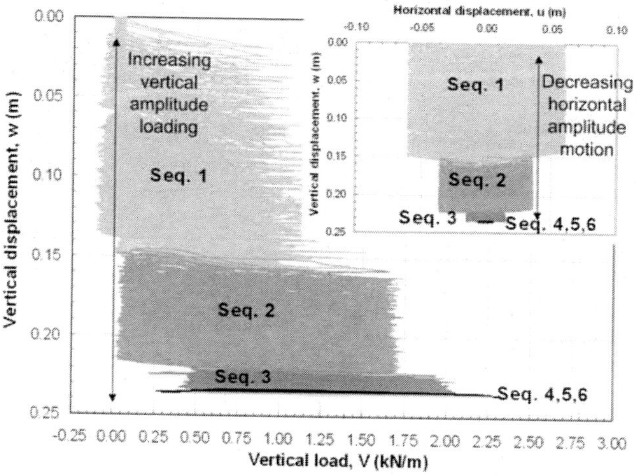

FIGURE 10: Dynamic lay simulation. Cyclic vertical loading and the associated pipe invert trajectory (after Gaudin & White, 2009)

Other important observations are the suction developed at the pipe invert during the first sequence of loading, where the pipe was lifted up from the soil (which potentially affects the stresses in the pipeline), and the dominant effect of the lateral motion amplitude compared to the cyclic vertical load. The first sequence, which featured the largest amplitude of lateral motion, resulted in a deeper embedment that the subsequent sequences, which featured a higher maximum vertical load but smaller lateral motion amplitudes. In other words, for these lay conditions and this soil type, the action of pushing the soil to either side of the pipeline during lateral motion (and concurrently remoulding it) has a dominant effect on the pipe embedment in comparison to any remoulding or penetration of the soil resulting from the vertical cyclic loading.

The results do stress the necessity to model accurately dynamic laying motion, in order to determine the pipe embedment, but also the remoulding of the soil resulting from the laying process, which are going to govern both on-bottom stability and lateral buckling. They demonstrate the capability of centrifuge modelling in replicating accurate laying motions and capturing particular soil behaviour features.

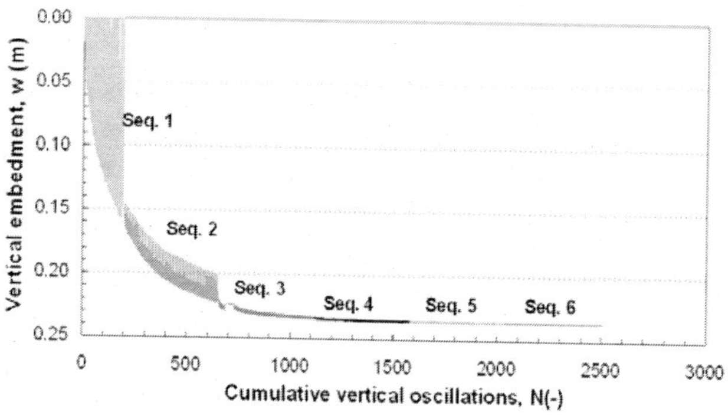

FIGURE 11: Accumulation of pipe embedment resulting from the dynamic lay simulation of Figure 32 (after Gaudin & White, 2009)

3.2 Example 5: Thermal Buckling

3.2.1 *Description of Pipeline Lateral Buckling*

To reduce the structural load resulting from thermal expansion, pipelines are permitted to buckle at targeted location, resulting in lateral displacements in order of 10-20 times the pipeline diameter (Bruton et al., 2006). In operation, as the pipeline experiences fluctuations in pressure and temperature, the pipe cycles back and forth across the same patch of seabed. Surface soil, swept ahead of the pipe on each cycle, then builds up into berms at the extremes of pipe displacement, as illustrated in Figure 12. Subsequent consolidation may increase the strength of the soil berms after disturbance. Pipe feeds axially into and out of the buckle with each cycle. On unloading, the pipeline attempts to return to the as-laid position but is prevented from doing so by both axial and lateral pipe-soil resistance generated by the berms.

Whilst lateral buckling reduces the axial load on the pipeline, it also generates bending moment in the buckling zone, which can ultimately lead to local bending failure. Design methods for buckling typically involve the definition of 'equivalent friction factors', which characterize the soil resistance to the lateral pipeline motion in the terminology of Coulomb friction. This simple approach is efficient to implement within the structural models of pipelines that are used to assess the in-service stresses and accumulating fatigue. However, it is important to note that the seabed is not a frictional surface and the soil resistance will not be proportional to the vertical pipe-soil load. Therefore, it is necessary to perform some form of geotechnical analysis – supported, if needed, by model testing – to estimate the limiting lateral and axial pipe-soil resistance. These values are then converted into 'equivalent' friction coefficients – which may vary with displacement or with cycle number – for implementation into the pipeline structural analysis.

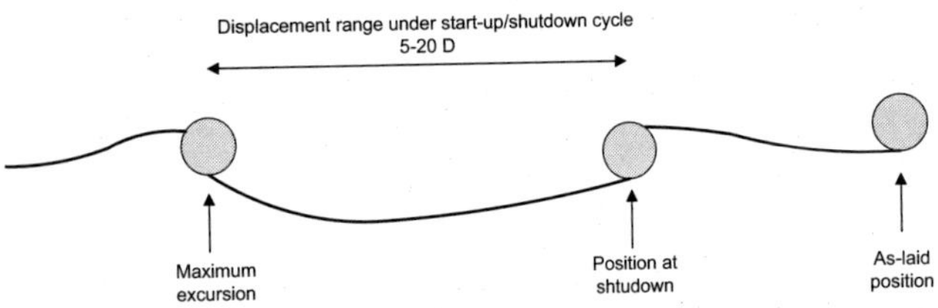

FIGURE 12: Typical cross profile at crown of lateral buckle on an operated pipeline (After Bruton et al., 2007)

Unlike on-bottom stability design, overestimating the friction factor is not conservative as both high and low equivalent friction factors may be more onerous in design. The determination of an accurate range of design friction factors is therefore pivotal for a safe and sound design. This requires the understanding of the soil behaviour at large displacements and through many cycles of loading, well beyond the point of failure, all features that can be particularly well investigated by centrifuge methods.

3.2.2 *Large Amplitude Lateral Pipe-Soil Interaction*

The general form of the pipe-soil interaction behaviour during lateral buckling in fine-grained soils has been investigated at UWA, through a series of project-specific industry studies and research projects using a test setup identical to the one presented in Figure 9. The modelling techniques developed have been presented in details in White & Gaudin (2008) and Gaudin & White (2009).

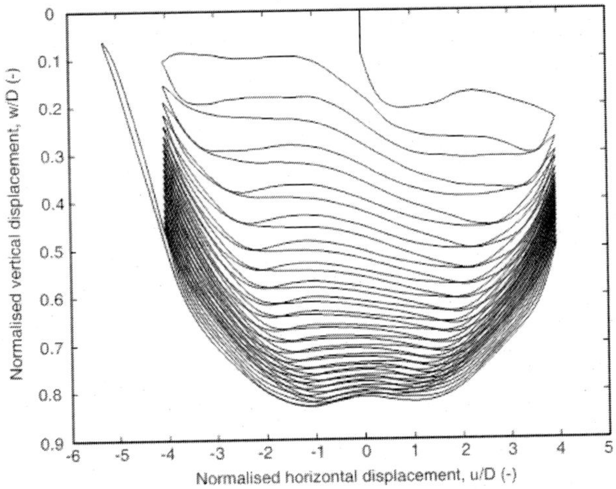

(a) Pipe invert trajectory

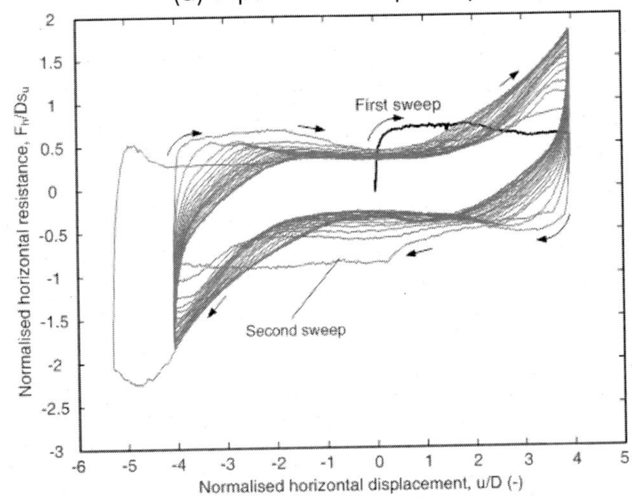

(b) Normalised horizontal resistance

FIGURE 13: A typical pipe-soil response normalized by the pipe diameter D during cyclic lateral movement under a constant vertical load (pipe weight) (Cheuk et al., 2008)

A typical lateral buckling pipe response on kaolin clay is presented in Figure 13, in terms of pipe trajectory and evolution of equivalent friction factor with lateral movement. The response is divided into an initial breakout, where the pipe moves away from the as-laid position, and a cyclic response as the pipe is swept back and forth under constant vertical load. The response presents key features that were subsequently incorporated into a schematic model used to describe pipe-soil interaction during lateral buckling (Cheuk et al., 2008).

- Breakout response: The response is characterized by a brittle peak (whose amplitude varies with initial embedment) followed by either a steady residual resistance for light pipes, or an increasing residual resistance for heavy pipes. The reason for this is the downward trajectory exhibited by heavy pipes, which results in the growth of the soil berm ahead of the pipe, leading to an increasing lateral resistance. For both pipes, the breakout peak partly from the loss of suction, and therefore tensile resistance, at the rear of the pipe, and also from the reducing embedment of the pipe as it rises towards the ground surface. This was directly observed in centrifuge tests reported by Dingle et al. (2008) and correlated to the pipe load displacement response.
- Cyclic response: The lateral response during the first sweep is characterized by a berm resistance and a cyclic residual resistance. For light pipes, as the pipe is swept across, it pushes forward a small berm of soil that has been scraped from the seabed mobilizing steady lateral resistance. Surface soil, swept ahead of the pipe on each cycle, builds into berm at the extremes of the pipe displacement, which are increasing in size after each cycle, providing increasing berm resistance. The response is slightly different for heavy pipes, where the pipe experiences increasing embedment and corresponding increase of the static berms.

3.2.3 *Development of Design Guidance and Empirical Correlations*

The observations shown in Figure 13 formed the basis of a pipe-soil interaction model described in Figure 14 in terms of equivalent friction factors defined at key locations along the sweep and varying with the number of cycles, in order to account for the contribution of the soil berms. This soil-structure interaction is difficult to capture in design without a pipe-soil interaction model that simulates the trajectory of the pipe, as well as the varying equivalent friction. Current theoretical solutions are insufficient to parametise fully the form of model shown in Figure 14. The initial breakout resistance might be assessed theoretically from published failure envelopes (Randolph & White 2008b) but the residual resistance is more challenging. It is controlled by the size of the soil berm ahead of the pipe, the strength of the soil within it, and the trajectory of the pipe. Models for the cyclic large-amplitude lateral need to account for the accumulation and deposition of berm material (White & Cheuk, 2008), but also for re-consolidation of the soil that has been remoulded and transported ahead of the pipe.

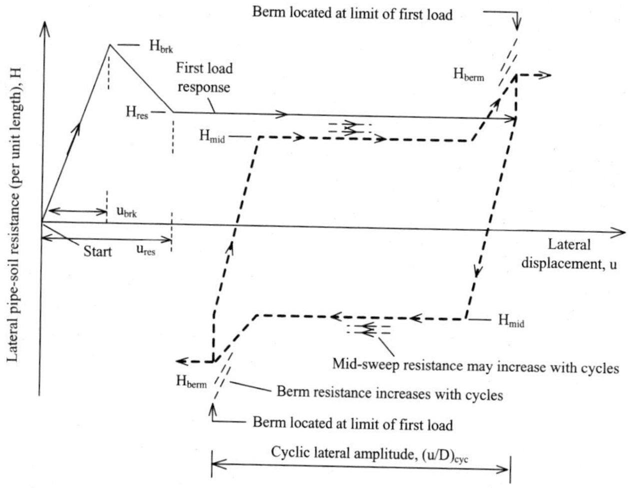

FIGURE 14: Idealized form of lateral pipe-soil interaction during buckling

The cyclic resistance is not only affected by the changing geometry. Depending on the soil type, pore pressure dissipation may occur during lateral sweeping, and is also likely to occur between start-up and shutdown events. This leads to reconsolidation of the disturbed soil within the berm, and also swelling of the unloaded seabed that is exposed by the scraping action of the pipe.

Given these complications, current design guidance uses correlations derived from physical model tests and it also recommends that project-specific physical model tests be performed if tighter bounding of the behaviour is required. One such correlation developed to assess the lateral breakout resistance of pipelines laid on fine-grained soils has been developed within the SAFEBUCK JIP, who collated results from 28 large-scale tests and 39 small-scale centrifuge tests. Using these results for the breakout resistance of shallowly-embedded pipelines, a correlation was developed to link the normalized breakout resistance, H_{brk}/s_uD, to the key geotechnical properties (intact undrained shear strength, su, unit weight, γ', and the vertical pipe-soil load, V), as well as the normalized pipe embedment, w/D. From dimensional analysis:

$$\frac{H_{brk}}{s_u D} = f\left(\frac{\gamma'}{s_u D}, \frac{w}{D}, \frac{V}{s_u D}\right) \quad ...(2)$$

A simple correlation was devised between these parameters, using theoretical solutions for guidance (e.g. Randolph & White 2008b). The correlation was calibrated to achieve the best agreement between the calculated and measured breakout resistance across all 67 measurements of breakout resistance in the database. During the calibration process, skew was eliminated with respect to each of the dimensionless groups in the above expression.

The details of the calibrated expression are confidential to participants. Essentially, the three dimensionless terms above were combined as additive components, with each contributing non-linearly to H_{brk}/s_uD according to a power law. The calibration process ensured that the mean ratio between the calculated and measured resistance was 1.0. The standard deviation of the ratio calculated/measured is 0.25.

The performance of the empirical correlation was assessed by White & Cheuk (2010) on a database consisting of 8 different studies. Despite some scatter, the correlation provided on average a correct assessment of the breakout resistance. It is important to note the guidance is skewed to the average conditions explored in the model test database. For example, the correlation is influenced by the particular pipe-soil bonding condition observed in these tests, as well as the particular distribution of soil strength that exists after the pipe has been penetrated into the seabed, and been left to consolidate in place. Given these considerations, the correlations should be used with caution within the range of conditions over which is has been calibrated. These conditions include:

- The range of pipe weight, V/s_uD, soil strength/weight ratio, γ'D/su, and embedment, w/D spanned by the database.
- The secondary soil properties (not considered in the correlation) that control the operative soil strength during breakout (which is likely to differ from the intact s_u). These secondary properties include the sensitivity (and hence the softening during pipe 'laying') and the consolidation parameters.
- The pipeline properties adopted in the model test database – including the pipe roughness (which is not considered in the correlation).
- The breakout processes simulated in the model test database – including the rate of breakout (which is not considered in the correlation).

This list of limitations highlights the value in supporting model test observations by theoretical analyses that can set the results within a robust framework. Such a framework allows more confident extrapolation. However, soil behaviour, and lateral pipe-soil interaction, is controlled by further properties in addition to those given in the above Equation, such as sensitivity, the consolidation characteristics and so forth. As a result, correlations and analyses based on idealized theory can often be significantly refined through well-planned physical modelling studies using a geotechnical centrifuge. The 8 project-specific studies performed at UWA in recent years to support pipeline developments worldwide are testament to this.

4. CONCLUSIONS

Centrifuge modelling techniques have played an important role in developing new design and assessment methods for jack-up spudcan foundations and pipeline-soil interaction, many of which have been incorporated into industry guidelines and practices. This paper provides an overview of recent contributions covering jack-up installation and extraction, and pipe dynamic laying and thermal buckling.

With the development of centrifuge technology, further opportunities will be created for solving the emerging issues faced by the offshore industry as they move into deeper water and more challenging seabed conditions. Some of the opportunities for the near future include amongst others: direct prediction of jack-up installations using penetrometer data, installation through footprints and sloped seabeds, spudcan behaviour in multi-layered and intermediate silty soils, pipeline behaviour through varying drainage conditions or steel catenary riser soil interaction.

The geotechnical centrifuge will continue to provide pivotal insights into spudcan and pipeline behaviour, as well as other offshore geotechnical issues. While the author advocates the use of centrifuge methods as a valuable and necessary tool to assist the offshore industry in developing and designing solutions for a wide range of geotechnical problems, it is also acknowledged that centrifuge modelling should not been considered as the unique tool to be used, but as a particular one which outcomes may be maximised if integrated into a global approach. It is believed that while technological and scientific developments will, without doubt, increase the utility of centrifuge methods, the most spectacular improvement will be in the understanding of the centrifuge methods contribution and their integration at key stages into a global design procedure, which will incorporate in-situ, numerical and analytical methods.

ACKNOWLEDGEMENTS

The paper presents an overview of centrifuge results and outcomes resulting from projects undertaken in collaboration with colleagues at COFS and already presented in various publications. The contribution of Prof. M. Cassidy, Prof. D. White, Dr B. Bienen and Dr M. Hossain, in establishing and formulating the results presented is gratefully acknowledged. Centrifuge modelling is an experimental technique, which rely upon the expertise and dedication of technical staff. The author would like to acknowledge the contribution of the technical team at COFS in carrying out successfully the projects presented in this paper. Progress of centrifuge techniques have greatly benefited from industry incentive and support. The contribution and financial support of the various industry partners involved in the projects presented in the paper are also acknowledged.

REFERENCES

Bienen B., Gaudin C., and Cassidy M.J. (2007) "Centrifuge tests of shallow footing behaviour on sand under combined vertical-torsional loading". International Journal of Physical Modelling in Geotechnics, 7, Issue 2, pp. 1-21.

Bienen B., Gaudin C., and Cassidy M. J. (2009) "The influence of pull-out load on the efficiency of jetting during spudcan extraction". Applied Ocean Research, 31, Issue 3, pp. 202-211.

Bruton D. A. S., White D. J., Cheuk C. Y., Bolton M. D., and Carr M. C. (2006) "Pipe-soil interaction behaviour during lateral buckling, including large amplitude cyclic displacement tests by the Safebuck JIP", Proc. Offshore Technology Conference, Houston, OTC17944.

Cathie D. N., Jaeck C., Ballard, J-C., and Wintgens J-F. (2005) "Pipeline geotechnics – State-of-the-art", Proc. Intern. Symp. on Frontiers in Offshore Geotechnics, Perth, Australia, pp 95-114.

Cassidy M. J., Byrne B. W., and Randolph M. F. (2004) "A comparison of the combined load behaviour of spudcan and caisson foundations on soft normally consolidated clay". Géotechnique, 54, Issue 2, pp 91-106.

Cheuk C. Y., White D. J., and Bolton M. D. (2007) "Large scale modelling of soil-pipe interaction during large amplitude movements of partially-embedded pipelines". Canadian Geotechnical Journal, 44, Issue 8, pp 977-996.

Cheuk C. Y., White D. J., and Bolton M. D. (2008) "Reply to Discussion by W.O. McCarron on 'Large scale modelling of soil-pipe interaction during large amplitude movements of partially-embedded pipelines'". Canadian Geotechnical Journal, 45, Issue 5, pp 744-749 (Original paper 44, Issue 8, pp 977-996).

Craig W. H., and Al-Saoudi N. K. S. (1981) "The behaviour of some model offshore structures", Proc. 10th Intern. Conf. on Soil Mechanics and Foundation Engineeting, Stockholm, Sweden, (2), pp 541-556.

Craig W. H. (1984). Preface. Proc. Intern. Symp. Application of Centrifuge Modelling to geotechnical Design, Manchester, UK, 1.

Craig W. H., and Chua K. (1990) "Deep penetration of spudcan foundations on sand and clay". Géotechnique, 40, Issue 4, pp 541-556.

Craig W. H., and Chua K. (1991) "Large displacement performance of jack-up spudcans", Proc. Int. Conf. on Centrifuge '91, Rotterdam: Balkema, pp 139-144.

Dean E. T. R., James R. G., Schofield A. N., Tan F. S. C., and Tsukamoto Y. (1993) "The bearing capacity of conical footings on sand in relation to the behavior of spudcan footings of jack-ups", Proc. Wroth Memorial Symp. Predictive Soil Mechanics, Oxford, pp 230-253.

De Catania S., Breen J., Gaudin C., White D.J. (2010) "Development of a multiple axis actuator control system", Proc. 7th Intern. Conf. on Phy. Model. in Geotechnics, Zurich, Switzerland, pp 325-330.

Dendani D., and Jaeck C. (2007) "Pipe-Soil Interaction in Highly Plastic Clays", Proc. 6th Intern. Offshore Site Investigation and Geotechnics Conf., London, pp 115-124.

Dingle H. R. C., White D. J., Gaudin C. (2008) "Mechanisms of pipe embedment and lateral breakout on soft clay". Canadian Geotechnical Journal, 45, Issue 5, pp 636-652.

Finnie I. M. S., and Randolph M. F. (1994) "Bearing response of shallow foundations in uncemented calcareous soil". Proc. Int. Conf. on Centrifuge '94, Rotterdam, Netherlands.

Gaudin C., and White D. J. (2009) "New centrifuge modelling techniques for investigating seabed pipeline behavior", Proc. of 17th Intern. Conf. of Soil Mech. and Geotech. Eng., Alexandria, Egypt, 1, pp 448-451.

Gaudin C., Cluckey, E. C., Garnier, J., and Phillips, R. (2010) "New frontiers for centrifuge modelling in offshore geotechnics", Proc. 2nd Intern. Symp. on Frontiers in Offshore Geotechnics, Perth, Australia, pp 155-188.

Gaudin C., Bienen B., and Cassidy M. J. (2011) "Investigation of the potential of bottom water jetting to ease spudcan extraction in soft clay". Géotechnique, 61, Issue 12, pp 1043-1054.

Gaudin C., Kong V., and Cassidy M. J (2012) "An overview of spudcan reinstallation near footprint", Offshore Technology Conference, Houston, USA.

Hill A. J., and Jacob H. (2008) "In-situ measurements of pipe-soil interaction in deep water", Proc. Offshore Technology Conf., Houston, OTC 19528.

Hossain M. S., Hu Y., Randolph M. F., and White D. J. (2005) "Limiting cavity depth for spudcan foundations penetrating clay". Géotechnique, 55, Issue 9, pp 679-690.

Hossain M. S., Randolph M. F., Hu Y., and White D. J. (2006) "Cavity stability and bearing capacity of spudcan foundations on clay", Proc. Offshore Technology Conf., Houston, OTC 17770.

Hossain M. S., and Randolph M. F. (2009) "New mechanism-based design approach for spudcan foundations on stiff-over-soft clay", Proc. Offshore Technology Conf., Houston, OTC 19907.

Hossain M. S., and Randolph M.F. (2010a) "Deep-penetrating spudcan foundations on layered clays: centrifuge tests". Géotechnique, 60, Issue 3, pp 157-170.

Hossain M. S., Randolph M. F. (2010b) "Deep-penetrating spudcan foundations on layered clays: numerical analysis". Géotechnique 60, Issue 3, pp 171-184.

InSafeJIP. (2009) Improved guidelines for the prediction of geotechnical performance of spudcan foundations during installation and removal of jack-up units, First year report. Authors: Osborne, J.J., Teh, K.L., Houlsby, G.T., Cassidy, M.J., Bienen, B., Leung, C.F.

InSafeJIP. (2010) Improved guidelines for the prediction of geotechnical performance of spudcan foundations during installation and removal of jack-up units, Proposed Industry Guidelines. Authors: Osborne, J.J., Teh, K.L., Houlsby, G.T., Cassidy, M.J., Bienen, B., Leung, C.F.

ISO. (2009) Petroleum and natural gas industries – Site-specific assessment of mobile offshore units – Part 1: Jack-ups, International Organization for Standardization, ISO 19905-1.

Jacob H., and Looijen P. (2008) "Development of a deepwater tool for in-situ pipe-soil interaction measurement and its benefit in pipeline analysis", Offshore Pipeline Technology Conference and Exhibition, Amsterdam, Netherlands.

Kim S. W. (1978) "Bearing capacity of footings on two-layered clays under inclined loads", MEng Thesis, Nova Scotia Technical College.

Lee K. K. (2009) "Investigation of potential punch-through failure on sands overlaying clay soils", PhD Thesis, The University of Western Australia.

Lee K. K., Randolph, M. F., and Cassidy, M. J. (2009) "New simplified conceptual model for spudcan foundations on sand overlying clay soils", Proc. Offshore Technology Conference, Houston, OTC-20012.

Leung C. F., Xie Y., Chow Y. K. (2006) "Centrifuge model study of spudcan-pile interaction", Proc. 16th Int. Offshore and Polar Engineering Conf. (ISOPE), San Francesco, pp 530-535.

Leung C. F., Xie Y., Chow Y. K. (2008) "Use of PIV to investigate spudcan-pile interaction", Proc. 18th Int. Offshore and Polar Engineering Conf. (ISOPE), Vancouver, pp 721-726.

Lund K. M. (2000) "Effect of increase in pipeline soil penetration from installation", Proc. Intern. Conf. Ocean Offshore Artic Eng. New Orleans, USA, OMAE2000/PIPE-5047

Murff J. D. (1996) "The geotechnical centrifuge in offshore engineering", Proc. Offshore Technology Conf., Houston, OTC 8265.

Noble Denton & Associates. (1987) Foundation fixity of jack-up units: a joint industry study, Noble & Denton Associates.

Osborne J. J., Teh K. L., Leung C. F., Cassidy M. J., Houlsby G. T., Chan N., Devoy D., Handidjaja P., Wong P., and Foo K. S. (2008). An introduction to the InSafe JIP. Proc. 2nd Jack-up Asia Conference. Singapore.

Osborne J. J., Houlsby G. T., Teh K. L., Bienen B., Cassidy M. J., Randolph M. F., and Leung C. F. (2009) "Improved guidelines for the prediction of geotechnical performance of spudcan foundations during installation and removal of jack-up units", Proc. Offshore Technology Conference, Houston, OTC 20291.

Okamura M., Takemura J., and Kimura T. (1997) "Centrifuge model test on bearing capacity and deformation of sand layer overlying clay". Soils and Foundations 37(1), pp73-88.

Purwana O. A., Leung C.F., Chow Y.K., and Foo K.S. (2006) "Breakout failure mechanism of jackup spudcan extraction", Proc. 6th Intern. Conf. of Physical Modelling in Geotechnics (ICPMG06), Hong-Kong, 1, pp667-672.

Purwana O. A., Quah M., Foo K. S., Leung C. F., and Chow Y. K. (2008) "Understanding spudcan extraction problem and mitigation devices", Proc. 2nd Jack-Up Asia Conference & Exhibition, Singapore.

Purwana O.A., Quah M., Foo K.S., Nowak S., and Handidjaja P. (2009) "Leg Extraction / Pullout Resistance - Theoretical and Practical Perspectives", Proc. 12th International Conference The Jack-Up Platform Design, Construction & Operation, London.

Randolph M. F. (2004) "Characterisation of soft sediments for offshore applications", Proc. 2nd Int. Conf. on Site Characterisation, Porto, 1, pp 209-231.

Randolph M.F., White D.J. (2008a) "Pipeline embedment in deep water: processes and quantitative assessment", Proc. Offshore Technology Conference, Houston, USA. OTC19128-PP.

Randolph M. F., White D. J. (2008b) "Upper bound yield envelopes for pipelines at shallow embedment in clay". Géotechnique, 58, Issue 4, pp 297-301.

Rowe P. W., and Craig W. H. (1981) "Applications of models to the prediction of offshore gravity platform foundation performance", Proc. Intern. Conf. on Offshore Site Investigation, London, pp 269-281.

Siciliano R. J., Hamilton J. M., Murff J. D., and Phillips R. (1990) "Effect of jackup spudcans on piles", Offshore Technology Conference, Houston, OTC 6467.

SNAME. (2008) Recommended practice for site specific assessment of mobile jack-up units. T&R Bulletin 5-5A, 1st Edition – Rev. 3, Society of Naval Architects and Marine Engineers, New Jersey.

Springman S. M., and Schofield A. N. (1998) "Monotonic lateral load transfer from a jack-up platform lattice leg to a soft clay deposit", Proc. Int. Conf. on Centrifuge '98, Rotterdam: Balkema, pp 563-568.

Teh K. L., Cassidy M. J., Chow Y. K., and Leung C .F. (2006) "Effects of scale and progressive failure on spudcan ultimate bearing capacity in sand", Proc. Int. Symp. On Ultimate States of Geotechnical Structures, Marne-la-Valèe, France, 1, pp 481-489.

Teh K. L., Leung C. F., Chow Y. K., and Handidjaja P. (2009) "Prediction of punch-through for spudcan penetration in sand overlying clay", Proc. Offshore Technology Conference, Houston, OTC 20060.

Teh K. L., Leung C. F., Chow Y. K., and Cassidy M. J. (2010) "Centrifuge model study of spudcan penetration in sand overlying clay". Géotechnique, 60, Issue 11, pp 825-842.

Wong P. C., Chao J. C., Murff J. D., Dean E. T. R., James R. G., Schofield A. N., and Tsukamoto Y. (1993) "Jack up rig foundation modelling II", Offshore Technology Conference, Houston, OTC 7303.

White D. J., Take W. A., and Bolton M. D. (2003) "Soil deformation measurement using particle image velocimetry (PIV) and photogrammetry". Géotechnique 53, Issue 7, pp 619-631.

White D. J. and Cheuk C. Y. (2008) "Modelling the soil resistance on seabed pipelines during large cycles of lateral movement". Marine Structures 21, Issue 1, pp 59-79.

White D. J., and Gaudin C. (2008) "Simulation of seabed pipe-soil interaction using geotechnical centrifuge modeling", Proc. 1st Asia-Pacific Deep Offshore Technology Conference, Perth, Australia.

White D. J. ,and Cathie D. N. (2010) "Geotechnics for Subsea pipeline and risers", Proc. 2nd Int. Symp. on Frontiers in Offshore Geotechnics. Perth, Australia.

White D. J., Hill A. J., Westgate Z. and Ballard J-C (2010) "Observations of pipe-soil response from the first deepwater deployment of the SMARTPIPE", Proc. 2nd Int. Symp. on Frontiers in Offshore Geotechnics, Perth, Australia, pp 851-85.

White D. J., Cheuk C. Y. (2010). SAFEBUCK JIP: Pipe-soil interaction models for lateral buckling design: Phase IIA data review. Report to Boreas Consultants (SAFEBUCK JIP), UWA report GEO 09497r2, pp 185. (currently confidential to JIP participants).

Xie Y., Leung C. F., Chow Y. K. (2006) "Effects of spudcan penetration on adjacent piles", Proc 6th Int. Conf. on Physical Modeling in Geotechnics, Hong Kong, 2, pp 701-706.

Xie Y., Leung C. F., Chow Y. K. (2010) "Study of soil movements around a penetrating spudcan", Proc 7th Int. Conf. on Physical Modeling in Geotechnics, Zurich, 2, pp 1075-1080.

Modelling of Improved Ground in Centrifuge

M. Kitazume, Y. Morikawa[1] and S. Nishimura[2]

Department of Civil and Environment Engineering, Tokyo Institute of Technology, Tokyo, Japan
[1] Geotechnical Engineering Division, Port and Airport Research Institute, Japan
[2] Division of Field Engineering for the Environment, Faculty of Engineering, Hokkaido University, Japan
E-mail: kitazume@cv.titech.ac.jp

ABSTRACT

Soft soil is often encountered when any type of infrastructure is constructed, in which large ground settlement and stability failure are anticipated. There are many soil improvement techniques developed against these problems. The behavior of improved ground is affected by many factors such as the properties of original ground and improved ground, interaction of original and improved grounds, and external loading conditions, etc. Many centrifuge model tests have been carried out on the research topics of ground improvement, which can be classified into two categories: investigation of properties' change and behavior of original ground during ground improvement work, and investigations of behavior of improved ground. The Port and Airport Research Institute carried out many centrifuge model tests on research topics related to several ground improvement techniques for many years. In this manuscript, several examples of centrifuge modeling regarding to the two categories which were carried out at the Institute are briefly introduced.

1. INTRODUCTION

It is an obvious truism that, structures should be constructed on good quality ground. The ground conditions of construction sites, however, have become worse than ever during recent decades throughout the world. We often encounter soft soil when any type of infrastructure is constructed, in which large ground settlement and stability failure are anticipated. Apart from clayey or highly organic soils, loose sand deposits under a water table can cause serious problems of liquefaction under seismic condition. In these cases, suitable soil improvement techniques are required to improve physical and mechanical properties of the soft soil in order to cope with these problems. There are many soil improvement techniques developed for these purposes.

Many numerical analyses and physical model tests have been conducted to investigate the ground behavior, interaction between original ground, improved ground and superstructure, and performance of improved ground. Based on these research results together with observations and experiences in field, design procedures for ground improvement techniques are established and brushed up.

The behavior of improved ground is affected by many factors such as the properties of original ground and improved ground, interactions between original and improved grounds, and external loading conditions, etc. In addition, the properties of original ground after improvement are not same as those before improvement because the original ground is subjected to disturbance and induced stresses to some extent due to execution of ground improvement in many cases: the smear phenomenon in vertical drain method is a typical example where the original soil is disturbed and its permeability is decreased due to installation of vertical drains.

Many centrifuge model tests have been carried out on the research topics of ground improvement, which can be classified into two categories: investigation on properties' change and behavior of original ground during ground improvement work, and investigation of behavior of improved ground. In the former, miniature ground improvement machines were developed to simulate actual ground improvement work in centrifuge as precisely as possible. The installation of sand drains, heavy tamping method by Mikasa *et al.*(1989), sand compaction pile method by Ng *et al.*(1998) and by Weber *et al.*(2006) are typical example of this category. The centrifuge modeling on ground heaving due to installation of compacted sand piles can be also included in this category. In the latter, quite large number of model tests were carried out for various ground improvement techniques. As the behavior of improved ground is affected by many factors as stated above, the rigorous idealization or simplification of the problem is inevitable in the centrifuge modeling and different materials and improvement processes from actual were adopted in many tests for the purposes.

The soil stabilization laboratory of the Port and Airport Research Institute, where the authors worked/work for, carried out many centrifuge model tests on research topics related on several ground improvement techniques for many years (Kitazume, 2009). In this manuscript, several examples of centrifuge modeling regarding to the two categories which were carried out in the centrifuge of the Institute (Kitazume and Miyajima, 1995) are briefly introduced.

2. INSTALLATION OF COMPACTION GROUTING

2.1 Introduction

Compaction grouting, an in-situ static compaction technique by means of grout injection, has been in use since the 1960s in the United States for settlement control and bearing capacity improvement. It has been increasingly adopted for improving the liquefaction resistance of loose sandy ground in more recent years (Coastal Development Institute of Technology, 2007). An increase in the liquefaction resistance of sand caused by compaction grouting is presumed to derive from three possible mechanisms. They are, (i) increase in the lateral confining stress, (ii) densification and (iii) reinforcement by hydrated and hardened grout piles. A design of grouting specifications should in theory be based on these mechanics. In practice, however, most existing design approaches are rule-of-thumb types, with the above factors only evaluated with SPT-*N* values before and after execution. A case for this approach in Japan is the accumulated data and confidence from the sand compaction pile method in the past. However, this practice, being highly empirical in nature, cannot be blindly extrapolated to conditions that fall outside past experience.

Detailed, systematic studies of ground condition changes due to compaction grouting have been limited in number. The study by Nichols and Goodings (2000) is one of the very few reported cases in which compaction grouting was simulated in a centrifuge. Other researchers mostly performed 1-*g* miniature model tests in calibration chambers (e.g. Shinsaka *et al.*, 2003; Yamazaki *et al.*, 2008; El-Kelesh and Matsui 2008). While the above studies were helpful in drawing an outline of ground behavior during compaction grouting, they typically focus on a single grout pile installation and offer little insight into the state of ground embraced by multiple piles. Understanding the process of stress changes by progressive, multiple grout pile formation is a necessary step from a basic study to a practical, quantitative study of the stabilization effects.

The in-flight apparatus capable of injecting three grout piles continuously was developed to investigate the stress changes caused by multiple grout pile installation (Nishimura *et al.*, 2012). Newly developed test apparatus and miniature injection techniques are described briefly in this section.

2.2 Experimental Setup

2.2.1 Grout Injection System

The whole assembly is shown in Figures 1 and 2. The adopted soil container was cylindrical, with an inner diameter of 540mm and a depth of 500mm. It had a rigid steel wall, whose inner was lubricated with silicone grease and a layer of latex membrane. The steel wall was reinforced by welded vertical ribs to endure the load from the hoist system atop. The hoist system consists of a steel frame carrying three linear-head motors diametrically deployed at 120° apart from each other. Each of them is driven independently, lifting three rams, each of which was clutching an injection rod. The horizontal locations of the rods were adjustable in the radial direction and the rod spacing could thus be changed while the triangle center was fixed at the container center. The motor gears were set to achieve a lift rate of 1.1 mm/sec.

FIGURE 1: View of centrifuge platform after apparatus setup

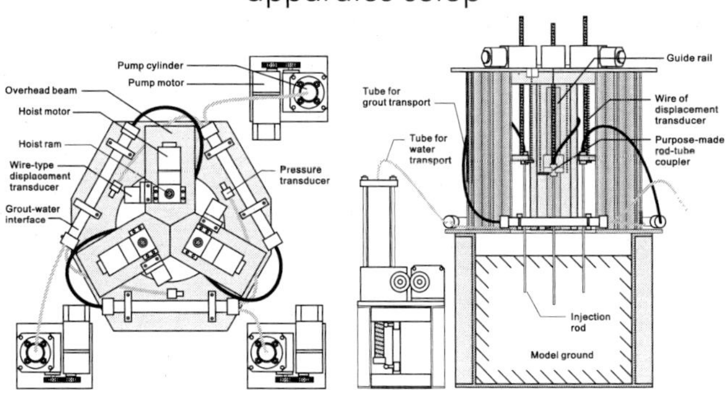

(a) Plan view (b) Vertical cross-section

FIGURE 2: Illustration of soil container and grout injection system

Three motor-driven hydraulic pumps were also mounted beside the container on the centrifuge. The pumps have a pressure capacity of more than 10 MN/m^2 and the injection rate can be preset at any value between 0 to 0.4 m*l*/sec. The injected volume was calculated by monitoring the piston displacement. The pumps were filled with water to generate water pressure, which was then applied to the grout stored in the interface cylinders mounted on the soil container rim. Each interface cylinder had a smooth piston inside to transmit the pressure with a minimum loss. The grout's consolidation and un-homogenization before injection, caused by the self-weight under centrifugal acceleration, was mitigated by designing the cylinders in a slender shape and laying them horizontally. The injection pressure from each interface cylinder was measured by the pressure transducer connected to the cylinder's water chamber.

The grout displaced from the interface cylinders was sent to the injection rods. They were made of stainless steel, with inner and outer diameters of 5 and 8mm, respectively, and a length of 45mm. Thick and stiff Synflex tubes were used for the water and grout transport between the pumps and the cylinders, and between the cylinders and the rods. The connection of the tubes was made with couplers produced by Swagelok Company, which had a straight, smooth inner with the same diameter as that of the Synflex tubes and the injection rods. This arrangement ensured that the grout's passage diameter was constant all along, reducing the possibility of clogging up. It also turned out essential to put in advance a small amount of high-viscosity silicone grease ahead of the grout's advancing front. This was to prevent a free-fall of the grout from the highest point in the tube under centrifugal acceleration at the initial stage of injection. Such free-fall led to instant segregation of the grout, and the segregated sand particles jammed the rod tip.

2.2.2 *Grout Preparation*

The mortar grout used in the compaction grouting technique is normally prepared by mixing a few different kinds of soil to adjust gradation such that it is sufficiently fine to flow through the tube and rod, while sufficiently coarse at the same time to displace the original ground without infiltrating into the void. To simulate a grouting process in the miniaturized centrifuge model, consideration must be given to the small tube and rod inner diameter. By trial and error of mixing and injection, the present study found a suitable mixture; the Soma Silica Sand #5, the Kawasaki

Clay, Portland cement and water at the ratio of 40%:60%:12%:50% by weight. A small dose of retardant (7% of the cement weight) was admixed to prevent the early hydration of cement before injection.

2.2.3 Ground Model and Grout Injection Procedures

A typical ground model is illustrated in Figure 3 in horizontal and vertical cross-sections. The ground was prepared by air-pluviating the Soma Silica Sand #5, targeting the relative density, D_r, of around 60%. Most of the test cases involved dry sand ground, while a few tests were performed with sand ground saturated with viscosity-adjusted fluid. The grout pile spacing, x, was set at 1.8, 2.4 or 3.0m with the fixed equivalent pile diameter, d, of 0.7m (quantities are quoted in the prototype scale in this section). The improvement ratio, a_s, defined as the pile's horizontal cross-section area per ground area, then varied according to x. The spacing adopted in practice so far in Japan has been 2.0m or less, with a corresponding improvement ratio of 5% or more.

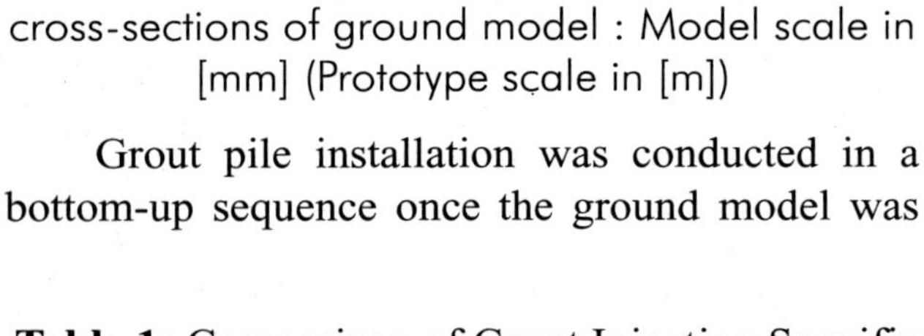

FIGURE 3: Example of vertical and horizontal cross-sections of ground model : Model scale in [mm] (Prototype scale in [m])

Grout pile installation was conducted in a bottom-up sequence once the ground model was subjected to a centrifugal acceleration of 30g. The subsequent procedures involved alternating stages of injection and rod-lifting, starting from a depth of 9.9m up to 2.4m, at 0.33m steps. The specifications of the injection are summarized in Table 1, which also indicates the model-to-prototype conversion ratios. The injection and lifting rates are both specified to be compatible with those commonly adopted in practice. Once a single grout pile was formed through Rod 1, the same procedures were repeated with Rod 2 and then Rod 3. Preliminary tests with pore water pressure probes confirmed that the excess pore water pressure was negligible at any time in the tests with saturated ground model.

2.3 Formation of Grout Piles and Injection Pressure

The grout piles formed during the centrifuge tests were observed by excavating the ground model after the centrifuge was stopped. A typical view of the formed piles is shown in Figure 4. They broadly resembled real grout piles formed in field, with no obvious discontinuity or tapering trend. The equivalent pile diameter after the tests, calculated from the length and the volume, was typically 0.60 to 0.63m, while the injected volume was controlled to make it 0.70m. This discrepancy was accountable by loss of the grout water during and after the injection, probably caused by radial consolidation. This consolidation might be exaggerated in the miniaturized centrifuge tests, in which the cement hydration time could not be reduced according to the similitude law. In field, the grout may start hardening before the consolidation is completed, while this is unlikely in the 1/30 scale model.

Table 1: Comparison of Grout Injection Specifications between Model and Prototype (factors with n indicate scale ratios of prototype to model)

Acceleration (g)	*Equivalent Pile Diameter, d (mm)*	*Bulb Volume (l)*	*Injection Rate (l/min.)*	*Vertical injection step (mm)*	*Time for injection per bulb (sec.)*
(n)	($1/n$)	($1/n^3$)	($1/n^2$)	($1/n$)	($1/n$)
1	700	128	20.7*	333	371
30	23.3	0.00475	0.0230	11.1	12.4

*Upper bound to commonly adopted rates in Japan is 30 *l*/min. (CDIT, 2007)

FIGURE 4: Grout piles formed in test case with x = 1.8 m and dry sand ground

An example of the time-histories of injection pressure, as measured by the pressure transducers is shown in Figure 5. The pressure loss between the injection rod tip and the water reservoir, caused by the piston friction and grout viscosity, was calibrated and found to be 100kPa at maximum. This was small in relation to the pressure magnitude involved in the tests, and no correction had been applied to the data shown here. It is noted in the pressure record in the figure that the pressure reached ultimate values quickly after resuming injection. The ultimate pressure, P_u, is plotted against depth for each test case in Figure 6. It is broadly linear above 8m depth, while later injection via Rods 2 and 3 tends to exhibit slightly higher P_u values, reflecting gradual stress build-up and increasing degrees of compaction.

The lines representing $P_u = 30\gamma_t z$ and $P_u = 50\gamma_t z$ (γ_t is the dry unit weight and z the depth), also shown in Figure 6, seem to bracket the experimental date. These factors, 30-50, are compatible with the commonly quoted field values of 20-100 (e.g. Shinsaka *et al.*, 2003). Shinsaka *et al.* (2003) found that the factor of only 8-10 was obtained in their miniaturized 1*g* model tests. It is thus suggested that reproducing the field stress level is an important factor in quantitatively studying the pressure and stress characteristics involved in compaction grouting. Below 8m in depth, the P_u values exhibited reverse, increasing-upward trends. This P_u-z curve shape may indicate combined effect of two mechanisms at simultaneous work. As the injection point moved upward, the previously compacted region beneath the current injection point functioned as a foundation, providing larger bearing capacity against subsequent injection (increasing-upward trend). At the same time, the overburden pressure, which also resisted the expansion of the grout bulb, decreased in proportion to the injection depth (decreasing-upward trend).

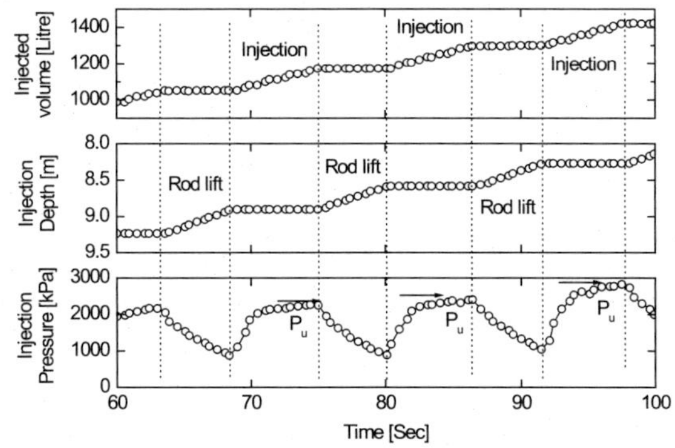

FIGURE 5: Time-history of injection operation and recorded injection pressure

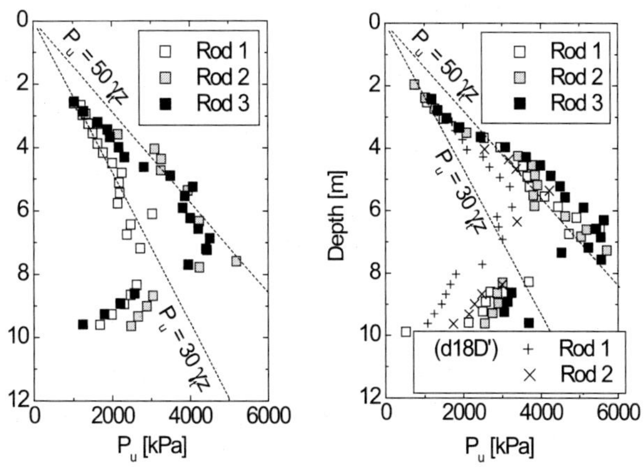

(a) x = 1.8 m (b) x = 3.0 m

FIGURE 6: Ultimate injection pressure, P_u, recorded against depth (d18D′, an unsuccessful case in which grout jammed in Rod 3, serves as a duplicate case of injection via Rods 1 and 2 in d18D, a dry sand case with x = 1.8 m)

2.4 Summary

Apparatus and experimental techniques for simulating compaction grouting were newly developed and confirmed to be successful, with relatively uniform grout piles formed in the centrifuge field. The recorded level of injection pressure was consistently within the range commonly encountered in field execution, suggesting that the centrifuge testing, by reproducing realistic stress levels in the ground models, was an appropriate approach to the problem. Following the success, this study investigated the stress regime changes caused by grout injection and proposed to take account of the new stress states as well as sand densification in evaluating liquefaction resistance improvement effect. The details are reported by Nishimura *et al.* (2012).

3. BEARING CAPACITY OF SAND COMPACTION PILE IMPROVED GROUND

3.1 Introduction

Sand compaction pile method (SCP) has been applied to improve the soft ground often found in the Japanese coastal area (Kitazume, 2005). When the method is applied to improve clay deposits, the improved ground is often called a composite ground. The behavior of the improved ground is influenced by many factors, including geometric conditions, initial and induced stress conditions in the ground. Also the in-situ sand piles can have scatter in their properties, and the surrounding ground is in many cases heavily disturbed due to the piles installation. There are two approaches in modeling the improved ground in the centrifuge; similar execution technique to the in-situ technique is adopted to make a model ground having similar properties to that in-situ, and a sort of ideal and simple model ground is prepared to avoid any influence factors due to the uncertainties in the field conditions. The authors adopted the latter approach in which frozen sand piles were prepared separately and installed into holes in the model ground in order to avoid the influence of soil disturbance effect due to pile installation. Here one of the studies is reported investigating the bearing capacity of the SCP improved ground with low improvement area ratio, a_s under a combination of vertical and horizontal loads (Terashi *et al.*, 1991).

3.2 Preparation of Model Ground

The setup of model ground is shown in Figure 7. A thick normally consolidated clay layer of Kaolin is reduced in scale and prepared in a strong specimen box which has the following inside dimensions: 30 cm deep, 10 cm wide and 50 cm long. All the model tests were carried out in the 50 *g* field. Therefore the prototype thus simulated in the strong box was an approximately 10 m thick alluvial clay deposit which was improved by large compacted sand piles with 1 m diameter. The material used for the sand pile and sand mound was Toyoura sand and the clay used was Kaolin clay. Both materials were selected because their characteristics were well known and also they were commercially available.

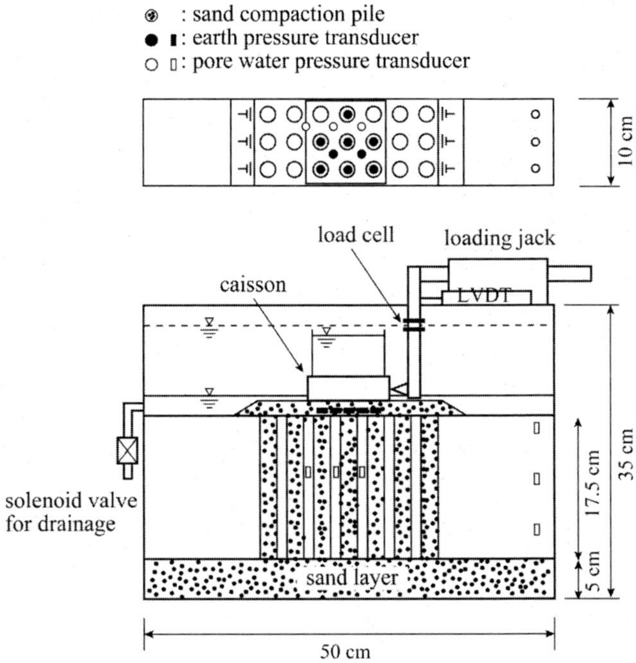

FIGURE 7: Setup of model ground (model scale)

A thick normally consolidated clay layer was prepared in a strong specimen box which had the inside dimensions of 30 cm deep, 10 cm wide and 50 cm long. The kaolin clay was thoroughly remolded at a water content of 120% which was sufficiently higher than its liquid limit. A drainage layer of Toyoura sand with 5 cm thickness was placed at the bottom of the strong box. Then the slurry of Kaolin clay was poured into the box. The preliminary consolidation was conducted under a vertical pressure of 10 kN/m^2 on the laboratory floor. Then the model ground was brought onto the swing platform for self weight consolidation under 50 *g* in order to prepare the normally consolidated clay ground with the thickness of 20 cm. Due to the pre-consolidation and the self weight consolidation, the completed model ground had thin layer of over-consolidated clay underlain by the thick normally consolidated clay. After the self weight consolidation, the centrifuge was once stopped for the preparation of improved ground on 1 *g* laboratory floor. Model compacted sand piles were manufactured following the procedure devised by Kimura *et al.* (1982) and installed into the model ground. Saturated Toyoura sand was poured into water filled tubes whose inner diameter was 20 mm. The sand and tubes were subjected to vibration until the specified density was attained. The sand piles thus prepared were then slowly frozen and both ends were trimmed. The thin-walled tubes with a 20 mm outer diameter were inserted into the clay ground at a regular rectangular pattern with a distance of 33 mm (model scale) which corresponded to a low a_s of 0.28. Then the clay inside the tubes was removed by a tiny auger to make holes. Finally frozen sand piles were inserted into the holes and left for gradual thawing. This procedure was different from actual but was adopted in the model ground preparation in order to avoid any soil disturbance effect due to actual pile installation. After the soil improvement, the strong box was mounted again onto the swing platform of the centrifuge for the loading test. In order to reflect the prototype loading condition of the breakwater or revetments to the model test, it was most appropriate to apply the vertical load component in advance to applying the horizontal load component. The vertical load component was applied by quick lowering of the water level. The horizontal component was applied immediately after that by means of the horizontal loading jack.

3.3 Test Results and Discussion

In the vertical loading test, the load increased with increasing settlement and neither peak nor the final constant load was observed. The bearing capacity of the improved ground was determined as a yield of the ground. The yield load of the

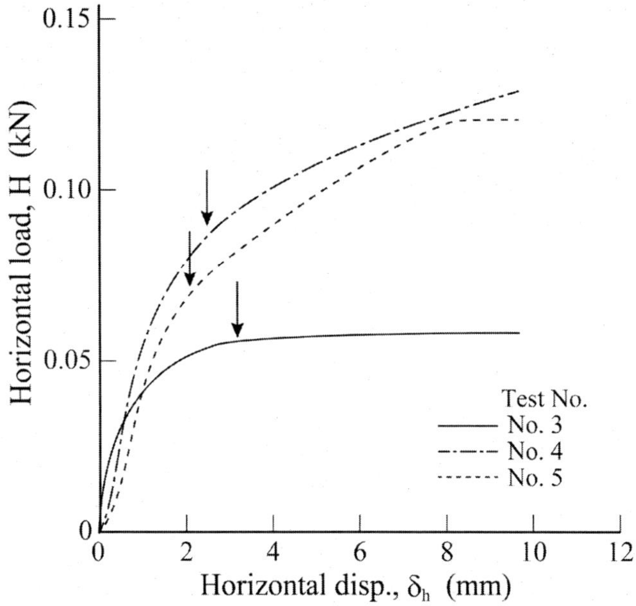

FIGURE 8: Horizontal load – displacement curves

ground was defined by the intersection of the initial tangent line of the curve and the tangent line at the straight portion of the curve at the larger settlement.

For the inclined load tests (see Figure 10 for the load inclination in each test), horizontal load H - horizontal displacement δ_h curves are obtained and shown in Figure8. In test No. 3 for the largest load inclination, the H increases with increasing δ_h but the load becomes constant after reaching a certain value. The bearing capacity for this particular case is therefore determined as this final constant value. Whereas the $H - \delta_h$ curves for the smaller load inclination do not show either peak or the final constant value. The bearing capacities for these cases are determined by the intersection of two tangent lines. As shown in the figure, it is known that the horizontal component of the bearing capacity is highly dependent upon the load inclination.

Figure 9 shows the photograph taken after the vertical load test (Test No. 1) to observe the deformation of sand piles directly. Shear planes and the sliding wedge are clearly observed in the central three rows of sand piles. The piles outside the wedge are deformed at their tops by the penetration of wedge but the overall improved ground does not reach the general shear failure.

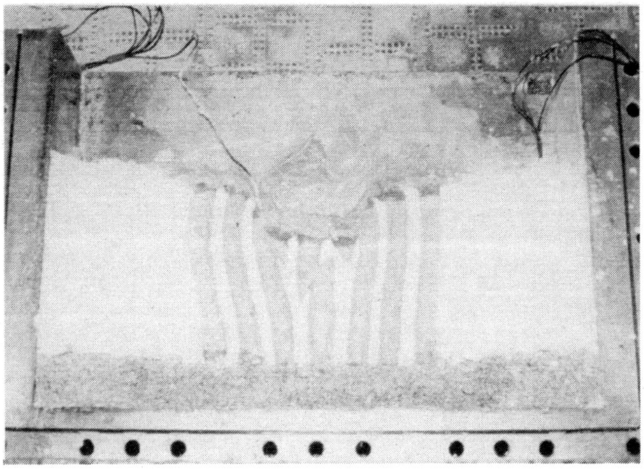

FIGURE 9: Failure mode of compacted sand piles (No. 1)

The vertical and the horizontal load components, V and H at yield or failure of the improved ground for all the tests are plotted in Figure10 to obtain the bearing capacity envelope on the V - H plane. The horizontal load which can be supported by the improved ground increases with the increasing vertical load. However, when the V reaches around a half of the vertical bearing capacity, the H becomes maximum and decreases with further increase in V level.

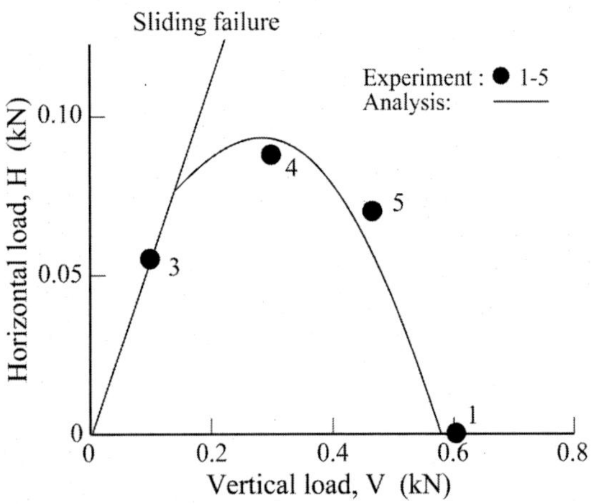

FIGURE 10: Bearing capacity envelope in V-H Plane

Several failure criteria for the composite ground have been proposed and employed in the routine design for many years in Japan. The bearing capacity is calculated by the Fellenius method of slip circle analysis combined by a shear strength expression and shown by a solid curve in Figure10. The curve gives the cigar shaped bearing capacity envelope in the V-H plane and is acceptable both qualitatively and quantitatively. Also calculated and shown by a solid straight line in Figure10 is the simple sliding failure of the foundation on the surface of sand mound. For this failure mode, the maximum horizontal load is calculated as a product of the effective weight of the foundation and the factor of friction of the sand mound. As is observed in the figure, the results obtained both by calculation and experiment are in perfect accordance for the sliding failure.

3.4 Summary

This series of test results was employed in the development of design formula of improved ground by the sand compaction pile method with low replacement area ratio. The centrifuge test results collaborated the analytical prediction of the improved ground's bearing capacity under combined loads, thus adding confidence to the predictive method.

4. STABILITIES OF GROUP COLUMN TYPE DEEP MIXING IMPROVED GROUND

4.1 Introduction

The Deep Mixing Method (DMM), a deep in-situ soil stabilization technique using cement and/ or lime as a binder was invented and developed by the Port and Harbour Research Institute (PHRI; the former body of PARI) to improve soft marine deposits (Coastal Development Institute of Technology, 2002). Due to quick strength increase and high shear strength of stabilized soil, DMM has become one of the most effective soil improvement techniques for marine and on-land constructions. A group column type improvement, where many columns are constructed in rows with rectangular or triangular arrangements, has been extensively applied to foundations of embankment or lightweight structures. A design procedure for the improved ground has been established in Japan, in which two failure patterns are assumed: external and internal stabilities (Public Work Research Center, 2004). In the external stability, the possibility of sliding failure is calculated, in which the DM columns and the clay between move horizontally on a stiff layer without any rearrangement of columns (i.e. a failure mechanism passes outside the improved region). In the internal stability analysis, rupture breaking failure is calculated by a slip circle analysis, in which the shear failure mode of DM columns is assumed (i.e. a failure mechanism passes inside the improved region).

The first author conducted a research project on the failure mechanism and stability of group column type improved ground subjected to embankment loading (Kitazume and Maruyama, 2006; 2007). In the centrifuge modelling of the external stability, acrylic pipes were used as model stabilized soil column instead of the mixture of soil and cement, in order to measure the axial force and bending moment induced in the pipe. In the internal stability, the mixture of soil and cement was used as model soil column to simulate its failure in the centrifuge. In both cases, a sort of ideal and simple model ground technique was adopted to avoid any influence due to the columns installation: the model columns prepared separately were installed into holes in the ground. In this section, the test results on the failure pattern of the improved ground and the criteria related to external and internal stabilities are briefly introduced.

4.2 Preparation of Model Ground

Figure 11 schematically shows a typical example of model ground setup, in which a normally consolidated clay ground of20cm thick and five rows of DM columns were modeled. In the model ground preparation, a drainage layer of Toyoura sand was made on the bottom of specimen box. Kaolin clay slurry was poured on the layer and then pre-consolidated one dimensionally by the vertical pressure of 9.8 kN/m^2 on a laboratory floor. After the self-weight consolidation at a 50 g acceleration field, the centrifuge was stopped once for preparation of improved ground on a laboratory floor. A thin walled tube with an outer diameter of 20 mm was penetrated into the clay ground. The clay inside the tube was then carefully removed using a tiny auger to make a hole, and a model DM column was inserted after removing the tube. Similar to the SCP model ground preparation, this procedure is different from the actual but is adopted in order to avoid any soil disturbance effect due to actual execution of constructing stabilized soil columns.

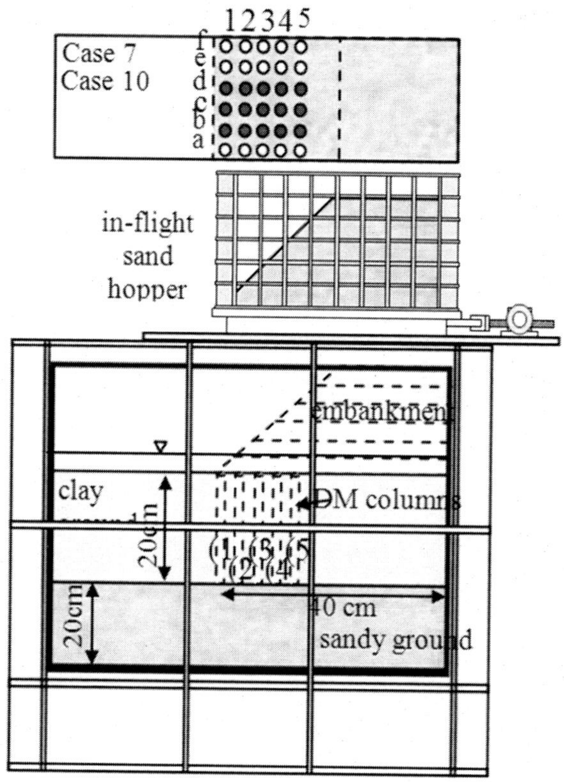

FIGURE 11: Model ground setup (Cases 7 and 10)

4.3 Test Procedure

The model ground thus constructed was brought up to a 50g centrifugal acceleration field, which corresponded to a 10m thick soft clay layer improved by DM columns of 1m diameter in a prototype scale. The model ground was allowed to consolidate again by the enhanced self-weight to minimize soil disturbance effect that might be induced during the model ground preparation. After the consolidation, the model embankment was constructed stepwise under almost undrained conditions by the in-flight sand raining device.

In the test series, two types of model column, 2cm in diameter and 20cm in length, an acrylic pipe and two different cement stabilized columns, were used as a DM column. A total of 11 model tests were carried out as summarized in Table 2.

The acrylic model column (named as A-column) was used to simulate soil column with very high strength in Cases 2 to 5 for investigating the external stability with bending moment measurements, while the cement stabilized columns (named as Tl-column and Th-column) were used in Cases 6 to 11 for investigating the internal stability with simulating rupture breaking failure. In order to detect the model column failure during loading, a carbon rod electric probe was embedded into the columns before hardening which are placed at the rows b, c and d, marked by red circles in Figure 11(Kitazume and Maruyama, 2007).

Table 2: Test Conditions and Major Test Results

	Improvement Condition				
	Width (cm)	*No. of Rows*	*Imp. Area Ratio, as*	*Mate-rial*	*qu (kN/ m2)*
Case 1	0	-	-	-	-
Case 2	8.6	3	0.28	A	-
Case 3	15.2	5	0.28	A	-
Case 4	21.8	7	0.28	A	-
Case 5	15.2	5	0.56	A	-
Case 6	8.6	3	0.28	Tl	425
Case 7	15.2	5	0.28	Tl	411
Case 8	21.8	7	0.28	Tl	391
Case 9	8.6	3	0.28	Th	1271
Case 10	15.2	5	0.28	Th	1290
Case 11	21.8	7	0.28	Th	1434

4.4 Test Results and Discussion

The measured embankment pressure and displacement curves for the external stability study (Cases 1 to 4) are shown in Figure12(a). In the figure, the vertical and horizontal axes show the embankment pressure, p_e, measured at ground surface and the horizontal displacement at the toe of embankment slope, δ_h, respectively. In the unimproved ground (Case 1), a relatively small horizontal displacement takes place as long as the embankment pressure remains at a very low level, but the displacement increases rapidly with further increase of embankment pressure. In the improved grounds with A-columns (Cases 2 to4), on the other hand, the horizontal displacement increases with increasing embankment pressure, but its magnitude is small compared to that in the unimproved ground. The magnitude of horizontal displacement becomes smaller when the improvement width increases.

The p_e and δ_h curves of Cases 7 and 10 for the internal stability are plotted in Figure12(b). In the figure, the letters beside the curves indicate the column ruptures, indicated by the ID number of the columns as shown in Figure11. In Case 7, Tl-1b column failed first at p_e of 26.2 kN/m^2, and Tl-2b, Tl-2d and Tl-3c failed at the same time. As p_e increased, the columns failed one by one in sequence from the forefront to the rearmost column. In Case 10, the forefront columns failed one by one at p_e of 34.2 to 50.2 kN/m^2. When p_e increased to 79.6 kN/m^2, Th-5b, Th-5c and Th-5d failed instead of the second, third and fourth

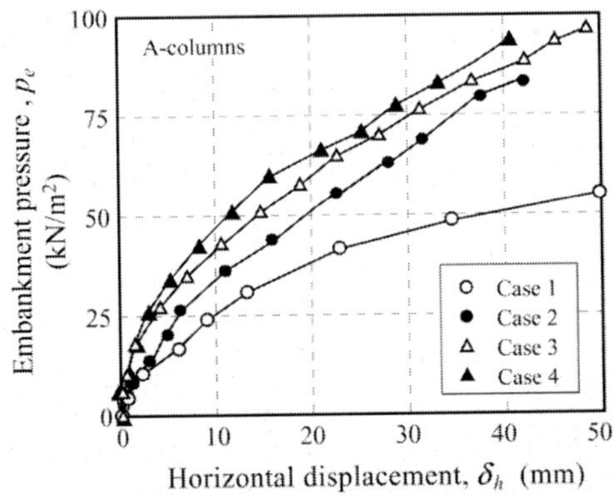

(a) for external stability study (Cases 1 to 4)

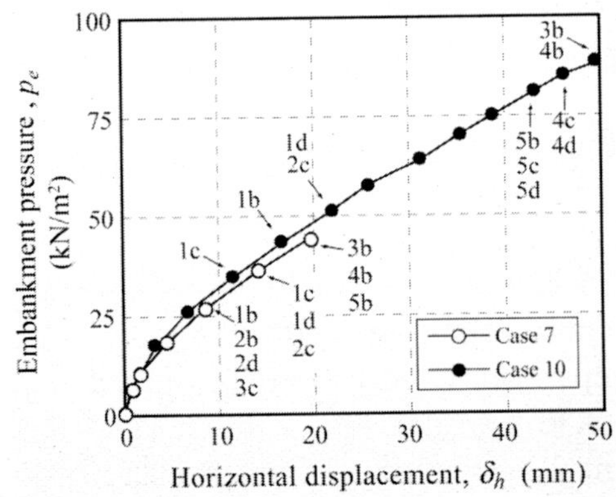

(b) for internal stability study (Cases 7 and 10)

FIGURE 12: Embankment pressure and horizontal displacement curves

row columns. After that, Th-4 and Th-3 failed in reverse sequence from the rearmost to the forefront column. It is of interest to note that p_e continually increased even after many columns failed.

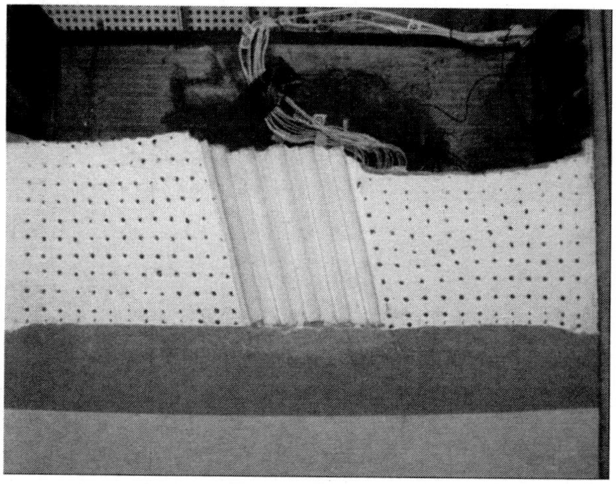

(a) Case 3

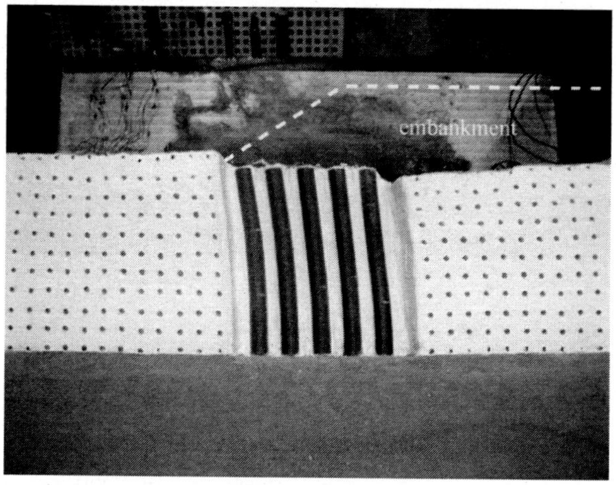

(b) Case 7 (b-line)

(c) Case 10 (c-line)

FIGURE 13: Failure pattern of improved ground

The DM columns after embankment loading in Cases 3, 7 and 10 are shown in Figure13. In Case 3, Figure13(a), all the columns tilted like dominos about their toe without any breaking failure. The inclination angle was the same throughout the columns, indicating that the improved area deformed uniformly as a simple shear failure. This failure pattern is quite different from that in the external stability analysis of the current design method where the sliding failure pattern is assumed.

In Case 7, Figure13(b), all the columns titled counterclockwise with bending failure. As the embankment loading was terminated at a relatively small embankment pressure to prevent heavy column failure, the tensile cracks cannot be observed clearly in the figure. According to Figure12(b), Tl-1b and Tl-2b failed first and then the other three columns, Tl-3b, Tl-4b and Tl-5b, failed at the same p_e of 43.9kN/m^2. In Case 10, Figure13(c), all the columns tilted counterclockwise with tensile cracks at two depths. The figure clearly shows that the column did not fail by shear failure mode but rather by bending failure mode. According to the detailed observation after the test, the bending failure took place at a shallow depth first and then at a deep depth.

According to the failure pattern observed in the model tests, a simple stability calculation was carried out. In the calculation of the external stability of clay ground between the columns were assumed to deform as a simple shear due to the unbalanced pressure of active and passive earth pressures acting on the side boundaries of improved area. For the calculation of the collapse failure pattern, the moment equilibrium at the bottom of improved area was analyzed. The calculated $p_{ef,collapse}$ for various stress concentration ratio, n and friction angle of embankment, ϕ_e are plotted in Figure14againstthe improvement width, D. The $p_{ef,collapse}$ increases almost linearly with D for all cases. The effects of ϕ_e and n are quite small on $p_{ef,collapse}$: the n value in particular has a negligible effect. In the figure, the model test results are also plotted. Although the calculated $p_{ef,collapse}$ still overestimates the test results for small D but well coincided them for large D. The calculation provides reasonable estimations compared with the experiments.

A series of calculations was carried out for various improvement widths and column strengths, and the relationship between the improvement width, D and the failure depth z is shown in Figure 15 for various q_u values. It is found that the z increases monotonically with increasing D, and with increasing q_u. However, the effect of q_u is not so dominant as compared to the shear failure pattern. In the figure, the model test results are also plotted. The calculation gives reasonable estimation of the failure depth, in which the calculations slightly overestimate the model test results for Cases 6 to 8 but underestimate for Cases 9 to 11.

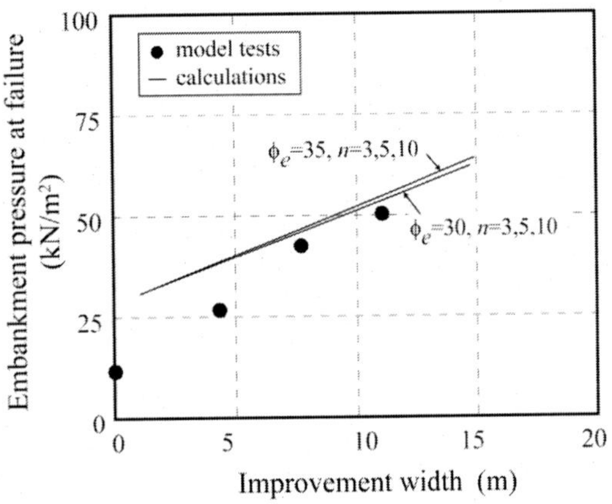

FIGURE 14: Comparison of calculations for collapse failure with model test results

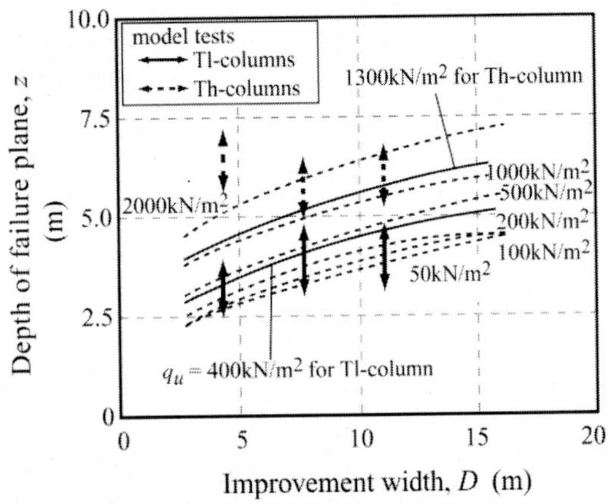

FIGURE 15: Relationship between the improvement width and the failure depth for bending failure mode

For the internal stability, a simple stability calculation was also proposed. In the calculation, all the DM columns were assumed to fail simultaneously with the bending failure mode and the improved area above a failure plane was assumed to deform as a simple shear.

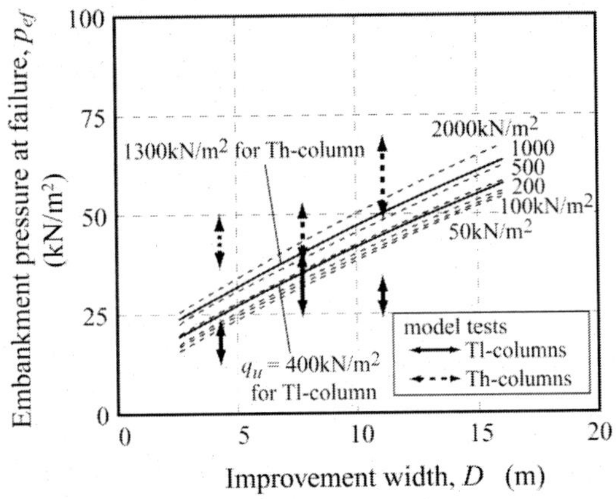

FIGURE 16 Embankment pressure at ground failure and improvement width

For the calculation, the moment equilibrium at the assumed failure plane was formulated. The $p_{ef,bending}$ is shown against D in Figure16. The figure shows the $p_{ef,bending}$ value increases with increasing D and q_u. However, the effect of q_u is relatively small. The model test results are also plotted in the figure. As compared with them, the calculations give reasonable estimation to the model tests.

4.5 Summary

According to the tests, it is found that the current design method overestimates the model test results in the external and internal stabilities, because inadequate failure patterns are assumed in the analyses. The proposed calculations based on the centrifuge model tests have relatively high applicability for evaluating the external and internal stabilities of the group column type DM improved ground.

5. EFFECT OF DEEP MIXING WALL SPACING ON LIQUEFACTION MITIGATION

5.1 Introduction

The deep mixing method has been adopted for many construction projects for improving the stability and reducing settlement. Recently, the grid type improved ground has also been applied for liquefaction prevention, in which a grid of stabilized columns functions to restrict generation of excess pore pressure by confining the soil particle movement during earthquake. The improvement effect of the technique was confirmed in the 1995 Kobe earthquake (Tokimatsu *et al.*, 1996). Experimental and numerical studies have been carried out to investigate the effect of grid spacing on pore pressure generation and liquefaction prevention (e.g. Namikawa *et al.*, 2007). Based on these studies, a quite simple guideline for the method was established. However, the current design practice of this method does not take into account the different seismic behavior at different depths but evaluates the possibility of liquefaction only at the mid-depth.

A series of centrifuge model tests was conducted to investigate the effect of grid spacing on generation of pore pressure and seismic response in a sand layer. In this research, the spacing of grid is the main concern and the failure of the wall is not priority. Therefore the model wall was prepared by Bakelite material instead of the mixture of soil and cement. The effect of grid spacing was assessed in the other centrifuge tests by taking into account the ratio of grid spacing to depth (Takahashi *et al.*, 2006).

5.2 Preparation of Model Ground

An example of model grounds is schematically shown in Figure 17, where two model grounds with different sizes were prepared in a specimen box for conducting many tests. The model grid was made of Bakelite panels with the thickness of 2 cm, as shown in Figure18. The model grid was fixed on the specimen box with bolts. Several model grids were prepared to perform parametric tests.

The sand material used in this study was Soma sand, whose U_c and D_{10} were 1.7 and 0.34 mm respectively (Takahashi *el al.*, 2006). Several accelerometers and pore pressure gauges were placed precisely at a depth of 4 cm and 10 cm from the ground surface. After filling the sand, the ground surface was carefully leveled by means of a vacuum, and then fully saturated by the percolation technique using Carbon dioxide gas and vacuum. The fluid used in this study was an aqueous solution of hydroxypropyl methylcellulose. The viscosity of fluid was controlled to be 25m²/s for the 25*g* centrifuge model test by changing its concentrations.

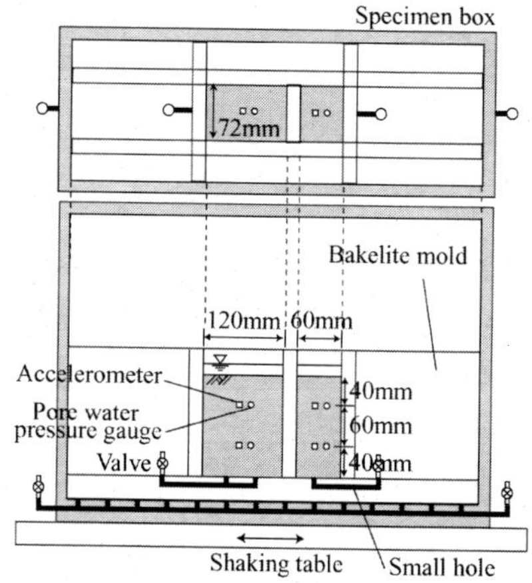

FIGURE 17: Schematic view of model ground

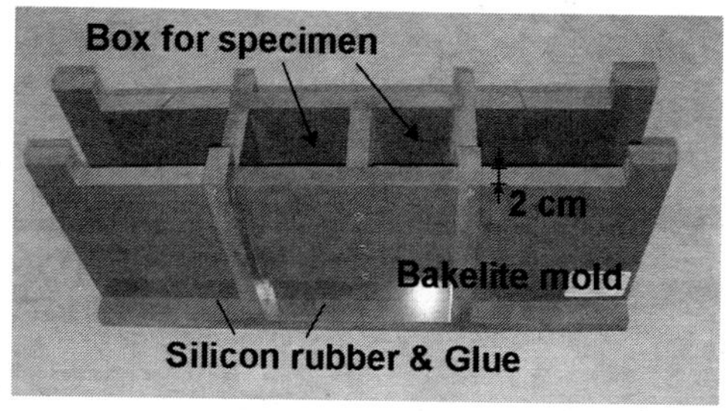

Figure 18: Bakelite panels

5.3 Test Procedure

Soon after reaching a centrifugal acceleration of 25 *g*, the model ground was subjected to seismic excitation of 50 sinusoidal waves with 4 Hz in the prototype scale. After confirming the dissipation of excess pore pressure generated in the previous excitation, the excitation level was increased stepwise until the model ground liquefied. A total of eleven model tests were carried out on the model ground prepared with various types of grid spacing, as summarized in Table 3.

Table 3: Test Conditions

Case	*Wall Spacing, L*	*Relative Density*	*Viscosity*
	(m)	*(%)*	*(m2/s)*
E1-1	1.5	49.7	23.6
E1-2	1.5	44.0	15.2

Case	*Wall Spacing, L*	*Relative Density*	*Viscosity*
	(m)	*(%)*	*(m2/s)*
E2-1	2.0	44.0	21.5
E2-2	2.0	41.7	13.0
E3-1	2.5	44.0	21.5
E3-2	2.5	41.7	13.0
E4-1	3.0	49.7	23.6
E4-2	3.0	44.0	15.2
M25-5	6.0	33.7	18.3
M25-6	6.0	53.0	20.9
M25-7	6.0	50.0	24.2

5.4 Test Results and Discussion

The relationships between the induced acceleration at a depth of 1.0 m(values are quoted in prototype scale in this subsection) and input excitation are plotted in Figure 19. The induced acceleration measured in various grid spacing initially increases almost linearly with increasing the input acceleration. The induced acceleration for the wall spacing, L of 2.0 m shows a sharp decrease at the input excitation of about 380 gal, which indicates that liquefaction took place in the ground. The model ground having the L of 1.5 m doesn't show sharp decreases in the acceleration even if the input acceleration exceeds about 500 gal. Similar phenomenon can be seen in the measurements at a depth of 2.5 m.

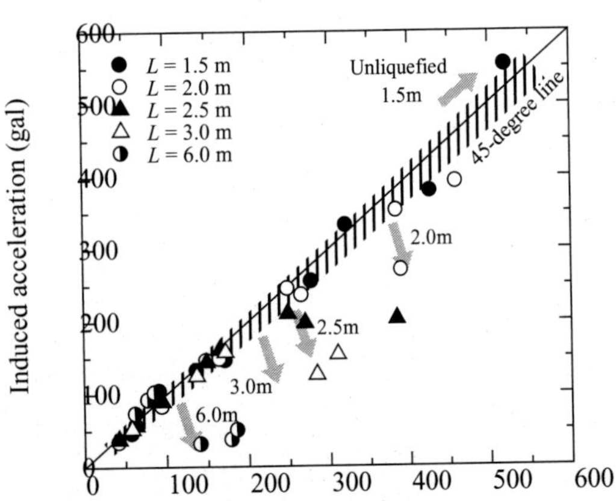

FIGURE 19: Relationship between induced acceleration and input excitation at a depth of 1.0 m

The relationship between the maximum excess pore pressure ratio, $\Delta u/\sigma'$ measured at a depth of 1.0 m and the input excitation is shown in Figure 20. As the input excitation increases, the $\Delta u/\sigma'$ increases gradually irrespective of the grid spacing. The $\Delta u/\sigma'$ increases to unity in the case where L is equal to or exceeds 2.0 m, and the input acceleration at that time decreases as increasing the L value. In the ground with L of 1.5 m, the $\Delta u/\sigma'$ does not increase to unity even when the input acceleration exceeds about 500 gal. According to the above discussion, the improvement effect on liquefaction prevention is highly influenced by the grid spacing.

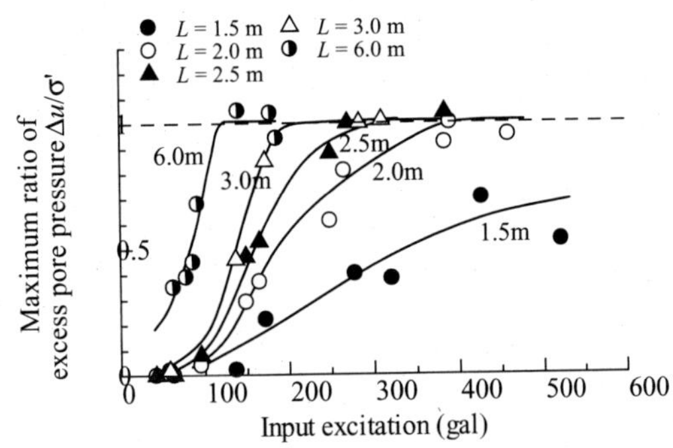

FIGURE 20: Relationship between maximum ratio of excess pore pressure and input excitation at a depth of 1.0 m

Figure 21 shows the relationship between the ratio of grid spacing to the grid depth, L/d, and the input acceleration. The d is the concerned depth measured from the ground surface. In the figure, the test data when liquefaction did or did not take place are plotted as a filled circle and an open circle, respectively. A gray-filled circle shows the excitation level where the ground liquefies to some extents during the excitation. Based on the plotted data, a liquefaction boundary can be drawn as the hatched portion in the figure. The boundary shows that the input acceleration necessary for liquefaction increases with decreasing L/d, very sharply when the L/d becomes less than about 2. This figure confirms that a large effect on liquefaction prevention can be expected when the L/d ratio becomes less than about 2. This figure also indicates that the liquefaction can be induced by a relatively small earthquake attack in the shallow depth of ground (i.e. having a small d value) even if the grid spacing is quite small. This suggests a limitation in the application of the grid type improvement for liquefaction prevention, and the shallow ground should be improved by the other technique.

5.5 Summary

Based on the accelerations and pore pressures measured in the model ground with five different grid spacing, it was confirmed that the improvement effect of liquefaction prevention was influenced by not only the grid spacing but also the magnitude of excitation and it also differs at different depth. In the present study, a new parameter, the ratio of grid spacing to depth, *L/d*, was proposed to evaluate the effect of grid improvement on liquefaction prevention.

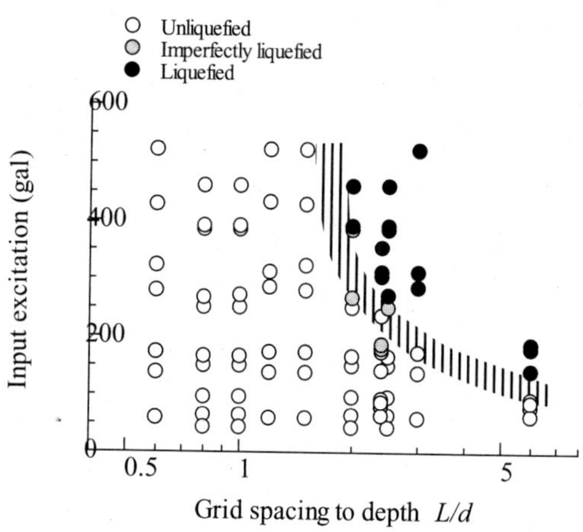

FIGURE 21: Relationship between input excitation and *L/d*

6. CONCLUDING REMARKS

In this paper, several examples of centrifuge modeling were briefly introduced, which include simulation of grouting installation and investigation of improved ground by the sand compaction pile method and the deep mixing method. In the first, a miniature installation machine was invented to simulate the actual grouting in centrifuge as closely as possible, and the stress development during the execution was measured and studied. In the other studies, the model ground was prepared with different material and different procedure from the actual execution in order to make an idealized and simple improved ground and to avoid any effect due to execution of improvement in field on the behavior of improved ground in model test. This approach has often been adopted in the Institute for performing parametric studies and to highlight the critical issue in the behavior of improved ground. This approach requires detail understanding and experiences in the field behavior for applying the model test results to actual conditions. The authors also recognize alternative approaches where a model ground is prepared in similar techniques and manners to the actual as precisely as possible and apply the test results to actual situations directly. Selection of the approaches is up to centrifuge modeler, but should be appropriately conducted accordingly considering the research target of model test. The authors should emphasize that detail understanding of behavior of actual improved ground and experiences of field ground improvement work are important, irrespective of the approach.

Finally the authors wish that this paper will be useful for performing centrifuge model tests relating to ground improvement techniques.

REFERENCES

Coastal Development Institute of Technology. (2002) The deep mixing method – principle, design and construction, A. A. Balkema, p123.

Coastal Development Institute of Technology. (2007) Technical manual of compaction grouting method as a countermeasure for liquefaction, p188 (in Japanese).

El-Kelesh, A. M. and Matsui, T. (2008) "Calibration chamber modeling of compaction grouting". Geotechnical Testing Journal, 31, Issue 4, pp 295-307.

Kimura, T., Nakase, A., Kusakabe, O., Saitoh, K. and Ohta, A. (1982) "Geotechnical centrifuge model tests at the Tokyo Institute of Technology", Technical Report No.30, Department of Civil Engineering, Tokyo Institute of Technology, pp 1-33.

Kitazume, M. (2005) "The sand compaction pile method", A. A. Balkema, p 232.

Kitazume, M. (2009) "Twenty Nine Years of experiences of physical modeling of geotechnical problems in port structures", International Journal of Physical Modelling in Geotechnics, 9, Issue 3, pp 1-19.

Kitazume, M. and Maruyama, K. (2006) "External Stability of Group Column Type Deep Mixing Improved Ground under Embankment Loading". Soils and Foundations, 46, Issue 3, pp 323-340.

Kitazume, M. and Maruyama, K. (2007) "Internal Stability of Group Column Type Deep Mixing Improved Ground under Embankment Loading". Soils and Foundations, 47, Issue 3, pp 437-455.

Kitazume, M. and Miyajima, S. (1995) "Development of PHRI Mark II geotechnical centrifuge", Technical Note of the Report of the Port and Harbour Research Institute, No. 817, p 33.

Mikasa, M., Takada, N. and Ohshima, A. (1989) "Dynamic consolidation test in centrifuge", Proceedings of the 11th International Conference on Soil Mechanics and Foundation Engineering, Vol. 2, pp 947-950.

Namikawa, T., Koseki, J. and Suzuki, Y. (2007) "Finite element analysis of lattice-shaped ground improvement by cement-mixing for liquefaction mitigation". Soils and Foundations, 47, Issue 3, pp 559-576.

Ng, Y. W., Lee, F. H. and Yong, K. Y. (1998) "Development of an in-flight sand compaction piles (SCPs) installer", Proceedings of the International Conference Centrifuge 98, Vol. 1, pp 837-843.

Nichols, S. C. and Goodings, D. G. (2000) "Physical model testing of compaction grouting in cohesionless soil". Journal of Geotechnical and Geoenvironmental Engineering, ASCE, 126, Issue 9, pp 848-852.

Nishimura, S., Takehana, K., Morikawa, Y. and Takahashi, H. (2012) "Experimental study of stress changes due to compaction grouting". Soils and Foundations, 51, Issue 6, pp 1037-1049.

Public Works Research Center. (2004) Technical manual on deep mixing method for on land works, p 334 (in Japanese).

Shinsaka, T., Zen, K., Sakamoto, K. and Yamazaki, H. (2003) "Experimental studies on the compaction effect of static grouting on sand", Proceedings of the Japan Society of Civil Engineers, No.764/III-67, pp 183-192 (in Japanese).

Takahashi, H., Kitazume, M., Ishibashi, S. and Yamawaki, S. (2006) "Evaluating the saturation of model ground by P-wave velocity and modeling of models for a liquefaction study". International Journal of Physical Modelling in Geotechnics, 6, Issue 1, pp 13-25.

Terashi, M., Kitazume, M. and Minagawa, S. (1991) "Bearing Capacity of Improved Ground by Sand Compaction Piles", Deep Foundation Improvements: Design, Construction, and Testing, ASTM STP 1089, Robert C. Bachus, Ed., American Society for Testing and Materials, 1990, ASTM, pp 47-61.

Tokimatsu, K., Mizuno, H. and Kakurai, M. (1996) "Building damage associated with geotechnical problems". Soils and Foundations, Special Issue 1, pp 219-234.

Weber, T. M., Laue, J. and Springman, S. M. (2006) "Centrifuge modelling of sand compaction piles in soft clay under embankment load", Proceedings of the 6th International Conference on Physical Modelling in Geotechnics, Hong Kong, pp 603-608.

Yamazaki, H., Kaneda, K., Adachi, M., Harada, Y., Yamada, K. and Takahashi, T. (2008) "Model tests on compaction grouting method", Proceedings of the Japan Society of Civil Engineers, Series C, 64, Issue 3, pp 544-549 (in Japanese).

Centrifuge Tests of Monopod and Tripod Bucket Foundations for Offshore Wind Turbine Tower

Y.W. Choo, D.J. Kim[1], S. Kim, J.H. Kim, D.S. Kim, S.H. Kim[1] and J.H. Choi[1]

Department of Civil and Environmental Engineering, KAIST, Daejeon, Republic of Korea
[1] *Hyundai Engineering and Construction Co. Ltd., Seoul, Republic of Korea*
E-mail: ywchoo@kaist.ac.kr, djkim@hdec.co.kr, kimsurin@kaist.ac.kr, zlawogus@kaist.ac.kr , dskim@kaist.ac.kr , shjee@hdec.co.kr, jhchoi@hdec.co.kr

ABSTRACT

Centrifuge load tests of monopod and tripod suction bucket foundations designed for the Western Sea in Korea were performed. The horizontal load and overturning moment for an offshore wind turbine were applied and monitored in the test, with the vertical load being simulated by self-weight of the bucket model. The load-rotation angle results for monopod and tripod foundations were compared. From the load-rotation angle curve, the initial stiffness and yielding point were determined and compared to study the behaviour of two foundation types.

1. INTRODUCTION

The market for offshore wind farm is expected to grow fast, and the design and installation of substructure and foundation are getting more important. As for the offshore wind farms located in the shallow (depth < 20m) water, the construction and installation of the substructure and foundation makes up about 1/4~1/3 of the offshore wind farm construction cost, and the portion is expected to increase because the turbine capacity is increasing from 2~3MW to 5MW or larger and the water depth of wind farms is also increasing over 30m (EWEA, 2008; EWEA, 2009a; EWEA, 2009b).

As a foundation for offshore wind turbine, the suction caisson foundation is being considered to be a highly competitive alternative to the conventional monopile or gravity based structure, because it has features suitable for the offshore construction such as quick installation, no heavy equipment for penetration and no hammering noise from driving.

Despite its promising prospect of engineering application, buckets foundations have been developed rather slowly so far. Available design experiences of monopod and tripod bucket foundations for offshore wind turbines are still very limited because the existing data obtained from laboratory model tests and field applications are insufficient. Furthermore, the soil-structure interaction and failure mode of bucket foundations under overturning moment have not been investigated thoroughly yet.

Compared to extensive previous works of bucket foundations on clay (Bransby and Randolph 1998; Gourvenec 2007), elaborated numerical analysis in cohesionless soils is relatively difficult. Previous works on bucket foundations installed in sand are mainly presented by the research group at the University of Oxford (Byrne and Houlsby 2003; Villalobos 2006). These are based on best-fitting results of small-scale laboratory model tests. Meanwhile, in Korea, projects for the development of offshore wind farms are under way at many sites and Western sea of Korea is considered as one of the most potential areas in Korea. As the soils in Western sea of Korea are silt or silty sand, the investigation on bucket foundations in silt and silty sand is much important.

In this paper, centrifuge load tests of monopod and tripod suction bucket foundation designed for the Western Sea in Korea were performed. The prototypes of the monopod foundation and tripod foundation were designed for a 3MW turbine. The dimensions of the steel monopod caisson were a diameter of 15.5m and a length of 10.5m. The prototype of the tripod foundation consisted of three bucket caissons with a diameter of 6.5m and a height of 8.0m. The distance between the centers of three bucket caisson is 26.9m. The site comprised of a dense silty sand layer above a thick sandy silt layer. The bedrock was over 40m below the seabed. The soils at the nearby seashore were collected and used to form the model soil profile in the container. The horizontal and overturning moment were applied at 33m from seabed and monitored in the test, with the vertical load being simulated by self-weight of the bucket model. The load-rotation angle relationships for monopod and tripod foundations were compared. From the load-rotation angle curve, the initial stiffness and yielding point were determined and compared to study the behavior of two types.

2. CENTRIFUGE MODELLING

2.1 Facility

Centrifuge tests were performed using the 240g-ton geotechnical centrifuge equipment at KAIST (Korea Advanced Institute of Science and Technology) in South Korea. The maximum capacity of this KAIST beam centrifuge, with a 5m radius, is 2400kg for up to 100g of centrifugal acceleration and 1,300kg at 130g of maximum centrifugal acceleration. The detailed specifications of the centrifuge equipment can be found in Kim et al. (2012). Figure 1 shows the centrifuge equipment with the testing system developed in this study.

FIGURE 1: Centrifuge equipped with a bucket foundation model

2.2 Prototype Seabed Conditions

Western sea of Korea is considered as one of the most potential areas for offshore wind farms in Korea. In this study, one site near Western coast in Korea, which is located near Wido island and Anma islands, was selected. The water level of the site is 15m deep. The seabed of the target site was basically composed of three layers: silty sand layer from seabed surface to 10m, ML and CL layer from 10m to 32m, silty sand from 32m to bedrock. The bedrock was at around 50m depth. Since the lengths of prototype tripod and monopod in this study were 8.0 m and 15.5 m, respectively, two layers from the surface to 32 m (which is about 2 times the length of monopod bucket foundation) were modeled in this study. The detailed information of the site is tabulated in Table 1.

Table 1: Conditions of Prototype Soil Layers

Item	SM layer		ML Layer	
Depth (GL)	0m ~ –4m	–4m ~ –11m	–11m ~ –20m	–
SPT N	N = 17 ~ 39	N = 50	N = 19 ~ 31	N = 10 ~ 19
Estimated D_r^*	70%		60%	

* D_r = Relative density

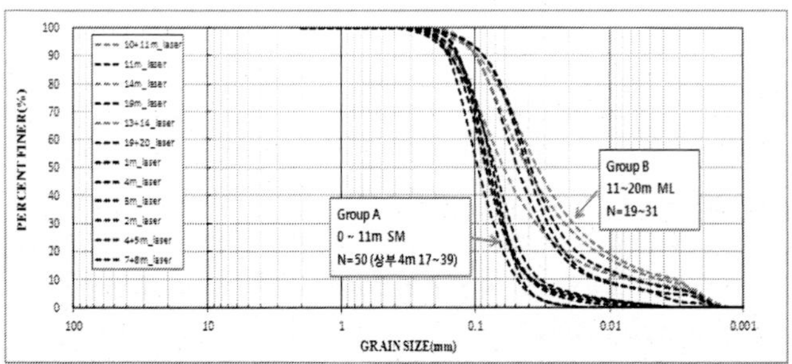

FIGURE 2: Grain size distributions of soils sampled the target site

Figure 2 presents grain size distributions of samples from the target site. As mentioned earlier, the soil deposit of the sites was classified into two groups: SM (0 m to 11 m) and ML (11 m to 32 m). In order to simulate the SM and ML layers of the site, model samples were collected from shores close to the target site. The final model soils were selected and their grain size distributions and basic properties are presented in Figure 2 and Table 2.

Table 2: Basic Properties of Model Soils

Item	*SM Layer*	*ML Layer*
Fine contents (passing #200)	47% (40% ~ 50%)	75%
Max. dry density*	1.65 (t/m^3)	1.77 (t/m^3)
Min. dry density**	1.20 (t/m^3)	1.14 (t/m^3)
Grain size (mm)	D_{10} = 0.045 D_{50} = 0.080 D_{60} = 0.095	D_{10} = 0.0055 D_{50} = 0.050 D_{60} = 0.060
c_u	2.11	10.91
G_s	2.67	2.64

* Head (2006), ASTM D 1557, KS F 2312
** ASTM D 4254

2.3 Preparation of Model Soil Layers

Based on in-situ SPT results, the prototype seabed was grouped into two layers: SM and ML layers. The relative densities of the prototype layers were estimated in Table 2 and they were adopted for target relative densities of the model layers.

Scaling ratio was 1/70. The final thickness of SM and ML model layer was 157 mm and 300 mm, corresponding to 11.0m and 21.0m in the prototype seabed. In order to prepare designed model layers having the pre-determined relative density, the model seabed were compacted by totally nine layers in turns. The lower ML layer was compacted by six sub layers with a height of 50 mm height each. The upper SM layer was compacted by three layers with a height of 52 mm. After dumpling soils into the container for each sub layer, in order to control the model density to the pre-determined value, the layer was compacted by dropping a 13.5kg cylindrical steel plate in a constant height of 500mm and then static pressure of about 800kPa was applied.

The resulting conditions of all centrifuge models are tabulated in Table 3. After all preparation of centrifuge models, water was filled very slowly from the model surface to saturate the model layers.

Table 3: Conditions of Centrifuge Soil Models

Test ID	*T1*		*T2*		*T3*	
Type	*Monopod*		*Tripod*		*Tripod*	
γ_d of Layers (t/m^3)	SM	ML	SM	ML	SM	ML
	1.50	1.44	1.55	1.44	1.51	1.49
D_r** (%)	72.4	58.4	82.7	58.4	75.8	66.3
Loading	Monotonic		Monotonic		Cyclic	

* γ_d = Dry density, ** D_r= Relative density

After all of testing system and model buckets were installed, a pilot spinning was performed with test-ready centrifuge models. In this pilot spinning, horizontal and moment load was not applied. This pilot spinning aimed to enhance the saturation of the model seabed and minimize the settlement of the seabed during the acceleration from 1g to 70g by inducing settlement in advance. After the pilot spinning, the status of the centrifuge models was checked.

Next to the pilot spinning, in order to evaluate seabed profile under inflight centrifugal acceleration, miniature cone penetration test and shear wave velocity measurement were performed during the second spinning prior to the loading test. The diameter of the miniature cone was 10mm. The cone was penetrated up to 300mm at a rate of 10mm/s in model scale.

The cone penetration results obtained at all centrifuge tests T1 to T3 are presented in Figure 3. For T2, the miniature cone was malfunctioned. The cone resistance q_c increased to approximately 10MPa with depth until a depth of 160mm in model scale (11.2m in prototype). After a depth 160mm, the q_c decreased about 5MPa or 8MPa, clearly showing two different layers. In addition, the q_c profiles were closely matched, showing constant model seabed conditions of all centrifuge tests.

The shear wave velocity was measured by a bender element array. The bender elements were installed in a fixed frame to stay in pre-assigned locations. The resulting shear velocity profile is shown in Figure 3. Similar to the qc profile, the V_s profiles of all centrifuge models are very closely matched and shows a sharp decrease at 11m, which was interpreted as a boundary of two layers. Based on the q_c and V_s analysis, the model soils were considered to be formed as designed.

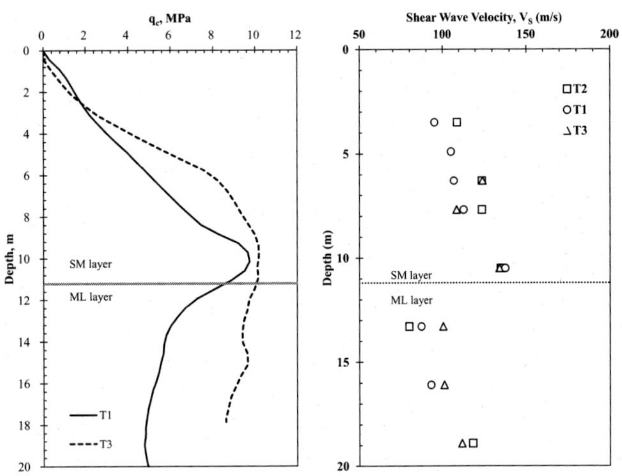

FIGURE 3: Cone tip resistance and shear wave velocity profiles

2.4 Modelling of Bucket Foundations

In this study, two types of prototype bucket foundations were selected: one is a monopod and the other is a tripod. Preliminary designs of the foundations were in Figure 4. Detail dimensions are summarized in Table 4. One of the main purposes of this study is to investigate the behaviours of two foundation types using centrifuge modelling.

The scaling ratio of 1/70 was used to model the prototype foundations. Outside dimensions were properly scaled as tabulated in Table 4. However, the thicknesses of the caisson side wall and the lid plate were selected with 1mm and 7mm. Because of the limitation in the model fabrication, thicker dimension was chosen for the side wall. Horizontal and overturning moment loads were applied through a tall tower rigidly connected on the top lid of the bucket foundation, probably resulting in extreme stresses on the lid. Therefore, the thickness of the top lid was selected

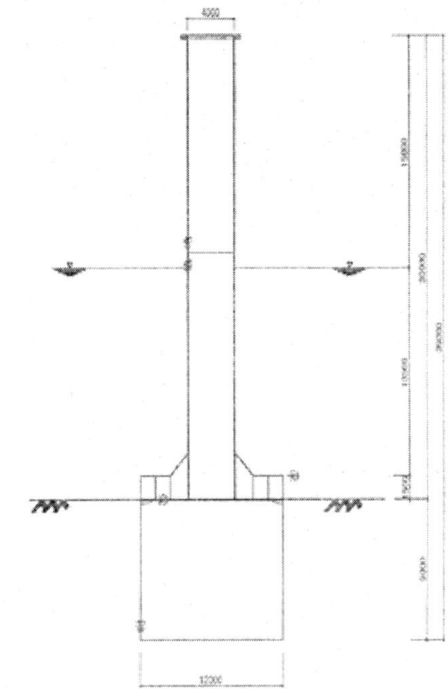

(a) Monopod

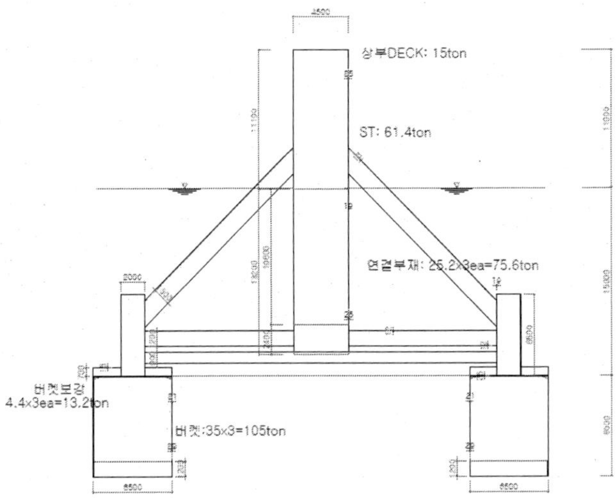

(b) Tripod

FIGURE 4: Designs of prototype bucket foundations

as a relatively thick value. In addition, thicker design dimensions in the model were expected to minimize the possibility of foundation structure deformation or breakage, leading to responses in the test mainly caused by soil behaviour. The dimensions are tabulated in Table 4 and Figure 5 shows the foundation models used for the tests.

Table 4: Specification of Prototype and Model Bucket Foundations

Item	*Monopod for 3MW*		*Tripod for 3WM*	
	Prototype	*Model*	*Prototype*	*Model*
Tripod CtC*	–		26.85 m	383.5 mm
Bucket dia. (D)	15.5m	221.4 mm	6.5 m	92.9 mm
Bucket length (L)	10.5m (0.68D)	150.0 mm	8.0 m (1.23 D)	114.3 mm
Side wall thickness	40 mm (0.26% D)	1 mm	25 mm (0.38% D)	1 mm
Top lid thickness	40 mm	7 mm	25 mm	3 mm
Weight	458 tonf		474 tonf	

* CtC denotes the distance between centers of buckets

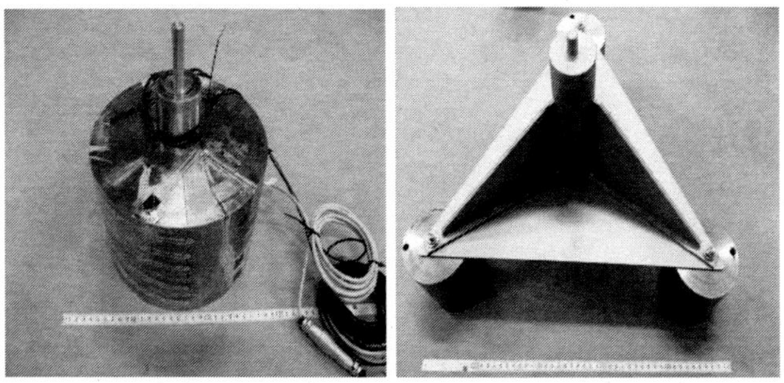

(a) Monopod (b) Tripod

FIGURE 5: Centrifuge model foundations

2.5 Simulation of Load Conditions

2.5.1 Horizontal and Moment Load

Offshore wind turbine towers are subjected to horizontal and moment loads simultaneously which is induced by wind load on nacelle, blade and tower and wave load on the substructure. Thus, a loading system for centrifuge tests was developed to simulate offshore wind turbine load conditions. The new loading system consisted of a linear actuator, an elevation spacer, a hinge, a clamp, and a Teflon collar as indicated in Figure 6. The linear actuator was capable of 10kN in model scale in a maximum rate of 8.3mm/s. The linear actuator adopted an electrical step motor capable of remote control and arbitrary loading pattern by displacement control.

The linear actuator was installed on top of the cylindrical soil container for stable support. In order to simulate simultaneous application of horizontal and moment load, horizontal load was applied at a predetermined height to a rigid tower rigidly connected on the top lid of the monopod or tripod as shown in Figure 6. The loadcell measured the load applied to the tower. This system provided horizontal and moment load on foundation system. The loadcell and clamp with the Teflon collar was installed at 471mm from seabed in model scale, simulating lateral load at 33m from seabed in prototype scale.

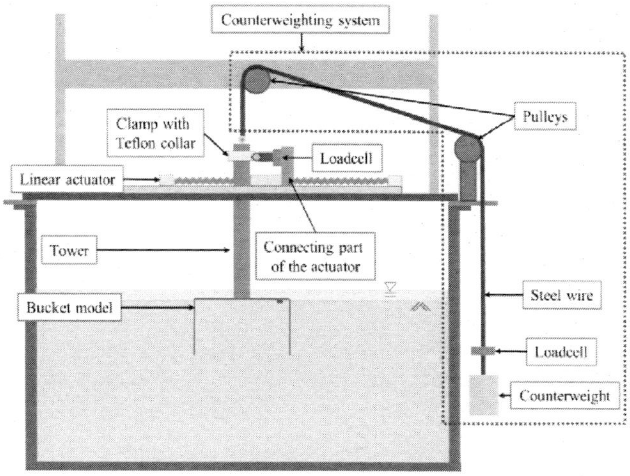

FIGURE 6: Schematic diagram of loading system

2.5.2 Vertical Load with Counter Weight System

The model foundations were heavier than the prototype because the thickness of the buckets and tower of models was thicker, resulting in more vertical load on seabed and foundation. Especially, resistance behaviour of the tripod was expected to be sensitively affected by the vertical load. Therefore, a counterweighting system was developed to adjust vertical load on the foundation. The schematic diagram of the counterweighting system is presented in Figure 6. The counterweighting system consists of a pair of pulleys, steel wire and a predetermined counterweight. A loadcell was attached in the middle of steel wire to measure the tension load in the wire. The weight of the counterweight was pre-determined to apply design vertical load to the foundation model and the system was designed to withstand 70g of centrifugal acceleration.

2.6 Instrumentation

The lateral load was measured by the loadcell installed between the extension rod of the actuator and the pin-jointed interface. The lateral load and moment induce lateral displacement as well as rotation. Therefore, displacement was measured at multiple locations on the tower to measure both lateral translation and rotation. Measurement locations are depicted in Figure 7 and tabulated in Table 5.

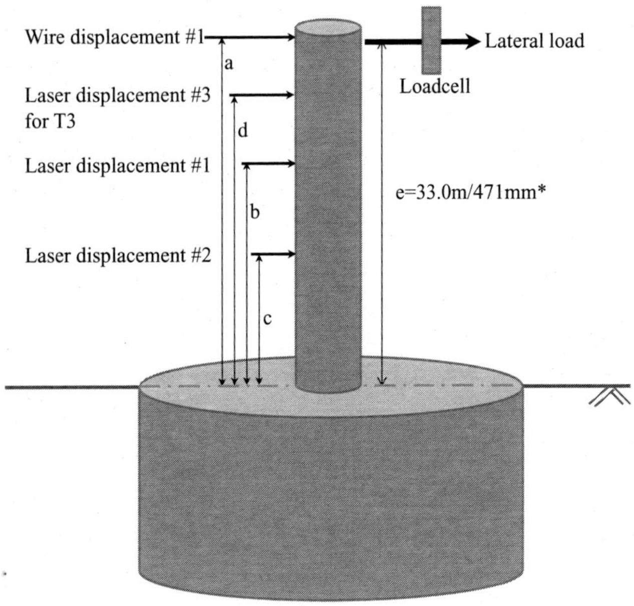

FIGURE 7: Measurement locations

Table 5: Measurement Locations

Test ID	*T1*	*T2*	*T3*
Wire #1 (a)*	33.1m (474 mm)**	32.3 m (461 mm)	33.2 m (474 mm)
Laser #1 (b)	17.2 m (246 mm)	20.0 m (285 mm)	20.8 m (297 mm)
Laser #2 (c)	8.0 m (114 mm)	12.6 m (180 mm)	13.4 m (192 mm)
Laser #3 (d)	–	–	25.1 m (359 mm)

* The letters denote the locations of sensors from the bucket lid in Figure 7.
** The values in () is model scale.

2.7 Experimental Procedure

The test program consisted of monotonic and cyclic loading tests. A moment load M can be achieved by applying a horizontal load H at a height M/H above the lid of the bucket. Recorded data at the target site was not sufficient to establish exact environmental loads. Thus, the loading condition was estimated based on the existing literature. The water depth of the site was 15m. The design loads on the top of bucket foundations were 3.48 MN of lateral load with 115MN-m of moment. Therefore, the lateral load in this study was applied at 33.0m (= 115MN · m/3.48MN) above the seabed. In the model, the clamp with the Teflon collar was attached at 471 mm above the seabed in model scale.

As summarized in Table 6, two types of loading were used in this study. For monotonic load, horizontal displacement was applied until yielding appeared. After the bucket foundations showed yielding behaviour, the controlled displacement was reversed until the horizontal load reached back to zero. After unloading, the horizontal load was applied again till 5 or 10 degree of rotation of the model was achieved.

T3 was performed with nine sequential cyclic loads. The nine sequential loads are summarized in Table 6 and they were sequentially applied to the tripod. Since the loading system used in this study was displacement control, the amplitude of displacement at the loading point was kept constant for each sequence and it is summarized in Table 6. The resulting M_{max} and M_{min} measured during the tests are tabulated in Table 6.

Table 6: Soil Conditions of Centrifuge Models

Sequence No.	D_{max} * *(mm)*	M_{max}/M_y **	M_{min}/M_y **	*# of cycles*
S1	0 ~ 0.7	0.39	–0.09	50
S2	–0.74 ~ 0.74	0.33	–0.47	
S3	0 ~ 1.23	0.48	–0.10	
S4	–1.23 ~ 1.23	0.62	–0.61	
S5	0 ~ 1.85	0.70	–0.16	
S6	0 ~ 4.71	1.34	–0.61	
S7	–1.23 ~ 1.23	0.62	–0.02	
S8	0 ~ 12.8	2.35	–2.33	
S9	0 ~	–	–	–

* D_{max} denotes the amplitude of displacement at the loading point and the plus denotes Direction 1
** M_{max} is the single amplitude of moment in the load cycle and M_y denotes the static capacity on the yield.

The loading direction for the tripod should be carefully selected because of the asymmetry of the tripod rather than the symmetry of the monopod. When the tripod is loaded with horizontal and moment loads, the tripod is governed by the pullout capacity of buckets under pullout loads. As shown in Figure 8, the resistance capacity of the tripod depends on loading direction because the number of buckets resisting pullout loads changes depending on the loading direction. The Direction 1 was determined to be the critical loading condition because only one bucket resists the applied load on the tripod. Therefore, the monotonic and one-way cyclic loads were applied in the Direction 1.

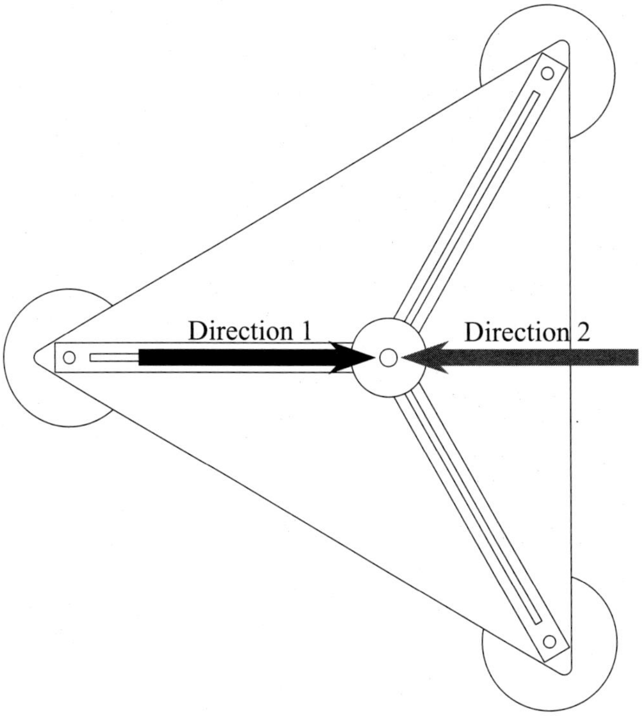

FIGURE 8: Loading directions on the tripod

3. TEST RESULTS

3.1 Behaviour under Monotonic Load

Figure 9 compares the resulting moment-rotation angle curves of the monopod and tripod. The two different bucket foundations showed clearly different behaviour. The stiffness of the tripod at the beginning of loading was stiffer than that of the monopod. The tripod showed almost linear behaviour before and after yielding and the yield point was relatively clear. The slope of the plastic range of the tripod (after the yielding point) became much lower than that of the elastic range. However, the yielding of the monopod was not clear and the loading-rotation angle curve of the monopod showed gradual decrease in the slope.

In order to define the point of yield, the method described by Villalobos (2006) was used. Straight lines were fitted to the initial stiff elastic section and the plastic section as shown in Figure 9. A horizontal line is drawn from the cross point to the load-rotation angle curve. The intersection was used to define a yield load which is denoted M_y. The M_ys for the tripod and monopod are 95 MN-m at 0.5° and 198MN-m at 3.4°, respectively. The yield load of the monopod was almost twice that of the tripod but the yielding rotation angle of the tripod was smaller than that of the monopod. It can be mentioned that the tripod is more efficient to control the rotation angle which is one of the most critical requirements for the wind turbine tower.

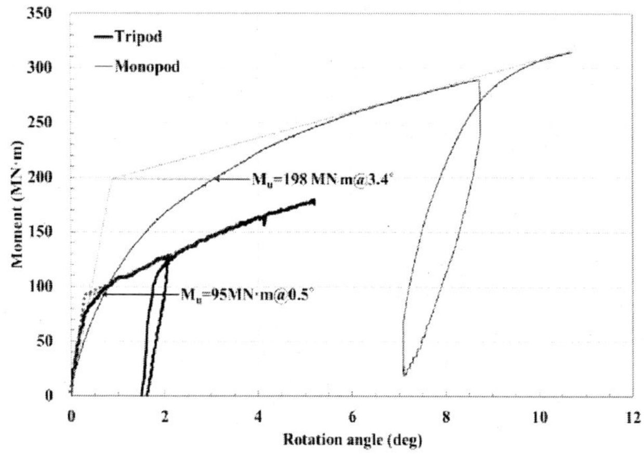

FIGURE 9: Moment-Rotation angle curve of T1 and T2

3.2 Behaviour under Cyclic Load

It is well-known that environmental loads in offshore areas are naturally cyclic under any weather conditions. In this study, various amplitudes of cyclic loads were applied to the tripod. Figure 10 presents the resulting moment-angle curves of the nine loading sequences and they compared with the monotonic loading curve of the tripod (T2).

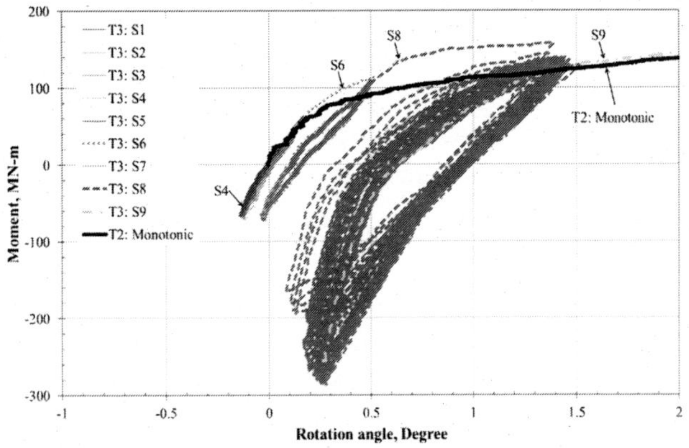

FIGURE 10: Moment-Rotation angle curves of T2 and T3

From S1 to S5, the amplitudes of cyclic load were within M_y and accumulated plastic deformations after 50 loading cycles were insignificant for both of one way and two way loading. On the contrary, as the amplitudes of cyclic load became larger than M_y (which are S6, S8), accumulated plastic deformations appeared. As the higher amplitude was applied, the larger was the accumulated plastic rotation angle. However, the accumulated plastic deformations became stabilized with the number of loading cycles. S9 (monotonic loading) after all of cyclic loads, interestingly, followed the plastic range curve with the same slope as the monotonic loading curve of T2.

Figure 11 plots the moment-rotation angle loops of S1 and S2 for 50 cycles. It is noteworthy to note that the slopes (stiffness) of the two sides (the plus rotation angle and the minus rotation angle) are different because the shape of the tripod is asymmetric. Under the plus rotation angle, one

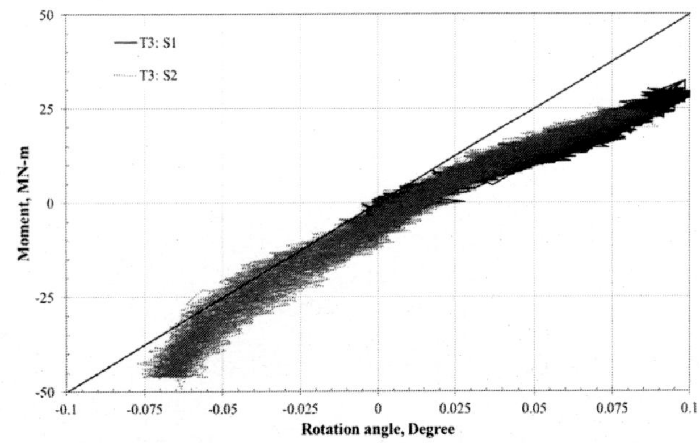

FIGURE 11: Moment-Rotation angle curves of S1 and S2 for T3

bucket is under pullout load but under the minus rotation angle two buckets are under pullout load.

4. CONCLUSION

This study aimed to investigate the resistance behaviour of the tripod and monopod bucket foundations designed for offshore wind turbine tower. In this study, three centrifuge model tests to simulate the monopod and tripod bucket foundations were successfully performed. Monotonic moment-horizontal combined load was applied to the monopod and tripod bucket foundations. In addition, cyclic moment-horizontal combined load was applied to the tripod. Interesting conclusions from the test results are as follows.

1. The tripod showed almost linear behaviour before and after yielding and the yield point was relatively clear. However, the slope of the load-rotation angle of the monopod creased gradually, so the yielding of the monopod was not clear.
2. The M_y for the tripod and monopod were 95 MN-m at 0.5° and 198MN-m at 3.4°, respectively. The M_y of the monopod was almost twice that of the tripod but the rotation angle of the tripod for the yield load was lower than that of the monopod.
3. Before the yield load, the tripod showed stiffer behaviour than the monopod.
4. The cyclic loading with amplitudes smaller than M_y resulted in insignificant accumulated plastic deformations for both of one way and two way loading types. On the contrary, the cyclic load higher than M_y induced significant accumulated plastic deformations. However, the accumulated plastic deformations became stabilized with number of loading cycles.

ACKNOWLEDGEMENT

This study was supported by a grant (10 CTIP E04) from the Offshore Wind-energy Foundation System (OWFS) program funded by the Korean Institute of Construction & Transportation Evaluation and Planning and the Korean Ministry of Land, Transport and Maritime Affairs.

REFERENCES

American Society for Testing and Materials (ASTM) (2007). Standard Test Methods for Laboratory Compaction Characteristics of Soil Using Modified Effort, D 1557-07.

American Society for Testing and Materials (ASTM) (2006). Standard Test Methods for Minimum Index Density and Unit Weight of Soils and Calculation of Relative Density, D 4254-00.

Bransby, M.F. and Randolph, M.F. (1998). "Combined loading of skirted foundations". Géotechnique, 48, Issue 5, pp 637-655.

Byrne, B.W. and Houlsby, G.T. (2003). "Foundations for offshore windturbines", Philosophical Transactions of the Royal Society of London, Series A: Mathematical and Physical Sciences, 361, Issue 1813, pp 2909-2930.

EWEA (2008). The Economics of Wind Energy, A Report by the European Wind Energy Association.

EWEA (2009a). Oceans of Opportunity, Harnessing Europe's Largest Domestic Energy Resource.

EWEA (2009b). Wind Energy – The Facts.

Gourvenec, S. (2007). "Failure envelopes for offshore shallow foundations under general loading". Géotechnique, 57, Issue 9, pp 715-728.

Head (2006). Manual of Soil Laboratory Testing, Vol. 1, Whittles Publishing, London.

Houlsby, G.T., and Byrne, B.W. (2000). "Suction Caisson Foundations for Offshore Wind Turbines and Anemometer Masts". Wind Engineering, 24, Issue 4, pp. 249-255.

Kim, D.S., Kim, N.R., Choo, Y.W., and Cho, G.C. (2012). "A newly developed state-of-the-art geotechnical centrifuge in South Korea". KSCE Journal of Civil Engineering (doi:10.1007/s12205-013-1350-5).

LeBlanc, C., Houlsby, G.T. and Byrne, B.W. (2010a). "Response of stiff piles in sand to long term cyclic loading". Géotechnique, 60, Issue 2 pp 79-90.

Senders, M. (2008). "Suction caissons in sand as tripod foundations for offshore wind turbines", PhD. Thesis, The University of Western Australia.

Villalobos, F.A. (2006). "Model Testing of Foundations for Offshore Wind Turbines", Ph.D Desertation, University of Oxford, UK.

Zhu, B., Byrne, B.W. and Housby, G.T. (2012). "Long term lateral cyclic response of suction caisson foundations in sand". ASCE, Journal of Geotechnical and Geoenvironmental Engineering (doi:10.1061/(ASCE)GT.1943-5606.0000738).

Geotechnical Solutions for Foundations on Slag Fills

K.S. Rama Krishna

L&T GeoStructure LLP, Larsen & Toubro Limited, Construction Division, Chennai
E-mail: ksrk@Lntecc.com

ABSTRACT

The Iron and Steel Industry in India is going through a major expansion. Slag, a by-product co-product of the iron and steel making process, is generally disposed of in the open areas around the existing steel plants resulting in large slag fills of thickness varying from a few meters to over 20m. These fills are heterogeneous in nature because of various other inorganic industrial wastes such as fly ash, cinders and other forms of debris are also dumped along with slag. Bored piles, which are generally chosen to support new plant structures and equipment located in the fill areas, when installed by conventional piling methods result in quality, cost and time issues. The author and his team carried out extensive laboratory and field tests on slag fill and presented three case studies to establish firmly that shallow foundations can be gainfully employed in lieu of bored piles for carrying heavy loads. Case study-1 presents successful implementation of open foundations as a solution after slag is excavated to required depths and filled back in well compacted layers by vibro-rollers. Case study-2 presents a foundation solution with vibro-replacement, for a plant to be set up on an ash pond area reclaimed using slag. Case study-3 presents successful field trials of vibroflotation to compact the slag fill in-situ when slag is overlying a competent original soil. Laboratory and field tests on slag and field footing load tests demonstrated high bearing capacity of compacted slag fills. Results of various laboratory and field tests and field trials are presented. The potential of centrifuge modeling technique to simulate the field tests and to understand the bearing capacity and settlement behaviour of shallow foundations on slag fills subjected to various degrees of compaction can be explored as a means to enable engineering decisions on alternative foundation solutions to conventional piling.

1. INTRODUCTION

1.1 Slag

Slag is a by-product of the iron and steel making process and it is produced during the process of pig iron manufacturing in blast furnace and steel manufacturing in steel melting shop (Fig. 1). Blast furnace slag (BFS) and steel furnace slag (SFS) are the two main types of slag. BFS is composed of the non-metallic components consisting essentially of silicates and aluminosilicates of calcium and other bases (IBM, 2002). SFS, is also a non-metallic product consists of calcium silicates and ferrites combined with fused and mineralogically combined oxides of iron, aluminium, manganese, calcium and magnesium. Both BFS and SFS have unique physical and chemical properties that make them particularly well suited to a variety of uses in construction and civil engineering projects (IRC, 2008).

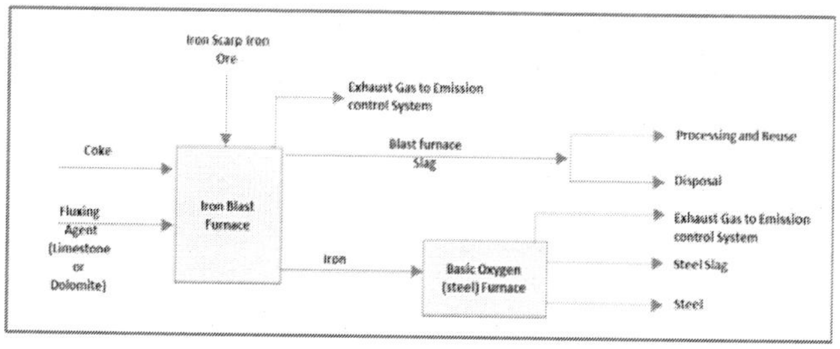

FIGURE 1: General process schematic of iron and steel production

It is gathered from published data that depending upon the content of iron in the ore feed, the amount of blast furnace slag generated ranges from 0.350 to 1.0 tonne per tonne of pig iron produced. Similarly, steel furnace slag is approximately 20% by mass of crude steel output.

Based on the cooling process, three types of blast furnace slag are generated: (1) air-cooled (solidification under ambient conditions), which finds extensive use in conventional aggregate applications; (2) expanded or foamed (solidified with controlled quantities of water, sometimes with air or steam), which is mainly used as a lightweight aggregate; (3) granulated (solidified by quick water-quenching to vitrified (glassy) state), which is mainly used in blended cement manufacture; and (4) pelletized (solidified by water and air-quenching in conjunction with a spinning drum), which is used both as a lightweight aggregate and in blended cement manufacture(NSG Report MF186-1). Blast Furnace slag does not expand whereas, Steel Furnace slag, due to un-hydrated lime (MgO or CaO) used in the steel making process shows a tendency to expand, which can be destructive to building and structures when used as fill or base materials. Expansive steel slag can be treated or "cured" before use as a construction aggregate.

Slag has several civil engineering and construction applications viz., as a structural base layer under pavements, floor slabs, foundations, etc. It has high angle of internal friction which is better than dense sands and gravel.

FIGURE 2: Typical heterogeneous slag fill site at TISCO Jamshedpur

Hence the slag material attracted researchers in utilizing the same for wide construction activities across the globe. In India huge quantities of slag has been disposed of into the open areas around the existing steel plant for several decades resulting in large areas of slag fill with depths ranging from a few meters to over 20 m. These are exposed to natural weather conditions of sun, wind and rain for all these years. Other inorganic industrial wastes like ash, cinders, used refractory bricks, iron and steel scrap and even construction debris has also been dumped in the same areas as a result of which slag has assumed a heterogeneous character (Fig. 2). Even though slag is dumped loosely over the years it has consolidated and gives an appearance of a well compacted and dense fill. Currently, the steel industry in India is undergoing a major expansion and new plants are being set up in the slag fill areas due to scarcity abundant land for plant expansion.

1.2 Conventional Bored Cast-in-situ Piles

Designers have been choosing bored cast-in-situ piles as the foundation system for most of the steel plant structures and critical equipment. Conventional tripod-winch-bailer/chisel method is used for drilling through slag and piles are usually terminated in the firm soil or bed rock below the slag. Typical pile diameters adopted are 550mm/600mm and 1000 mm. In the case of piles of diameter 550 and 600mm the pile bore in slag is generally protected from collapse using a temporary steel casing. However for 1000mm diameter piles permanent steel liners are lowered in small length cans as the pile bore is advanced with chisel and bailer in slag. The joint between successive cans is achieved by field welding to form a full length liner up to the firm virgin soil below. In both cases the steel casing follows the boring as it would not be possible to push the casing ahead of boring due to the friction of the slag material. This obviously entails a borehole of bigger diameter than the outer diameter of the casing. The pile is required to be terminated in the Mica Schist layer lying below the silty sand layer. The process is extremely slow and tedious besides producing a borehole diameter larger and irregular compared to the designed pile diameter. It was reported that boring was taking a significantly long period (about 7 days for 550 mm dia. bored pile of 20m depth). As a result the concrete consumption was very high and sometimes even as high as two times the theoretical volume.

The above method suffers many shortcomings for the following reasons:

1. Any percussion method of boring such as dropping bailer/chisel would dislodge otherwise stable walls of the pile bore resulting in collapse. Bentonite would not serve the purpose as it can easily escape through the voids in the BFS fill. Permanent steel liner makes piling very expensive besides damaging the good shaft friction that otherwise would have resulted between the pile and surrounding slag (concrete pile would have taken the shape and profile of the drilled bore). Majority of the pile capacity would therefore come from the bottom virgin soil layers – a highly inefficient foundation system.
2. Bailer or chisel can get easily worn out because of the abrasive nature of the BFS material.
3. Advancing a temporary casing and withdrawing the same is also very difficult due to high friction and abrasive nature of BFS fill. In the process of extraction, further collapse of BFS material from pile bore wall is expected.

1.3 Alternatives to Piling in Slag

Recognizing the excellent geotechnical properties of slag the author and his team have undertaken various laboratory and field tests and field trials on slag to find alternative foundation solutions to commonly employed bored piles using the above conventional method. Three case studies are presented in the following chapters to firmly establish that shallow foundations with proper treatment of heterogeneous slag fill would be the right and efficient foundation system.

2. CASE STUDY-1: SHALLOW FOUNDATIONS FOR A PLANT AT TISCO, JAMSHEDPUR

2.1 Subsoil Profile at Sinter Plant Area

In the year 2005 M/s Tata Iron and Steel Company, TISCO, Jamshedpur proposed to set up unit-4 of Sinter Plant, called SP-4, in an area filled with loosely dumped slag. The fill was gradually built over several decades. The ground level was varying from RL +175.0 m to +164.0 m. A detailed subsoil investigation was carried out to explore the suitability of shallow foundations as a replacement to conventional bored cast-in-situ piling. It comprised of boreholes drilled up to 30.5 m with SPT tests conducted at regular intervals, pressuremeter tests which met with no success due to sharp edges of slag and irregular profile of borehole wall, shallow trial pits of 1.0m × 1.0m × 1.0m deep at different elevations to obtain field density, moisture content, specific gravity and particle size distribution, deep trial pits below the expected depth of foundations, logging of excavated faces, plate load test (0.75m × 0.75m size) and footing load tests (1.5m × 1.5m and 1.5 × 3.0m size). Laboratory tests comprised of gradation tests and large size shear box tests (300mm × 300mm). The subsoil profile at SP-4 is shown in Fig. 3. The stratification of subsoil layers is described below:

Layer-1: Dense to very dense, silvery/dark grey slag combined with other inorganic waste materials such as fly ash, cinders, refractory bricks, wire ropes, structural steel pieces, etc. Slag thickness varied from 15 to 25m. SPT tests generally met with refusal. The size, shape and gradation of slag suggest that the SPT N-values in this layer are not dependable. In other words a high SPT-N would not directly co-relate to a very well compact and dense slag layer and we need further evidence of the condition of the fill. A physical examination of several exposed walls of excavations in the site including a 9.0m deep nearly vertical (3V:1H) excavation in slag however revealed that the fill is generally in a very compact and dense state but for the occasional presence of loose pockets and large size voids in the fill. Arching of slag could be seen above these loose pockets and voids. This is probably due to the high friction and interlocking characteristics of slag as revealed by the laboratory shear box tests and field density tests (Table 1).

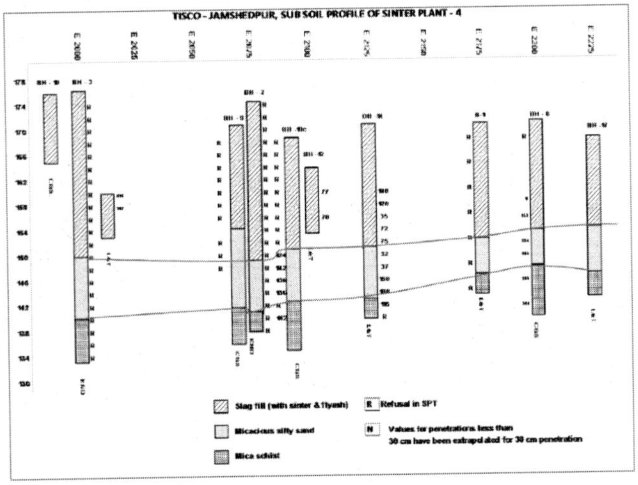

FIGURE 3: Subsoil profile at Sinter plant 4 area, TISCO Jamshedpur

Layer-2: Dense to very dense, light brownish yellow Micaceous Silty Sand. Layer thickness was varying from 6m to 13m. SPT tests generally met with refusal. Top of this layer is the original ground.

Layer-3: Highly to moderately weathered brownish yellow Mica Schist rock. SPT tests generally met with refusal.

The ground water table was well below the slag layer at the time of investigation. It is to be noted that as slag has very high degree of well-connected voids, any amount of rain water would quickly drain out in to the lower areas around the plant. Several years of exposure to free draining rain water must have stabilized the slag from any harmful effects of hydration and expansion.

The field density tests on slag using water replacement method (Fig. 4) indicated that the dry unit weight was ranging from 11 to 18.4 kN/m^3. The gradation tests indicate that the slag was well graded within the gravel range, as shown in Fig. 5. Results of large shear box tests (300 mm × 300 mm) have shown that the slag has an inter-particle friction angle of around 44 degrees.

Table 1: Summary of the Laboratory and Field Density Test Results

Sr. No.	*Test*	*Unit*	*Results*
1	Field density test: Unit weight	kN/m^3	11.0–18.4
2	Shear box test*:		
	(a) Apparent cohesion	kN/m^2	80
	(b) Friction angle	degree	44

* Size of shear box used: 300mm × 300mm.

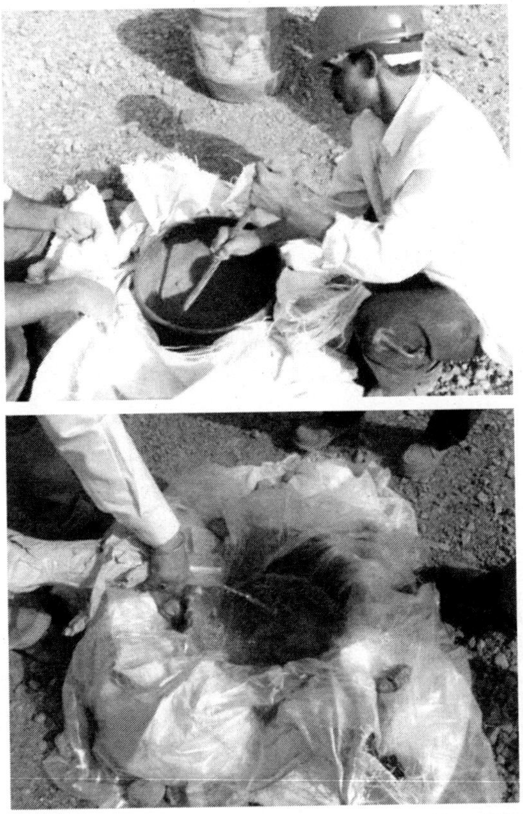

FIGURE 4: Field density test as per IS 2720 Part33-1971

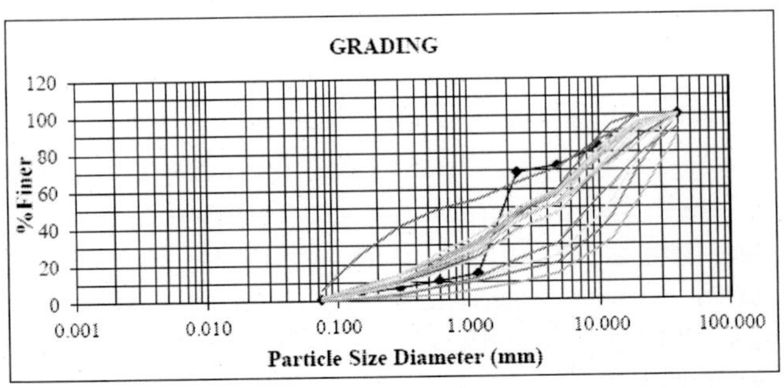

FIGURE 5: Grain size distribution of slag

2.2 Plate Load and Footing Load Tests

As the slag fill is generally in a compact and dense state, it was decided to carry out plate load test and footing load tests on the fill at different elevations to ascertain the bearing capacity and settlement characteristics of the heterogeneous fill.

- PLT-1: A 750 mm × 750 mm plate at +159.8 m level (4.5 m below the FGL (+164.3) to be tested to failure.
- FLT-1: A 1500 mm × 1500 mm footing at +158.3 m level (6.0 m below the FGL (+164.3) to be tested up to 450 kN/m2 (Fig. 6).
- FLT-2: A 1500 mm × 1500 mm footing at +160.3 m level (4.0 m below the FGL (+164.3) to be tested up to 450 kN/m2
- FLT-3: A 1500 mm × 3000 mm × 100 mm footing at +160.3 m level (4.0 m below the FGL) to be tested up to 450 kN/m2 in an area where the fill has high percentage of ash and cinders.

FIGURE 6: Footing load test arrangement

The load settlement curve of Plate load Test-1 with 750 mm × 750 mm plate is presented in Fig. 7. There was no bearing capacity failure. Settlement at a failure load is taken as 10% of least plate dimension is 75 mm. Even at a high load intensity of 2200 kN/m^2 the settlement of the plate was only observed to be 12.0 mm.

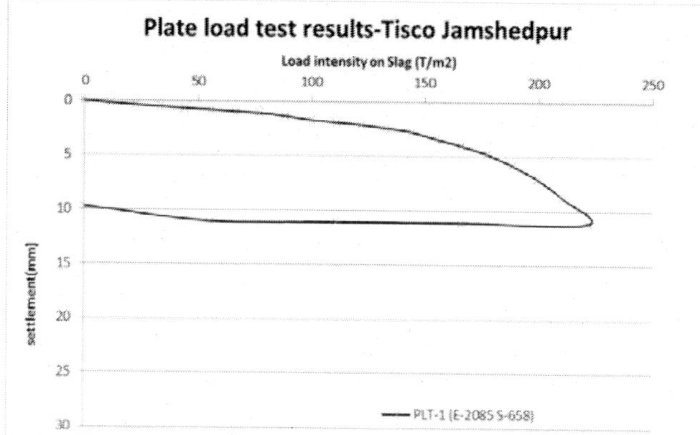

FIGURE 7: Load settlement curve of PLT-1

Based on footing load tests FLT1, FLT2, the settlements corresponding to the maximum applied load of 43.4 kN/m^2 are 7.56 mm and 6.89 mm respectively. Both the test results are plotted in Fig. 8.

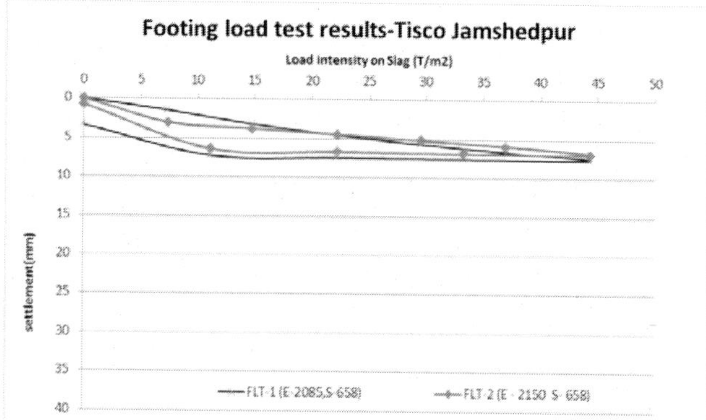

FIGURE 8: Footing load test results of FLT1 and FLT2

The interpretation of the footing load test results for allowable bearing capacity has been carried out as per the procedure outlined in IS: 1888–1982 Clause 5.3. Based on the pressure verus settlement curves for FLT -1 and FLT - 2, the maximum pressure for 150 mm (10% of the width of the footing) settlement for a footing size of 1.5 m is 1150 kN/m^2. The bearing pressure was restricted to 550 kN/m^2 with a factor of safety of 2. The corresponding settlements for 3 m, 4.5 m and 6 m footing sizes are 12, 13.5 and 15 mm, respectively. The bearing capacity is a function of width of the footing. The calculated value of 550 kN/m^2 is for 10% of the footing size of 1.5 m. Therefore the value of allowable bearing pressure will be greater than 550 kN/m^2 for larger footing sizes. The bearing capacity does not govern the design. For the footing sizes of 3 m, 4.5 m and 6 m, the corresponding pressures for total settlements of 10 mm, 15 mm, 25 mm and 40 mm have been calculated using extrapolation. The results of limiting values are given in Table 2.

One more footing load test, FLT 3 was conducted in an area, where the slag contains higher percentages of ash and cinders, to understand the behaviour of shallow foundations in an apparently weak fill area. The footing load test indicated that the settlement corresponding to the maximum applied load of 443.4 kN/m^2 is 31.94 mm. The results of are given in Fig. 9.

Table 2: Interpretation of Footing Load Test Results

Footing Size	*Total Settlement (mm)*	*Pressure (kN/m^2)*	
		FLT-1	FLT-2
3	10	500	510
	12	550	550
4.5	10	475	475
	13.5	550	550
6	10	450	470
	15	550	550

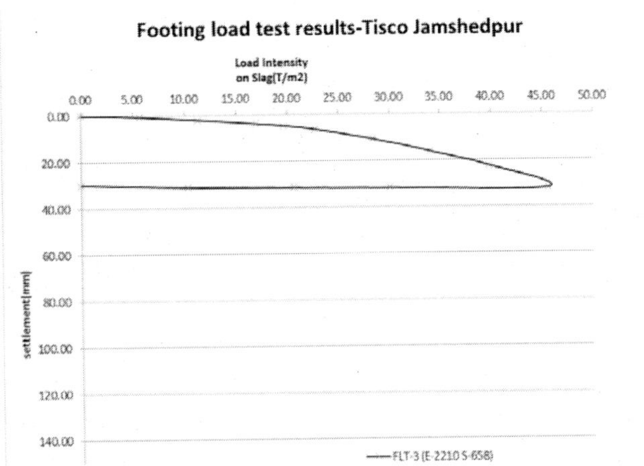

FIGURE 9: Load settlement curve of footing load test-3

The site was graded from the level of +171.0 to +163.5 m, i.e. 7.5 m of slag overburden was removed which is equivalent to a load intensity of 135 kN/m^2. In other words the slag fill below +163.5 m was subjected to a preload of 135 kN/m^2 and this explains the good results of plate load test.

It was therefore decided to go ahead with shallow foundations.

While the fill itself is in a very compact condition, there could be some local variations in stiffness due to heterogeneous nature of the fill. In such an event, the rigid footings for the plant, which are in the size range of 5 to 6 m, were expected to take care of the variations. As an abundant precaution and to confirm the required performance of the foundations, settlement observations were carried out during the construction and equipment erection phase and the footings performed well.

However as a further precaution it was also decided to excavate and remove any weak/undesirable material at or near the founding level, fill back with selected slag and compact the surface with a 10 T vibro-roller. To determine the number of passes required by a 10 T vibro-roller for obtaining maximum compaction and field density, field trials of roller compaction were carried out as per the layout given in Fig. 10. Field

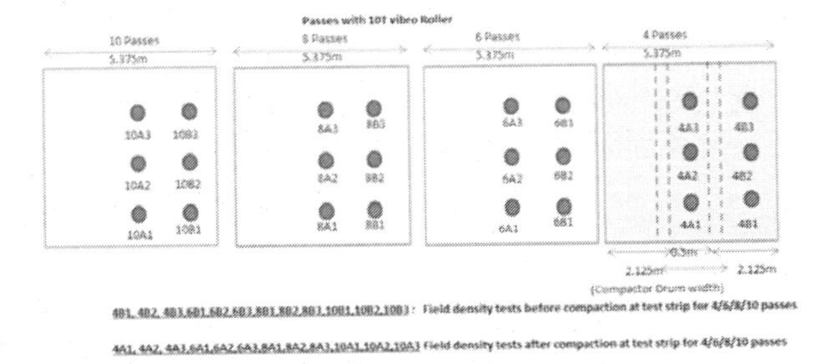

FIGURE 10: Trial compaction layout

Table 3: Number of Roller Passes Versus Field Density Before and After Roller Compaction

No.of Passes		Field Density Test Id Mark.	Initial Volume of Reading with Ring Only Vj (cc).	Final Volume of Reading With Ring & Cavity Vf (cc).	Volume of Cavity, V = Vf - Vj (cc).	Weight of Wet Material Ww (g).	Wet Density γ= Ww/V (g/cc)	Weight of "surface dry stones" of size > 60mm Wr (g)	Volume of "surface dry stones" of size > 60mm Vr (cc)	Moisture content w (%)	Dry Density γd = (Ww-Wr)/((V-Vr)(1+w/100)) (T/m3)	Average dry density (T/m^3)	% increase in Dry density	Remarks
1		2	3	4	5	6	7	8	9	10	11	12	13	14
4	Before Compaction	4B1	10674	17591	6717	18603	2.77	0	0	6.50	2.60	1.90	0.5%	Sample 4B1 not considered due to wide variation
		4B2	9520	21860	12340	26320	2.13	0	0	5.90	2.01			
		4B3	10937	21414	10477	20122	1.92	0	0	4.52	1.84			
	After Compaction	4A1	10500	15530	5030	9660	1.90	0	0	7.06	1.86	1.94		
		4A2	9500	13010	3510	7868	2.24	0	0	6.04	2.11			
		4A3	9840	15314	5474	10894	1.99	0	0	8.25	1.84			
6	Before Compaction	6B1	13231	25684	12453	14842	1.19	0	0	8.10	1.10	1.23	66.5%	
		6B2	12400	25140	12740	15981	1.25	0	0	8.57	1.15			
		6B3	9732	16317	6585	10256	1.56	0	0	8.22	1.44			
	After Compaction	6A1	16760	31500	14740	32665	2.22	0	0	8.45	2.04	2.05		
		6A2	12802	24871	12069	26101	2.30	0	0	8.10	2.15			
		6A3	6906	25162	18256	38958	2.13	0	0	5.26	2.03			
8	Before Compaction	8B1	13261	30000	16739	30600	1.83	0	0	8.23	1.69	1.74	23.0%	
		8B2	13020	24300	10600	23944	2.22	0	0	8.81	2.04			
		8B3	8837	33000	24063	37734	1.57	0	0	5.93	1.48			
	After Compaction	8A1	11988	23964	11975	28039	2.34	0	0	9.40	2.14	2.14		
		8A2	12560	24972	12812	29045	2.27	0	0	8.57	2.09			
		8A3	14256	27500	10796	25387	2.35	0	0	6.85	2.20			
10	Before Compaction	10B1	12000	25584	13584	26802	1.97	0	0	6.83	1.85	1.67	18.2%	Sample 10A2 not considered due to wide variation
		10B2	9141	23581	14440	24475	1.69	0	0	7.02	1.58			
		10B3	9684	23380	13778	23678	1.72	0	0	8.95	1.58			
	After Compaction	10A1	12941	22588	9747	21000	2.16	0	0	11.23	1.94	1.97		
		10A2	10456	24666	5205	19834	3.81	0	0	8.17	3.51			
		10A3	12206	24517	12312	26843	2.20	0	0	9.52	2.01			

density tests were conducted before and after compaction for 4, 6, 8 and 10 number of passes of the roller to establish the optimum number of passes. As shown in Table 3, and Fig. 11, eight (8) number of roller passes resulted in maximum field density.

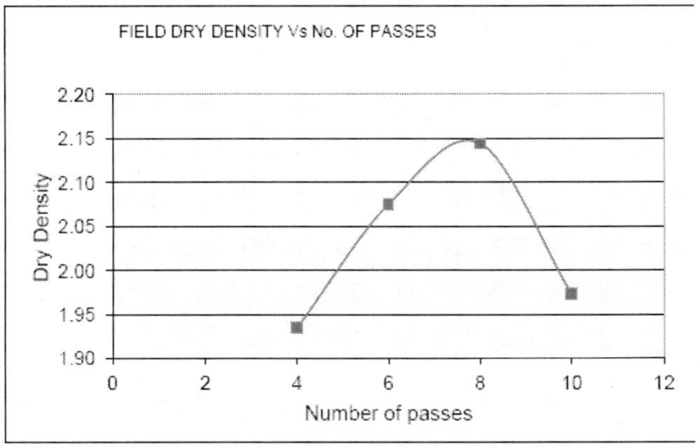

FIGURE 11: Number of roller passes vs field density

2.3 Steps Followed During Execution of the Project

Based on the above field investigations, data interpretations and the bearing capacity analyses, the following steps were followed in the design and construction phase of the project:

- For the footing sizes and permissible settlements, corresponding bearing pressure values were adopted from Table 2.
- The minimum width and depth of the footing was 1.5m and 1.0m from the finished grade level of the site respectively. The depth of the footing was increased to suit functional requirements of equipment and structures.
- When loose pockets of cinder, ash, etc., were encountered at the founding level, the same was removed and replaced by well-graded slag in layers of 300mm and well compacted by 10T vibro-roller.
- The foundation construction was sequenced in such a way that the lower level foundations were cast first and the higher-level foundations are cast later.
- The entire area of SP-4 was with provided with sufficient surface drainage system.
- Monitored settlements of critical foundations during the construction and erection of equipment and they all performed well.

The above step-wise procedure led to a systematic and trouble free construction of all foundations and the early completion of the project.

3. CASE STUDY-2: VIBRO-REPLACEMENT TRIALS AT TATA BLUE SCOPE SITE, BARA, JAMSHEDPUR

3.1 Introduction

Tata Blue scope Ltd proposed to set up a metal coating and coil paint plant at Bara, near Jamshedpur. The site was part of an ash pond reclaimed in part by dumping slag generated by TISCO steel plant. Client's consultants proposed bored cast-in-situ piles for the proposed plant and equipment foundations. Depending upon the area and load imposed by the structures various pile diameters, viz., 550 mm, 750 mm and 1000 mm were proposed. The total number of piles was around 8000 with a pile length of up to 20 m.

L&T carried out a preliminary study of various options to minimize and/or replace the piles with a value engineered foundation system and at the same time create a sound and uniform substrata below the entire plant area so that not only plant structures, buildings and equipment rest on good foundations even the roads and other utilities would perform without any distress in the design life time of the plant. The major task was to find a technology that would enable reinforcing the ash layer in situ with slag and compact the heterogeneous slag fill. M/s BSP-UK, M/s Menard Soltraitement-France and M/s Keller Ground Engineering Pvt. Ltd.- India were invited to independently study the feasibility of punching slag columns in to the ash layer as a reinforcement by either dynamic compaction(by Menard) or rapid impact compaction(RIC by BSP UK) or install stone columns using vibro-replacement technique(Keller). Based on various discussions, it was concluded that both dynamic compaction and rapid impact compaction techniques had the limitation of punching slag columns in to the ash layer to a maximum depth of 6.0m provided the water table was lowered to increase the efficiency of the energy imparted and the ground lowered by 1.5 m below the proposed finished grade level. Because of the inherent properties of slag such as high friction, interlocking and abrasive nature, and having no known precedence in slag, M/s Keller wanted to establish the feasibility of creating stone columns before submitting a firm proposal for the ground improvement of the entire plant area. Permission was obtained from the client to carry out the field trials with the Keller vibro-replacement technique and follow it up with field density and footing load tests.

3.2 Topography of the Site

The plan area and ground levels at the proposed plant site are shown in Fig. 12. The total area was divided in to four zones A1, A2, A3, and A4 depending on the existing ground levels. As per the plan, the entire plant area was to be brought to one uniform level by cutting slag at the higher levels and filling the lower level areas.

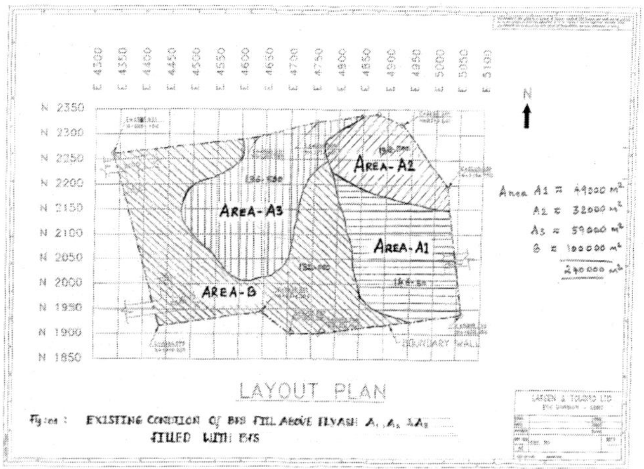

FIGURE 12: Topography of the site

3.2 Sub Soil Description

The sub soil consisted of varying thickness of slag (Fig. 13). After site grading works the ground was to be brought to +136.0 m level. After this the thickness of slag layer would be varying from 2 m to 5 m. The gradation of a typical slag sample is shown in Fig. 14. The top slag layer is followed by ash of about 5 to 6 m thickness with SPT N ranging from 0 to 9, sandy clay of 2 to 3 m thickness with SPT N ranging from 25 to more than 50 and quartz and gneiss rock layers.

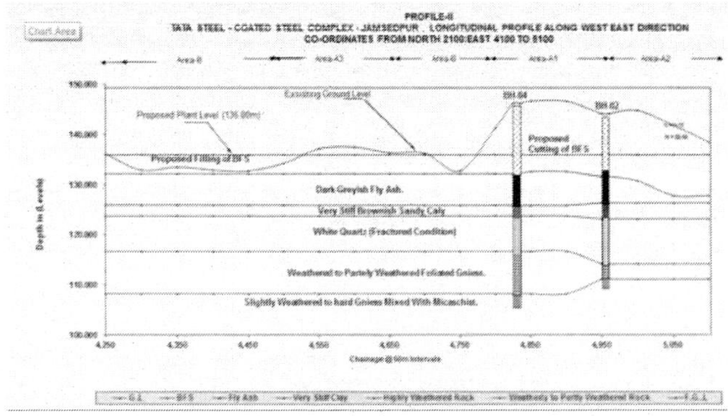

FIGURE 13: Typical subsurface profile at Tata Blue Scope at Bara, Jamshedpur

FIGURE 14: Typical gradation of slag

It can be noticed that the ash layer was sandwiched between two compacted/hard layers, i.e., the slag on top and stiff to hard sandy clay layer at the bottom. The ash layer is considered to be a week layer with un-established strength and settlement characteristics. Going by the low SPT values there was a need to improve this layer in order to avoid or minimize piling.

3.3 Vibro-Replacement Concept

The vibro-replacement technique proposed by Keller (Keller Report, 2006) was a result of further development of vibro-compaction process to cater to the improvement of pure silts/clays or mixed deposits of silts/clays/sand. This technique introduces coarse grained material such as slag as a load bearing element, called stone column, into the soft soil. To form a vibro-replacement point the vibrator with its follower tubes is placed over the selected point by means of a suitable supporting rig (crane). Out of the two vibro-replacement methods, viz., wet method and dry method, wet method was adopted in the field trials is briefly explained below. The essential equipment for this process is a depth vibrator – a long, heavy tube enclosing eccentric weights, driven by an electric motor. A typical vibrator is shown in Fig. 15. In wet method, high pressure water jets existing at the tip of the vibrator assist penetration of the vibrator in to the soil.

FIGURE 15: Field trials – vibro-replacement and vibro-compaction

To form a vibro-replacement point the vibrator with its follower tubes is placed over the selected point by means of a suitable supporting rig (crane). After starting the motor the vibrator is lowered into the ground. The oscillating vibrator with its extension tubes sinks rapidly in to the soil under its own weight. With a series of up and down motions of the vibrator and with the help of water jets an annular space is created up to the designed depth. Aggregates (slag in this case) are then transported in using a loader, deposited around the probe point and allowed to fall into the annular space. A constant water flow is maintained to ensure that the stone backfill is washed to the tip of the vibrator and the soft in situ soils are flushed out. The vibrator is slowly withdrawn in steps of 0.7 to 1.0 m to allow the stone to fall in to the hole. The vibrator is then lowered back in to the hole between 0.7 to 0,8 m, thereby creating 0.2 to 0.3 m length of stone column. The action of the vibrator compresses backfill material radially in to the surrounding soil and also compacts the slag material in the annular space, resulting in a column of diameter which is dependent upon the relative stiffness of the in situ soil – the softer the soil the larger the diameter of column. The average column diameter can be determined by monitoring the slag consumption during installation. This compaction procedure continues until the full length of the column has been constructed. Upon completion of the installation process, a compacted column of slag is left in the ground, surrounded by a soil/slag matrix of increased density. The column and the in situ soil form an integrated system having low compressibility and high shear strength. The excess pore water pressure in the surrounding soil can be dissipated through the column, which also acts as a vertical drain. The settlement expected for the treated soil is reduced while the rate of settlement is increased when compared with the untreated soils.

The degree of improvement (settlement reduction and increased shear strength) achieved by the vibro-replacement process depends on the soils being treated, the diameter of column installed and the installation spacing. Vibro-compaction method is explained separately in Case study-3. Vibro-replacement points may be arranged so as to accommodate varying combinations of load and foundation configurations. The spacing between the columns is determined in accordance with the design criteria and the in situ soil conditions.

3.4 Field Trial

In order to assess the performance of ground improvement solution and to check the penetrability of the vibrator in the top slag layer, 16 Nos of trial stone columns and 13 Nos of trial penetrations were carried out. As per the trial layout, 16 Nos of vibro stone columns were installed from the ground level of +136 m at the spacing of 2 m × 2 m c/c up to the depth of hard strata (N >15). The layout plan is shown in Fig. 16. The backfill material used for the stone columns was of locally available slag of size 100 mm down to 12 mm. Hence, no external material was required which would result in substantial saving of cost. The analysis of the penetrability in the slag, fly ash, diameter of the stone column formed in the slag and fly ash are calculated and summarized in Table 4 below. The total time required for installation of each columns including penetration, washing and building columns varies from 21 min to 47 minutes The average time was about 30 minutes.

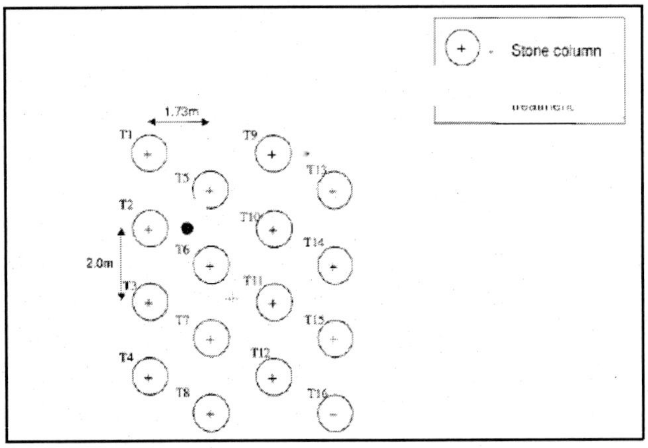

FIGURE 16: Trial layout of stone columns

Table 4: Time Summary for Trial Columns (Keller Report, 2006)

S.NO	Date	Probe No	Time (min) for			Depth (m)			Rate of (m/min)		Quantity			Diameter		
			Total	Penetration	Building columns	Slag	Fly ash	Total	Penetration	Column building	Slag	Fly ash	Total	Slag	Fly ash	A.Dia
1	27/5/06	T1	24.32	4:30	15:00	3.90	6.87	10.77	2.39	0.72	2.40	7.20	9.60	0.89	1.16	1.07
2	27/5/06	T2	26:42	4:45	16:30	3.90	6.81	10.71	2.25	0.65	2.40	8.00	10.40	0.89	1.22	1.11
3	27/5/06	T3	26:12	7:00	11:00	3.90	7.13	11.03	1.58	1.00	2.40	10.00	12.40	0.89	1.34	1.20
4	25/5/06	T4	32:00	6:30	14:00	3.90	7.38	11.28	1.74	0.81	3.20	5.60	8.80	1.02	0.98	1.00
5	27/5/06	T5	25:08	4:30	16:30	3.90	6.94	10.84	2.41	0.66	2.40	6.80	9.20	0.89	1.12	1.04
6	26/5/06	T6	30:00	5:45	12:30	3.90	6.94	10.84	1.89	0.87	3.20	8.00	11.20	1.02	1.21	1.15
7	26/5/06	T7	23:00	5:15	12:10	3.90	7.13	11.03	2.10	0.91	1.60	7.84	9.44	0.72	1.18	1.04
8	25/5/06	T8	47:00	4:30	14:15	3.90	6.81	10.71	2.38	0.75	3.20	6.80	10.00	1.02	1.13	1.09
9	27/5/06	T9	26:40	3:30	16:45	3.90	7.13	11.03	3.15	0.66	2.40	7.20	9.60	0.89	1.13	1.05
10	26/5/06	T10	26:00	4:00	15:00	3.90	7.32	11.22	2.81	0.75	2.40	5.60	8.00	0.89	0.99	0.95
11	26/5/06	T11	29:00	4:00	19:00	3.90	7.00	10.90	2.73	0.57	4.80	8.80	13.60	1.25	1.27	1.26
12	26/5/06	T12	21:00	4:15	12:15	3.90	7.06	10.96	2.58	0.89	1.60	5.60	7.20	0.72	1.00	0.91
13	27/5/06	T13	37:14	4:00	22:00	3.90	6.94	10.84	2.71	0.49	3.20	7.60	10.80	1.02	1.18	1.13
14	26/5/06	T14	47:00	3:00	15:30	3.90	7.26	11.16	3.72	0.72	1.60	7.20	8.80	0.72	1.12	1.00
15	26/5/06	T15	22:00	4:00	14:15	3.90	7.19	11.09	2.77	0.78	3.20	8.00	11.20	1.02	1.19	1.13
16	26/5/06	T16	22:00	4:15	12:15	3.90	6.94	10.84	2.55	0.88	1.60	5.60	7.20	0.72	1.01	0.92
Average			29:06	4:37	14:56	3.90	7.05	10.95	2.48	0.76	2.60	7.24	9.84	0.91	1.14	1.07

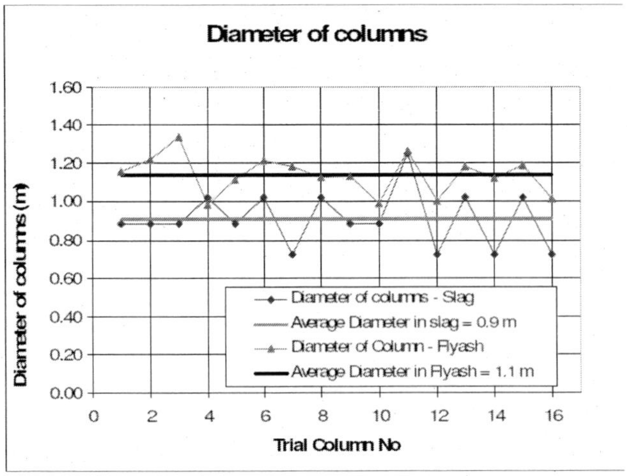

FIGURE 17: Diameter of the column formation due to vibro-compaction (Keller Report, 2006)

Diameter of the stone columns formed in the fly ash was varied from 1.0 m to 1.3 m and the average diameter of the stone column formed is about 1.1 m. The diameter formed in the slag varied from 0.7 to 1.2 m and the average is about 0.9 m. Diameter was calculated based on the approximate slag consumption in slag portion and fly ash portion. The graphical presentation is given in Fig. 17. The trial firmly established that vibro-replacement could be employed for ground improvement of both slag and the ash layers below. Due to the prevailing conditions at the site, load tests on the treated ground were not conducted. However, another opportunity was created in another slag filled site where footing load tests were carried out after the slag layer which was overlying a firm layer was improved by vibro-compaction with the help of M/s Keller.

4. CASE STUDY 3: VIBRO-COMPACTION AT IISCO, BURNPUR

4.1 Introduction

A major expansion of ISCO-steel plant at Burnpur in West Bengal, India was underway where thousands of bored cast-in-situ piles were envisaged in a vast area largely covered by slag. A major opportunity was foreseen to introduce alternative foundation solutions. Permission was obtained from client to carry out field trials in order to study the possibility of shallow foundations on the heterogeneous slag fill improved by vibro-compaction. The following chapter presents the results of vibro-compaction trials, footing load tests, laboratory and field tests.

4.2 Soil Profile

The plant is located in a vast area filled mainly with slag and occasionally with heterogeneous materials such as unbroken slag skulls, steel scrap, wire ropes, rails, sleepers, refractory bricks, etc. (Fig. 18). The fill whose thickness varied from 8 to 24 meters with reference to the terrace level (+130.00 m) was followed by 2.5 m thick dense sand layer (SPT N = 39), 2.0 m thick hard clay layer (SPT N = 41) and sand stone in that order. Water table was present at a depth of 18 m below ground level. Since borehole data was not available in the area allotted for field trials, a borehole was drilled with the help of rotary piling rig to check the thickness of the slag at the trial area allotted by client for field trials. It was found that slag was up to 11.0 m below the existing ground level at this location.

FIGURE 18: Slag skulls and other obstructions removed from the slag fill

4.3 Problems Faced during Piling

Bored piling was executed using the hydraulic rotary drilling rigs and tripods. Huge obstructions were encountered in the form fused slag chunks, slag skulls, etc. As a result the pile drilling productivity came down drastically. Many times piling operations were stopped to allow excavators to excavate to the required depth to remove these heavy obstructions and fill back with slag to allow further piling work. Boring tools were also wearing out fast and with frequent necessity to change the teeth of the auger there was further dent in the drilling productivity.

4.4 Vibro-Compaction Technique

Vibro-compaction is a technique (Keller Report, 2009) that rearranges/compacts the particles of non-cohesive soils into a denser state by inducing vibration. The essential equipment for this process is a depth vibrator - a long, heavy tube enclosing eccentric weights, driven by an electric motor. A typical vibrator is shown in Figure 19. The vibrator is connected to a source of electric power and a high-pressure water pump. Extension tubes are added as necessary, depending on the treatment depth, and the whole assemblage is suspended from a crane. With the electric power and water supply switched on, the vibrator is lowered into the ground. The combination of vibration and high-pressure water jetting causes liquefaction of the soils surrounding the vibrator, which assists in the penetration process. When the required depth is reached, the water pressure is reduced and the vibrator pulled up in short steps. With the inter-particle friction temporarily reduced, the surrounding soil particles then fall back below the vibrator and, subjected to vibratory energy, are rearranged into a denser state. This process is repeated back up to the ground level, leaving on completion, a column of very dense material surrounded by material of enhanced density. The degree of compaction achieved at a particular point depends on the properties of the soil being treated, the amount of time spent at each compaction step and the distance from the vibrator. Vibro-compaction is suitable for treating granular material with a fines content of less than 10–15%. The spacing of probes is designed to ensure that the zones of influence overlap sufficiently to achieve minimum requirements throughout the treated area. Generally, the effect of the compaction becomes visible at the ground surface in the form of a cone-shaped depression. The depression formed around the vibrator or the extension tubes is continually in-filled with granular materials, which is either imported or obtained from the natural granular deposits at the site. Water required for the penetration and compaction process is obtained either by direct pumping from nearby water source or ground water using well points.

FIGURE 19: Depth vibrator (Vibroflot) setup and slag feeding

4.5 Vibro-Compaction Trials

Encouraged by the fact that vibroflot, as it is popularly known, (depth vibrator) could easily penetrate the thick slag layer at Bara, Jamshedpur, M/s Keller agreed to the proposal by M/s Larsen & Toubro Ltd to carry-out vibro-compaction trials. The allotted trial area was divided in to three zones (Fig. 20) to try three different spacings of vibro-compaction points (2.0, 2.5 m and 3.0 m c/c in a square grid) with the aim of obtaining the optimum spacing for achieving maximum field compaction.

4.6 Field Observations

4.6.1 Time Required for Compaction

During the vibro-compaction process, the vibrator was able to penetrate through the slag to its full depth of about 11.5 m and the time taken for the penetration was generally 4 to 10 minutes (average 5 minutes). The time required for the compaction ranged from 11 to 29 minutes (average 20 minutes). The overall time required for each probe including penetration and compaction varied from 16 min to 35 minutes and average 25 minutes and considering the equipment movement, the overall time for each probe was 30 minutes. The average time is inclusive of time for removal of some localized obstructions, break down in water supply during operations, power obstructions etc. Penetration and compaction time for 2.0 m. 2.5 m and 3.0 m spacing vibro points are shown in Table 5.

Table 5: Average Time for Different Trial Spacings

Spacing (m)	*No. of Vibro Points*	*Average Time for Each Point (min)*
2.0	50	24
2.5	32	27
3.0	32	23

4.6.2 Ground Subsidence after Vibro-Compaction

After vibro-compaction, the ground subsidence was observed to be lowered by 1.63 m, 1.37 m, 0.88 m for 2.0 m, 2.5 m and 3.0 m spacing respectively (Fig. 21). The details are given in Table 6.

Table 6: Volume of Slag Material Consumed for Each Trial Area

Sl. No	Spacing	No of Vibro Points	Theoritical .Area Covered (m2)	Reduced Level (m)		Av.Level of subsidence (m)	Actual Area Covered (m2)	Total volume of slag subsided (m3)
				Av. Before Compaction	Av. After Compaction			
1	2.00	50.00	200.00	129.83	128.20	1.63	224.47	290.41
2	2.50	32.00	200.00	129.83	128.46	1.37	199.96	215.09
3	3.00	32.00	288.00	129.65	128.77	0.88	207.09	157.33
	Total	114	688					

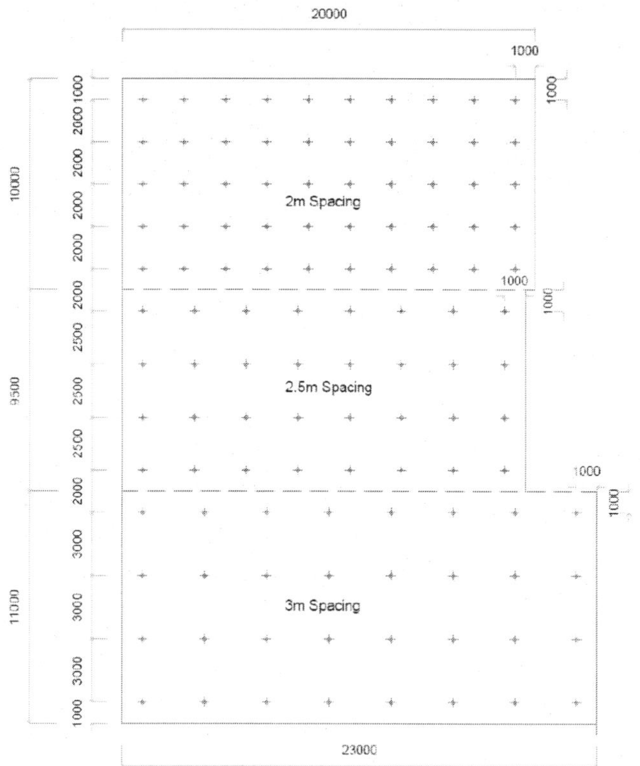

FIGURE 20: Trial layout pattern for vibro points

FIGURE 21: Ground subsidence measurement at trial area

4.7 Laboratory Tests

Slag samples before and after vibro-compaction were collected at different locations within the trial area and demarcated as 2.0 m, 2.5 m, and 3.0 m spacing. Samples and laboratory tests conducted for grain size distribution, relative density, strength parameters from large direct shear test (300 × 300 mm square), swelling properties etc. The test results are given in Fig. 22 and Table 7. The slag material in the trial location had 65% to 70% gravel and 10% to 14 % fines. This shows that the properties of the slag material are in line with the requirements for vibro-compaction. The specific gravity was in the order of 2.5 with zero swell index.

Shear box tests were carried out on three samples collected from the three zones in the trial area. Each sample was compacted to two different densities to match with the before and after vibro-compaction at site. The angle of internal friction was 40 to 46 degrees when tested at a dry unit weight of 16 kN/m^3, whereas it is 53 to 59 degrees when tested at a dry unit weight of 18.5 kN/m^3. The results are shown graphically in Fig. 23. This shows that the slag material has higher angle of internal friction than that of sand and bearing capacity failure of foundations resting on slag fill is unlikely.

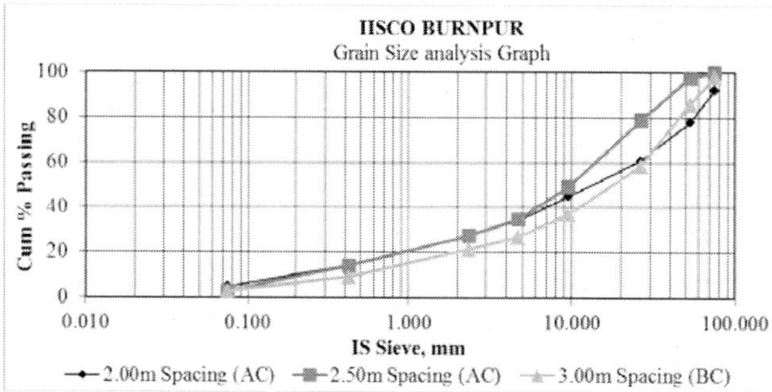

FIGURE 22: Grain size distribution curve

Table 7: Laboratory Test results

LAB SAMPLE ID	SWELLING INDEX	GRAIN SIZE DISTRIBUTION*						SPECIFIC GRAVITY	SAMPLE DRY DENSITY		DIRECT SHEAR GRAVEL (UU)		
		GRAVEL	SAND			SILT	CLAY					Lower side	Higher Side
			COARSE	MEDIUM	FINE				(Lower side)	(Higher Side)	C	ϕ	ϕ
	%	%	%	%	%	%	%		g/cc	g/cc	Kg/cm²	Degrees	Degrees
SM-755/B/01	NIL	66	7	13	9	5		2.51	1.60	1.87	–	40	53
SM-755/B/02	NIL	65	8	13	12	2		2.48	1.60	1.85	–	46	54
SM-755/B/03	NIL	73	6	12	7	2		2.66	1.60	1.85	–	51	59

4.8 Field Density Tests

Prior to the vibro-compaction, five numbers of field density tests were carried out in each trial area with a total of 30 no's covering the total trial area. 15 Nos of tests are carried out prior to the vibro-compaction and 15 Nos of tests post to the vibro-compaction. It was observed that prior to the compaction the field density values had a range from 12 to 17 kN/m^3 with an average being 14.6 kN/m^3 for 2.0 m spacing, 14.4 kN/

m³ for 2.5m spacing and 14.3 kN/m³ for 3.0 m spacing and field density values after compaction having a range of 16 to 26 kN/m³ with an average being 19.9 kN/m³ for 2.0 m spacing, 18.7 kN/m³ for 2.5 m spacing and 18.5 kN/m³ for 3.0 m spacing. Percentage increase in field density after compaction is 26% for 2.0 m spacing, 23% for both 2.5 m spacing and 3.0 m spacing that of before compaction. The consolidated field density test results are shown in below Table 8 and Fig. 24. Over all there is an increase in field density from 15 kN/m³ to 19 kN/m³ resulting in a corresponding change in void ratio from 0.8 to 0.5.

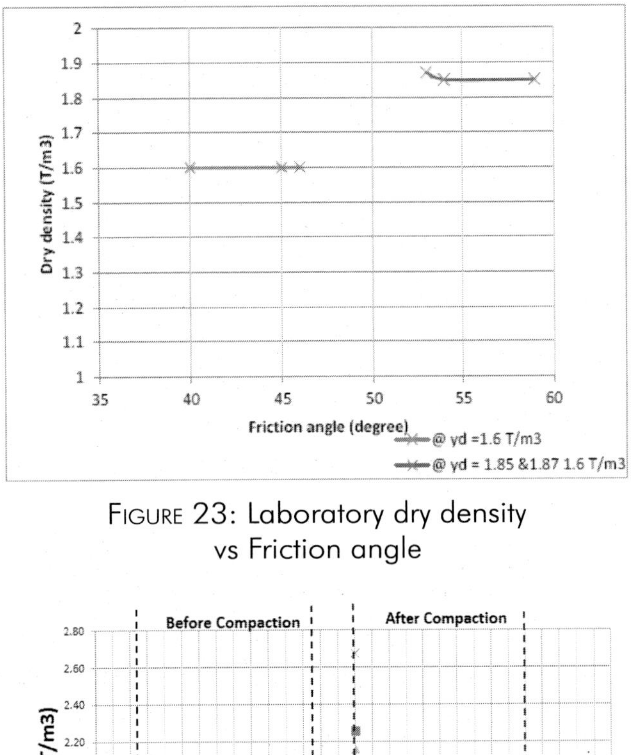

FIGURE 23: Laboratory dry density vs Friction angle

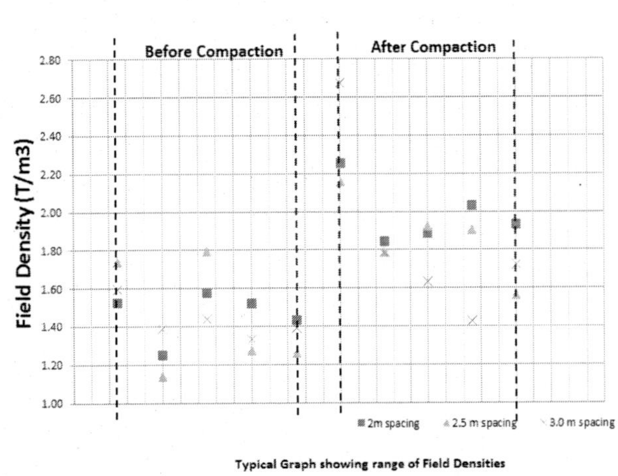

FIGURE 24: Field density test results before and after Vibro-compaction

4.9 Footing Load Tests

The footing load tests were carried out at varied spacing as per IS 1888. The setup is shown in Fig. 25. The analysis and interpretations for each footing load test data are described here. The safe bearing capacity (SBC) values have been extrapolated to full size footings between 1.5 to 5m from the load test data carried out on 1.3 × 1.3 × 0.6 m size footing. Fig. 26 shows the consolidated graph with 2.0 m, 2.5 m, 3.0 m spacing test results.

All the three tests showed similar load-settlement behavior while the footing load test in 2 m spacing area showed better behaviour for the given settlement. Each unloading stage indicated significant permanent settlement and the reload curve is almost flat. This shows the benefits of preloading in case higher loads need to be sustained by the footings. For example at a load intensity of 600 kN/m² the settlement of the footing for 2.5 m spacing of vibro-compaction point is in the range of 20 mm. When the same was unloaded and reloaded to 600 kN/m² the settlement would be about 5.0 mm. For structures and equipment sensitive to settlement preloading would give much better benefits in terms of cost and time instead of when the same was supported on piles and pile cap.

Table 8: Field Density Test Results

Sl no	Identity	Description	Trial no	Wet density (γ = Ww/V)	Moisture content (w%)	Dry density (γd = (Ww-Wr)(V-Vr)) / (1+ w%)	Average Dry density (γd)	% increase
1	2m spacing Trial area	Before compaction	FD-BC-1	1.64	3.50	1.53	1.46	26%
2			FD-BC-2	1.33	3.20	1.25		
3			FD-BC-3	1.62	2.60	1.58		
4			FD-BC-4	1.57	2.90	1.52		
5			FD-BC-5	1.48	3.10	1.44		
6		After compaction	FD-AC-1	2.41	6.85	2.25	1.99	
7			FD-AC-2	1.99	7.69	1.84		
8			FD-AC-3	2.07	9.38	1.89		
9			FD-AC-4	2.18	7.50	2.03		
10			FD-AC-5	2.09	8.10	1.94		
11	2.5 m spacing Trial area	Before compaction	FD-BC-1	1.79	2.60	1.74	1.44	23%
12			FD-BC-2	1.19	4.10	1.14		
13			FD-BC-3	1.87	3.50	1.79		
14			FD-BC-4	1.31	3.00	1.28		
15			FD-BC-5	1.38	2.40	1.26		
16		After compaction	FD-AC-1	2.34	8.15	2.16	1.87	
17			FD-AC-2	1.96	9.65	1.79		
18			FD-AC-3	2.06	7.61	1.92		
19			FD-AC-4	2.09	9.82	1.90		
20			FD-AC-5	1.73	10.60	1.56		
21	3.0 m spacing Trial area	Before compaction	FD-BC-1	1.71	5.10	1.60	1.43	23%
22			FD-BC-2	1.45	3.60	1.39		
23			FD-BC-3	1.47	2.10	1.44		
24			FD-BC-4	1.36	2.30	1.33		
25			FD-BC-5	1.43	3.10	1.39		
26		After compaction	FD-AC-1	2.89	8.10	2.67	1.85	
27			FD-AC-2	1.92	7.69	1.79		
28			FD-AC-3	1.79	9.88	1.63		
29			FD-AC-4	1.54	7.50	1.43		
30			FD-AC-5	1.86	8.10	1.72		

Assuming a conservative friction value of 45 degrees from laboratory test results the safe bearing capacity for a footing of 2.5 m × 2.5 m size would be 600 kN/m² (as per IS-6403-1981). Based on shear and settlement criteria and a 2.0m spacing for vibro-compaction, shallow foundations of sizes ranging from 2.5 m × 2.5 m to 5.0 m × 5.0 m, resting at 2.0 m depth below FGL

FIGURE 25: Footing load test setup

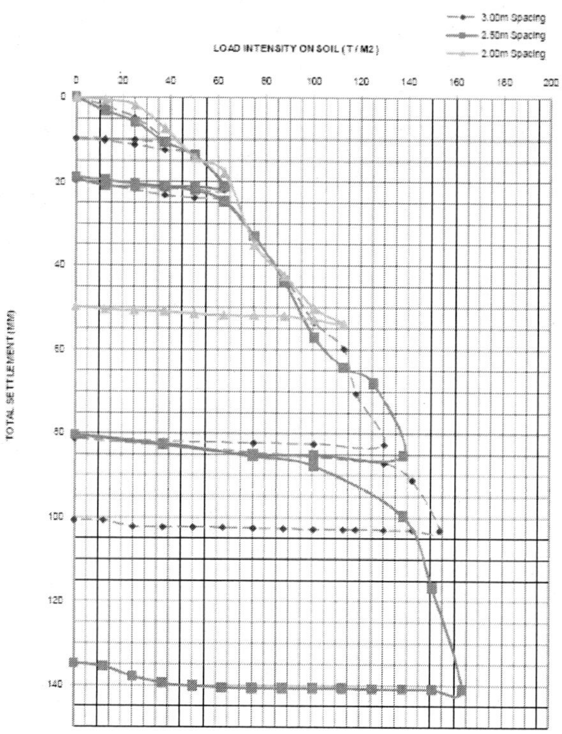

FIGURE 26: Field Load Test Results for 2.0m, 2.5m, 3.0m Vibro Probe spacings

can be safely designed for 300 kN/m² for 10 mm settlement, 400 kN/m² for 25 mm settlement and 450 kN/m² for 40 mm settlement. The combined approach of vibro-compaction, preloading of foundation can yield very safe and high bearing capacities leading to significant savings in time and cost when compared to bored piling. This will definitely address the recent emphasis of carbon footprint, environmental sustainability, etc. In the opinion of the author even if structures are made to rest on piles and pile cap the external load would be transferred from pile cap directly to the slag in contact with it making the piles redundant.

5. CONCLUSIONS

Based on several field and laboratory tests, field trials, discussions with specialist companies and experts shallow foundations in slag fills are recommended as a very effective and sound alternative to conventional piling solutions currently being followed. Because slag fills are observed to heterogeneous certain ground improvement methods as recommended below may be considered as a prerequisite to shallow foundations.

- When slag fills are deposited over firm original soils, it would be cheaper to excavate slag up to the effective depth of the footing and fill back up to founding layer with selected slag in layers of specified thickness and well compacted using vibro-rollers. 10 ton vibro-roller with 6 to 8 passes and 300 mm thick compacted layer could be considered for estimation and planning purposes. Since each site can be different field compaction trials shall be considered as mandatory.
- Preloading of footings on slag would significantly improve allowable bearing pressures by reducing the post construction settlement.
- In case original soil on which slag is deposited is week, slag columns formed by vibro-replacement technique would be an ideal solution and the resulting improved layers would minimize or totally eliminate the need for pile foundations.
- If heterogeneous slag is deposited on a firm original ground, vibro-compaction would be an ideal alternative to piling. It would create a uniformly compact and competent founding

layer for all plant structures, equipment and even roads and utilities.

- When heterogeneous slag with thickness of 6 to 8m is resting on firm original ground rapid impact compaction technique would offer a faster and competitive solution.

Further geotechnical studies on slag may be taken up by an independent institution to recommend broad guidelines that would benefit the clients and the country in completing the projects faster and cheaper.

In the present study, an attempt has been made to establish a procedure to construct shallow foundations on un-conventional fill sites like slag fills through field tests. The analysis and interpretation of results revealed that the slag fills were found to exhibit good load carrying capacities and has a capacity to support different structural loadings. It can be noted that executing field load tests is tedious and not encouraged in practice by clients due to cost, time and resource various obstructions and difficulties. However, considering the practical difficulties and time delay in performing field tests, the adoption of centrifuge modelling technique can be a viable option. In a centrifuge testing, behaviour of prototype structures can be studied as reduced scale model of 1/N times by subjecting to N times acceleration due to gravity in a controlled environment (N = scale factor or g-level). Moreover, the adoption of centrifuge modelling technique can also help in assessing load-bearing characteristics of slag fills under long-term conditions (say for about 2 to 3 years) under different compaction conditions. In order to achieve this, it may be appropriate to adopt the following step-wise procedure:

- Characterization and modelling of slag fill characteristics (limited to one or two types)
- Development of a centrifuge model test package for assessing load carrying capacity-settlement characteristics with and without treatment options (say at g-levels of the order of 20-50),
- Assessing load carrying capacity-settlement characteristics of slag fills with time,
- Performing parametric study to evolve guide-lines for designing appropriate foundations on slag fills.

The adoption of above-procedure can help in the application of centrifuge modelling technique to address issues relevant to foundations on slag fills. This can lead to considerable economics in reducing foundation construction costs, project time reduction and a viable methodology to have a geotechnical solution for constructing foundations on slag fills.

ACKNOWLDGEMENTS

The author would like to express his sincere thanks and gratitude to the management of M/s TISCO-Jamshedpur, M/s IISCO-Burnpur and M/s Larsen & Toubro Limited for giving opportunity to the author and his team for carrying out various field and laboratory studies and field trials; to Prof. V.S. Raju for his valuable guidance in the studies carried out at Sinter Plant SP4 project; to Dr. V.R. Raju and his team from M/s Keller Ground Engineering India Pvt. Ltd for enthusiastically participating in the field trials at Bara and Burnpur; to all his colleagues in L&T who have worked with passion and dedication during all the field and laboratory investigations and field trials. Thanks are also due to Prof. B.V.S. Viswanadham (who happened to be a student of Prof. V.S. Raju), Department of Civil Engineering, IIT Bombay for bringing-out the relevance of centrifuge model testing to evaluate the long-term load carrying capacity-settlement characteristics of slag fills. Finally the author wishes to acknowledge the support provided by Mr. P. Suresh Kumar, who tirelessly assisted in the preparation of this manuscript.

REFERENCES

IBM (2002). Indian Bureau of Mines Report.

IRC (2008). Industrial Resources council report.

BIS-1888 (1982). Method of Load Tests on Soils.

BIS-2720 Part33 (1971). Methods of tests on Soils - Determination of the density in-Place method by the ring and water replacement method.

BIS-6403 (1981). Code of Practice for Determination of Bearing Capacity of Shallow Foundations.

USGS (2002). Minerals information-Minerals year book – Iron and Steel Slag.

Studies on Jack-up Spudcan Foundations

C.F. Leung

Department of Civil and Environmental Engineering, National University of Singapore, Singapore
E-mail: ceelcf@nus.edu.sg

ABSTRACT

Mobile jack-up rigs are often used for offshore energy explorations. This paper provides a summary of centrifuge model studies conducted at the National University of Singapore to investigate problems concerning jack-up spudcan foundation including punch through, extraction difficulties, spudcan footprint interaction and spudcan pile interaction. Tests employing full model spudcans were conducted to evaluate the severity and mechanism of the problems under different configurations. Tests using half-cut model spudcans placed behind transparent window were employed to examine the underlying fundamental mechanism with emphasis on the effects of soil movements due to spudcan penetration and extraction on the problems understudy.

1. INTRODUCTION

Jack-up rigs are commonly employed for offshore oil and gas explorations in relatively shallow waters with seabed up to 150 m deep. Despite being accidents prone, jack-up rigs are by far the most popular rig employed for offshore explorations particularly for marginal oil/gas fields as the rigs can be moved from one site to another. Singapore is a major fabricator of jack-up rigs and its yards currently fabricate about two-thirds of jack-up rigs worldwide. Fig. 1 shows a typical jack-up rig in operation. The plan of a modern jack-up rig is of triangular in shape with a spacing of about 50 m between the legs, though the spacing can be larger for rigs employed in deeper waters. Each of the three legs (Fig. 2) is typically supported by a shallow spudcan foundation of size ranging from 10 m to 25 m in diameter. Rigzone.com (2012) reported that in August 2012, there were 556 jack-up rigs utilized or under construction worldwide. Asia is a very active region for offshore explorations with Southeast Asia leading the pack employing 109 jack-up rigs (63% of all offshore rigs in the region), followed by Far East Asia with 45 jack-up rigs and South Asia with 38 jack-up rigs. The utilization rate is high in Asia with about 83% of jack-up rigs operating. Depending on the rig size, the rate for the operation of a jack-up rig is up to US$150,000 per day.

FIGURE 1: Mobile jack-up rig
(Source: www.gustomsc.com)

FIGURE 2: Schematic of jack-up with adjacent permanent jacket platform (source: www.offshore-technology.com)

Owing to mobility of jack-up rigs, spudcans are not designed for a site-specific seabed condition. McClelland et al. (1981) reported that jack-up rigs are accident prone and a high percentage of the accidents were due to problems in the spudcan foundations. They also noted that great difficulties when extracting spudcans in deep soft clay seabed. Osborne and Paisley (2002) noted that it is fairly common for punch through failure to take place during spudcan installation and preloading in Southeast Asia due to its peculiar subsurface conditions under the sea beds.

In some instances, a jack-up rig may return to an old site to drill additional wells or maintain old wells. Stewart and Finnie (2001) reported that the existence of old footprints due to former jack-up activities may be hazardous for the new installation as an existing footprint in soft soil can exceed 10 m in width and depth. The new installation of jack-up spudcan foundation at a certain distance from the footprint centre may induce excessive stress on the jack-up legs or slewing of the rig into the old footprint. The uneven soil bearing resistance at and around a footprint during a new spudcan installation is believed to be a significant factor for the problem of interaction between the spudcan and an existing footprint. The interaction between the jack-up spudcan foundation supporting a mobile jack-up rig and adjacent pile foundations supporting a nearby platform (Fig. 3) can be a major concern during the installation and extraction of the spudcan (see Fig. 2).

With the rapid growth of oil and gas works in many parts of the world, the National University of Singapore (NUS) officially launched the

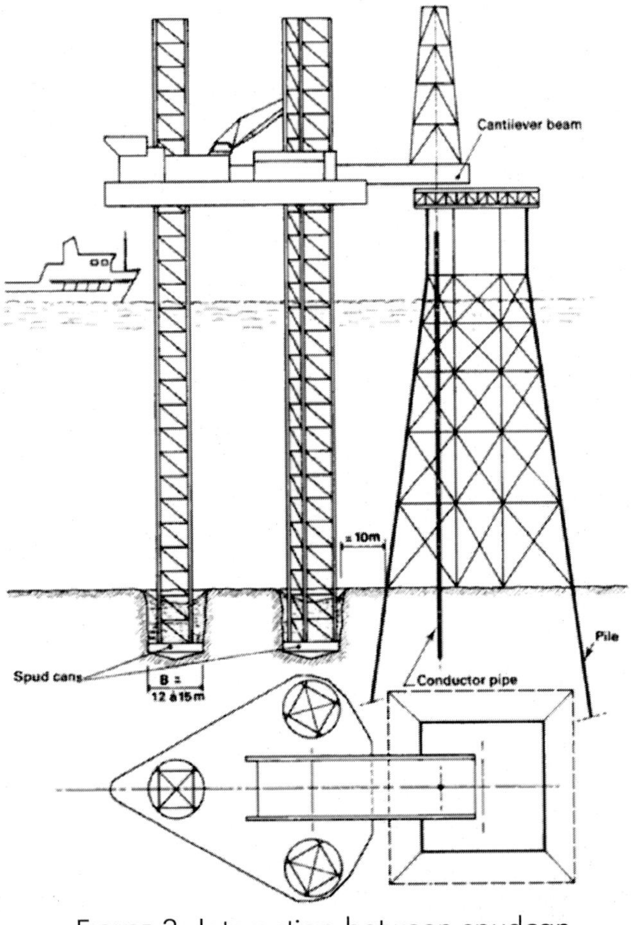

FIGURE 3: Interaction between spudcan foundation of jack-up rig and pile foundation of jacket platform (top side elevation view; bottom plan view) (after Le Tirant and Pérol 1993)

Centre for Offshore Research and Engineering (CORE) in 2004 to spread head various offshore research studies across different disciplines. One of the CORE's research activities is to investigate the behaviour of jack-up spudcan foundation. Extensive centrifuge model studies had been performed at the National University of Singapore for the past 10 years to investigate the above mentioned spudcan jack-up foundation problems. A review of the experimental setup, results and findings are presented in this paper.

FIGURE 5: Full spudcan test setup

2. CENTRIFUGE MODEL SETUP

Two series of centrifuge model tests were conducted with one employing a full size model spudcan (Fig. 4a) while the other using a half-cut model spudcan (Fig. 4b). Figs. 5 and 6 show the experimental setup of full and half spudcan tests, respectively. The half spudcans are placed behind a transparent window of the model container with beads placed on the soil to facilitate image processing analysis on the photographs taken during the experiments. The full spudcan tests aim to evaluate the severity and mechanism of the problems under different configurations while the half spudcan tests examines the underlying fundamental mechanism with emphasis on the effects of soil movements due to spudcan penetration and extraction on the problems under study.

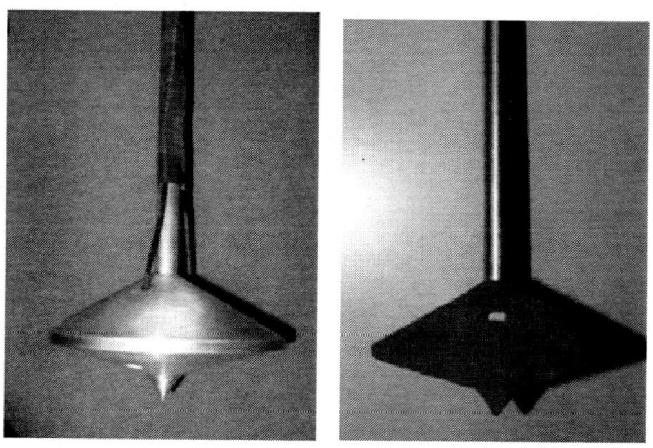

(a) (b)

FIGURE 4: (a) Full and (b) half-cut model spudcans

FIGURE 6: Half spudcan test setup

The clay used in the tests is constituted from dry kaolin clay with a liquid limit of 80%, plastic limit of 35% and a specific gravity of

2.6. The dry clay powder is thoroughly mixed with de-aired water at 1.2 times the liquid limit of the clay. To prepare the soil specimen, dense sand is first rained into the model container to form the bearing and the bottom drainage layer. This is followed by the clay slurry which is first subjected to a consolidation pressure of 20 kPa at 1g and subsequently to self-weight consolidation at 100g in the centrifuge. At the end of in-flight consolidation, the clay has achieved a minimum of 90% consolidation with a thickness of about 260 mm. For tests in soft normally consolidated clay, the strength profile has an undrained shear strength of 5 kPa for the upper 4 m (prototype scale) clay crust followed by a linear increase in strength of 1.43 kPa/m depth for the underlying normally consolidated clay.

The preparation for the half spudcan test is essentially similar as that for full spudcan tests. To facilitate image processing, beads were placed on the clay during preparation at 1g (Fig. 7). During a test, high resolution photographs were taken regularly to capture the change in the trend of soil movement patterns. After the test, Particle Image Velocimetry (PIV) technique was employed to obtain the displacements of a prescribed grid of fine patches using GeoPIV8 software (White and Take, 2002) from a sequence of high resolution photographs taken during the test. A more precise displacement fields could be obtained by applying close-range photogrammetry technique as has been described in White et al. (2003).

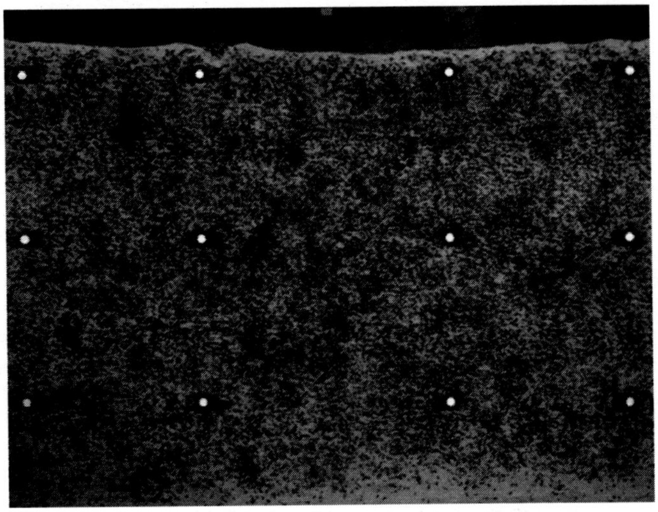

FIGURE 7: Beads placed on clay for image processing

When installing a spudcan in sand overlying clay, a sudden rapid penetration may occur if the underlying weak soil is not strong enough to resist the relatively large applied load that exceeds the maximum bearing resistance developed in the upper sand layer. As such, when a spudcan is installed in stiff over soft soil profile, rapid leg penetration may occur due to failure of the underlying soft soil. This is a hazardous situation whereby one spudcan leg rapidly penetrates while the other two legs stay stationary inducing severe stresses or even failure of the jack-up rig structure. This situation is commonly termed as punch-through hazard. Fig. 8 shows the phenomenon of punch through put forward by Young et al. (1984).

3. STUDIES ON SPUDCAN PUNCH THROUGH

During installation, a spudcan is progressively loaded from the seabed to the desired depth so as to achieve the required bearing resistance against vertical and environmental loads. The spudcan penetration depth in clay could be up to 2 times its diameter giving rise to very significant displacement of surrounding soil (Craig and Chua, 1990) and collapse of cavity above the penetrating spudcan (Hossain et al., 2005); whereas for penetration in sand, the spudcan foundation evokes a progressive failure mechanism during its limited penetration.

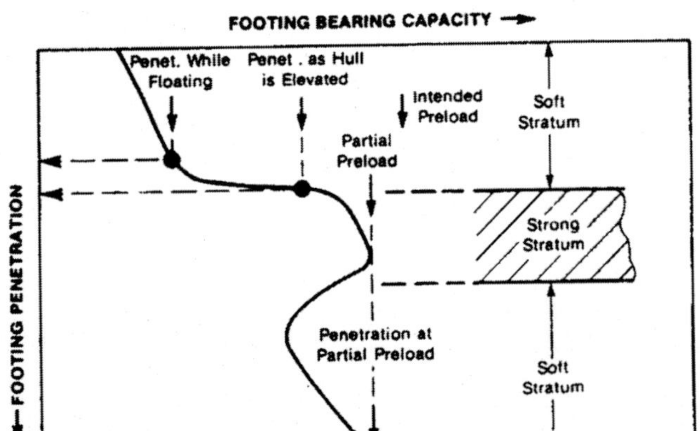

FIGURE 8: Schematic of punch through hazard (after Young et al. 1984)

The hazard of punch through in sand overlying clay has been studied by Teh et al. (2008, 2010). The occurrence of sand overlying clay is not uncommon in many parts of world. As an illustration, selected locations of seabed with the presence of sand overlying clay are highlighted in Fig. 9.

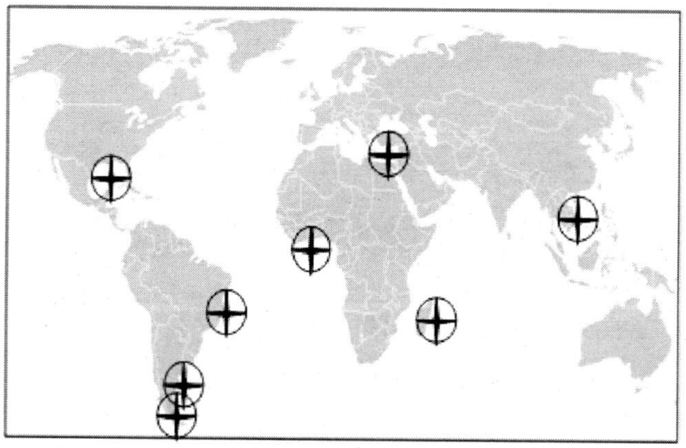

FIGURE 9: Selected locations with sand overlaying clay seabed

The experimental set-up and dimensions of the 100-mm diameter model spudcan are shown in Fig. 10. The upper sand layer and the underlying soft clay were formed using Toyoura sand and kaolin clay, respectively. In some tests, pore pressure transducers (PPTs) and potentiometers were installed to investigate the pore pressure distribution within the underlying clay and the ground surface settlement during spudcan penetration, respectively. The clay slurry was first consolidated at 100g. The sample was then brought to rest where sand was rained onto the consolidated clay layer. This was followed by saturating the upper sand layer before the layered soil sample was reconsolidated at 100g. Once the underlying clay achieved over 90% of degree of consolidation under self-weight and sand surcharge, the spudcan penetration test commenced at 100g. The soil strength profile of the sample was evaluated by performing miniature piezocone penetration test.

The spudcan was installed using load control. In the present study, the drainage condition of the soil was checked using the drainage criterion proposed by Finnie (1993) such that the upper sand layer was loaded under drained condition whereas the underlying clay behaved undrained during the spudcan penetration. The spudcan penetration load was measured by a load cell attached between the spudcan leg and the loading actuator; and the corresponding penetration was measured using a potentiometer. All the discussions in the subsequent sections are presented in prototype scale, unless stated otherwise.

The spudcan penetration load-depth profile obtained from a typical centrifuge test employing a full spudcan is shown in Fig. 11. The penetration load, Qv, which is measured by the load cell, includes the applied load, the submerged weight of spudcan and leg, and the weight of backflow soil if any. The spudcan penetration depth, d, is measured from the lowest elevation of the spudcan widest cross-sectional area with respect to the original ground surface. Qv begins to register when the spigot of the spudcan penetrates beneath the ground surface. After the spudcan penetrates to $d = -0.8$ m, the load increases with d linearly until $d = 0$ where the spudcan widest cross-sectional area is in full contact with the ground. For $d > 0$, the load keeps increasing with d at a steeper gradient before reaching a maximum value, $Q_{v,max}$, at $d = 1.1$ m. An immediate reduction in Qv is then recorded causing the spudcan to penetrate rapidly until a resistance identical to $Q_{v,max}$ is reached at a greater depth. During the rapid spudcan penetration, a second peak load with magnitude smaller than $Q_{v,max}$ is recorded. A linear increase in Qv is then observed when the spudcan penetrates into the underlying clay (i.e. $d \geq H$) reflecting the linearly increasing strength profile of the clay.

During spudcan penetration, the responses of the surface potentiometers were monitored. The potentiometer closest to the spudcan records surface settlement as the penetrating spudcan compresses the underneath sand and causes the neighbouring sand falling into the void space formed by the spudcan advancement. On the other hand, ground heave is noted at locations further away from the spudcan. For a deeply embedded spudcan, the sand back-flow forms a concave deformed surface with a diameter of about 2 times and a maximum depth of 0.55 time of spudcan diameter.

All the pore transducers embedded in the underlying clay layer show an instant build-up

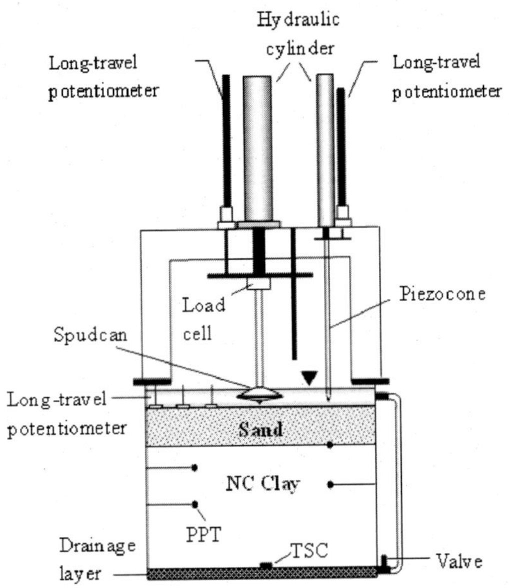

(a) Side elevation

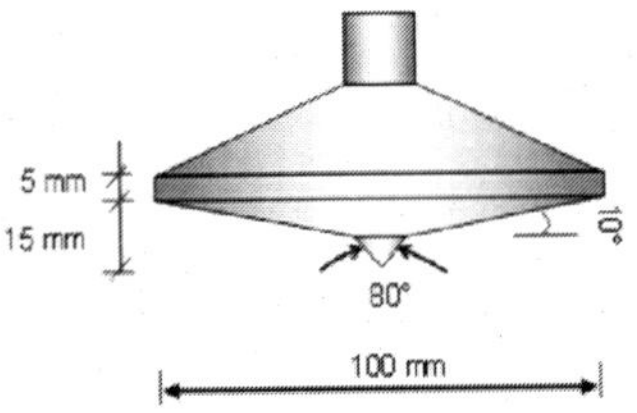

(b) Model spudcan

FIGURE 10: Centrifuge model setup of sand overlying clay (after Teh et al. 2010)

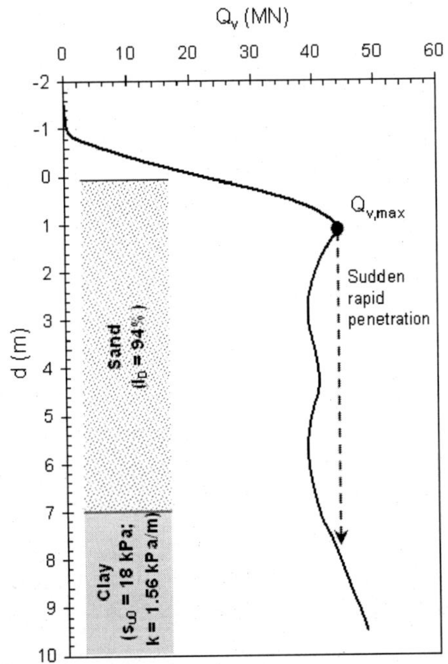

FIGURE 11: Typical test results of spudcan penetration in sand overlying clay

of excess pore pressure as soon as the spudcan records its first penetration resistance, indicating an immediate stress distribution phenomenon. A sand plug of similar height as the upper sand layer is observed underneath the penetrating spudcan, see the results observed from a half spudcan test shown in Fig. 12. This is similar to that observed by Craig and Chua (1990). The presence of sand plug causes the bearing pressure to be distributed to the surrounding soil. When the sand plug travels past the elevation of a pore pressure transducer, a reduction in pore pressure is noted.

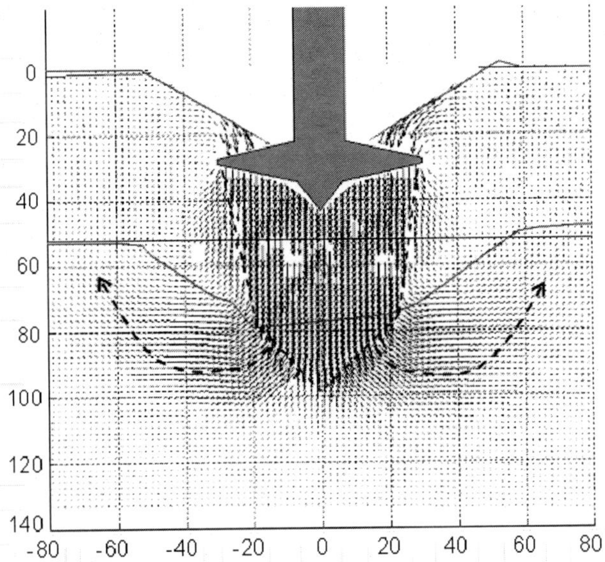

FIGURE 12: Sand plug beneath advancing spudcan (after Teh et al. 2008)

4. STUDIES ON SPUDCAN EXTRACTION

As mentioned earlier, the penetration of the spudcan in soft clay can be deep. A jack-up rig may be extracted and moved to other operating locations. It has been reported that the extraction of a deeply buried spudcan can be very difficult due to soil consolidation after spudcan installation. In situations with deep penetrations, typically 2 times spudcan diameter in soft clay, the legs may be stuck on an attempted extraction and require considerable effort to be retrieved. This is not economical for the drilling operator due to loss of opportunities. In view of this, a centrifuge model study was carried out to examine the causes of difficulties in extracting a spudcan and the results are reported by Purwana et al. (2005).

The first test series involves experiments using full spudcans. The model spudcan was extensively instrumented using total and pore pressure transducers placed at various places at the top and base of the spudcan. The changes of stresses above and below the spudcan as well as the development of pore pressure in the soil are monitored during spudcan installation, jack-up operation period and spudcan extraction. Purwana et al. (2005) established that the uplift resistance Q comprises the resistance of the soil above the spudcan Qt and suction at the spudcan base Qb. The latter is found to increase significantly with time and a major component of the breakout force for spudcans with long operation period at a given site.

The test results are summarized in Fig. 13. Altogether six centrifuge model tests had been carried out covering a prototype jack-up operation period ranging from a very short duration of 1 day to a relatively long duration of 850 days (2.3 years). The measured spudcan uplift resistance consists of the uplift resistance beneath the spudcan, the self-weight of the spudcan and the suction between the soil and the spudcan base. While the self-weight of the spudcan remains constant for the six tests, it is evident from Fig. 13 that the uplift resistance above the spudcan does not increase significantly with increase in jack-up operation period. On the other hand, the suction between the spudcan base and the soil increases significantly with increase in jack-up operation period. This indicates that as the soil beneath the spudcan base reconsolidates after spudcan installation, the suction at the spudcan base increases with time revealing the longer a jack-up operates in a particular field, the harder to extract the spudcan particularly in soft soil with a deeply buried spudcan.

To enhance the understanding of spudcan extraction mechanism, half-cut model spudcan tests were also carried out. The results were reported by Purwana et al. (2006). By conducting image processing on a series of high resolution photographs captured in the tests, it is evident that three distinct phenomena exist throughout the extraction event. The initial condition is where the central plane of the model spudcan is embedded inside the textured clay prior to extraction. The subsequent extraction generates movements of the soil surrounding the spudcan. At the top of the spudcan, the overlying soil is compressed and the compression progressively propagates to the surface as the spudcan is further pulled. In contrast, the underlying soil is in an extension state as it moves up along with the spudcan. After a certain spudcan upward displacement, the underlying soil stops moving further upward and the movement of a soil column above the spudcan was predominant.

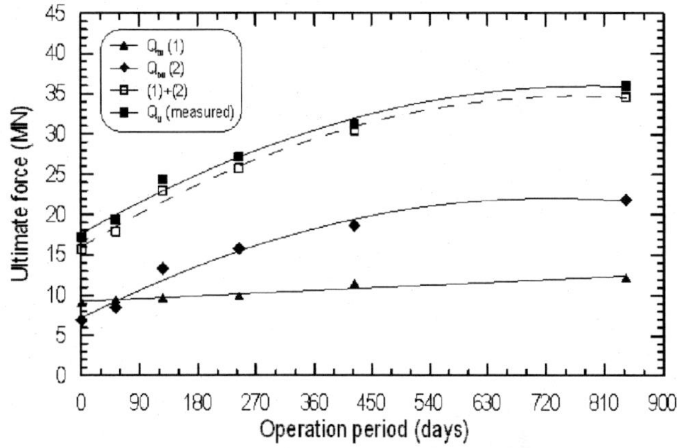

FIGURE 13: Composition of breakout force for spudcan extraction (after Purwana et al. 2005)

Upon further extraction, the area beneath the spudcan is observed to be filled with non-texture soil. This indicates the flows of the surrounding soil to beneath the spudcan base. The soil flow continues until a stage where a cavity beneath the spudcan is formed in which water can enter. The water gap progressively develops as the spudcan lifts while the footprint is still able to maintain the shape. When the spudcan is virtually pulled out of the soil surface, an abrupt failure of the cavity occurs and the surrounding surface soil flows into the footprint. The key moment of spudcan breakout is shown in Fig. 14 revealing that as the spudcan breakouts during uplift, there would be separation of soil represented by the zone of non-textural soil just beneath the spudcan base. This illustrates that the suction at the spudcan needs to be overcome before the spudcan can break out of the soil.

5. STUDIES ON SPUDCAN FOOTPRINT INTERACTION

After the spudcans have been extracted when a jack-up is moved offsite, depressions would be left by the previous spudcan installations and these depressions are commonly termed as footprints. Fig. 15 shows the footprints left by previous spudcans at a site. Another jack-up may visit the same site some months or years later. The installation of new spudcans (represented by the continuous black circles in Fig. 15) in existing footprints may cause a spudcan leg to slide into a footprint due to uneven soil conditions in the seabed. Such sliding movements not only cause structural damage and unsatisfactory positioning of the jack-up unit, but also injury to personnel and at worst lead to catastrophic failure (SNAME, 2008). As such, the interaction between spudcan installation and existing footprint needs to be evaluated.

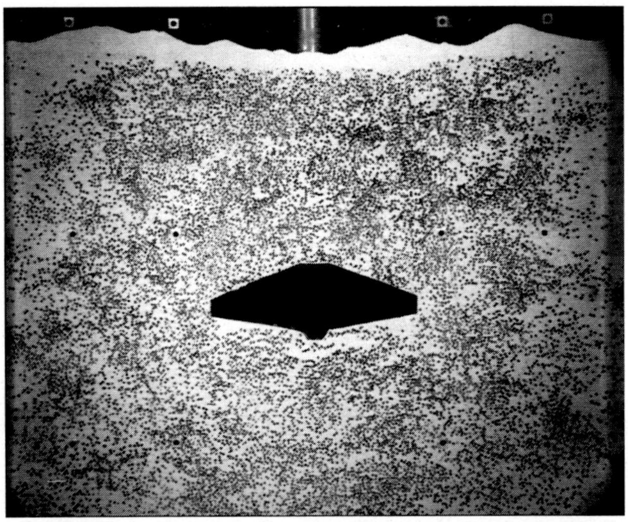

FIGURE 14: Separation of soil from spudcan base at breakout of spudcan

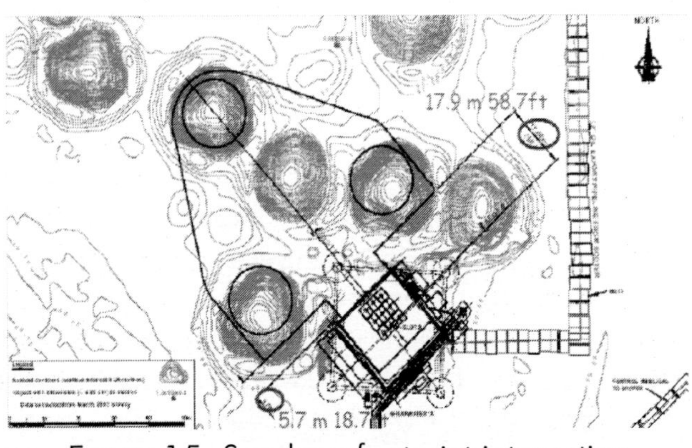

FIGURE 15: Spudcan footprint interaction (plan view)

A centrifuge model study has been performed to examine the spudcan footprint interaction problem. In the tests, the model spudcan leg was kept fixed and instrumented with three levels of strain gauges (Fig. 16). In such setup, the spudcan would not be able to slide into the footprint. Instead, there would be severe lateral force and moment induced on the leg and the magnitudes denote the degree of severity of spudcan footprint interaction. Gaudin et al. (2012) reported centrifuge test results that if the top of the spudcan leg is allowed to rotate, the induced moment and lateral force reduce considerably if the leg is allowed some degree of freedom. It should be noted that the rotation of the spudcan leg is also a concern at a site as the installation of jack-up could be off target as a result of spudcan leg rotation.

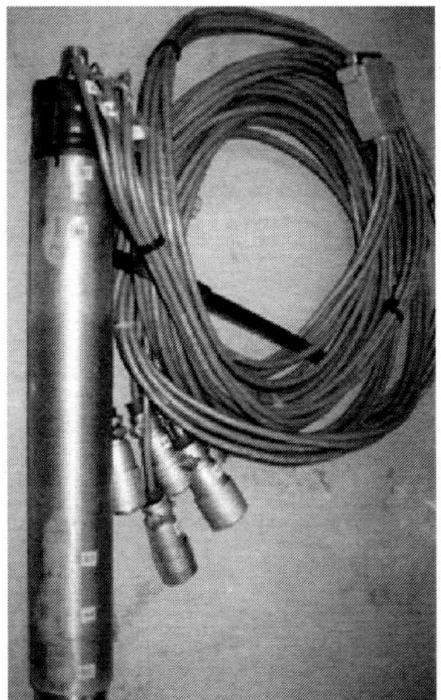

FIGURE 16: Instrumented spudcan leg to monitor induced lateral force and moment on spudcan leg

Gan et al. (2008) presented the results of centrifuge model tests on spudcan footprint interaction. The load penetration of spudcan installation and reinstallation at 0.5 spudcan diameter offset of the footprint is shown in Fig. 17. It is evident that due to the presence of cavity in the footprint, the reinstallation of spudcan requires a smaller load. As such, for the same preload, the

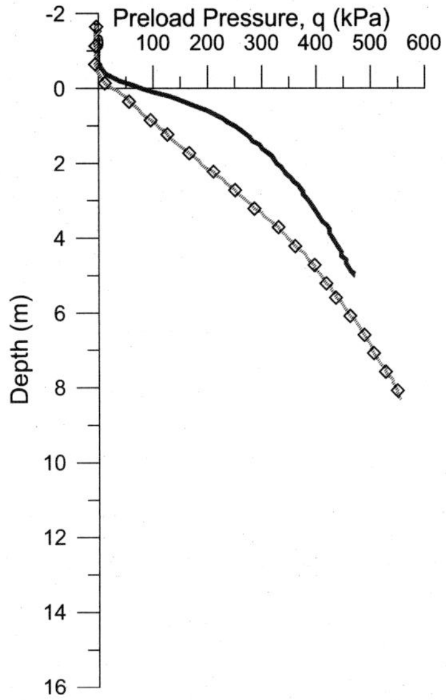

FIGURE 17: Load-penetration curve of spudcan penetration (right curve) and spudcan re-penetration after footprint is formed (left curve with symbols)

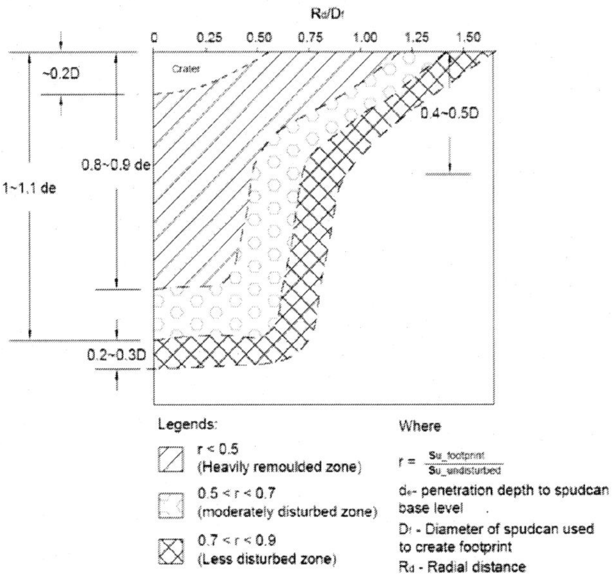

FIGURE 18: Shear strength variations beneath spudcan footprint

penetration of the spudcan is considerably deeper as compared to first time installation. To further examine the problem, the shear strength variations beneath a footprint are investigated and the generalized results obtained from a number of tests are presented in Fig 18. Gan et al. (2012) further presented the results of time effects on spudcan footprint interaction involving different times of spudcan extraction of first spudcan installation representing different jack-up operation periods and different elapsed times denoting the revisit of a jack-up rig to a site at various time frames

6. STUDIES ON SPUDCAN FOOTPRINT INTERACTION

Significant soil movements would be induced during the installation of the spudcan as the spudcan diameter typically ranges from 10 to 25 m which is very large. Figs. 2 and 3 illustrates that the large soil movements due to spudcan installation may induce severe stress on the comparatively small pile foundation supporting the adjacent permanent jacket platform. SNAME (2008) recommends that spudcan-pile interaction should be analysed if the clearance between the spudcan and the pile is less than one spudcan diameter.

A centrifuge model study has been conducted to examine the interaction between spudcan and adjacent pile foundation. The experimental setup for the full model spudcan study is shown in Fig. 19. The test results are presented in detail in Xie et al. (2012) who reported that the induced lateral pile behaviour is significant as compared to the induced axial pile behaviour. From safety view point of the jacket platform, the induced pile bending moment is deemed to be most critical. Fig. 20 shows the variations of pile bending moment profiles with clearance between spudcan and pile at spudcan penetration depths of 5 m and 20 m. In these tests, a 40-m long fixed head pile embedded in 37 m of soft clay and 3 m of sand at the bottom is modelled. Though the observed bending moment profiles are similar, the maximum induced pile bending moment increases significantly with decreasing spudcan-pile clearance, as expected. The magnitude of maximum induced moment increases with increasing spudcan penetration depth.

In order to provide a preliminary guidance for offshore geotechnical engineers to assess the severity of spudcan pile interaction in soft clay, the test results are summarized in chart form and presented by Leung et al. (2012). A typical chart on the variation of clearance factor with spudcan-pile clearance at different spudcan penetration depths is

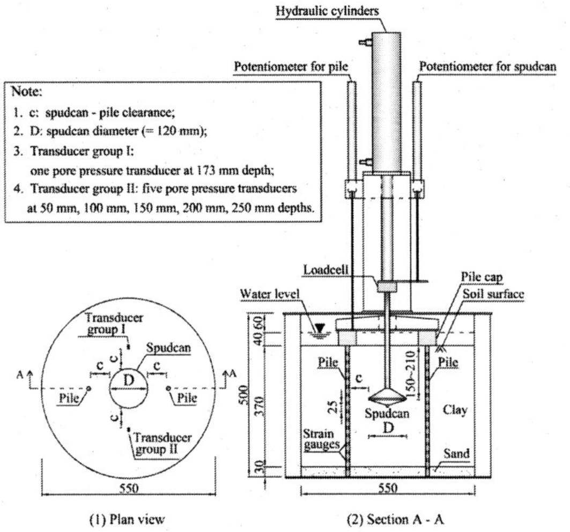

FIGURE 19: Experimental setup of spudcan-pile interaction study (after Xie et al., 2012)

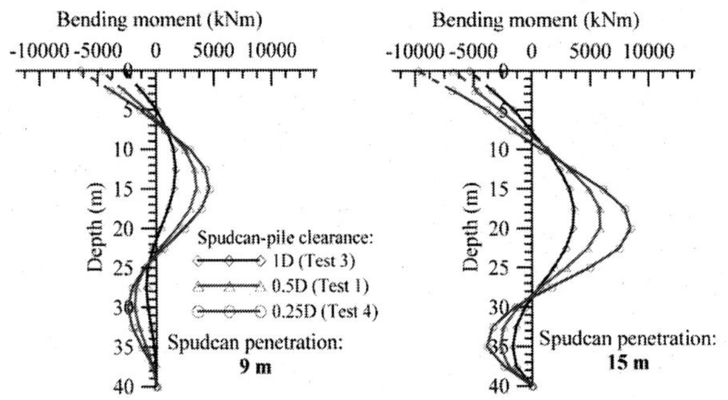

FIGURE 20: Induced pile bending moment profiles at different spudcan-pile clearances (after Xie et al., 2012)

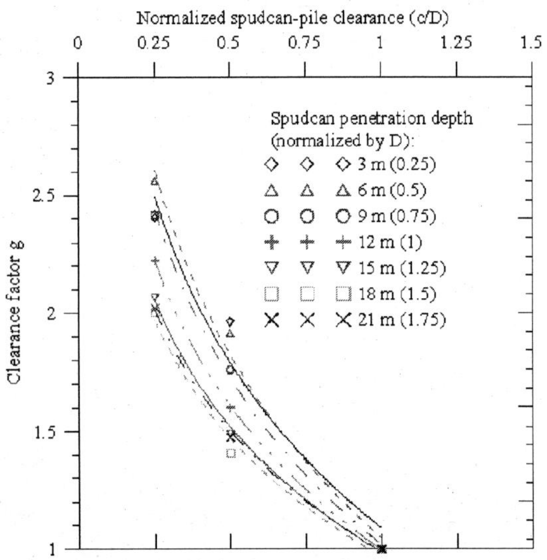

FIGURE 21: Clearance correction factor for maximum induced pile bending moment at pile head for different spudcan pile clearances

shown in Fig. 12. It can be seen that the maximum induced pile bending moment at spudcan-pile clearance of 0.25 spudcan diameter is over 2 times that at 1 spudcan diameter clearance. It is observed that the spudcan penetration depth corresponding to the occurrence of maximum bending moment varies with spudcan-pile clearance. The bending moment reaches a maximum at an early penetration stage if the pile is closer to spudcan. Factors such as total pile length, pile socket length, distinct soil layering, pile diameter and rigidity are also reported by Leung et al. (2012).

7. CONCLUDING REMARKS

This paper provides a summary of findings on centrifuge model studies conducted on full and half model spudcans to investigate various jack-up foundation problems including punch through, extraction difficulties, spudcan footprint interaction and spudcan pile interaction. The key findings are summarized as follows.

a. When installing a spudcan in sand overlying soft clay, a sand plug would be formed beneath the spudcan base. This phenomenon is different from that provided in traditional shallow foundation theory on layered soil. It has been further identified that the load-penetration response of the spudcan in sand overlying soft clay consists of three critical components: resistance at the seabed, maximum resistance where punch through would occur and the sudden rapid penetration depth whereby the reduced resistance would increase back to the punch through resistance.

b. The main cause of difficulties in extracting a deeply buried spudcan in soft clay has been identified as the suction developed between the soil and spudcan base. The suction increases as the soil beneath the spudcan reconsolidates after spudcan installation. As such, the longer the operation period of a jack-up rig at a site, the more difficult it would be to extract a deeply buried spudcan.

c. The study on spudcan footprint interaction reveals that the re-penetration of a spudcan at a site with existing footprint would depend on the size of the crater on the ground surface as well as the change in soil strength due to previous spudcan installation. As such, the duration of previous jack-up operation and the elapsed time for the jack-up rig revisit play a role in the interaction between the re-penetration of a spudcan and existing spudcan footprint.

d. SNAME (2008) recommends further examination if the penetrating spudcan is less than 1 spudcan diameter away from the nearest jacket platform piles. The results of the centrifuge model study establish that the lateral pile behaviour is more severely affected by spudcan penetration as compared to the axial pile behaviour. The induced bending moment on the pile which is located at 0.25 spudcan diameter clearance from the spudcan would be more than double of that on a pile located at 1 spudcan diameter away.

ACKNOWLEDGEMENTS

The author wishes to extend his sincere thanks to Prof Y K Chow, Executive Director of CORE and co-investigator of the studies, for his valuable contributions to the projects. In addition, the great contributions from former PhD students Dr Okky Purwana, Dr TEH Kar Lu, Dr GAN Chengti and Dr XIE Yi to the studies presented in this paper are gratefully appreciated. Last but not the least, the able assistance of the laboratory staff of Geotechnical Centrifuge Laboratory is acknowledged.

REFERENCES

Craig, W. H., and Chua, K. (1990) "Deep penetration of spudcan foundations on sand and clay". Géotechnique, 40, Issue 4, pp 541-556.

Finnie, I. W. S. (1993) "Performance of shallow foundations in calcareous soils", PhD thesis, University of Western Australia.

Gan, C. T., Leung, C. F., and Chow, Y.K. (2008) "A study on spudcan footprint interaction", Proc. of 2nd BGA International Conference on Foundations, Dundee, pp 861-872.

Gan, C. T., Leung, C. F., Cassidy, M. J., Gaudin, C., and Chow Y. K. (2102) "Effect of time on spudcan-footprint interaction in clay". Geotechnique, 62, Issue 5, pp 401-413.

Gaudin, C., Kong, V., and Cassidy, M.J. (2012) "An overview of spudcan reinstallation near a footprint", Proc Offshore Technology Conference, Houston, OTC 23050.

Hossain, M. S., Hu, Y., Randolph, M. F., and White, D. J. (2005) "Limiting cavity depth for spudcan foundations penetrating clay". Géotechnique, 55, Issue 9, pp 679-690.

Le Tirant, P., and Pérol, C. (1993) "Stability and operation of jackups", Paris: Technip.

Leung, C. F., Tho, K. K., Chow, Y. K. Xie, Y., Wong, P. C., and Purwana, O. A. (2012) "Experimental and numerical studies of spudcan pile interaction", Proc Offshore Technology Conference, Houston, OTC 23053.

McClelland, B., Young, A. G., and Remmes, B. D. (1981) "Avoiding jack-up rig foundation failures", Proc. Symposium on Geotechnical Aspects of Coastal and Offshore Structures, Bangkok, Vol 1, pp 137-157.

Mirza, U. A., Sweeney, M., and Dean, A. R. (1988) "Potential effects of jack-up spud can penetration on jacket piles", Proc Offshore Technology Conference, Houston, OTC 5762.

Osborne, J. J., and J. M. Paisley. (2002) "SE Asia jack-up punch-throughs: The way forward?", Proc. Int. Conf. Offshore Site Investigation and Geotechnics, London, pp 301-306.

Purwana, O. A., Leung, C .F., Chow, Y. K., and Foo, K. S. (2005) "Influence of base suction on extraction of jack-up spudcans". Geotechnique, 55, Issue 10, pp 741-753.

Purwana, O. A., Leung, C. F., Chow, Y. K., and Foo, K. S. (2006) "Breakout failure mechanism of jack-up spudcan extraction", Proc 6th International Conference on Physical Modelling in Geotechnics, Hong Kong, Vol 2, pp 667-672.

SNAME. (2008) Guidelines for Site Specific Assessment of Mobile Jack-up Units. Society of Naval Architects and Marine Engineers, Technical and Research Bulletin 5-5A, New Jersey.

Stewart, D. P., and Finnie, I. M. S. (2001) "Spudcan-footprint interaction during jack-up workovers", Proc International Society of Offshore and Polar Engineers (ISOPE), Vol. 1. pp 61-65.

Teh, K. L., Cassidy, M. J., Leung, C. F., Chow, Y. K., and Quah, M. (2008) "Revealing the bearing capacity mechanism of a penetrating spudcan through sand overlaying clay". Geotechnique, 58, Issue 10, pp 793-804.

Teh, K. L., Leung, C. F., Chow, Y. K., and Cassidy, M. J. (2010) "Centrifuge model study of spudcan penetration in sand overlying clay". Geotechnique, 60, Issue 11, pp 825-842.

White, D. J., and Take, W. A. (2002) GeoPIV: Particle Image Velocimetry (PIV) software for use in geotechnical testing. Cambridge University Engineering Department, Technical Report D-SOILS-TR322.

White, D. J., Take, W. A., and Bolton, M. D. (2003) "Soil deformation measurement using particle image velocity (PIV) and photogrammetry". Géotechnique, 53, Issue 7, pp 619-631.

Xie, Y., Leung, C. F., and Chow, Y. K. (2012) "Centrifuge modeling of spudcan-pile interaction in soft clay". Geotechnique, 62, Issue 9, pp 799-810.

Young, A. G., Remmes, B. D., and Meyer, B. J. (1984) "Foundation Performance of Offshore Jack-up Drilling Rigs". Journal of Geotechnical Engineering, ASCE, 110, Issue 7, pp 841-859.

Geotechnical Challenges in Indian Offshore

R.K. Ghanekar

Institute of Engineering and Ocean Technology (IEOT), Oil and Natural Gas Corporation Limited, Panvel, Navi Mumbai

E-mail: rkghanekar@gmail.com

ABSTRACT

Offshore Geotechnics in India till now is mainly focused on application for development of shallow water offshore oil and gas facilities. Till very recent times, Oil and Natural Gas Corporation Ltd. (ONGC) had virtual monopoly on offshore exploration and production. The area of focus too had been limited primarily to shallow waters of western Indian offshore. The structures in shallow waters have been fixed offshore platforms for production and processing, Jack-ups for exploration and work-over, and, a large pipeline network. With economic liberalization and increasing energy demand in India, attention is now being given to exploration and production of hydrocarbons in deep waters. In addition, it is anticipated that economical exploitation of shallow water marginal fields, offshore wind, wave and tidal energy will soon attract greater attention. With this changing scenario, new structural concepts requiring new (at least to Indian offshore) foundation concepts are bound to register increasing presence in Indian offshore posing special challenges to geotechnical engineers. This paper presents in brief the geotechnical problems and challenges in present Indian scenario and anticipated challenges in near future. Many of these problems would require theoretical, numerical and physical modeling for their solution. The paper also briefly reviews the physical modeling efforts in India focused on offshore environment and highlights the need for industry-academia interaction in this context.

1. INTRODUCTION

From the first discovery of oil in Indian offshore in 1974 till very recent times, Oil and Natural Gas Corporation Ltd. (ONGC) had virtual monopoly on offshore exploration and production. ONGC, having been the dominant player in Indian offshore, a bias towards oil and gas industry and ONGC perspective is perhaps inevitable while discussing the offshore geotechnical issues. Private players too are now engaged in exploration and production of hydrocarbons from offshore fields including deep water areas.

This paper tries to present in brief the geotechnical problems and challenges in present Indian scenario and anticipated challenges in near future. The paper does not attempt to focus exclusively on physical modeling aspect but tries to present an overview of challenges - present and future. The paper also addresses the (much clichéd) issue of need for interaction of industry and academia – indigenous and foreign – which however, in author's opinion, is extremely important and needs serious attention if the industry has to obtain competent technical solutions to its problems economically and in optimum timeframe. The reason for addressing this issue here is two unrelated and independent developments related to physical modeling in Mumbai itself which could address the offshore problems – establishment of a model testing facility by IEOT, ONGC and a centrifuge facility at Indian Institute of Technology Bombay (IITB), Mumbai.

2. CHALLENGES IN SHALLOW INDIAN WATERS

Traditionally, foundations for fixed offshore structures in India have been open ended driven steel pipe piles. Most of the Jack-up units operating in Indian waters are three legged structures with individual spudcan at each leg. Required soil characterization, foundation design and field performance of these foundations appear to be satisfactory in an overall perspective. However, there still are some issues which need more attention.

2.1 Soil Characterization

Soils in western Indian offshore are known to be calcareous in nature. Methodology for taking this into account while characterizing the soil has evolved over time with experience. Uncertainty in identification of cementation in calcareous/ carbonate sands, its variability in extent and strength is still a concern.

CPTU is the main in-situ tool currently being used in Indian offshore. Full flow penetrometers for soft sediments for better characterization needs to be included in the normal soil investigations. Use of seismic CPTU also needs to be examined for soil characterization especially for cemented sediments.

2.2 Driven Piles

a. Driveability aspect and sporadic cases of pile refusal still concern the engineers.
b. Requalification of old platforms or life extension issues now are of primary concern where codal changes over the years demand increased environmental design loads to be considered. This raises the issue of adequacy of existing pile foundations. Examination of time effect on load bearing capacity of driven piles has hence become very important.

2.3 Jack-ups

a. Jack-up spudcan–footprint interaction while redeployment of jack-ups near fixed platforms for work-over operations.
b. Identification and mitigation of potential punch-through situations is another aspect which needs more detailed investigation.

2.4 Other Issues

With high oil prices prevailing, exploitation of marginal fields, hitherto unattractive, is now becoming commercially viable. Extent of economic benefit of developing such areas for production is also dependent on the reduction in the cost of structures and foundations. Soil characterization, comparative evaluation of different foundation concepts, their analysis and design for sub-sea structures in such marginal fields are aspects which may have to be dealt with regularly in near future.

Foundations for wind farms with their special requirements are also seen as an important issue to be addressed in near future.

3. CHALLENGES IN DEEP WATERS

Exploration and exploitation of hydrocarbons in Indian deep waters has been initiated comparatively recently. Deep water environment poses new problems to geotechnical engineers. Research is still continuing worldwide on many of the special geotechnical aspects of deep and ultra deep water environment.

Deep waters require floating and sub-sea structures which in turn require anchoring systems, analysis and design of which is markedly different compared to traditional pile foundations. Very soft soils generally found in deep waters pose additional challenge in their characterization. It is known from limited experience in deep waters in the eastern Indian offshore, that very soft clayey soils exist up to a significant depth below the seabed. In certain deltaic areas under-consolidation too is suspected. Soil Characterization, design and installation of foundations in such areas require more sophistication compared to those in shallow water areas. Deep water temperature and pressure regime along with very soft surface sediments generate new problems for pipeline stability and design.

Based on the reported experience of offshore industry worldwide in deep and ultra deep waters, some of the important challenges which will have to be faced in India in near future are:

3.1 Soil Characterization

a. Deployment of special soil investigation systems for sampling and in-situ testing
b. Use of new in-situ testing tools in very soft soils e.g. full flow penetrometers
c. Advanced laboratory testing, interpretation of in-situ test results and integration of data
d. Greater emphasis (compared to onshore) on use of geophysical surveys and integration of data with geotechnical data especially in view of deep water geo-hazards
e. Estimation of stiffness characteristics and cyclic strength of the very soft surface sediments

3.2 Analysis and Design of Shallow Foundations for Sub-sea Structures for Compression and Uplift e.g. for Pipeline End Manifolds (PLEM) and Terminals (PLET)

3.3 Analysis and design of anchoring systems for floaters and their comparative evaluation for application

a. Drag anchors
b. Suction anchors
c. Suction Embedded Plate Anchors (SEPLA)
d. Dynamically penetrating anchors (Torpedo Anchors)

3.4 Geo-hazard Assessment and their Impact on the Offshore Facilities Including Pipelines

a. Submarine slides and mud-movement
b. Effects of gas hydrate degeneration / dissociation
c. Effect of shallow gas

3.5 Pipeline Design

a. Pipeline – seabed interaction
b. Trenching
c. Pipeline walking
d. Pipeline lateral buckling
e. Catenary risers – soil interaction

Research on many specific aspects of the above shallow and deep water challenges is being carried out worldwide using both numerical and physical modeling. It is felt that solutions to many of the geotechnical problems of Indian offshore should emerge through indigenous efforts.

4. INDIAN SCENARIO – PHYSICAL MODELLING FOR OFFSHORE APPLICATION

As far as author's knowledge goes, very little attention has been given to physical modeling for offshore problems in India in recent times. Some early work on model testing was performed in IIT Delhi, in the area of offshore foundations on the following aspects:

a. Behaviour of submarine pipelines on soft seabed.
b. Behaviour of piles in calcareous soils.
c. Plate anchors, piled anchors and super-piles in soft clays for floating offshore structures.

Institute of Engineering and Ocean Technology (IEOT), ONGC started work on developing a model testing facility with the help of Norwegian Geotechnical Institute (NGI) during 1998-2000. Testing was carried out on a model of suction foundation in marine sand to study the behaviour of the foundation under static and cyclic loading conditions. Subsequently, IEOT developed a better facility (Figures 1 and 2) for model testing to address offshore foundation problems with technical support from COFS, Australia in the year 2000.

This facility was mainly aimed at model testing for foundations e.g. suction caissons, bucket foundations. However, a wide variety of studies can be carried out using the facility with some modifications and limited fabrication works. The model testing facility is housed in a shed of size 15m × 6 m × 4.3m (L × W × H) which is equipped with a 3 Ton EOT Crane, 10

Bar compressed air supply, office cabin and other facilities. The equipments of the facility are briefly described below.

- **Test Tanks:** Three sets of test tanks of diameter 1000 mm and height 1200 mm with arrangements for drainage and connections for transducers.
- **Sample Preparation Equipment:**
  - **Clay** – hydraulic consolidation frame (90 Ton maximum) and a piston and air pressure (50 kPa maximum) arrangement, slurry mixer of 1 m^3 capacity vacuum pump for mixing under vacuum, slurry pump to transfer slurry from mixer to test tanks, facility for applying suction at the bottom of the sample.
  - **Sand** – sand rainer for preparing loose to medium dense samples using a sieve of 2.5 mm opening and a winch for lifting to control the density, vibrator for preparing dense samples.
- **Loading Actuator:** a stepper motor driven, multi-axis actuator for applying both vertical and horizontal loads up to 10 KN, inclined loads in 5° increments for up to 45° maximum inclination, 10 mm/sec maximum speed of loading
- **Syringe Pump:** A software controlled syringe Pump of 6 litre capacity for suction installation of the caissons driven by a stepper motor
- **Investigation Tools:** A penetrometer with interchangeable tips for Cone penetrometer, T-Bar and Ball penetrometer tests with load cell capacity of 700 N (corresponding to about 9 MPa for a cone of 10 mm diameter), a hand shear vane
- **Instrumentation:** Required, load cells, displacement transducers, inclinometers, earth pressure transducers, pore pressure transducers etc.
- **Data Acquisition System:** Computer based data acquisition system with required data cards and the amplifier.

Due to some unavoidable reasons, only a limited amount of work could be taken up using

FIGURE 1: Set-up for a test

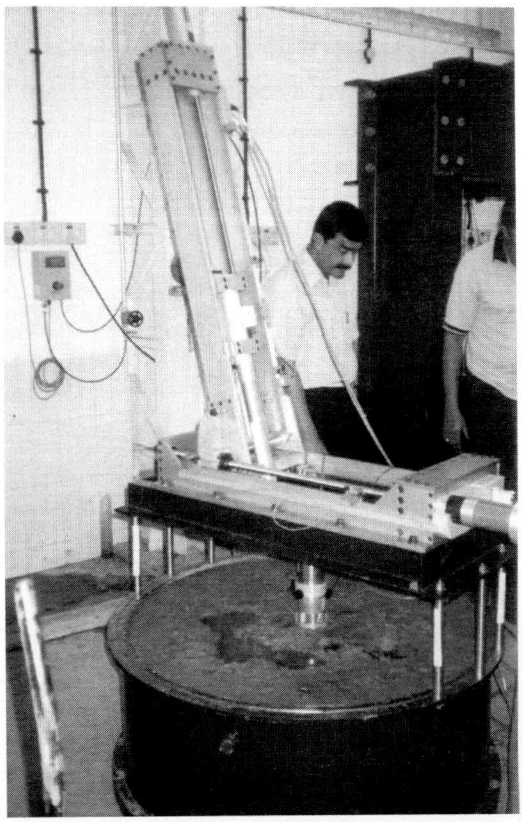

FIGURE 2: Inclined loading being applied on a model

the facility so far. There is a potential to revive this facility in collaboration with reputed academic institutes in India and address the geotechnical issues concerning Indian offshore industry.

IITB, Mumbai has an excellent centrifuge facility which could also be utilized for addressing offshore geotechnical problems in collaboration with offshore industry.

5. INDUSTRY – ACADEMIA INTERACTION

A dynamic and fruitful interaction between industry and academia has been lacking in the field of offshore geotechnical engineering in India due to many reasons. However, we are now in a transition period where industry is moving from shallow to deep waters and from "only hydrocarbons" to "hydrocarbons and alternative energy sources". Hence, this appears to be the most opportune time to correct the situation.

In author's opinion, there is a need to develop a Center of Excellence in the area of Offshore Geotechnics in India having capabilities in theoretical, numerical and physical modeling. The role models for such an entity can be for example Norwegian Geotechnical Institute (NGI), Norway and Center for Offshore Foundation Systems (COFS), Australia.

In many aspects of offshore geotechnical engineering, the world seems to have moved much ahead - hence - attracting talent from abroad for such a facility must not be ruled out. The funding could come partly from Indian companies working in Indian offshore and partly from Indian Government perhaps through reputed Indian Academic Institutions like IITs, Indian Institute of Science (IISc), Bangaluru.

Joint Industry Projects focusing on the problems being faced by Indian offshore industry can be referred to such a facility. Research sponsorship for Indian students by industry is another possibility. Exchange of personnel between industry and academia is another way of developing better interaction and understanding of needs of each other.

6. CONCLUSIONS

A brief review of the present and anticipated future geotechnical challenges in Indian shallow and deep waters has been presented. Many aspects of these problems can be addressed through physical modeling techniques. As reported in literature, research is continuing in this direction throughout the world. A brief review of work on physical modeling in India aimed at solving offshore geotechnical problems is presented.

Need of the hour is to develop and sustain indigenous expertise in the field of Offshore Geotechnics through interaction between industry and academia – both Indian and foreign. Establishment of a Center of Excellence in Offshore Geotechnics in India is a possibility which must be explored. There is no doubt that potential for this to fructify exists in India.

NOTE: *The views expressed in this paper are author's own views and do not necessarily represent the views of his organization.*

REFERENCES

Institute of Engineering and Ocean Technology (IEOT), ONGC. (2000) Model Testing of Suction Caisson in Marine Sand, IEOT Report No. IEOT-NGI/GEOTECH/WP-5/MT.

Joer, H. A., and Randolph, M. F. (2001) "Model testing for Deep Water Foundation Systems: Testing Procedures", University of Western Australia, COFS. Report (No. Geo:01262) prepared for IEOT, ONGC.

Datta M., Director, PEC University of Technology. (2012) "Personal communication".

K-water Geotechnical Centrifuge Facility

N.R. Kim, D.H. Shin, D.S. Park and U.S. Jung

Infrastructure Technology Center, K-water Institute, Daejeon, Korea

E-mail: namryong@kwater.or.kr; shindh@kwater.or.kr; fulgent@kawter.or.kr; usjung@kwater.or.kr

ABSTRACT

One of the largest geotechnical centrifuges in the world is being installed at K-water Institute in Daejeon, Korea for research purposes on mainly experimental performance verifications of dams. The 800 g-tons maximum capacity geotechnical centrifuge has 8.0 m platform radius and can be operated up to 150 g centrifugal acceleration. An experimental model with maximum weight of 8,000 kg can be loaded in 2.0 m (L) × 2.0 m (W) × 2.0 m (H) large platform. The centrifuge is equipped with fluid rotary joint for hydraulic supplies, fiber optic rotary joint for network connections, sliprings for electrical power supply and signal transmission, and on-board data acquisition system. A large scale uniaxial servo hydraulic in-flight earthquake simulator is also developed in order to simulate dynamic test for geotechnical systems. It has 1.6 m (L) × 0.8 m (W) experimental platform and can provide maximum 22 g horizontal acceleration in sinusoidal vibration load to 1,500 kg payload. An in-flight uniaxial loading system as well as miniature transducers will be prepared for general physical modelling purpose.

1. INTRODUCTION

Reduced-scale physical modelling in geotechnical engineering is often conducted within a large centrifuge in order to provide correct scaling of the self-weight stresses (Schofield, 1980; Taylor, 1995) and it is proven that this technique can provide various advantages in the analysis of geotechnical systems (Kim et al., 2012a).

Physical modelling technique has advantages in performance evaluation of large dams. Fill dams generally have various zoning in the body and complicated boundary conditions in the interfaces, therefore the mechanical behaviour of these characteristics of the dams cannot be sufficiently represented by numerical methods and the global performance may result differences from prototype. Physical modelling using geotechnical centrifuge can be an alternative tool to verify the performance of dams in a reasonable manner. Since major dams in the world which require critical performance verification due to safety issue are sometimes higher than 100 m, the centrifuge for this test must be large enough to represent the prototype according to precise scaling law and to simulate the global behaviour appropriately.

There are currently four geotechnical centrifuge facilities in Korea, however it is found that the capacities of these equipment have limitations on simulation of large dams currently operated in Korea. In this context, K-water (Korea Water Resources Corporation) decided to build a large scale geotechnical centrifuge facility as a part of dam safety research program. Considering preliminary experimental research on evaluating performance of dams at several centrifuge facilities, K-water decided to build an 8.0 m platform radius and 600 g-tons capacity geotechnical centrifuge which is one of the largest in the world. The installation of the centrifuge will be completed by the end of 2012 and the operation will start from 2013 for experimental research for dam safety as well as general geotechnical engineering research.

The purpose of this article is to introduce a new geotechnical centrifuge which is being installed at K-water. Details of the main centrifuge and the large scale in-flight biaxial earthquake simulator are presented. Furthermore, various equipment for experiment and model preparation are introduced.

2. CENTRIFUGE FACILITY AT K-WATER

In 2006, K-water institute has constructed a new testing facility for large geotechnical centrifuge and installed an 8.0 m platform radius, 600 g-tons centrifuge which is the second largest centrifuge in the world in terms of platform radius and capacity (Ha *et al.*, 2006). However, K-water decided to install other type of geotechnical centrifuge from a different supplier and made a contract with ACTIDYN SYSTEMES SA (Elancourt, France) on June 2011.

The building is a three-story reinforced concrete and steel structure building with a total of 1,596 m^2 of floor space for office and laboratory facilities. The centrifuge chamber and model preparation area are located in the 1st floor at the same level considering easy access to the chamber and conveniences of working process. The centrifuge chamber also has an access at the top roof for access to be used for installing heavy weight equipment such as earthquake simulator or maintenance purpose. For safety reasons, the reinforced concrete wall of the chamber was designed as 1 m thickness and furthermore the opposite side from the working area is surrounded in the ground. The control room and research area are located at 2nd floor and it has direct access to the top of the centrifuge chamber. The plan view of the building is shown in Figure 1.

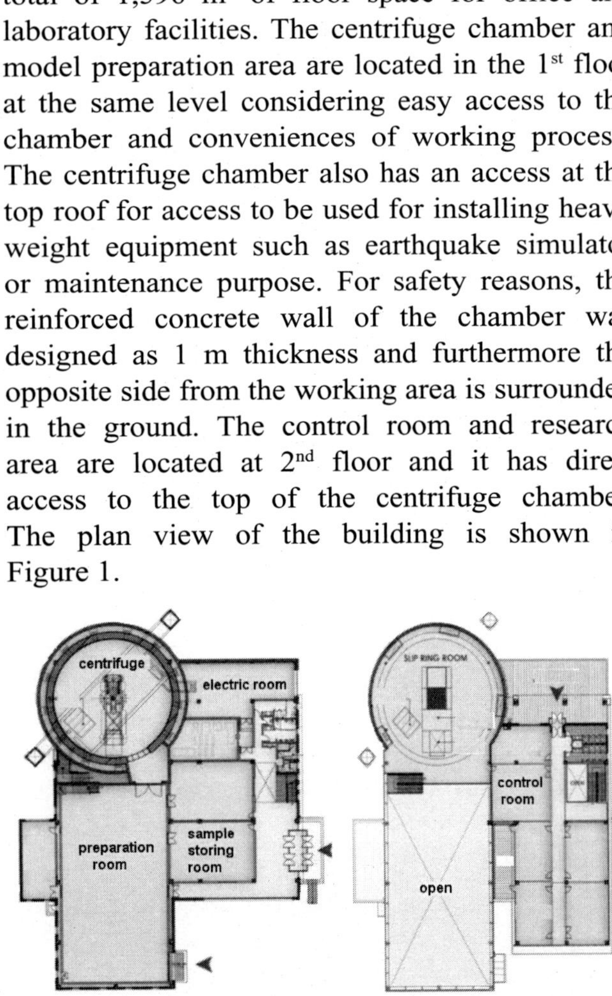

FIGURE 1: Layout of the Centrifuge Facility

The basement area underneath the chamber is prepared for electrical power supply to the main system and installation of driving system (i.e. motor, gear reducer and braking resistor). The centrifuge building was partially modified to accommodate a new geotechnical centrifuge hence a new concrete pedestal for anchoring the large centrifuge and the pad for equipment in the basement are newly constructed.

3. DETAILS OF THE EQUPMENT

3.1 800 g-tons Geotechnical Centrifuge

The geotechnical centrifuge of K-water is manufactured by ACTIDYN SYSTEMES SA and its installation will be completed by the end of 2012. The centrifuge is designed as asymmetric arm type and it is mainly balanced by massive counterweights at the opposite side of the platform. This type of design has advantages of users' convenience on balancing and reduces aerodynamic drag force during operation hence the electrical power consumption can be reduced.

The platform radius of the centrifuge is 8.0 m and the maximum capacity is 800 g-tons. The centrifuge is one of the largest in the world in terms of both platform radius and capacity. Considering total weight of earthquake simulator, the maximum payload of the centrifuge was decided as 8,000 kg. A model payload with maximum weight can be accelerated up to 100 g, and the maximum centrifugal acceleration of the system at nominal radius of 7.5m is restricted as 150 g with 3,800 kg payload, which is enough to simulate most of large dams appropriately. Detailed technical specifications of the machine are listed in Table 1 and the conceptual design of the centrifuge is described in Figure 2.

In this centrifuge, signal slip rings and a fiber optic rotary joint are installed at the top of the centrifuge for data transmission. Brush type sliprings for 12 electrical signal channels and 8 high frequency video channels are prepared. Local network is provided in the centrifuge on-board system by Ethernet connection via fiber optic rotary joint therefore users can easily configure and access on-board control system. Especially, a data acquisition system is mounted at the on-board cabinet and controlled remotely, so users can avoid noise problems in data collection.

At the bottom of the centrifuge, there are power sliprings for electrical power supply to the on-board systems and fluid rotary joints along the rotating axis beneath the driving system. Total number of passages in the fluid rotary joint is 10; 6 for hydraulic oil pressure for earthquake simulator, 4 for pneumatic air or water supply lines.

The centrifuge driving system consists of 3 major systems; the power drive system, an AC brushless variable speed induction motor, and a gear reducer in the room below the centrifuge chamber (Figure 2). The motor and gear system operates according to users' command and rotates the vertical drive shaft which connects entire system of the centrifuge.

The centrifuge is equipped with two different balancing mechanisms. Users move the primary counterweights which consist of five 10 ton metal plates to a certain location corresponding

Table 1: Technical Specifications of K-water Centrifuge

Item	*Specification*
Platform radius	8.0 m
Max. capacity	800 g-tons
Max. acceleration	150 g at 7.5 m nominal radius
Payload	8,000 kg up to 100 g 5,000 kg up to 120 g 3,800 kg up to 150 g
Payload dimensions	2.0 m (L) × 2.0 m (W) × 2.0 m (H)
Power consumption	1,200 kVA
Signal sliprings	12 channels for electrical signals 10 lines for electrical power supply 8 lines for high frequency video
Fluid rotary joints	6 passages for hydraulic oil supply 4 passages for air or water supply
Fiber optics	1 passage of 1 GHz FORJ
In-flight balancing	Fine adjustment by hydraulic actuator

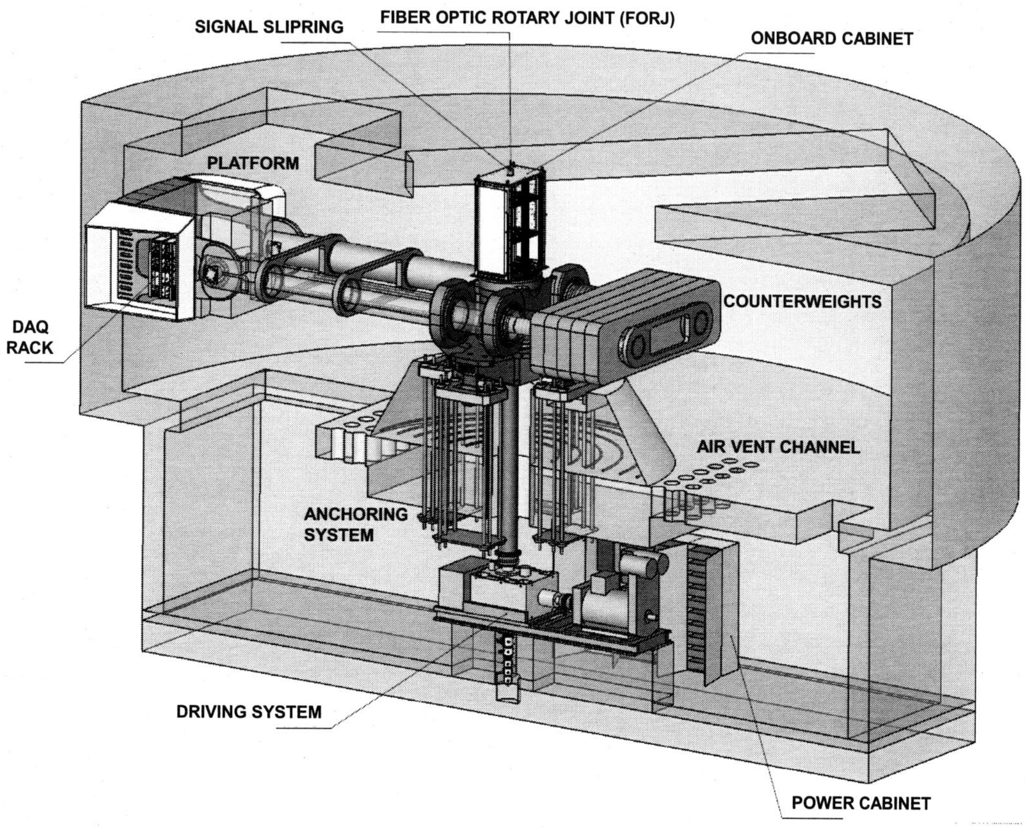

FIGURE 2: 800 g-tons Geotechnical Centrifuge at K-water

to the weight of payload using motor and gear system before operation. Fine adjustment for unbalance load during operation can be reached by automatic balancing system and the unbalance load is monitored using the strain gages at two metal springs among four anchoring systems. Two cylindrical metal counterweights are installed inside the arm tubes and it moves by hydraulic actuators and this mechanism adjusts the unbalance load.

3.2 Centrifuge Data Acquisition System

The Centrifuge Data Acquisition (DAQ) system main frame is installed in the on-board cabinet near the rotating axis and it is directly connected to local network via gigabit fiber optic rotary joint. This configuration of centrifuge DAQ system is common in most of beam type centrifuge cases while signal slip rings had been used for data transmission in early years of centrifuge testing.

The Centrifuge DAQ System is developed by combining brand-new National Instruments PXI and SCXI series to condition signals from a wide variety of transducer types, and to ensure precise timing and synchronization. The configuration of this system is described in Figure 3. This kind of system is widely used in many centrifuge facilities due to its good flexibility to configure the system as user's requirements. This system basically consists of a PXIe Express chassis, PXIe-8133 quad-core embedded controller, five PXIe-6341 X-Series data acquisition modules, one PXI-6733 analog output module, and one PXI-2565 relay module therefore the configuration of this system can be expanded to cover 288 input channels. Particularly, the storage unit of the controller module was selected as a

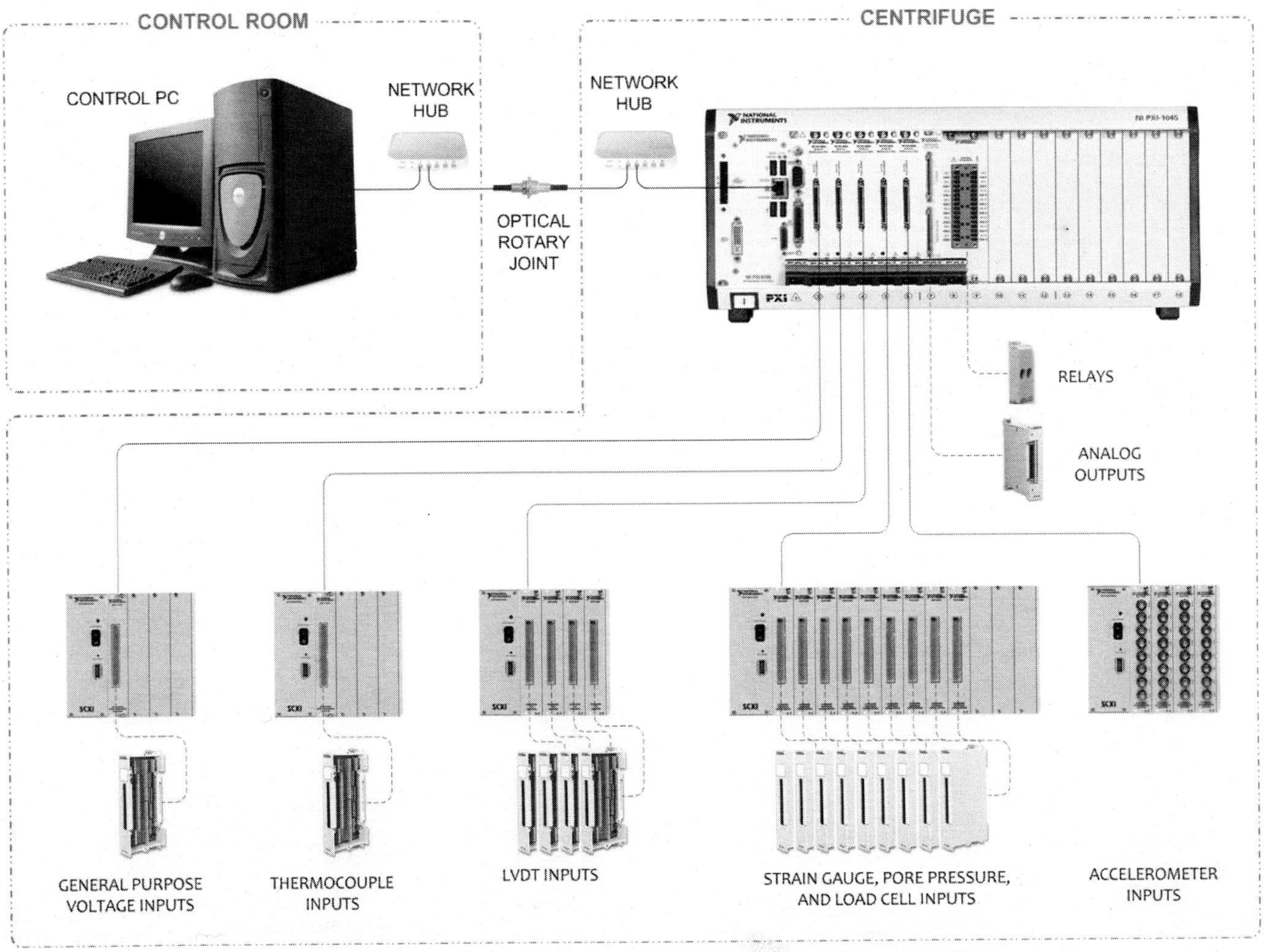

FIGURE 3: The Configuration of Centrifuge Data Acquisition System

solid-state hard drive because disk type hard drive sometimes do not work under high g environment. Each PXIe-6341 data acquisition module is cabled to a SCXI signal conditioning chassis which is located at the experimental platform (Figure 1). Any combination of SCXI-1000 4-slot chassis and SCXI-1001 12-slot chassis may be used. Signal conditioning modules include the SCXI-1102C, SCXI-1520, SCXI-1531, and SCXI-1540. Supported sensors include accelerometers, quarter-, half-, and full-bridge strain gages, pore pressure sensors, LVDTs, thermocouples, and analog voltages. Any SCXI module can be installed in any slot of any SCXI chassis. The PXIe-6341 modules perform high-speed analog-to-digital conversion of the conditioned signals. The PXI-6733 analog output module and PXI-2565 relay module control a variety of equipment such as shaker tables, cameras, pumps, and valves.

3.3 Earthquake Simulator

Recently, lots of research on geotechnical earthquake engineering using centrifuges has been undertaken throughout the world. In-flight shaking tables for geotechnical centrifuges are mostly operated using servo hydraulic systems. Because of the scaling law, the frequency characteristics of ground input motion for the shaking tables have to be N times of prototype ground motion. Therefore, the maximum operating frequency range of those shaking tables is usually up to over 300Hz.

Table 2: Technical Specifications of K-water Earthquake Simulator

Item	*Specification*
Type	Electo-Hydraulic Servo System
Payload dimensions	1.6 m (L) × 0.8 m (W) × 0.8 m (H)
Max. Payload mass	1,500 kg (including container)
Centrifuge acc. range	10 to 100 g
Peak shaking force	500 kN
Max. stroke	5.0 mm
Max. shaking velocity	1.1 m/s
Max. shaking level	60 g without payload 22 g sine motion with full payload
Frequency range	20 to 350 Hz
Acceptable input motion	Earthquake record, sine, random.

The earthquake simulator of K-water is a self-balanced uniaxial shaking table. This system is basically designed for operation in 100g centrifugal acceleration and the maximum shaking acceleration level is 60g without any payload mass.

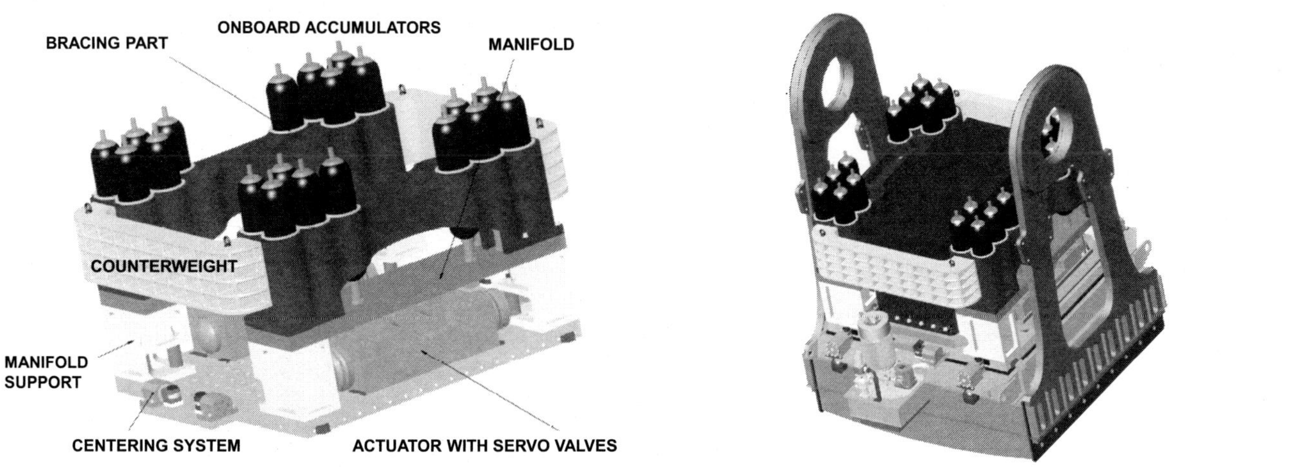

(a) Details of the earthquake simulator

(b) Earthquake simulator equipped in the centrifuge platform

FIGURE 4: Conceptual Design of the Earthquake Simulator for K-water Geotechnical Centrifuge

Experiments using this system must be designed according to proper scaling principles considering the frequency response and allowable simulating acceleration capacity, because the response of this system depends on both payload and frequency of input motion. The specifications of the system are listed in Table 2.

The earthquake simulator has been designed and operated based on dynamic balancing principle (Perdriat et al., 2002; Kim et al., 2012b). There are counterweights with similar mass of the experimental model at the table, and dynamic balancing is achieved by reciprocal actuation of the model and the balancing counterweight. In this system, the dynamically balanced motion is obtained by close loop control of its parallel pair of actuators. It is known that it offers better performances than any known type of mechanical bearing guidance associated with symmetry of construction. The dynamic balancing is important not only for simulating accurate input motion but for reducing the risk of damaging mechanical part of the centrifuge itself.

For the control of the earthquake simulator, the digital matrix controller acts as an outer loop controller that compensates for the acceleration (force) loops errors and for the dynamic properties of the model/exciters under test. A white noise signal with a small magnitude is used to evaluate system dynamics (gain and phase) at the first stage as measured from accelerometer attached to the table, and stores the resulting matrix of transfer functions. The digital controller software generates command signals (actuators drive signals) that are a complex combination of the desired simulator commands (reference vibration profiles) and the measured dynamic response (control point measurements). These command signals result in identical motion profile for each actuator of a pair as long as the test system exhibits predominantly linear behavior.

3.4 Miscellaneous Items

K-water geotechnical centrifuge facility prepared various equipment to be frequently used for physical modelling. Several containers including large rectangular container with a window on one longitudinal direction is prepared for static test of large dams, and an equivalent shear beam type container which has the interface to be attached to the earthquake simulator is prepared to provide appropriate boundary conditions for dynamic test. Stress and strain analyses using finite element analysis were conducted for these containers to verify its safety and the deformation was limited to ensure rigidity of the boundaries.

A penetrometer is also developed to provide vertical loading or movement in a certain direction during the test. It can apply vertical load up to 100 kN downward and 50 kN upward. It is useful system to provide unidirectional loading to a model and can be used conveniently. The designs of the loading system and the ESB container are shown in Figure 5.

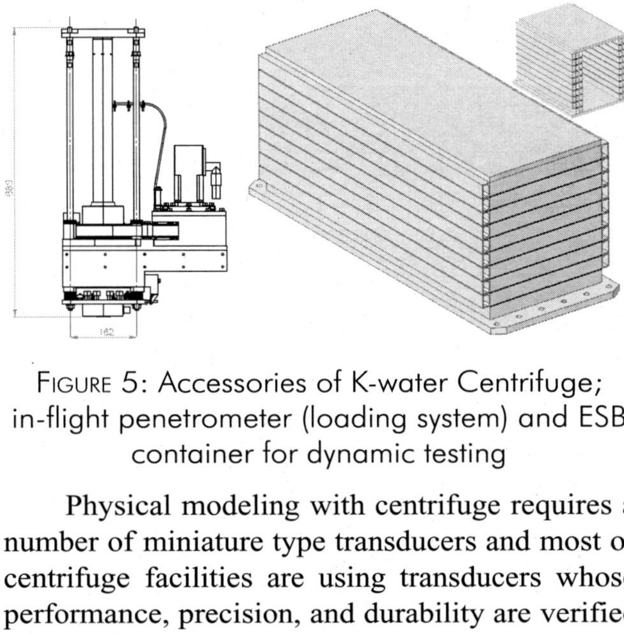

FIGURE 5: Accessories of K-water Centrifuge; in-flight penetrometer (loading system) and ESB container for dynamic testing

Physical modeling with centrifuge requires a number of miniature type transducers and most of centrifuge facilities are using transducers whose performance, precision, and durability are verified under high g environment. K-water geotechnical centrifuge facility selected transducers considering these criteria. High speed camera which enables remote control via Ethernet local network is also prepared to capture high speed video for dynamic event during the test. The accessories which K-water prepared for geotechnical centrifuge facility are briefly described in Table 3.

4. CLOSURE

A brief introduction of the new K-water geotechnical centrifuge facility has been

Table 3: Major Accessories or Equipment Related to K-water Geotechnical Centrifuge

Item	*Purposes*	*Specification*
Penetrometer and loading system	Moving or penetrating experimental model in vertical direction Applying vertical load	100 kN vertical loading (push down) 50 kN up-lift loading 400 mm stroke
Vacuum clay mixer	Mixing clay slurry removing air trap	Total volume : 200 liter
Automatic sand rainer	Preparing homogeneous/uniform dry sand model by fluviation method	Coverage area: 1,786mm (L) ×1,830 mm (W) Vertical movement: 1,200 mm Max. horizontal speed: 14 m/min
Containers	Soil containers for geotechnical modelling Large rectangular container with window Rectangular container with window (available for the earthquake simulator) Equivalent shear beam (ESB)	Outside : 1,800mm (L) × 800 mm (W) × 800 mm (H) Inside : 1,600mm (L) × 600 mm (W) × 600 mm (H) Outside : 1,600mm (L) × 600 mm (W) × 600 mm (H) Inside : 1,400mm (L) × 400 mm (W) × 400 mm (H) Outside : 1,600mm (L) × 600 mm (W) × 600 mm (H) Inside : 1,440mm (L) × 440 mm (W) × 500 mm (H)
High speed camera	To capture a video inside the model for dynamic event	Phantom v9.1 (Vision Research) Max. resolution : 1,632 × 1,200 @ 1,016 fps Normal frame rate : 960 × 480 @ 3,906 fps Max. frame rate : 153,846 fps
Transducers	ICP accelerometers LVDTs Laser displacement sensors Miniature pore pressure transducers	Up to ±500 g acceleration, 1 to 10,000 Hz frequency range Linear displacement up to ±2.54 mm Linear displacement up to ±5 mm at 48 mm reference distance Up to 10 bar water pressure range

presented focusing on state-of-the-art equipment for physical modelling. This centrifuge will be able to cover most of geotechnical modelling applications, especially for large earth structure, with advantage of its capacity. Furthermore, 1.8m long platform of the earthquake simulator will provide new opportunities for large structures, and conveniences in physical modelling, for example, by preparing multiple structural models at the same time for soil structure interaction study.

After complete installation of the centrifuge by the end of 2012, K-water will start a research program of physical modeling study on the performance evaluation of infrastructure related with water resources such as large dams. The K-water geotechnical centrifuge facility will also provide the opportunities of experimental research using its infrastructure in the field of physical modeling to geotechnical engineering communities in the world. It will be also available for joint research or cooperation.

REFERENCES

Ha, I. S., Seo, M. W., Jung, W. S., Kim, H. S., and Park, D. S. (2006) "Development of a large scale geotechnical centrifuge in KOWACO", Physical Modelling in Geotechnics – 6th ICPMG `06, Hong Kong, pp 135-139.

Kim, D. S., Kim, N. R., Choo, Y. W., and Cho, G. C. (2012a) "A newly developed state-of-the-art geotechnical centrifuge in Korea". KSCE Journal of Civil Engineering, DOI 10.1007/s12205-013-1350-5

Kim, D. S., Lee, S. H., Choo, Y. W., and Rames, D. (2012b) "Self-balanced earthquake simulator on centrifuge and dynamic performance verification". KSCE Journal of Civil Engineering. (Accepted).

Perdriat, J., Phillips, R., Nicolas Font, J., and Huntin, C. (2002) "Dynamically balanced broad frequency earthquake simulation system", Physical Modelling in Geotechnics - ICPMG `02, Newfoundland, Canada, pp 169-173.

Schofield, A. N. (1980) "Cambridge geotechnical centrifuge operation". Géotechnique, 20, Issue 3, pp 227-268.

Taylor, R. N. (1995). Geotechnical centrifuge technology, Blackie Academic, London, UK.

Modeling Reinforcement-Soil Interactions under Oblique or Transverse Forces – A Review

B. Umashankar, C. Hari Prasad and M.R. Madhav

Department of Civil Engineering, Indian Institute of Technology, Hyderabad
E-mail: buma@iith.ac.in, hariprasadiith@gmail.com, madhavmr@gmail.com

ABSTRACT

The pullout resistance of the reinforcement is an important parameter in the design of reinforced earth structures. The existing design procedures consider the pullout resistance due to only axial pull. However, the kinematics of failure clearly establishes that the reinforcement is displaced obliquely. In this paper, a review on the physical modeling of sheet reinforcement subjected to an oblique force applied at one end of the reinforcement is presented. The review ranges from simple physical models that consider linear normal stress-deformation of the fill material and rigid plastic behavior of shear stress-displacement at soil-reinforcement interface to more sophisticated models that consider nonlinear behavior for both normal stress-deformation of the fill and shear stress-displacement of soil-reinforcement interface. The effects of various parameters on the reinforcement-soil interaction are also highlighted. Review shows that the pullout resistance of the reinforcement is higher when subjected to the oblique force compared to that for an axial pullout.

1. INTRODUCTION

Reinforced earth or soil structures (reinforced retaining walls, embankments, subgrades beneath the pavements, etc.) are becoming increasingly popular mainly due to their (a) flexibility that enables to withstand large displacements, and (b) ability to perform better under seismic conditions. For proper design of any reinforced structure, soil-reinforcement interaction should be determined. The interaction mechanism can broadly be classified into two types: direct shear mechanism and pullout mechanism (Figure 1). The interface shear strength and shear stress-displacement response are evaluated by the modified direct shear test. The resistance offered by the reinforcement due to axial pullout can directly be determined from the pull-out test. Pull-out resistance of the reinforcement is an important parameter for the internal stability design of reinforced soil walls. In internal stability, localized mobilization of the reinforcement force (or tension in the reinforcement) and direction in the vicinity of the failure surface is considered. The localized mobilization of reinforcement force is dependent on the kinematics of failure of reinforced structure. The kinematics of failure dictates that the failure surface intersects the reinforcement obliquely (Figure 2). Response of reinforcement to an oblique force is a combination of responses due to transverse and axial forces. Most available theories and design methods however evaluate the pullout capacity of the reinforcement considering only axial pull-out mechanism (Jewell 1992; Sobhi and Wu 1996) and do not consider oblique pull or displacement.

Several experimental studies using direct shear tests have been performed to establish the influence of reinforcement on the improvement in the strength of the soil. For example, Figure 3 shows the pattern of deformation of reinforcement near the shear zone as proposed by Juran et al. (1982), Gray and Ohashi (1983), and Shewbridge and Sitar (1988).

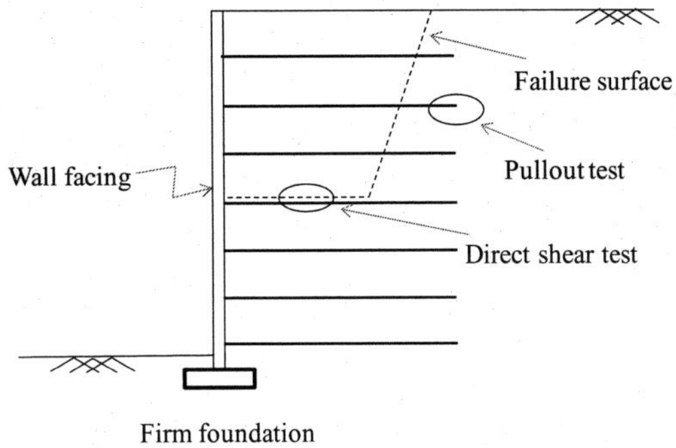

FIGURE 1: Soil-reinforcement interaction

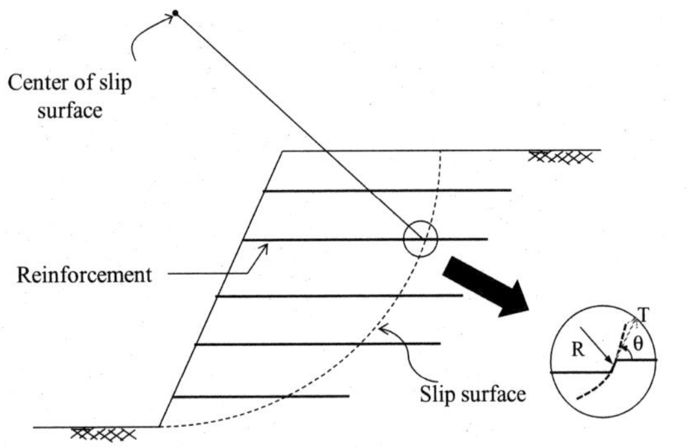

FIGURE 2: Kinematics of reinforced embankment

Bergado et al. (2000) have carried out large scale direct shear tests with inclined geotextiles and hexagonal wire mesh. The shear stress-displacement response of reinforcement with inclinations of 30°, 45°, 60° and 90° with the horizontal were obtained along with the deformation patterns.

In this review, the available physical models to study the reinforcement-soil interaction due to an oblique/transverse downward pull at the end of reinforcement is presented.

1.1 Pullout Mechanism of Oblique End Force

Oblique force is summation of horizontal and vertical forces. Analysis of reinforcement is fairly straight forward under the action of axial pull. In the case of an axial pull, the normal stress on the reinforcement-soil is the same as the gravity stresses and consequently the shear resistance mobilized at the interface is proportional to these normal stresses. However, under the action of transverse downward pull, the soil beneath the reinforcement deforms transversely leading to an increase in the normal stresses due to this deformation. Hence, the shear resistance mobilized could be considerably different in case of reinforcement subjected to a transverse force.

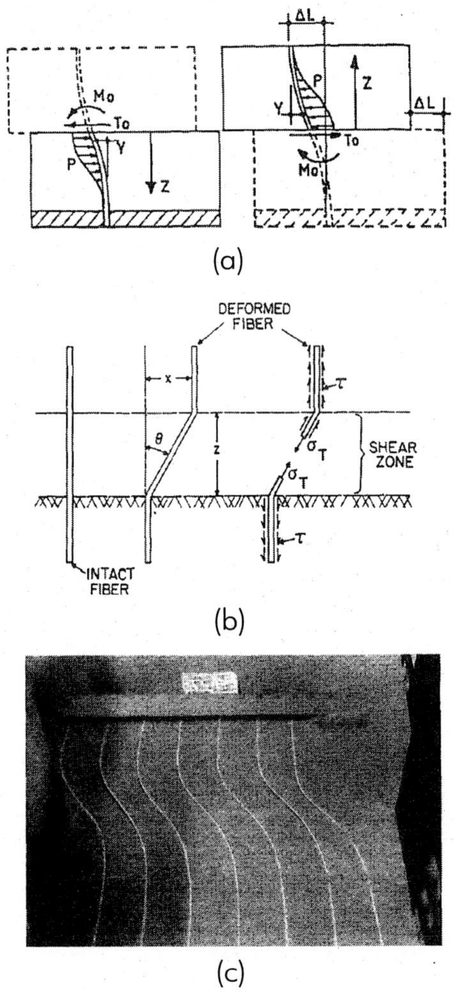

FIGURE 3: Proposed deformation patterns: (a) after Juran et al. (1982), (b) after Gray and Ohashi (1983), and (c) after Shewbridge and Sitar (1989)

2. REVIEW ON AVAILABLE MODELS

Madhav and Umashankar (2003a, 2003b, 2003c)

Madhav and Umashankar (2003a, 2003b, 2003c) did pioneering work in modeling the reinforcement-soil interaction when subjected to a transverse force. A sheet reinforcement of length L embedded at a depth D_e from the ground surface and subjected to a transverse force P at front-end of the reinforcement was considered.

Unit weight of the fill material is γ and the reinforcement-soil interface friction angle is ϕ_r (Figure 4). Both extensible and inextensible reinforcements were considered.

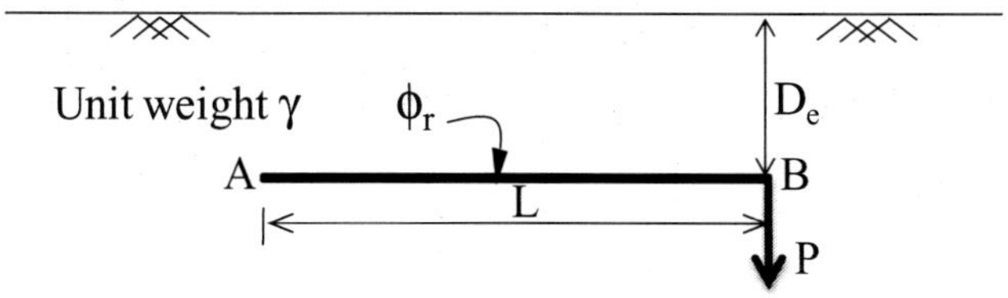

FIGURE 4: Definition sketch of reinforcement with transverse end force

For a transverse pull, P, or a transverse displacement, w_L, at the end of the reinforcement, Figure 5 shows the deformed shape of the reinforcement and the forces acting on the reinforcement. The normal and shear forces on the top and bottom surfaces of the reinforcement are assumed to act on its projected length. Figure 6 shows the forces acting on an infinitesimally small element of the reinforcement of length Δx. q_t and q_b, and τ_t and τ_b are the normal and shear stresses acting on the top and bottom surfaces, respectively, of the reinforcement. T and $T + \Delta T$ are the tensions developed in the reinforcement at x and $x+ \Delta x$, respectively. q_t is the overburden stress at embedment depth of the reinforcement and q_b is the mobilized normal stress that will be a function of the normal stress-deformation (q vs w) characteristics of the fill material.

Madhav and Umashankar (2003a and 2003b) modeled the behavior of inextensible reinforcement considering linear and non-linear (hyperbolic) responses for the normal stress-deformation response of the fill material, and the mobilized shear stress-displacement along the top τ_t and bottom τ_b of the reinforcement. These responses can be expressed mathematically as:

Normal Stress-Deformation Responses

$$q_b - q_t = k_s w \text{ ....Linear response (Figure 7a)} \quad (1)$$

and,

$$q_b - q_t = \frac{k_s w}{1 + \frac{k_s w}{q_{ult}}} \text{ .... Non-linear response (Figure 7b)} \quad (2)$$

Shear Stress-Displacement Responses

$$\left.\begin{aligned} \tau_t &= q_t \tan\phi_r \\ \tau_b &= q_b \tan\phi_r \end{aligned}\right\} \text{ ......Rigid plastic response (Figure 8a)} \quad (3)$$

$$\left.\begin{aligned} \tau_t &= \frac{k_\tau u}{1 + \frac{k_\tau u}{(\tau_{max})_t}} \\ \tau_b &= \frac{k_\tau u}{1 + \frac{k_\tau u}{(\tau_{max})_b}} \end{aligned}\right\} \text{ ......Non-linear response (Figure 8b)} \quad (4)$$

where, w is the normal displacement of the fill material, u is the horizontal displacement of the interface, k_s and k_τ are the initial slopes of q versus w and τ versus u curves, τ_{max} is the maximum shear stress that can be mobilized along the interface and q_{ult} is the ultimate bearing resistance of the soil.

The governing differential equations can be derived by taking the equilibrium of forces in the horizontal and vertical directions for the element (Figure 6). In this review, model considering the linear normal stress-deformation of the fill material and rigid plastic response of the interface-fill for an inextensible case is given in detail.

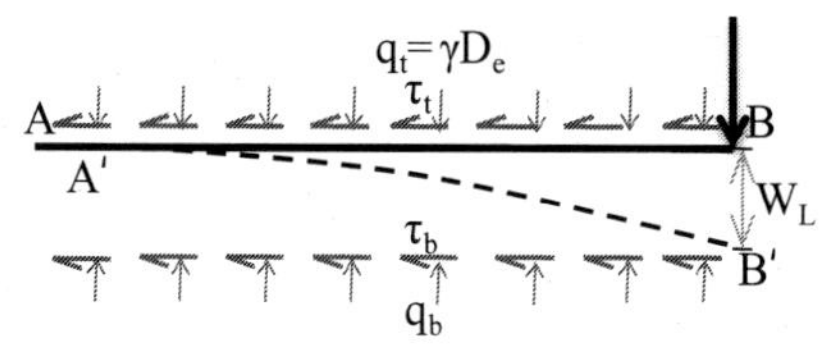

AB- original position; A'B'- deformed position

FIGURE 5: Deformed profile of the reinforcement

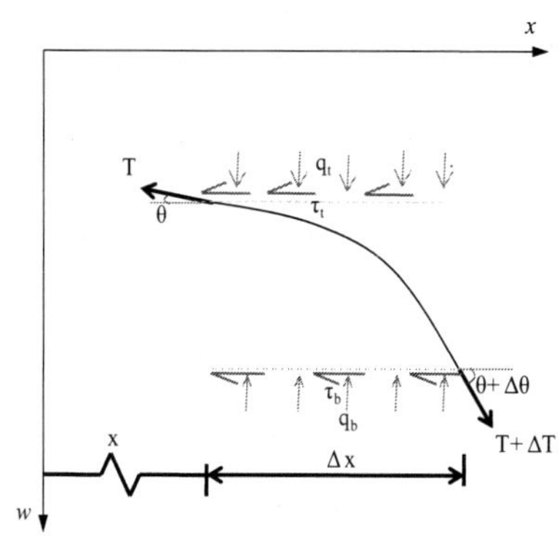

FIGURE 6: Forces acting on an element of reinforcement

The front-end displacements assumed to be in the range 0.001L-0.01L. Hence, the inclination of

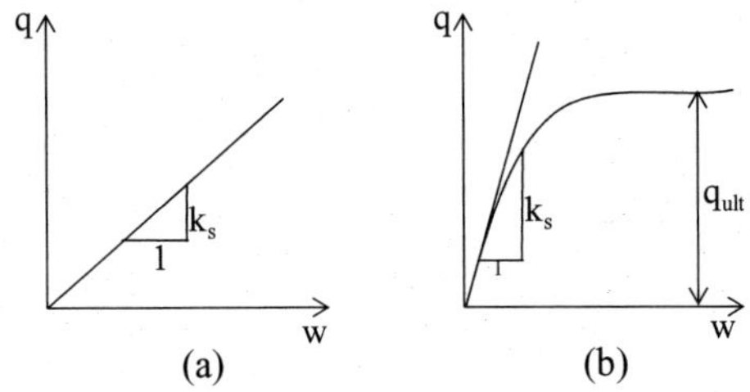

FIGURE 7: Linear and nonlinear normal stress-deformation responses of fill material

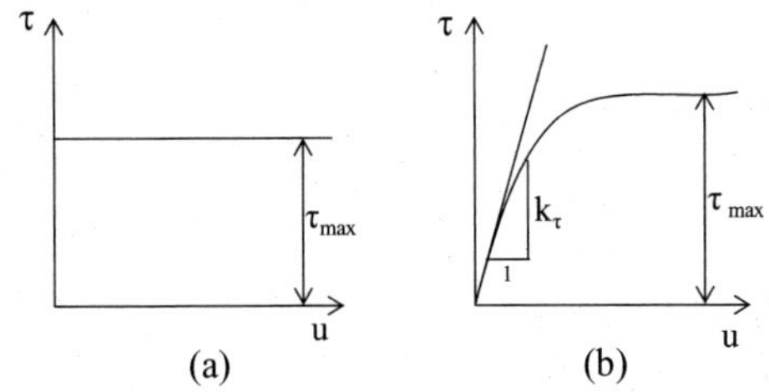

FIGURE 8: Rigid plastic and nonlinear shear stress-displacement response of reinforcement-soil interface

the reinforcement, θ, can be assumed to be small (i.e. cos θ → 0 and sin θ → 0).

The equilibrium equations can be simplified to

$$\frac{dT}{dx} = (q_t + q_b)\tan\phi_r = (k_s w + 2\gamma D_e)\tan\phi_r \tag{5}$$

$$-T\frac{d^2w}{d^2x} + k_s w = 0 \tag{6}$$

The applied transverse load, P, is obtained from the vertical equilibrium of forces and can be given as

$$P = \int_o^L k_s w dx \tag{7}$$

The above equations in normalized form become

$$\frac{dT^*}{dx} = (\mu W_L W + 2)/2 \tag{8}$$

$$-T^*\frac{d^2w}{d^2x} + \frac{\mu W}{2\tan\phi_r} = 0 \tag{9}$$

and,
$$P^* = \mu W_L \int_o^1 W dX \tag{10}$$

where, $X = x/L$, $W = w/w_L$, $T^* = T/T_{max}$ ($T_{max} = 2\gamma D_e L \tan\phi_r$, the axial pullout capacity), and $P^* = P/\gamma D_e L$, $\mu = k_s L/\gamma D_e$, a relative fill stiffness factor.

The boundary conditions in non-dimensional form becomes: at $X = 0$; $T^* = 0$ and $dW/dX = 0$ and at $X = 1$; $W = 1.0$. As the coupled equations [Eqs. (8) and (9)] cannot be solved analytically, a finite difference approach is adopted.

Madhav and Umashankar (2003c) modeled the response of extensible reinforcement of stiffness, J, subjected to transverse end force. Rigid plastic behavior of the soil interface is considered. The main difference between the behavior of inextensible and extensible reinforcements lies in the fact that the shear stresses are mobilized over the entire length of an inextensible reinforcement, whereas the shear stresses are mobilized only over the active length of an extensible reinforcement. This active length x_o is calculated by equating the extended length, $L_\varepsilon\left(= x_o + \int_{L-x_o}^{L}\frac{T(x)}{J}dx\right)$ obtained from the tension developed along the reinforcement to the length of reinforcement obtained from the change in geometry, L_g, and given by

$$L_g = \int_{L-x_o}^{L}\sqrt{1+\left(\frac{dw}{dx}\right)^2}\,dx \tag{11}$$

Governing equations (8) and (9) are solved using the finite difference scheme. The displacement and tensions corresponding to nodes at which x_o is attained are the actual displacement profile and the tensions developed along the reinforcement.

Madhav and Manoj (2004)

Madhav and Umashankar (2003a, 2003b, 2003c) evaluated the response of inextensible and extensible reinforcement assuming small deformations (of the order of 1% of the length of reinforcement) and hence the formulation is applicable only to very small inclinations θ_L at the reinforcement end. Madhav and Manoj (2004) extended the model to incorporate the inclination of the reinforcement corresponding to large displacements.

The governing equations for this case become

$$\frac{dT}{dx} = (k_s w + 2\gamma D_e)\tan\phi_r\cos\theta + k_s w\sin\theta \tag{12}$$

$$-T\frac{d\theta}{dx} = (k_s w + 2\gamma D_e)\tan\phi_r\sin\theta + k_s w\cos\theta \tag{13}$$

For inextensible reinforcement,

$$\frac{d\theta}{dx}=\cos^2\theta\frac{d^2w}{dx^2} \tag{14}$$

For extensible reinforcement,

$$\tan\theta=\frac{dw}{(dx+du)}=\frac{dw}{dx}\left(\frac{1}{1+(du/dx)}\right) \tag{15}$$

and,

$$\left(\frac{du}{dx}\right)=\left(\frac{T}{J}\right)=\varepsilon_x \tag{16}$$

Substituting for du/dx and differentiating with respect to x,

$$\frac{d\theta}{dx}=\frac{\cos^2\theta}{1+(T/J)}\left[\frac{d^2w}{dx^2}-\frac{dw}{dx}\frac{dT}{dx}\frac{1}{(T+J)}\right] \tag{17}$$

Shahu (2007) and Shahu and Hayashi (2009)

Shahu (2007), and Shahu and Hayashi (2009) studied the response of inextensible and extensible reinforcements subjected to an oblique end force (Figure 9). Shahu (2007) considered linear normal stress-deformation of the fill material and rigid plastic response of the reinforcement-soil interface, whereas Shahu and Hayashi (2009) considered nonlinear response for both normal stress-deformation and shear stress-displacement of the interface. The governing equations were derived by applying the equilibrium equations on the actual deformed shape of the reinforcement (Figure 10).

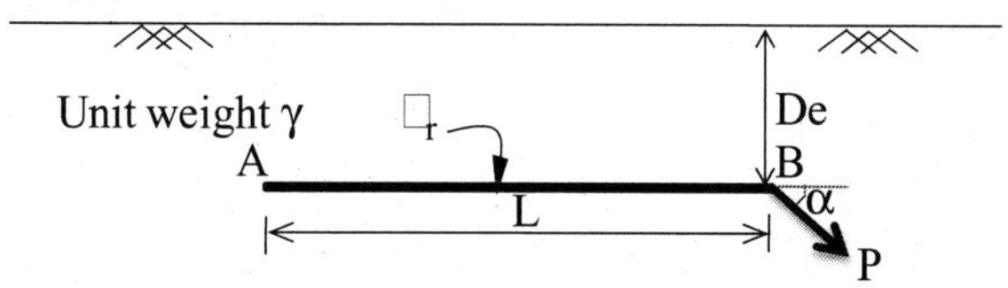

FIGURE 9: Definition sketch of reinforcement with an oblique end force

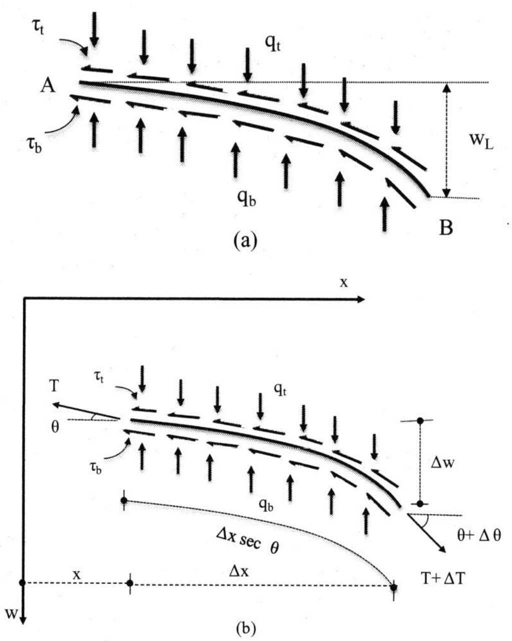

Figure 10 Forces on (a) reinforcement, and (b) infinitesimal element

Based on the force equilibrium of an element, Shahu (2007) derived the following governing equations:

$$\frac{dT}{dx}=k_s w\tan\theta+\left(k_s w+2\gamma D_e\right)\tan\phi_r \tag{18}$$

$$-T\cos^2\theta\frac{d^2w}{d^2x}+k_s w=0 \tag{19}$$

Patra and Shahu (2012)

In this model, the shear stiffness of the fill material/subrgrade is considered by representing the subgrade as a two parameter Pasternak model in place of Winkler springs. Figure 11 shows the model representation. The stresses on an infinitesimal element are shown in Figure 12. Governing equations for this model can be derived as

$$\frac{dT}{dx}=\left(k_s w-GH\frac{d^2w}{dx^2}+2\gamma D_e\right)\tan\phi_r+\left(k_s w-GH\frac{d^2w}{dx^2}\right)\tan\theta \tag{20}$$

$$-T\cos^2\theta\frac{d^2w}{d^2x}=k_s w-GH\frac{d^2w}{dx^2} \tag{21}$$

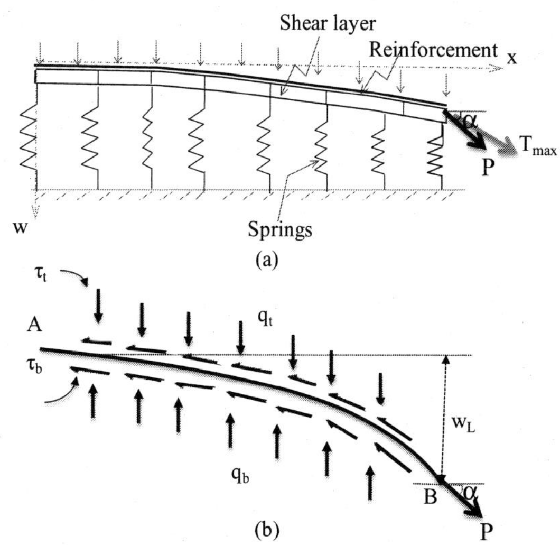

Figure 11 (a) Pasternak model representing fill material, and (b) forces acting on reinforcement

This model is found to provide a more realistic response of the pullout behavior. This is because as the shear stiffness increases, the displacement profile of the reinforcement becomes more uniform and bending of the reinforcement reduces. Normal stresses are, hence, reduced leading to smaller values of maximum reinforcement tension mobilized within the reinforcement.

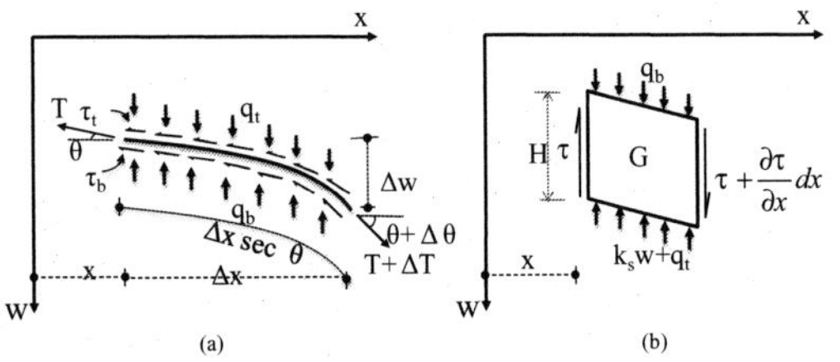

FIGURE 12: Forces acting on an infinitesimal element of (a) reinforcement, and (b) shear layer

3. CONCLUSIONS

1. The kinematics of downward transverse pull increases the normal stresses and consequently the shear resistance mobilized on the bottom surface of the reinforcement. The axial component of the pullout force due to transverse pull is somewhat greater than the purely axial pullout capacity of the reinforcement. This paper presents the review on generalized models for the reinforcement subjected to an oblique pull.
2. The pullout capacity of reinforcement is found to increase as front-end displacement, relative ground stiffness factor and obliquity increases.
3. The increase in shear stiffness factor of the fill material, G* (= $GH/\gamma DL$), from 0 to 50, the maximum tension mobilized in the reinforcement and the oblique pull out force are found to decrease.

REFERENCES

Bergado, D. T., Teerawattanasuk, C. and Long, P. V. (2000) "Localized Mobilization of Reinforcement Force and Its Direction in the Vicinity of Failure Surface". Geotextiles and Geomembranes, 18, Issue 5, pp 311-331.

Gray, D. H., and Ohashi, H. (1983) "Mechanics of Fiber Reinforcements in Sand". Journal of Geotechnical Engineering, ASCE, 109, Issue 3, pp 335-353.

Jewell, R. A. (1992) "Keynote Lecture: Links between the testing modeling and design of reinforced soil", In: Proceedings of the International Symposium on Earth Reinforcement Practice, Fukuoka Japan, 2, pp 755–772.

Juran, I., Shaffiee, S., Schlosser, F., Humbert, P., and Guenot, A. (1982) "Study of Soil-Bar Interaction in the Technique of Soil Nailing", Eighth European Conference on Soil Mechanics and Foundation Engineering, pp 513-516.

Madhav, M. R., and Umashankar, B. (2003a) "Analysis of Inextensible Sheet Reinforcement Subject to Transverse Displacement/Force: Linear Subgrade response". Geotextiles and Geomembranes, 21, Issue 1, pp 69-84.

Madhav, M.R., and Umashankar, B. (2003b) "Analysis of Inextensible Sheet Reinforcement subject to Downward Displacement/Force: Nonlinear Subgrade Response". Geosynthetics International, 10, Issue 3, pp 95-102.

Madhav, M. R., and Umashankar, B. (2003c) "Response of Extensible Reinforcement to Transverse Pull/ Displacement: Linear Subgrade Behavior".Indian Geotechnical Journal, 33, Issue 4, pp 397-421.

Madhav, M. R., and Manoj, T. P. (2004) "Reinforcement-Soil Interactions Under Transverse and Oblique Forces", Proc., 12th Asian Regional Conf. on Soil Mechanics and Geotechnical Engineering, C.F. Leung et al., eds.,World Scientific, Singapore, pp 1139–55.

Patra.S, and Shahu J. T. (2012) "Pasternak model for oblique pullout of inextensible reinforcement". Journal of Geotechnical and Geoenvironmental Engineering,doi: 10.1061/(ASCE)GT.1943-5606.0000720.

Shahu, J. T. (2007) "Pullout Response of Inextensible Sheet Reinforcement Subject to Oblique End-Force". Journal of Geotechnical and Geoenvironmental Engineering,133, Issue 11, pp 1440–48.

Shahu, J. T., and S. Hayashi. (2009) "Analysis of Extensible Reinforcement Subject to Oblique Pull". Journal of Geotechnical and Geoenvironmental Engineering,135, Issue 5, pp 623–634.

Shewbridge, S .E., and Sitar, N. (1989) "Deformation Characteristics of Reinforced Sand in Direct Shear". Journal of Geotechnical and Geoenvironmental Engineering, 115, Issue 8, pp 1134-1147.

Sobhi, S. and Wu, J. T. H. (1996) "An Interface Pullout Formula for Extensible Sheet Reinforcement". Geosynthetics International, 3, Issue 5, pp 565–582.

Physical Modelling of Jointed Rockmass Under Polyaxial Stress State

K.S. Rao and R.P. Tiwari[1]

Department of Civil Engineering, IIT, Delhi
[1] Department of Civil Engineering, Rewa Govt. Engineering College, Rewa, M.P.
E-mail: raoks@civil.iitd.ernet.in

ABSTRACT

The strength of rock mass probably is the first engineering parameter that a geotechnical engineers may be keen to establish before the start of any mining and civil engineering project. The strength of a rock mass at site is generally influenced by joint geometry and the stress state that it experiences. Field studies show that stress conditions around underground structures will be generally true triaxial or polyaxial, i.e. σ1>σ2>σ3. The rock mass behavior in such field conditions can effectively be simulated in laboratory by conducting large-scale physical modeling. In this study the most commonly occurring joint configurations (three joint sets) were created on rock mass models in the laboratory and tested under polyaxial stress compression. The model rock was a sand-lime block with low strength and medium deformability. The joint configurations were varied to assess the influence of joint geometry on the stress-strain curve, failure mechanism and strength anisotropy. Similarly, on each type of geometry, σ2/σ3 ratio was also varied to study the effect of intermediate principal stress on engineering behavior of rock mass. The maximum enhancement of 310% in strength due to increase in σ2/σ3 ratio was observed, which was corresponded to joint inclination, θ = 60°. Based on testing results the polyaxial strength criterion was evolved similar to the von-Mises failure envelope.

1. INTRODUCTION

One of the major problems confronting designers of engineering structures in rock is that of estimating its strength. The rock mass usually made up of an interlocking matrix of discrete blocks and failure of such rock mass may be dominated sliding along joint or rotation of block element. The rock slopes and deep underground structures are always subjected to higher confining pressure except at the boundary of the tunnel faces. In case of rock slopes, at the surface and in tunnel openings near the face, only σ1 and σ2 exist and the third stress σ3 is negligible or zero, so biaxial stress condition prevails. Further, at deep depths, rock slopes, foundations and tunnel openings or caverns experience all the three sets of stresses and triaxial or polyaxial condition is prominent. The strength responses of rock mass can be estimated through numerical modeling, in-situ and laboratory large-scale testing and physical model studies. The numerical methods need experimental results for their validity before they can be used and the insitu and laboratory methods are practically very difficult and time consuming. So, physical model studies are useful in such situations.

Numerous strength criteria have been developed in the past and some important contributions to mention are Franklin (1971), Jaeger and Cook (1979), Hoek and Brown (1980), Rao (1984), Yu (1983), Amadaei (1988), Sheorey (1997), Singh et al. (1998), Haimson and Chang (2000) and Yu et al. (2002), though each one is generally suitable only for a certain type of materials and stress levels. Kulatilake et al. (1993, 1994, 1997) based on laboratory model study observed that rock mass may exhibit different failure modes and internal distribution of stresses within rock mass can be highly complex. The combined effect of joint geometry was expressed through fracture tensor component

(Kulatilake et al., 1993). Jing (1992) performed experimental study and observed anisotropy and stress dependency on strength and deformability characteristics of rock joints. Further, Kulatilake et al. (2003) developed a non- linear rock mass failure criterion in different failure modes for biaxial loading conditions incorporating the fracture tensor component. Colmenares and Zoback (2002) suggested failure criterion using polyaxial test data for different rocks. Thus, few strength criteria are available in literature which include the effect of intermediate principal stress without much considering the anisotropy of rock mass, amongst them the Mohr-Coulomb and Hoek and Brown criteria are the most widely used strength criteria for rock.

To assess the effect of intermediate principle stress, $\sigma 2$ and angle of discontinuity in rock mass on its stress-strain curve, failure mechanism and failure strength, an extensive rock mass model testing was carried out in true triaxial stress state ($\sigma 1 > \sigma 2 > \sigma 3$) and discussed in this paper. Rock is heterogeneous and quite often discontinuous, i.e. rock masses are typically jointed. These rock joints are mechanical discontinuities of geological origin. These discontinuities may be in the form of joints, faults, bedding planes or other recurrent planar fractures, which may be filled with gouge material. The mechanical and engineering properties of these joints depend upon whether these joints are clean or filled, open or close. The presence of infill or gouge material in these joints reduces the shear strength.

Sources of infill material include products of weathering or overburden washed into open joint, water conducting in discontinuities, precipitation of minerals from the ground water, by-products of weathering adulterations along joint walls, crushing of parent rock surfaces due to tectonic and shears displacements, and thin seams deposited during formation. In general, infill materials may consist of partially loose to completely loose cohesionless soil or fine grained clay. Normally fine-grained clays are more frequently found as fillers and are more troublesome in terms of structural stability. Thickness of the infill material varies from micrometers to several meters and it plays an important role in shear behaviour. In tectonically crushed zones, the infill thickness may exceed several meters.

2. ROCK MASS TESTING

The different joint geometries were formed in rock mass specimens with the help of model rock and testing, which were performed under triaxial and true triaxial stress conditions.

2.1 Model Material Characterization

The model blocks were manufactured in single batch to ensure the uniformity in material mixture and engineering properties to give reproducible results. The material was composed of 22% hydrated lime and 78% sieved sand (passing 600 micron sieve) mixed with 22% water (by weight) of hardness less than 330 mg/l. The mixture was then poured in casts of size 23 cm × 10.5 cm x 6.9 cm and pressed under pressure of 39.6 MPa. The blocks so formed were put in an autoclave for 4 hours at 180°C and then cured in air for about three weeks. The physical and engineering properties of the model material were determined in the laboratory as per the suggested methods of ISRM (1981) and are shown in Table 1 (Tiwari and Rao, 2004). The model rock can be classified as 'EM' (Low strength and medium modulus) on Deere and Miller's (1966) classification chart.

Table 1: Material Properties of Sand – Lime Block (after Tiwari and Rao, 2004)

Property	Range of values	Representative value
Dry density, γd (KN/m3)	16.10-17.80	16.86
Specific gravity, G	2.66-2.67	2.66
Absolute porosity, n (%)	32.70-39.30	36.50
UCS (intact cyl.), σ_{ci} (MPa)	11.10-15.90	13.50
Tensile strength, σ_{tb} (MPa)	1.63-2.17	1.89
Failure strain UCS, ε (%)	0.43-0.63	0.53
Poisson's ratio, ν	0.18-.0.27	0.25
Modulus, E_{t50} (MPa)	4782-4866	4866
Internal friction angle, ϕ (°)	-	44.70
Cohesion, c (MPa)	-	3.29
Joint friction angle, ϕ_j (°)	-	36.80
Deere-Miller's classification	-	EM

2.2 Geometry of Specimens Tested

Most frequently occurring joint geometries in rock mass at site was created in model rock. The specimens were prepared by cutting model rock into small cubes of size 2.5 cm and arranging the small cubes to form the test specimen of size 15 cm x 15 cm x 15 cm having three sets of orthogonal joints. The joint set-I was continuous and inclined at various angles θ = 0°, 20°, 40°, 60°, 80°and 90° with x- axis. The joint set-II was perpendicular to set-I and was kept at stepping (s) = 0.5 of width (b) of small block. The set-III was always continuous (see Fig. 1). Each specimen was given a unique identification number *viz.* A60/0.31/1.62 which indicates Type-A specimen of θ = 60° and is acted by principal stresses, σ_3 = 0.31 MPa and σ_2 = 1.62 MPa along y and x-axis respectively.

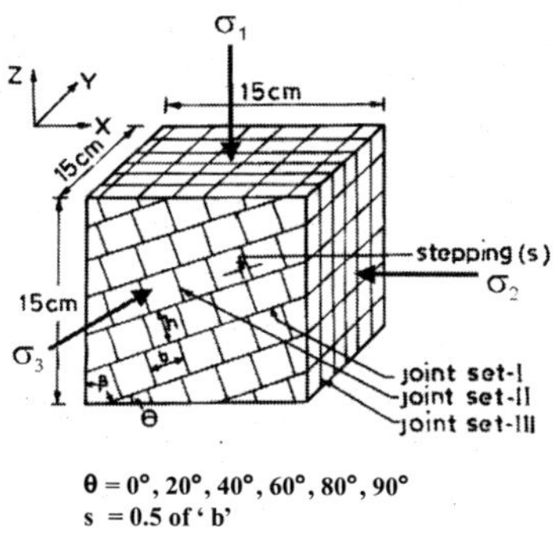

FIGURE 1: Key sketch of model tested (after Tiwari and Rao, 2004)

2.3 Testing Details

The True Triaxial System (Rao and Tiwari, 2002) was used for testing of rock mass specimens (see Fig.2). The system consists of a 1000kN capacity vertical frame, a biaxial frame of 300kN capacity fitted with two pairs of hydraulic jacks and platens, constant confining pressure unit for applying, monitoring and maintaining horizontal stresses (σ_2 and σ_3) on specimen faces, eight channel data acquisition system and personal computer to record all load and deformation data. The Teflon sheets smeared with high vacuum silicon grease were to be used for reducing friction between specimen and platen faces. Pairs of Teflon sheets with silicon grease were subjected to direct shear tests for determining the interface friction angle. It was observed that the friction angle between Teflon sheet interfaces is negligible (3.4°) as compared to joint friction angle (36.8°) of the model rock. Hence, the Teflon sheet was selected to use between specimen face and platen for achieving friction free and uniform loading. It is expected that Teflon sheet may undergo compression under loading hence, net deformation of specimen along any face was obtained by subtracting the compression of Teflon sheet from total deformation of the combined specimen and Teflon sheet assembly. The compression testing of Teflon sheet smeared high vacuum silicon grease was carried out and the normal stress vs compression curve of sheets was plotted and used for correction of deformation in each specimen.

FIGURE 2: True triaxial system used in present study (Rao and Tiwari, 2002)

The prepared rock mass specimens were first oven dried at 108°C and then air dried at room temperature for 30 days in rock mechanics laboratory before testing. The cured specimen was carefully transferred on the vertical loading platens of the true triaxial system and all vertical and horizontal platens were moved slowly to contact the specimen faces. The friction free and uniform stress on all six faces was ensured by putting 0.5 mm thick pair of Teflon sheets between specimen and platen and adjusting the alignment of horizontal jacks by movement of horizontal frame in vertical and horizontal planes using screw levers. The final micro adjustment of platens on specimen lateral faces was done by moving the platens vertically and horizontally with the help of mini screw levers fixed on horizontal biaxial frame.

After the final adjustment of specimen and testing system, all three independent principal stresses were applied initially in quasi-isotropic conditions (i.e. $\sigma_1 > \sigma_2 = \sigma_3$) and upon reaching the pre-determined level of σ_2 and σ_3, the σ_1 was increased monotonically strained controlled at rate of 0.75×10^{-3} cm/s as shown in Fig. 3 from $\sigma_1 = \sigma_2$ to a level when the specimen yielded and continued into the post failure zone until residual strength of specimen was achieved. The vertical deformation of specimen was recorded through an LVDT attached at vertical hydraulic jack and its tip touching the base of lower loading platen. The horizontal deformation of specimen along σ_3 directions was measured through LVDT's fitted at pair of hydraulic jacks on either side of specimen faces. The sum of deformations of both LVDT's was taken as average deformation of specimen in σ_3 directions. Similarly, LVDT's fitted in σ_2 directions were also used to measure the lateral deformation along σ_2 axis. The vertical loading was recorded using a vertical load cell fitted on top of specimen. The lateral loads to apply σ_2 and σ_3 on lateral specimen faces were measured through two load sensors fitted in the hydraulic lines of horizontal jacks. The signal from the five LVDT's and the three load cells/sensors were continuously transferred to data acquisition system, which finally transferred them to a PC as a notepad file. Strain controlled vertical loading was applied at a rate such that failure of specimen occurred in 10-20 minutes. Table 2 explains the details of testing programme.

2.4 Interpretation of Results

Figures 4 and 5 show the typical stress strain curves obtained from the rock mass testing results. The curve shows deviotoric stress ($\sigma_1 - \sigma_3$) plot against axial strain (ε_1), lateral strain along σ_2 (ε_2), lateral strain along σ_3 (ε_3), and volumetric strain. Rock mass specimens (with $\theta = 0°$, 20°, 80° and 90°) exhibit strain softening behavior with a steep elastic portion up to almost the peak and gradual fall in stress after failure (Fig. 4). Ductile or strain hardening behavior was seen for specimens with $\theta = 40°$, 60°), showing short elastic range followed by a continuous increase in stress and decrease in tangent modulus with increasing strain up to failure and beyond peak, stress increase is continued.

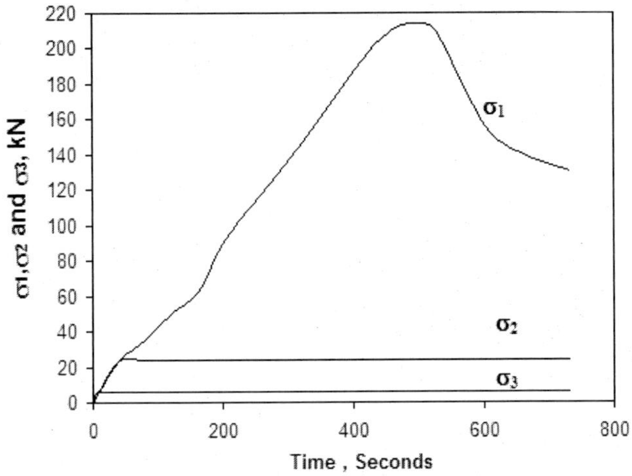

FIGURE 3: Typical Load Vs Time Plot for Complete Testing of a Specimen

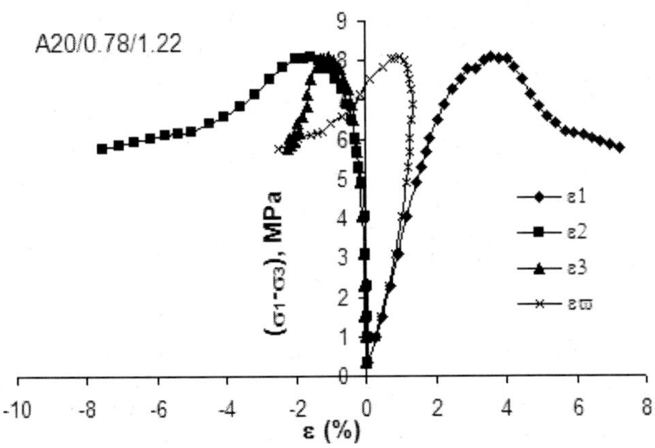

FIGURE 4: Typical stress-strain curves for A20 specimen

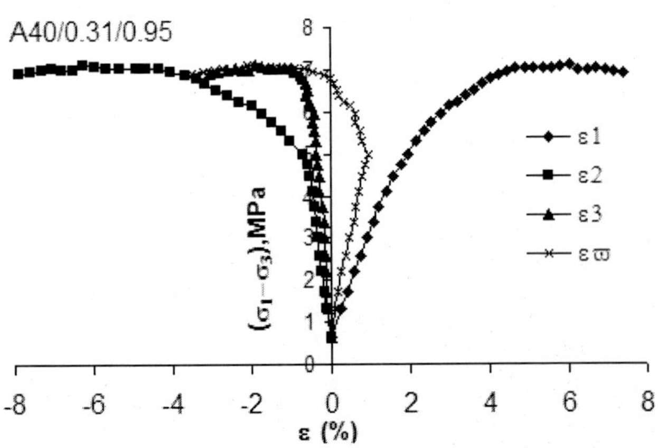

FIGURE 5: Typical stress-strain curves for A40 specimen

Table 2: Details of Testing Program

Test cases	θ ° and s details	σ_3 MPa	σ_2 MPa	σ_2 σ_3 Axis
TypeA (54)	s = 0.5 θ =0, 20, 40, 60, 80 , 90	0.31	0.31, 0.59 0.95, 1.22, 1.62	σ_2 -x σ_3 -y
		0.78	0.78, 1.22, 2.24	
		1.22	1.22	

The specimens also have shown different failure pattern depending upon joint configuration, stress ratio, and stress orientation. Shearing of intact material and joints was observed in case of specimens with θ = 0 °, 20°, 80° and 90°. The shear planes developed on YZ or σ_2 face dips along σ_3 directions and fracture dip increases with increasing σ_2 as shown in Figs. 6 and 7. In rock mass shearing was seen in θ = 60°. The shifting of failure mode from sliding and dilation along joint to shearing of intact material with specimen with θ = 40°, joint dilation and shearing of some blocks occurred whereas sliding of small blocks along joint mixed with increasing σ_2/ σ_3 ratio was also observed.

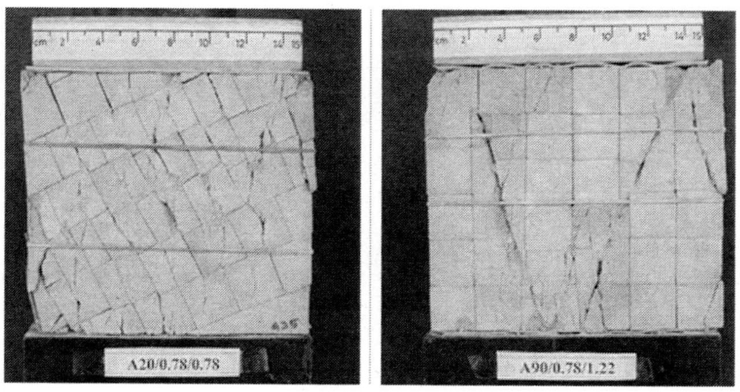

FIGURE 6: (a) Shearing of intact blocks and joints (b) Shear planes developed on the face at which is σ_2 is applied

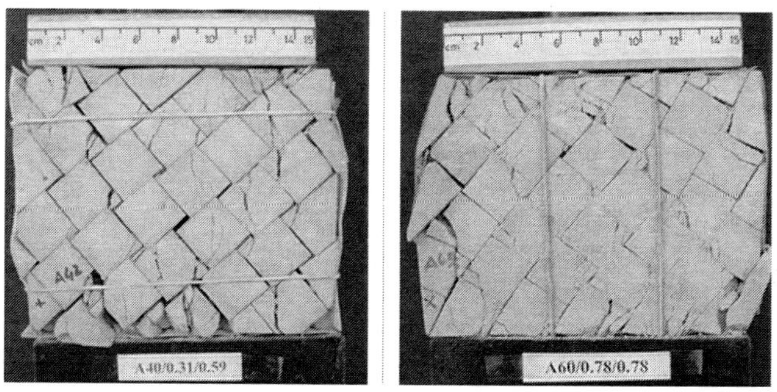

FIGURE 7: (a) Shearing of intact material and joint dilation (b) Bulging of specimen and rotation of blocks

2.5 Influence of Horizontal Stress Ratio

The failure strength increases with increasing σ_2/ σ_3 ratio at each joint inclination θ of rock mass specimens. Table 3 shows that at σ_2/σ_3 ratio equal to 1, failure strength is highly influenced by joint configuration of rock mass and is minimum (1.84 MPa) at θ = 60 ° and maximum (7.28 MPa) at 90°. Further more Table 3 and Fig. 8(a) show that the rate of increase is maximum for θ = 60 ° and strength reaches 7.53 MPa corresponding to σ_2/ σ_3 = 5.2. In case of θ = 90 °, rate of increase is the lowest, i.e. from 7.28 to 9.18 MPa only. In all joint inclinations, the rate of increase in strength at failure is higher initially which declines gradually with increasing σ_2/σ_3 but in no case lower than the strength corresponding to triaxial stress state.

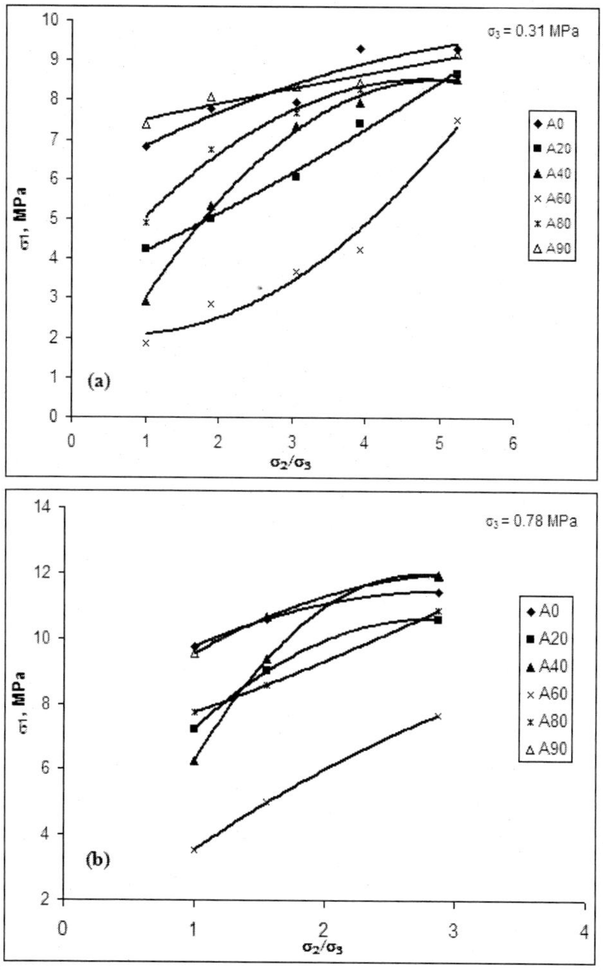

FIGURE 8: Effect of intermediate principal stress on failure strength of rock mass at σ_3 (a) 0.31 MPa and (b) 0.78 MPa

The above observation suggests that the effect of joint geometry on strength of rock mass is dominant at low confinement level only whereas as stress level increases, the effect of joint reduces gradually and strength response is similar in rock mass with different geometry. The Fig. 8(b) shows the similar behaviour in strength of rock mass at $\sigma_3 = 0.78$ MPa case too.

Table 3: Failure Strength σ_{1f} Values for Rock Mass Specimens

σ_3 MPa	σ_2 MPa	σ_{1f} (MPa) Inclination, θ (°)					
		0	20	40	60	80	90
0.31	0.31	6.81	4.23	2.92	1.84	4.91	7.28
0.31	0.59	7.78	5.01	5.32	2.85	6.76	8.68
0.31	0.95	7.65	6.05	7.34	3.67	7.69	8.34
0.31	1.22	9.32	7.41	7.96	4.24	8.27	7.33
0.31	1.62	9.32	8.68	8.54	7.53	8.54	9.18
0.78	0.78	9.76	7.23	6.26	3.54	7.73	9.54
0.78	1.22	10.60	9.01	9.38	4.99	8.59	10.65
0.78	2.24	11.45	10.60	11.91	7.66	10.86	11.97
1.22	1.22	12.51	9.32	9.13	4.93	9.36	13.13

2.6 Influence of Joint Geometry

The failure strength, deformation modulus and modulus ratio of rock mass are significantly influenced by joint configuration or inclination of critical joint set-I. The variation of engineering parameters with β (= 90° – θ) is known as anisotropy. The anisotropy ratio R_c is defined by the following equation:

$$R_c = \frac{\sigma_{1(90)}}{\sigma_{1(\min)}} \tag{1}$$

where, σ_{190} is strength value corresponding to β = 90°. The $\sigma_{1\min}$ is the minimum values of strength near β = 30° to 50°. The anisotropy in engineering parameters of rock mass can be classified using Table 4. The anisotropy ratio R_c for engineering parameters reduces with increase in σ_2/σ_3 value.

Figures 9(a) and (b) show the variation of strength with change in joint inclination, θ at different σ_2/σ_3 ratio. Anisotropy ratio, R_c reduces from 4.0 to 1.2 when σ_2/σ_3 is increased from 1.0 to 5.2 at $\sigma_3 = 0.31$ MPa. Similarly at higher confining

Table 4: Classification of Anisotropy

R_c Range	Description	Symbol
1.0 – 1.1	Isotropic	I
> 1.1 – 2.0	Low	L
> 2.0 – 4.0	Medium	M
> 4.0 – 6.0	High	H
> 6.0	Very High	VH

Table 5: Anisotropy Ratio R_c for Failure Strength

σ_3, MPa	σ_2/σ_3	Anisotropy ratio, R_c value	Symbol
0.31	1.0	4.0	H
0.31	1.9	2.8	M
0.31	3.1	2.3	M
0.31	3.9	2.0	L
0.31	5.2	1.2	L
0.78	1.0	2.7	M
0.78	1.6	2.1	M
0.78	2.9	1.6	L
1.22	1.0	2.7	M

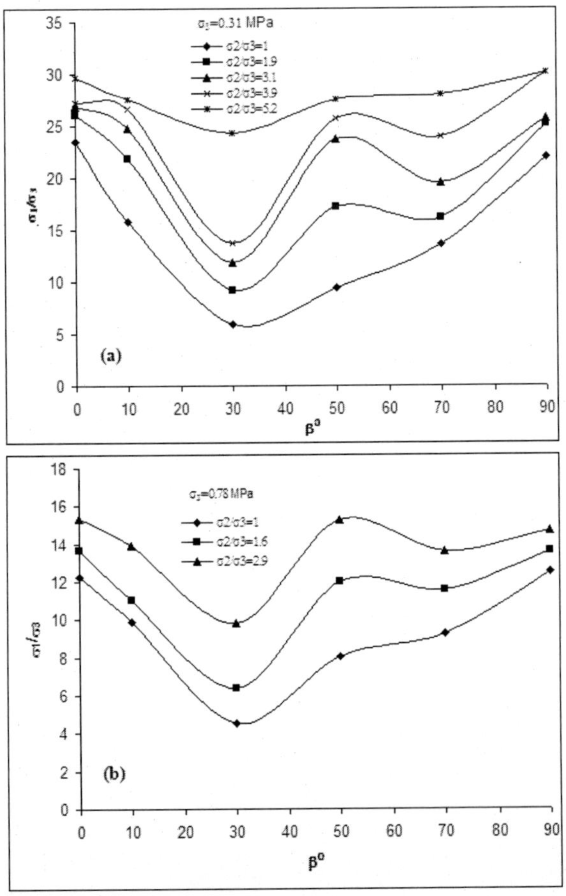

FIGURE 9: Strength anisotropy in rock mass under polyaxial compression at (a) $\sigma_3 = 0.31$ MPa and (b) $\sigma_3 = 0.78$ MPa

stress $\sigma_3 = 0.78$ MPa, R_c reduces from 2.7 to 1.6 with increase in σ_2/σ_3 from 1.0 to 2.9 as shown in Table 5. It can be explained from Table 5 that anisotropy in strength is high at lower confinement and diminishes (low anisotropy value) with increase in σ_2/σ_3 value.

3. SUGGESTED STRENGTH CRITERION

The extensive test results were obtained from present experimental study. The rock mass specimens failed due to both brittle failure and plastic yielding of specimens. The results were used to develop following true triaxial strength criterion in the pattern of generalized von Mises theory (Mogi, 1971).

$$\tau_{oct} = f(\sigma_{oct}) \tag{2}$$

where, τ_{oct} is the octahedral shear stress, i.e. $1/3[(\sigma_1-\sigma_2)^2 + (\sigma_2-\sigma_3)^2+(\sigma_3-\sigma_1)^2]^{1/2}$ at the point when σ_1 attains its peak value $\sigma_{1,peak}$ and σ_{oct} is the mean normal stress acting on failure plane and *f* is monotonically increasing function. It is based on the assumption that yielding occurs in the entire body of specimen and such behaviour in most of the specimens of present study may be assumed. Further, the octahedral shear stress is a function of the full octahedral normal stress, $\sigma_{oct.}$ The σ_{oct} is taken equal to $(\sigma_1+\sigma_2+\sigma_3)/3$. Based on the results of laboratory testing, τ_{oct} and σ_{oct} were calculated

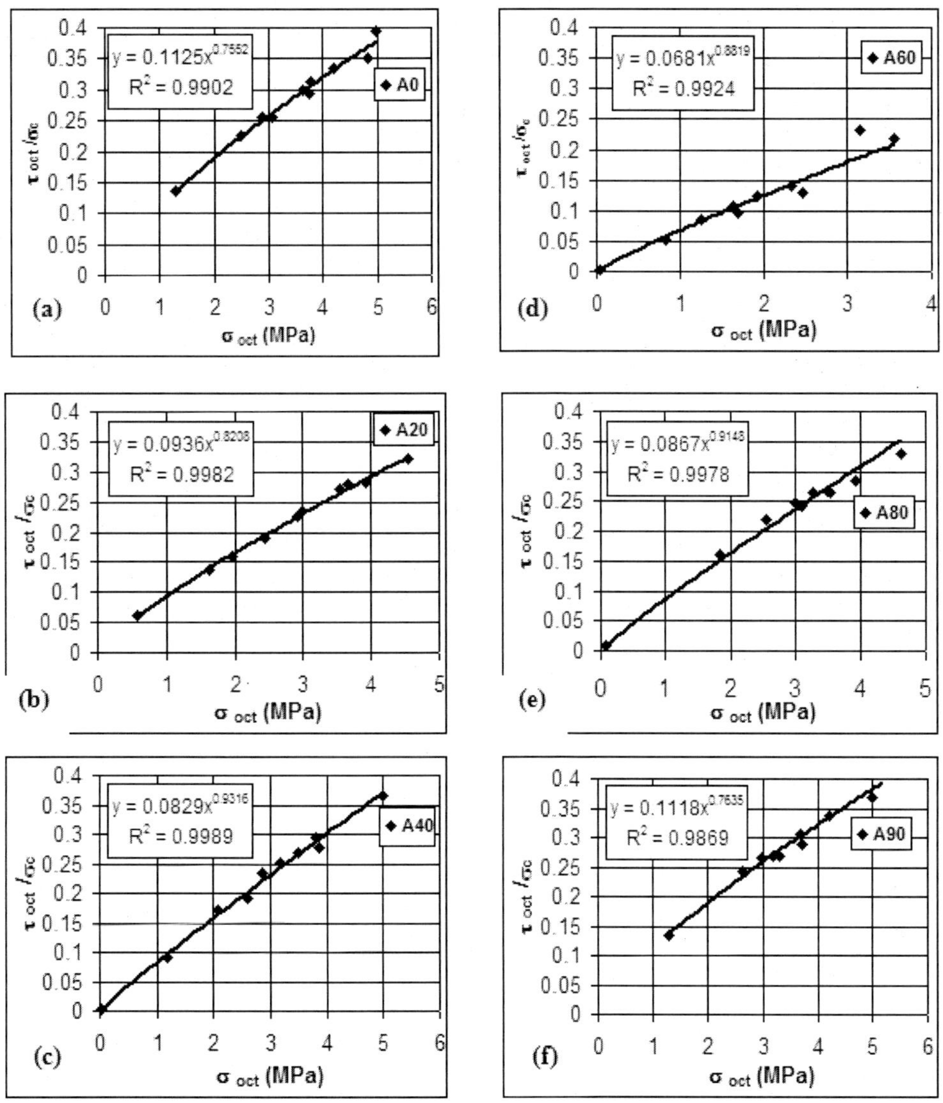

FIGURE 10: Polyaxial strength envelopes for rock mass specimens at inclination, θ (a) 0° (b) 20° (c) 40° (d) 60° (e) 80° and (f) 90°

for each θ value. The curves were fitted as shown in Fig. 10 of this paper and generalized equation of strength criterion was developed using curve fitting of least square method. The strength criterion developed is presented in Eqn. (3)

$$\tau_{oct} = D_m \sigma_{ci} \sigma_{oct}^{e_m} \tag{3}$$

The D_m and e_m are true triaxial parameters and are function of joint geometry. D_m and e_m systematically vary with joint configuration for rock mass specimens. The σ_{ci} is the uniaxial compressive strength of intact model material. The criteria have been derived based on the assumption that joints are smooth (ϕ_j =36.8°) and the extent of interlocking level (s = 0.5) is medium.

The strength envelopes for all specimens are plotted from the results of rock mass testing under polyaxial compression as shown in Fig. 10. Strength parameters D_m and e_m at different inclination (θ) are estimated through these envelopes. The parameters D_m and e_m for rock mass specimen is listed in Table 6. It can be seen from this Table 6 that parameters D_m is minimum at inclination, θ equal to 40° and maximum at θ = 90°. Similarly parameter e_m is the maximum at θ = 40° and the minimum at 90°. The data of D_m and e_m given in table 6 were plotted against (40°-θ) or (2π/9-θ). The equation of curve between D_m and (40°-θ) is represented by Eq.(4). Similarly curve between e_m and (40°-θ) is represented by Eq.(5) using curve fitting method of least square.

$$D_m = 0.08\left[\cos\left(\frac{2\pi}{9} - \theta\right)\right]^{-0.8} \tag{4}$$

$$e_m = 0.59\exp\left[0.40\cos\left(\frac{2\pi}{9} - \theta\right)\right] \tag{5}$$

Table 6: Polyaxial Strength Parameters D_m and e_m for Rock Mass

θ (°)	Strength parameters	
	D_m	e_m
0	0.11	0.76
20	0.09	0.82
40	0.08	0.93
60	0.07	0.88
80	0.09	0.91
90	0.11	0.76

Table 7: Comparison of Model Parameters for Rock Mass

θ (°)	Mohr-Coulomb model		Drucker-Prager model		Proposed model	
	c_m	ϕ_m	α	k	D_m	e_m
0	0.93	44.0	0.35	1.00	0.11	0.76
20	0.52	43.6	0.34	0.56	0.09	0.82
40	0.15	40.5	0.32	0.17	0.08	0.93
60	0.23	33.4	0.26	0.27	0.07	0.88
80	0.78	42.1	0.33	0.86	0.09	0.91
90	0.95	44.1	0.33	0.86	0.11	0.76

The parameters D_m and e_m can be interrelated by the following equation:

$$e_m = 0.32 D_m^{-0.41} \tag{6}$$

The Eqn. (3) can be applied in field with good accuracy for prediction of strength of rock mass by making input of relevant parameters. The parameters needed are uniaxial compressive strength, σ_{ci} of intact cylindrical rock core; insitu stresses (σ_1,σ_2 and σ_3 state and joint geometry (θ° and s) of rock mass obtained from field scan line survey.

4. PREDICTION OF MODEL PARAMETERS

The proposed strength model along with Mohr-Coulomb and Drucker-Prager models is presented in terms of stress invariants in principal stress space. The strength parameters these models are estimated for rock mass geometries used in the present study. The variation of model parameters with joint inclination θ is interpreted here.

4.1 Mohr-Coulomb Model

Mohr-Coulomb equation can be written in terms of stress invariants as below:

$$F = \frac{I_1}{3}\sin\phi_m + \frac{\sqrt{J_2}}{\sqrt{3}}\cos\left(\theta_D + \frac{\pi}{3}\right)\sin\phi_m + \sqrt{J_2}\sin\left(\theta_D + \frac{\pi}{3}\right) - c_m\cos\phi \tag{7}$$

where, F is yield function, I_1 is first invariant of stress tensor, J_2 is second invariant of deviatoric stress tensor. In a stress tensor with three principal stresses σ_1,σ_2, and σ_3, the stress invariants I_1, I_2 and I_3 will be equal to (σ_1+ σ_2 + σ_3), ($\sigma_1\sigma_1$ +$\sigma_1\sigma_1$ +$\sigma_1\sigma_1$) and ($\sigma_1\sigma_1\sigma_1$) respectively. The J_1 will be zero and J_2 will be taken equivalent to $1/3(I_1^2 - 3I_2)$. The θ_D is lode's angle and is defined as:

$$0 \le \theta_D = \frac{1}{3}\cos^{-1}\left(\frac{-3\sqrt{3}}{2}\cdot\frac{J_3}{J_2^{1/2}}\right) \le \frac{\pi}{3} \qquad (8)$$

in which J_3 is the third invariant of deviatoric stress tensor and will be equal to $\frac{1}{27}\left(2I_1^3 - 9I_1I_2 + 27I_3\right)$.

Mohr-Coulomb yield surface in principal stress state is irregular hexagonal pyramid with its axis coinciding with hydrostatic stress axis (i.e. $\sigma_1 = \sigma_2 = \sigma_3 = 0$). The deviatoric stress section on π-plane is an irregular hexagon.

The parameters c_m and ϕ_m are calculated from triaxial test on rock mass specimens. The Table 7 shows that c_m and ϕ_m vary systematically with inclination, θ and are the minimum corresponding to inclination ranges $\theta = 40°$ to $60°$.

4.2 Drucker-Prager Model

The Drucker-Prager model is modification over Mohr-Coulomb's model. The failure surface is a right circular cone. The cone will circumscribe the hexagonal pyramid of Mohr-Coulomb criterion. The model can be written as Eqn. (9) when represented by stress invariants in principal stress space.

$$F = \alpha I_1 + \sqrt{J_2} - k \qquad (9)$$

where, α and k are material constants. The stress path and range of stress in present rock mass testing are such that all data lie near to compressive meridian (i.e. $\theta_D = 60$ °). The Drucker-Prager model parameters α and k are calculated using Eqns. 10 and 11.

$$\alpha = \frac{2\sin\phi}{\sqrt{3}(3-\sin\phi)} \text{ and } k = \frac{6c\cos\phi}{\sqrt{3}(3-\sin\phi)} \qquad (10)$$

Tensile Meridian

$$\alpha = \frac{2\sin\phi}{\sqrt{3}(3+\sin\phi)} \text{ and } k = \frac{6c\cos\phi}{\sqrt{3}(3+\sin\phi)} \qquad (11)$$

Table 7 again, shows that the parameters α and k are minimum for joint inclination in the range $\theta = 40°$ to $60°$.

4.3 Proposed Model

The proposed model is an important modification over Drucker-Prager model. The results of true triaxial tests were plotted between octahedral normal and octahedral shear stress. The meridian section plot revealed that the failure surface is non-linear in contrast to linear surface of Ducker–Prager model. Thus, in the present study a model was proposed as shown in Eqn. 12, which incorporates non-linearity parameters (also known as true triaxial strength parameters) D_m, and e_m when represented in terms of stress invariants in principal stress space.

$$F = \sqrt{\frac{2J_2}{3}} - D_m\left(\frac{I_1}{3}\right)^{e_m}\sigma_{ci} \qquad (12)$$

The strength parameters D_m and e_m for different joint inclinations θ are estimated using the strength envelopes shown in Fig. 10 for rock mass. Table 4 shows that these parameters vary with joint inclination, θ similar to Mohr-Coulomb and Drucker-Prager model parameters.

In above Eqn. (12), stress invariants J_2 and I_1 can be estimated from insitu stress values and parameters D_m and e_m from joint pattern of rock mass at site. The σ_{ci} is obtained through laboratory tests. Thus, this equation can be effectively used for estimation of rock mass strength at site under polyaxial stress compression.

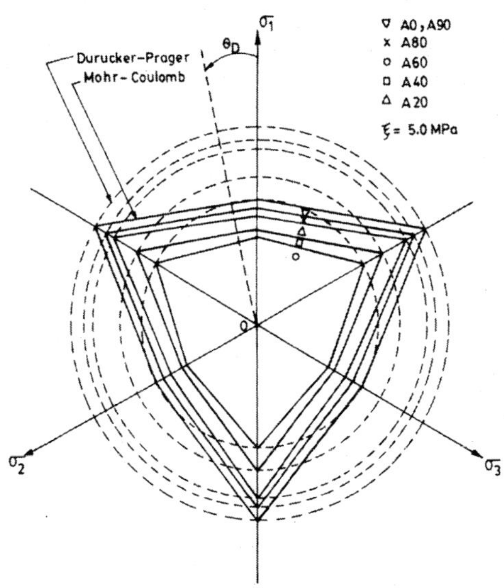

FIGURE 11: Deviatoric section of Mohr-Coulomb and Drucker- Prager failure envelopes at hydrostatic stress $\xi = 5.0$ MPa

5. STRENGTH MODEL PLOTS

The failure envelope plots of polyaxial rock mass testing results are made using all three strength models. The shape of strength envelope can be clearly visualised by plotting meridian section and deviatoric section in octahedral plane. All the three models are pressure dependent, two parametric models. For plotting the sections of failure envelopes, the two parameters needed are radius, ρ of deviatoric section and different levels of hydrostatic stress, ξ. The radius, ρ of deviatoric section increases linearly or non-linearly with increase in hydrostatic stress, ξ depending upon the specific strength model used.

The hydrostatic stress ξ is estimated using the following Eqn. (13).

$$\xi = \frac{I_1}{\sqrt{3}} \tag{13}$$

where, I_1 is first invariants of stress tensor and will be equal to $(\sigma_1 + \sigma_2 + \sigma_3)$ for stress path used in the present study.

The radius of deviatoric section ρ is estimated using the following Eqn. (14).

$$\rho = \sqrt{2J_2} \tag{4}$$

Alternatively, expressions for estimating ρ using all the three models can be written as following:

5.1 Mohr-Coulomb Failure Surface

In tensile meridian ($\theta_D = 0°$),

$$\rho_t = \frac{2\sqrt{6}\, c\cos\phi - 2\sqrt{2}\,\xi\sin\phi}{3+\sin\phi} \tag{15}$$

In compression meridian ($\theta_D = 60°$)

$$\rho_c = \frac{2\sqrt{6}\, c\cos\phi - 2\sqrt{2}\,\xi\sin\phi}{3-\sin\phi} \tag{16}$$

5.2 Drucker-Prager Failure Surface

Since section of failure surface is circular, so radius ρ will be same in tensile and compressive meridians as following:

$$\rho_t = \rho_c = \sqrt{2}\,k - \left(\sqrt{6}\,\alpha\right)\xi \tag{17}$$

5.3 Proposed Failure Surface

In this case also section is circular so radius ρ will be same in tensile and compressive meridians and be expressed as:

$$\rho_t = \rho_c = \sqrt{3}\,.\,D_m.\left(\frac{\xi}{\sqrt{3}}\right)^{e_m}.\,\sigma_{ci} \tag{18}$$

Thus, failure surface plots can be made for all three models at each joint inclination, θ of rock mass specimens using ρ and ξ values.

6. MODEL COMPARISON

The radius of failure envelope is calculated using the strength models and is compared with the

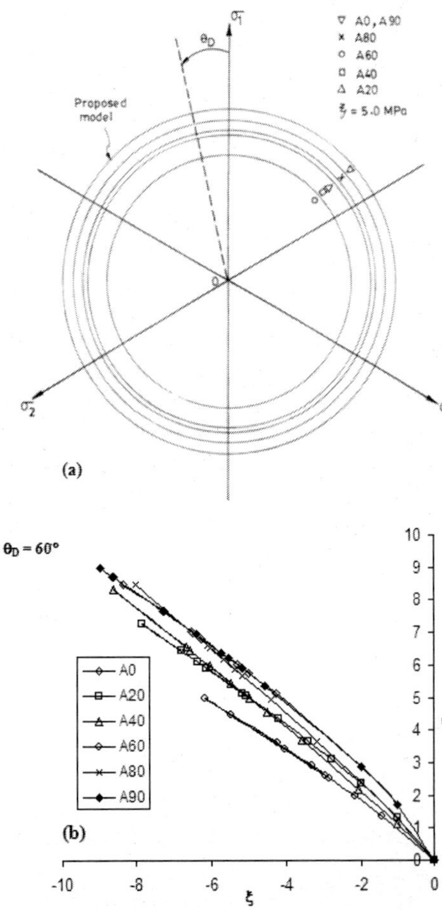

FIGURE 12: Failure surfaces of proposed model in octahedral plane for rock mass specimens (a) Deviatoric section (b) Meridian section

experimental results. The radius, ρ estimated at inclination θ = 0 °, 20°, 40°, 60°, 80° and 90° is listed in Tables 8 to 13. The lode angle θ_D for experimental values is lying near to 60°, i.e. in compression meridian. It is very clear from tables that in comparison to Mohr-Coulomb and Drucker-Prager models, the proposed model back predicts the results more closer to the experimental values at all joint inclinations, θ. If one sees the deviatoric section in Fig.11, it can be observed that Mohr-Coulomb and Drucker-Prager failure surfaces are irregular hexagonal pyramid and circular cone respectively. The proposed model is represented by a failure surface, which is circular cone with curvilinear surface (Fig.12).

Further, it can be seen from the Figs.11 and 12 that the radius of deviatoric section is the minimum for joint inclination θ = 60 ° and the maximum at θ = 0°, 90° in all strength models.

7. DISCUSSION AND CONCLUSIONS

In present study, the rock mass specimens having three joint sets were tested in true triaxial system for their strength response. The joint inclinations were varied to account the effect of joint geometry on strength response. The results clearly depict the enhancement in strength of block mass models due to increase in σ_2. The enhancement in strength is maximum at the weakest joint pattern of θ = 60° and comparatively lower at 0° and 90° which inferred the fact that soft rocks are more susceptible to enhancement than hard rocks. The specimen may yields before the failure takes place and follows a criterion, which is in pattern of octahedral-shear stress series of strength criteria. The main conclusions drawn from the present study can be summarized as in Table 8 and 9.

Table 8: Predicted and Experimental Values of ρ in Rock Mass Specimens (A0)

Confining stresses (MPa)		Hydro-static Stress (MPa)	Lode angle (°)	Radius, ρ in Deviatoric section of failure surface				
				Predicted values				Exp. values
				Mohr-Coulomb model		Drucker-Prager model	Proposed model	
σ_3	σ_2	ξ	θ_D	ρ_t	ρ_c	ρ_{DP}	$\rho_{proposed}$	ρ_{exp}
0	0	2.24	60.0	2.08	3.33	3.33	3.12	3.16
0.31	0.31	4.28	60.0	3.17	5.07	5.07	5.12	5.31
0.31	0.59	5.00	58.1	3.65	5.69	5.69	5.76	5.98
0.31	0.95	5.31	55.7	3.71	5.95	5.95	6.03	5.99
0.31	1.22	6.26	54.8	4.21	6.75	6.75	6.83	7.01
0.31	1.62	6.49	52.3	4.34	6.95	6.95	7.02	6.88
0.78	0.78	6.53	60.0	4.36	6.98	6.98	7.05	7.33
0.78	1.22	7.27	57.7	4.75	7.61	7.61	7.65	7.85
0.78	2.24	8.34	52.8	5.32	8.53	8.53	8.49	8.18
1.22	1.22	8.62	60.0	5.47	8.77	8.77	8.71	9.21

Table 9: Predicted and Experimental Values of ρ in Rock Mass Specimens (A20)

Confining stresses (MPa)		Hydro-static stress (MPa)	Lode angle (°)	Radius, ρ in deviatoric section of failure surface				
				Predicted values				Exp. values
				Mohr-Coulomb model		Drucker-Prager model	Proposed model	
σ_3	σ_2	ξ	θ_D	ρ_t	ρ_c	ρ_{DP}	$\rho_{proposed}$	ρ_{exp}
0	0	1.01	60.00	1.03	1.65	1.65	1.35	1.42
0.31	0.31	2,79	60.00	1.98	3.16	3.16	3.12	3.20
0.31	0.59	3.41	56.96	2.30	3.68	3.68	3.67	3.73
0.31	0.95	4.22	54.16	2.73	4.36	4.36	4.36	4.45
0.31	1.22	5.16	53.24	3.23	5.15	5.15	5.15	5.46
0.31	1.62	6.12	51.63	3.73	5.96	5.96	5.92	6.37
0.78	0.78	5.07	60.00	3.18	5.08	5.08	5.08	5.27
0.78	1.22	6.35	57.28	3.86	6.16	6.16	6.11	6.55
0.78	2.24	7.85	52.08	4.65	7.43	7.43	7.27	7.49
1.22	1.22	6.78	60.00	4.09	6.52	6.52	6.44	6.61

- The typical three joint sets found in rock mass at site were physically simulated in laboratory using low strength model material. The block mass was tested under polyaxial stress compression at different low stress levels.
- The significant enhancement in strength (310%) was seen due to increase in σ_2 in rock mass geometry with θ = 60 °. The enhancement is the lowest at θ = 90°.
- The strength values show anisotropy with joint inclination, θ (= 90°-β) with minimum value at 60°. This anisotropy reduces with increase in σ_2/σ_3 ratios.

Table 10: Predicted and Experimental Values of ρ in Rock Mass Specimens (A40)

Confining stresses (MPa)		Hydro-static stress (MPa)	Lode angle (°)	Radius, ρ in deviatoric section of failure surface				
				Predicted values				Exp. values
				Mohr-Coulomb model		Drucker-Prager model	Proposed model	
σ_3	σ_2	ξ	θ_D	ρ_t	ρ_c	ρ_{DP}	$\rho_{proposed}$	ρ_{exp}
0	0	0.06	60.00	0.19	0.29	0.29	0.09	0.09
0.31	0.31	2.04	60.00	1.18	1.83	1.83	2.18	2.13
0.31	0.59	3.59	57.15	1.96	3.04	3.04	3.68	3.98
0.31	0.95	4.96	55.28	2.65	4.11	4.11	4.98	5.49
0.31	1.22	5.47	53.75	2.91	4.51	4.51	5.45	5.90
0.31	1.62	6.04	51.48	3.19	4.96	4.96	5.97	6.25
0.78	0.78	4.51	60.00	2.42	3.76	3.76	4.55	4.47
0.78	1.22	6.56	57.39	3.46	5.37	5.37	6.46	6.85
0.78	2.24	8.61	53.07	4.49	6.97	6.97	8.31	8.55
1.22	1.22	6.67	60.00	3.51	5.45	5.45	6.56	6.46

Table 11: Predicted and Experimental Values of ρ in Rock Mass Specimens (A60)

Confining stresses (MPa)		Hydro-static stress (MPa)	Lode angle (°)	Radius, ρ in deviatoric section of failure surface				
				Predicted values				Exp. values
				Mohr-Coulomb model		Drucker-Prager model	Proposed model	
σ_3	σ_2	ξ	θ_D	ρ_t	ρ_c	ρ_{DP}	$\rho_{proposed}$	ρ_{exp}
0	0	0.06	60.00	0.29	0.42	0.42	0.09	0.09
0.31	0.31	1.42	60.00	0.89	1.29	1.29	1.37	1.25
0.31	0.59	2.16	54.23	1.21	1.76	1.76	1.99	1.97
0.31	0.95	2.84	49.67	1.51	2.19	2.19	2.53	2.52
0.31	1.22	3.33	47.22	1.72	2.49	2.49	2.91	2.91
0.31	1.62	5.46	50.19	2.66	3.85	3.85	4.49	5.43
0.78	0.78	2.94	60.00	1.55	2.25	2.25	2.61	2.25
0.78	1.22	4.03	54.54	2.03	2.95	2.95	3.44	3.27
0.78	2.24	6.16	48.38	2.97	4.29	4.29	4.99	5.12
1.22	1.22	4.25	60.00	2.13	3.09	3.09	3.61	3.03

Table 12: Predicted and Experimental Values of ρ in Rock Mass Specimens (A80)

Confining stresses (MPa)		Hydro-static stress (MPa)	Lode angle (°)	Radius, ρ in deviatoric section of failure surface				
				Predicted values				Exp. values
				Mohr-Coulomb model		Drucker-Prager model	Proposed model	
σ_3	σ_2	ξ	θ_D	ρ_t	ρ_c	ρ_{DP}	$\rho_{proposed}$	ρ_{exp}
0	0	0.14	60.00	0.84	1.33	1.33	0.21	0.20
0.31	0.31	3.18	60.00	2.42	3.81	3.81	3.67	3.75
0.31	0.59	4.42	57.79	3.05	4.81	4.81	4.93	5.15
0.31	0.95	5.16	55.51	3.44	5.42	5.42	5.68	5.78
0.31	1.22	5.65	54.00	3.69	5.82	5.82	6.17	6.16
0.31	1.62	6.04	51.48	3.89	6.13	6.13	6.56	6.25
0.78	0.78	5.36	60.00	3.54	5.58	5.58	5.88	5.67
0.78	1.22	6.11	57.13	3.93	6.19	6.19	6.62	6.21
0.78	2.24	8.00	52.29	4.91	7.73	7.73	8.47	7.71
1.22	1.22	6.80	60.00	4.29	6.76	6.76	7.31	6.65

Table 13: Predicted and Experimental Values of ρ in Rock Mass Specimens (A90)

Confining stresses (MPa)		Hydro-static stress (MPa)	Lode angle (°)	Radius, ρ in deviatoric section of failure surface				
				Predicted values				Exp. values
				Mohr-Coulomb model		Drucker-Prager model	Proposed model	
σ_3	σ_2	ξ	θ_D	ρ_t	ρ_c	ρ_{DP}	$\rho_{proposed}$	ρ_{exp}
0	0	2.23	60.00	2.09	3.35	3.35	3.11	3.15
0.31	0.31	4.56	60.00	3.33	5.34	5.34	5.36	5.69
0.31	0.59	5.17	58.18	3.66	5.87	5.87	5.91	6.23
0.31	0.95	5.54	55.88	3.85	6.18	6.18	6.22	6.31
0.31	1.22	5.74	54.13	3.96	5.34	5.34	6.39	6.29
0.31	1.62	6.41	52.14	4.32	6.92	6.92	6.95	6.77
0.78	0.78	6.40	60.00	4.31	6.91	6.91	6.94	7.15
0.78	1.22	7.29	57.74	4.79	7.68	7.68	7.67	7.88
0.78	2.24	8.64	53.11	5.51	8.83	8.83	8.73	8.61
1.22	1.22	8.98	60.00	5.69	9.12	9.12	8.98	9.72

- A new polyaxial strength criterion was developed which incorporates UCS of intact rock, Joint inclination θ and surrounding stresses σ_1, σ_2, and σ_3. The σ_2 is included in this criterion while calculating stress invariants I1 and deviatoric shear stress invariants J2. The deviatoric section of failure envelope in octahedral plane is circular cone with root J2 radius. The meridian section is curve linear surface.
- The new strength criterion is modification over Drucker–Prager model with inclusion of non-linearity. This criterion considers the effect of anisotropy by way of including the angle of discontinuities of rock mass.

REFERENCES

Amadei, B. (1988). "Strength of a regularly jointed rock masses under biaxial and axisymmetrical loading conditions". Int. J. Rock Mech. Min. Sci. & Geomech. Abstr. 25, pp. 3-14.

Colmenares, L.B. and Zoback, M.D., (2002). "A statistical evaluation of intact rock failure criteria constrained by polyaxial test data for five different rocks". Int . J. Rock Mech. Min. Sci., 39, Issue 6, pp. 695-729.

Deere, D.U., Miller, R.P., (1966). "Engineering classification and index properties for intact rock". Technical Report No. AFNL-TR-65-116. Air Force Weapons Laboratory, New Mexico.

Franklin, J.A., (1971). "Triaxial strength of rock materials," Rock Mech., 3, pp86-98.

Haimson, B. and Chang, C., (2000). "A new true triaxial cell for testing mechanical properties of rock and its use to determine rock strength and deformability of westerly granite". Int. J. Rock Mech. Min. Sci. 37, pp. 285-296.

Hoek, E. and Brown, E. T., (1980). "Empirical strength criterion for rock masses". Jl. Geotech. Engng. Div., ASCE, 106(GT9), pp. 1013-1035.

ISRM (1981). "Rock characterization, testing and monitoring". ISRM Suggested Methods, Brown E.T. (ed.), Pergamon Press.

Geotechnical Challenges Posed by Sikkim Earthquake (2011)

Chandan Ghosh

Geohazards Divn., National Institute of Disaster Management, Ministry of Home Affairs, Govt. of India
E-mail: cghosh24@gmail.com

ABSTRACT

The Sikkim Earthquake-2011 caused more than 350 landslides, damaged more than 34000 buildings, disrupted lifeline services including roads and bridges, water and power supplies and vital communication hubs. Government of Sikkim sought an estimated central assistance of Rs 3500 crore from national Calamity Contingency Fund. About $200 million damage was caused in Tibet (China), and slightly higher in Eastern Nepal in addition to losses in Bhutan with around 6000 buildings damaged. It called for immediate responses and relief measures all throughout the region. Though the life loss was not much, the impact of this earthquake in terms of direct and indirect economic loss and setback to developmental activities of the affected area, including multipurpose hydroelectric projects; needs thorough assessment. This must ensure assimilation of information into a consolidated report with details of economic value of the infrastructure, physical and economical losses. While planning for post-earthquake reconstruction strategy it is important to go for sustainable management of natural resources by consolidating institutional and individual capacities in human, technical, physical, and financial terms in order to ensure integrated management of the affected areas. This paper highlights some of the important issues concerning to India's scientific and geo-political temperament that still continue to undermine the much advocated "paradigm shift" from "relief centric" to "mitigation centric" attitudes.

1. INTRODUCTION

The Mw 6.9 earthquake (Focal depth 19.7KM as per USGS) occurred on September 18, 2011 at 6.11PM in the Sikkim-Nepal border (27.723°N, 88.064°E, 72km NW of Gangtok) region (Fig. 1). The main shock was followed by a few significant aftershocks located by the national seismological network, maintained by India Meteorological Department (IMD) and some other scientific organizations of the country such as NGRI, IIT-Roorkee, IIT-Kharagpur and others. IMD, the nodal organization of the country, has been maintaining 20 broadband stations in the NE region but on the fateful day, it recorded 2 aftershocks of M5.0, M4.5 within 2hrs and another of M4.2 at 3:21AM on September 19. National Disaster Rescue Force from Kolkata and other places were flown to different affected areas for rescue and relief works. Army, ITBP, Police and other forces, local public and NGO's

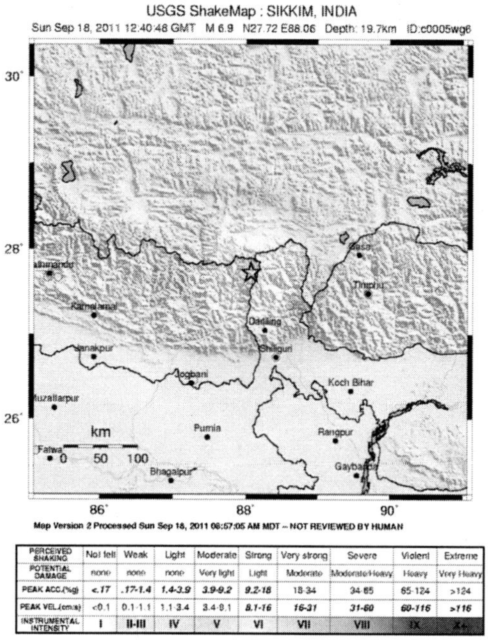

PERCEIVED SHAKING	Not felt	Weak	Light	Moderate	Strong	Very strong	Severe	Violent	Extreme
POTENTIAL DAMAGE	none	none	none	Very light	Light	Moderate	Moderate/Heavy	Heavy	Very Heavy
PEAK ACC.(%g)	<.17	.17-1.4	1.4-3.9	3.9-9.2	9.2-18	18-34	34-65	65-124	>124
PEAK VEL.(cm/s)	<0.1	0.1-1.1	1.1-3.4	3.4-8.1	8.1-16	16-31	31-60	60-116	>116
INSTRUMENTAL INTENSITY	I	II-III	IV	V	VI	VII	VIII	IX	X+

FIGURE 1 Shake map prepared by USGS within 10 minutes from the earthquake event

also joined the operation. Kathmandu, the capital of Nepal, experienced two aftershocks of M4.8 but had no serious impact. The earthquake was widely felt in all the North Eastern states of India, West Bengal, Bihar, Uttar Pradesh, Haryana, Rajasthan, including capital city Delhi. The earthquake tremors coupled with heavy rain fall caused rock fall and consequent blockade of important road corridors in Sikkim and Darjeeling Hills.

Utilizing high speed computer and communication technology, India, having more than 120 seismic stations deployed across the country, needs a functional coordination and protocol among agencies like IMD, DST, WIHG, NGRI, IITs, ...in order to create near real time post-earthquake information dissemination for the disaster managers, policy makers and community.

Historical and instrumentally recorded data on earthquakes show that the Sikkim and adjoining area lies in a region prone to be affected by moderate to great earthquakes in the past.

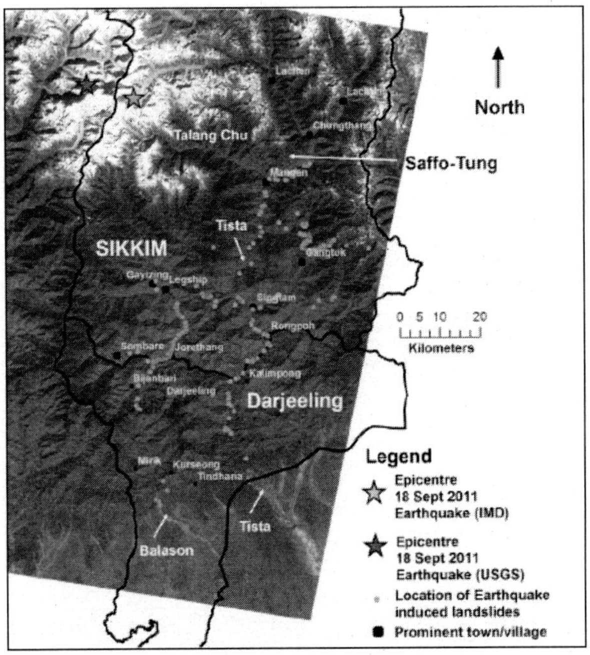

FIGURE 2: Map showing locations of 18 September 2011 Sikkim earthquake induced landslides observed along important road corridors in Sikkim-Darjeeling Himalaya (in background multi-spectral IRS P-IV Satellite's LISS III image of 2005 was used). (Chakraborty et al, 2011)

This region has experienced relatively moderate seismicity in the past, e.g. 12 January, 1965, 19 November, 1980, 14 February 2006, 31 December, 2006, 20 May, 2007 and it has experienced 18 earthquakes of magnitude 5 or greater over the previous 35years. Some of the landslides spots identified are shown in Fig. 2 (Chakraborty, et. al., 2011).

According to the Indian Meteorological Department (IMD), the earthquake was of Magnitude M6.8 occurred at 18.11 hours IST on 18th September 2011 with its epicenter (27.7°N, 88.2°E, Focal depth 10km) lying near the Sikkim-Nepal border, about 68 km northwest of Gangtok. As per the USGS estimates, the epicenter of this earthquake (Mw6.9, Focal depth 19.7 km, 27.723°N, 88.064°E) lies slightly west of the location shown by IMD. In above respect Indian seismological studies must try for understanding tectonic and fault activity mapping of NE region more intensive manner to address discrepancy among scientific communities.

Eight north eastern states – Assam, Meghalaya, Mizoram, Tripura, Nagaland, Arunachal Pradesh, Manipur and Sikkim – are considered by seismologists to be the sixth most earthquake-prone belt in the world. This region faces M5+ earthquakes every 4 months. The earthquake recurrence intervals between the largest main shocks may vary from several decades to several hundred years. Delineating a potential zone and predicting future seismic hazards in this region is urgently required so that appropriate measures to minimize the destructive effects can be reaped by administration/district officials dealing with emergencies/responses.

Most of the landslides/slope failure is triggered due to rainwater infiltration and/or poor drainage that build up pore water pressures in the ground. The September 2011 earthquake has triggered huge landslide along NH55A, at Tindharia, Kalimpong, which is about 100km from the epicenter. On 19th July 2012, heavy rainfall caused further slides in this area (Fig. 3). Post-Sikkim Earthquake damage assessment has brought out the attention of the disaster managers of the country of the different sectors, i.e. governance, the environment, disaster risk

Table 1: Distribution of Indian Earthquakes in the Last 110 Years, Indicating One M5+ Earthquake Occurring in Less than Every 4 Months in the North East Region

Seismic Regions in India	*No. of Earthquakes of Magnitude*				*Return Period*
	5.0-5.9	*6.0-6.9*	*7.0-7.9*	*8.0+*	
Kashmir & Western Himalayas	26	7	2	1	2.5-3 yrs.
Central Himalayas	68	28	4	1	1 yrs.
North East India	200	129	15	4	< 4 months
Indo-Gangetic Basin and Rajasthan	14	6	-	-	5 yrs
Cambay and Rann of Kutch	4	4	1	1	20 yrs.
Peninsular India	31	10	-	-	2.5-3 yrs.
Andaman & Nicobar	80	68	1	1	< 8 months

management, socio-infrastructure and cross-cutting sectors (youth, gender, vulnerable persons, and employment). The earthquake induced instability in the hill slopes has certainly enhanced landslide vulnerability and it is going to be a cause of major concern during monsoon. Therefore, disaster mitigation efforts need to be directed towards the identification of vulnerable regions, with emphasis on preventive steps as well as for planning relocation of settlements wherever necessary.

(a) Slides on 18/09/2011

(b) Slides on 19/07/2012

FIGURE 3: The Sikkim earthquake of 18th Sept. triggered landslides at Tindharya and nearby Paglajhora area, snapping road cum Heritage Darjeeling Toy Train link. No mitigation measures were taken since then thus resulting into further slides on 19th July 2012 due to heavy rainfall. The railway establishment, a heritage site, is now at stake

Though country is having wealth of information and expertise on the landslides mitigation techniques such as reinforced earth retaining wall using wide variety of geosynthetics, no consensus has yet reached among the West Bengal-PWD, Northern Frontier Railway, MoRST and design/construction agencies of the country. It is learned that a massive gabion wall of height more than 100m has been sanctioned to mitigate this slide. It is high time now to form a consortium, consisting of GSI, Geotechnical experts dealing with reinforced earth wall construction and design, West Bengal-PWD, NHAI, Northern Frontier Railway and MoRST and decide appropriate measures keeping mind to the quality worksmanship, construction feasibility and cost-economy.

2. EARTHQUAKE EFFECTS AND RESPONSE

The earthquake claimed 63 human lives in Sikkim, including 10 at the Teesta Stage III Hydroelectric power project site, 719 persons injured and also caused substantial loss to livestock (Table 2). It shook for 30-40 seconds at intensity MSK-VII, triggered landslides, and caused extensive damages to houses (Fig. 4) and infrastructures, mainly roads, water supply, transmission towers etc. in the state of Sikkim and North Bengal districts such as Siliguri, Darjeeling, Kalimpong, New Jalpaiguri. Lighter tremors (MM IV–III) spread southward through populous regions, with these motions reported in the Patna capital of Bihar and as far southwest as Bihar Sharif. The earthquake was felt in Nepal, India, Bhutan, Bangladesh and China. Tremors were felt in Assam, Meghalaya, Tripura, parts of West Bengal, Bihar, Jharkhand, Uttar Pradesh, Rajasthan and Delhi of India. In Tibet, the earthquake was felt in Shigatse and Lhasa. Wide scale damages were also reported from neighbouring countries like Nepal, Bhutan, Tibet and Bangladesh.

Figure 4: The only 3 storey newly constructed private building near Lumsey Govt. Primary School, Gangtok that completely collapsed during September 18, 2012 earthquake. The approach road to this building was so narrow and in bad shape that no emergency vehicle (not even a bicycle/Riksha) can reach there.

Though some of the Govt./public buildings that suffered severe damages have either been repaired or being retrofitted, Gangtok city administration has to promulgate some policy instruments in place to take up such cases as shown above. In above case, the owner of the house possibly decided to do debris demolition on his own means and terms. Therefore, the country needs a functional policy to take up collapsed structure demolition and compensation for the private buildings when these are constructed in municipality area guided by building-bye-laws, building code and existing development control rules.

Table 2: Earthquake Effect on Establishments and Livestock in Sikkim State

Sr. No.	*Type of Element*	*Number of Elements*	*Degree of Damage/ Loss*
1	Human Lives	63	Killed
		710	Injured
2	Houses	34159	Collapsed / damaged
3	Primary Schools	416 out of total 779	Damaged
	Junior High Schools	192 out of total 348	Damaged
	Secondary and Sr. Secondary Schools	151 out of total 195	Damaged
4	Hospitals and Health Centres	377	Damaged
	Primary Health Centres	350	Require recons-truction
	Community Health Centres	25	
	Other important health buildings	2	
5	Water Supply Schemes	1529 (1484 in rural areas and 44 in urban areas)	
6	Roads	3230km	
	State Roads	178.71 km	
	District Roads	744.77 km	
	Other district roads	1289 km	
	Urban Roads	132.44 km	
	PMGSY Road	884.89 km	
	Bridges	314	
	Culverts	7081	

Sr. No.	Type of Element	Number of Elements	Degree of Damage/ Loss
	Rural bridges & culverts	360	
	Foot bridges	380	
7	Agrl. Crops	7500 Hectares	
8	Minor Irrigation Channels	204	
	Flood Control Works	365	
	Urban Drainage	168 km	
9	Community Assets		
	Panchayat Ghar	60	
	Rural marketing Centres	8	
	Community Toilets	155	
	Intra-village roads	156	
	Anganwadi centres	875	
	Village Cooperatives	49	
10	Religious buildings	259	
11	RCC/Pucca House	432 houses	100%
		2965 houses	Major
		481 houses	Minor
12	Kutcha House	643 houses	100%
		5398 houses	Major
		456 houses	Minor
13	Livestocks killed	307 Cows	
		210 Goat	
		232 Bull	
		250 Pig	
		410 yak	
		800 Sheep	
14	Vehicles	231 (including bus, trucks, taxis, and government vehicles)	

Source: Resident Comm. Office, Sikkim State, New Delhi.

The Secretariat building, Sikkim Press building, Education, Health and Police Headquarters in Gangtok suffered severe damage. More than 34,000 buildings in Sikkim, such as multi-storey RCC buildings, monasteries, hospitals, schools, Gumpas, Panchayat Ghars, public health centres have suffered various degrees of damages. Urban water supply and drainage schemes in Gangtok, irrigation channels, PMGSY – village roads and rural water supplies in various districts have also been adversely affected. National Highways, State Highways including the rural connectivity in the Chungthang, Lachung and Lachen area have been snapped. Several aftershocks accompanied with incessant rainfall were causing further distress in the relief and rescue operations. While BRO cleared debris at more than 20 spots along NH31A upto Gangtok within 20 hrs, followed by road to Mangan, the North District HQ., via Dikchu in 3 days, road to Nathu-la, the Chinese border by 2nd October, and from Mangan to Chungthang by 18th October, it took more time to allow traffic upto Lachung and to further North.

Early rescue operations included four teams of National Disaster Response Force, who have been rushed to Sikkim and five more teams were being sent from Kolkata. A group of 14 tourists were rescued by the army from north Sikkim. The army had deployed 72 columns including infantry troops, combat engineers, four Dhruv and five Cheetah helicopters. Many of the areas in North Sikkim including Chungthang, Lachen, Lachung and Dzongu were inaccessible without the services of choppers and helicopters. Army, Air force, Teesta Urja and State Department of Tourism were using helicopters to rescue people trapped in these inaccessible areas and provide relief to them from Ringhim helipad at Mangan in North Sikkim district. It was reported that roads so cleared for traffic had not been found fit for carrying heavy machineries (Fig. 5). to many Hydroelectric Power (HEP) project sites, thus resulting in inordinate delay in their target achievements. Fig. 6 shows the portending danger in the heart of Gangtok city wait.

FIGURE 5: On December 19, 2011, 8 people killed and 3 injured in a bridge collapse at Rangchang Khola, about 18 kms from Singtam towards Dikchu on the Singtam-Mangan highway. The incident took place when a 48-wheeler heavy trailer carrying transformer of Teesta Urja Project to its project site in North Sikkim was attempting to cross the bridge span.

It was urgently required to conduct detail damage as well as vulnerability assessment survey of all these establishments in the affected areas with particular motivation towards the development of national policies/strategies for repair, strengthening and retrofitting measures.

Figure 6: Several of these buildings at the heart of Gangtok city were damaged/cracked, yet there were no systematic attempt to document them following the earthquake.

This area was subjected to earthquake ground vibration intensity MSK-VI, which according to Indian seismic code (IS: 1893-2002) is the minimum that such buildings at Seismic Zone –IV, should be able to resist.

Hardly any of these buildings appeared to face sudden collapse due to the fateful earthquake on 18th September, 2011. But propensity of their existence without a plausible resilience in hilly terrain warrants for detail vulnerability assessment followed by repair, retrofitting and re-strengthening measures.

Paucity of plain land forced Gangtok to have multistory buildings on steep slopes. Living in such housing establishments is itself a nightmare. Despite the country taking phenomenal stride in the application of Incident Response System (IRS) for the last 5 years (under USAID program) there could be hardly any well defined response system in place (as seen in the picture), which can cater the complexity of emergency operation inflicted by another earthquake of higher intensity.

The understanding of the resilience in the above respect is not to be presumed that the establishments as above are quite apt enough to face intensity of VII or more. It is rather a matter of chance that hilly terrain responded favorably, for which Indian expertise in the seismic microzonation of Sikkim must have to relook into their assumptions and presumptions so that more realistic guidelines are prepared keeping in view the damages incurred.

Report presented in this form contains category of items in numerals only (Table 2). It has hardly any supporting foot prints of the damages/ loss with areal maps, categorisation of damage patterns, graphs, figures, etc. It may be noted that allocation of financial aids/funds becomes more meaningful when damage/loss patterns are critically examined by experts. Therefore, post-disaster damage & loss documentation must be given utmost priority.

3. SEISMIC ACTIVITY IN SIKKIM REGION

The Sikkim region falls between Nepal and Bhutan and comprises the lesser active part of the 2500 km stretch of the active Himalayan belt. The seismic hazard scenario of Sikkim Himalaya appears to be quite underestimated considering its zone IV status in the seismic zonation map of India. This region is surrounded by Great earthquake occurrences during the past, namely the 1934 Bihar- Nepal earthquake (M 8.3) to the west, the 1897 Shillong earthquake (M 8.7) to the southeast and the 1950 Assam earthquake (M 8.7) to the east. The effect of topography and vulnerable establishments made the problem more complex.. The most significant earthquake that occurred in its neighbourhood is the 1934 Bihar-Nepal border earthquake of M 8.3 to the west, that caused intensity VIII damage in Sikkim Himalaya. In 1988 the region again witnessed an M 6.6 earthquake with intensity VI in Gangtok. The major tectonic features traversing the Sikkim Himalaya are the well-defined Main Boundary Thrust (MBT) and the circular overturned Main Central Thrust (MCT) to the north.

Seismic hazard in the Sikkim region is further accentuated due to the presence of sedimentary rocks and loose soil in the region, which tend to amplify the seismic ground motion. Nath *et al. (2008)* have indicated that, in general, site amplification in the Sikkim region increases from northeast to southwest along the Tista and Gangtok lineaments, peaking to a maximum of 1.6 at Singtam, southwest of Gangtok. This is attributed to the presence of thick sedimentary rocks to the south, at the Himalayan foothills.

Due to M5.4 Sikkim earthquake in 2006, most of the damage was observed in Phodong and Theng. In Gangtok, maximum damage was reported for government establishments like the Police Headquarters, school buildings and some old constructions. The Raj Bhavan, in Gangtok, was severely damaged, which later on got retrofitted. The 2011 earthquake brought attention of the policy makers, scientists and various stake holders in earthquake management of the country. Due to this earthquake too, several facilities, such as, transportation infrastructure (road networks, bridges, tunnels, culverts, retaining walls, village footpaths), energy infrastructure (generation plants, electrical grid, substations, transformers and local distribution), water management infrastructure (drinking water supply, drainage systems, irrigation and flood control systems) got severely affected.

4. EARTHQUAKE INFORMATION DISSEMINATION

India Meteorological Department (IMD) is the nodal agency of Government of India responsible for monitoring seismic activity in and around the country. The operational task of the department is to quickly estimate the earthquake source parameters immediately on occurrence of an earthquake and disseminate the information to all the user agencies including the concerned State and Central Government agencies responsible for carrying out relief and rehabilitation measures. The earthquake information is also transmitted to public information channels, press, mediam, etc. and posted on IMD's Website. An Earthquake Operational Centre is functioning on a 24×7 basis at IMD Headquarters in New Delhi, which has the operational responsibility of keeping a round-the-clock watch of seismic activity in and around the country The earthquake information is also transmitted to public information channels, press, media etc. and posted on IMD's Website. An Earthquake Operational Centre is functioning

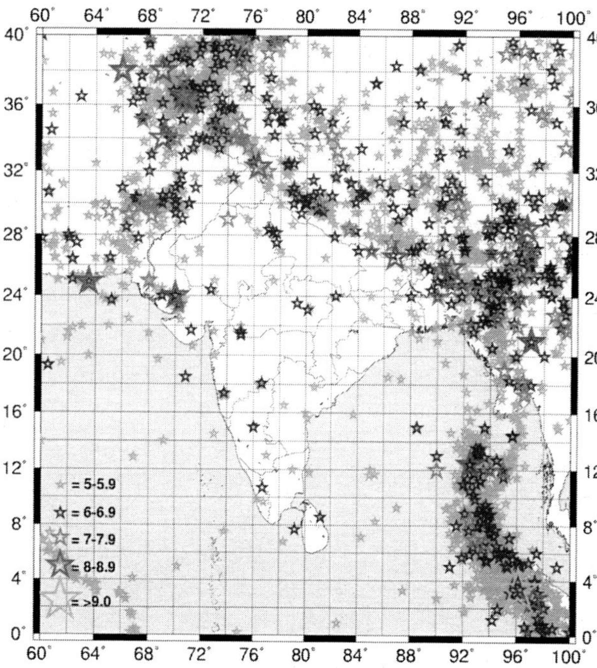

FIGURE 7: Seismicity of India (1505 to March, 2010) (M>5.0) (*Source:* IMD)

on a 24X7 basis at IMD Headquarters in New Delhi, which has the operational responsibility of keeping a round-the-clock watch of seismic activity in and around the country. The responsibilities of the Centre include retrieval of waveform data from remote field stations either in real time or through VSAT / dial up communication facilities, data analyses and quick dissemination of earthquake information to various user agencies including State and Central Government departments dealing with relief and rehabilitation measures. Fig. 7 shows the location of $M>5.0$ earthquakes in India and adjoining region.

IMD data in above form bears hardly any significant aspects of seismicity for the entire country. Some attempts (as shown in Fig. 8) have been made for clasterisation of seismic events but prediction exercise all over the world is being discouraged. Therefore, in India, more emphasis must be given for forecasting scenarios and active fault mapping.

Recent advances in seismic sensor technology, data acquisition systems, digital communications, and computer hardware and software make it possible to build reliable real-time earthquake information systems. Such systems provide a means for modern urban regions to cope effectively with the aftermath of major earthquakes and, in some cases; they may even provide warning, seconds before the arrival of seismic waves (Kanamori, 2003). Today, real-time systems are capable of providing basic earthquake information within minutes, and in the future this could be reduced to tens of seconds. In some cases, facilities will receive this information even before ground shaking begins (early warning). This may allow for clean emergency shutdown or other protection of systems susceptible to damage, such as power stations, transport and computer systems. Moreover, quantitative measures of building performance in earthquakes can only be achieved if shaking during damaging earthquakes is recorded with spatial density sufficient to characterize the ground motion. Several real-time earthquake information systems have been developed in Japan, Mexico, Taiwan and the United States.

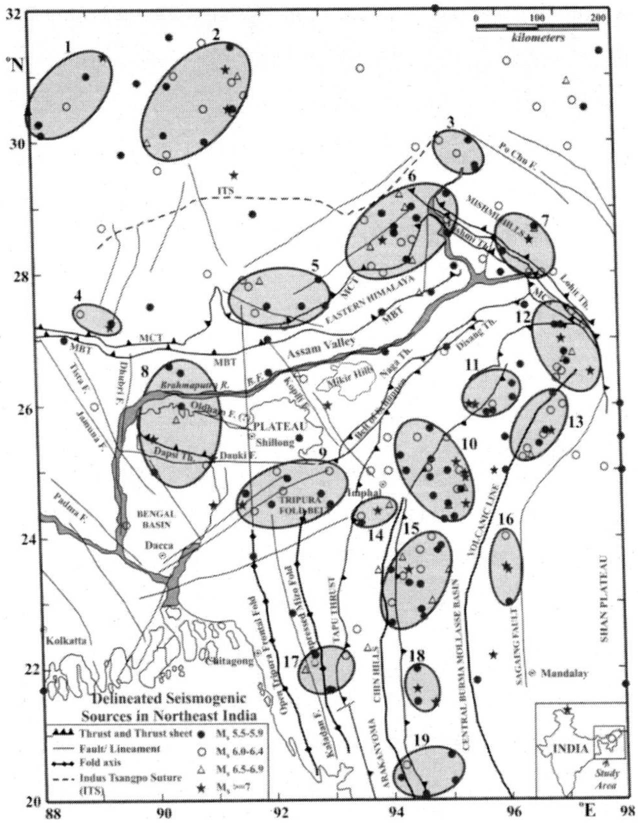

FIGURE 8: Spatial distribution of earthquake with $Ms \geq 5.5$ occurred during the period 1906-2008 in northeast India Himalaya and its adjoining regions.

In the NE region 19 seismogenic sources have been delineated in relation to the major tectonic features of the area and are considered to test the validity of Time-and Magnitude Predictable model

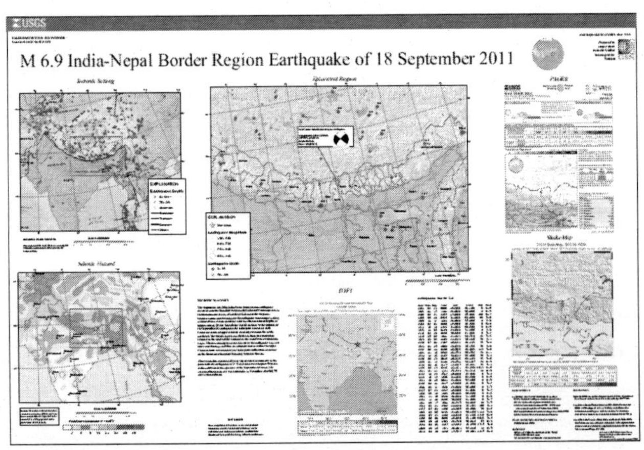

FIGURE 9: Using network of seismic stations USGS brings out above type of information within few minutes after any earthquake, which is widely refereed and appreciated

and to estimate the probability of future seismic hazards in each of the delineated seismogenic sources. The seismogenic sources are enclosed by shaded elliptical areas (Panthi et al., 2011).

Earthquake monitoring and information dissemination requires streamlining of the all the agencies working in India and bring out post-earthquake information through national media and web.

FIGURE 10: Portending danger of built infra structures buildings in the heart of the city, Gangtok – way forward to disaster resilience and preparedness

Buildings in capital city Gangtok are standing on steep slope with little or no measures for foundation protection. Almost same is the situation in all hilly states of India, mostly in seismic zone –V. They are laid one adjacent to others with no spaces in between for free vibration during earthquake shaking. Utilities are mostly lying in tattered mode. Jhora training works are massive yet overly inconsistent with the reality of landslips triggered during monsoon.

Buildings are having openings at random, soft storeys and stiffness variation knows no bounds, drainage measures on the natural slopes are way behind with the needs, roads are facing subsidence and they are being filled up without looking into proper cross drainage works, rock shooting are happening with warning sign, so many high rise buildings with defined beams and columns frames but with ill configured geometry in plan and elevation; no breathing space in between buildings; narrow road with sharp bends; breast wall with weep holes but flow clogged; Recent earthquake with MSK-VI shook all these buildings for 30 sec. Many buildings showed pounding effect; some buildings showed failure in the load bearing members at level where soft openings were there.

FIGURE 11: Controling measures around the Bagthang water falls, in the form of buttressed retaining walls are no way effective in preventing landslides. The road has no culvert or x-drains. Boulders are constant threat to unwary travellors

FIGURE 12: Landslides caused severe trouble to dwindling infrastructures services and utilities. Once again restoration of the same has hardly any technology in place

Road blocks due to earthquake induced landslides cleared in hours by BRO but their sustainability and timeline to normalcy is point of serious concern to the proponents of various industry and power projects in the region (Figures 11-12).

5. CONCLUSIONS

One of the prime requirements of the society is the forecasting of earthquake related hazards with reasonably high degree of probability which eventually help mitigating expected hazards that would save loss of lives and property. But it is not an easy task to predict precisely the time, magnitude and location associated with future earthquakes with any accuracy, despite many efforts being made worldwide by scientific community. Understanding earthquake generation processes in the NE region is the key to evaluate future seismic hazards in space and time domains. The major tools for such evaluation are the systematic analysis of the observed phenomena before and after the earthquakes and use of statistical tools for analysis of seismicity data considering different models with probabilistic approach. Any wrong prediction of earthquake may pass the wrong message to the public and hence it would be even more hazardous than predicting the seismic hazards. Nevertheless, several researchers are attempting to find suitable earthquake generation models and searching various earthquake precursors for reliable prediction of future seismic hazards.

The Sikkim earthquake caused damage in a large area, including Gangtok, state's capital and several areas in Northern Hill districts of West Bengal. Grave concerns are there with type of vulnerabilities of roads on hill slopes as well as portending scenarios of road widening works, multi-storey buildings juxtaposed on steep slopes with little or no breathing space and dwindling lifeline infrastructures in hilly terrain. Regions in the immediate vicinity of the epicentre, located close to the Sikkim–Nepal border are occupied by high mountains bordering southern Tibet. Accessibility to these regions was further restricted due to the landslides and rock falls that blocked the roads and mountain passes. Several buildings, mostly multi-storied, were damaged in Chungtang, many developed only minor cracks. Several hydroelectric power projects under construction, have met with setback as the road leading to the project site has not been repaired to carry heavy machineries.

Attempts have been made to highlight important lessons with suggestions for betterment, in respect of post-earthquake information dissemination, understanding of seismogenicity and augmenting forecasting ability, activation of Incident of Response System (IRS), documentation of damage and losses by independent expert team, capacity building of road and building construction officials in the modern earthquake resistant technologies, etc. Some salutary ideas and comments have been narrated at the footnote of each illustration.

ACKNOWLEDGEMENT

The author would like to acknowledge the company of Dr Surya Parkash, Associate Professor, NIDM, while on tour to Sikkim from 22-26 Sept., 2011, Mr Prabhakar Rai, Dy-Secretary, LR&DM, Govt. of Sikkim; Mr Ashoke Kumar, DM Cell, Sikkim; DCs of East and North District, Mr Arvind Kumar, Resident Commissioner, Sikkim House, Delhi; Mr Bhadauria, Secretary, LR&DM; Mr G.C. Khanal, Mr Tashi & Mr Liutel, Geology & Mines; Er Chowbey and Er Rohan Sonkar, BRO, 87-RCC; Wg. Cdr. Praful Rao, STH, Kalimpong; Vikrant Mahajan, SPHERE India, and many others who helped gaining access to various earthquake affected areas in Sikkim and Silliguri. Last but not the least, gratitude to Dr Satendra, the Executive Director, NIDM, who provided all helps and encouragement in accomplishing the mission.

REFERENCES

Applied Technology Council. (2005) Field manual: Post-earthquake safety evaluation of buildings, ATC 20-1, second edition. Redwood City, CA.

Applied Technology Council. (1989) Field manual: Post-earthquake safety evaluation of buildings, ATC-20-1. Redwood City, CA.

Applied Technology Council, (1995). Addendum to the ATC-20 post-earthquake building safety evaluation procedures, ATC-20-2. Redwood City, CA.

Applied Technology Council. (1996) Cases studies in rapid post-earthquake safety evaluation of buildings, ATC-20-3. Redwood City, CA.

Agnastopoulos, S.A., Moretti, M. (2008) "Post-earthquake emergency assessment of building damage, safety and usability. Part 1: Technical issues". Soil Dyn. & Earthq. Engin., 28, Issue 3, pp 223-232.

Bilham, R., Gaur, V.K. and Molnar, P., (2001) "Himalayan seismic hazard". Science, 293, pp. 1442-1444.

Bureau of Indian Standards (BIS). (2002) Criteria for earthquake resistant design of structures: General provisions and buildings, IS 1893:2002 (Part I).

Chakraborty, I, Bhattacharya, D. and Bora, A. (2011) "Aftermath of the 18th September 2011 Sikkim Earthquake", Report, Engineering Geology Division, Geological Survey of India, Eastern Region, Kolkata.

De, R. and Kayal, J.R., (2003) "Seismotectonic model of the Sikkim Himalaya: Constraint from microearthquake surveys". Bull. Seismol. Soc. Am., 93, pp. 1395-1400.

De, R., (2000) "A micro-earthquake survey at the MBT zone: Sikkim Himalaya". J. Geophys, 21, pp. 1-8.

Dunn, J.A., Auden, J.B., Ghosh, A.M.N. and Roy, S.C., (1939), 'The Bihar-Nepal earthquake of 1934". Geol. Surv. India Mem., 73, p. 280.

Ghosh, C. and Parkash, S. (2011) "Sikkim earthquake 2011 – a roadmap for resurrection", A report submitted to the Ministry of Home Affairs, Govt. of India, New Delhi (100 page).

Guha, S.K. and Bhattacharya, U. (1984) "Studies on prediction of seismicity in northeast India", Proc. World Conf. on Earthquake Engineering, San Francisco, USA, July, 21-27.

Gupta, H.K. and Singh, H.N. (1986) "Seismicity of northeast India region: Part II: Earthquake swarm precursory to moderate magnitude to great earthquakes". Jour. Geol. Soc. India, 28, pp. 367-406.

Gupta, H.K. and Singh, H.N. (1989) "Earthquake swarm precursory to moderate magnitude to great earthquakes in northeast India region". Tectonophysics, 167, pp 255-298.

Gupta, H.K. and Singh, V.P. (1982) "Is Shillong region, northeast India, undergoing dilatancy stage precursory to a large earthquake?". Tectonophysics, 85, pp. 31-33.

Kayal, J.R., (2001) "Microearthquake activity in some parts of the Himalaya and tectonic model". Tectonophysics, 339, pp. 331-351.

Kayal, J.R. (1987) "Microseismicity and source mechanism study: Shillong Plateau, Northeast India", Bull. Seismological Society of America, 77, Issue 1, pp. 184-194.

Kayal, J.R. (1996) "Earthquake source process in Northeast India: A review". Him. Geol., 17, pp. 53-69.

Kayal, J.R. (1998) "Seismicity of Northeast India and surroundings – development over the past 100 years". Jour. of Geophysics, 19, Issue1, pp 9-34.

Kharshing, A.D., Khattri, K.N., Moharir, P.S. and Chander, R. (1986) "Neotectonics in Shillong massif and neighboring regions", Proc. Int. Symp. Neotectonics in South Asia, India. Feb. 18-21, pp 407- 416.

Khattri, K.N. (1987) "Great earthquakes, seismicity gaps and potential for earthquake disaster along the Himalayan plate boundary". Tectonophysics,138, pp. 79-92.

Khattri, K.N. (1993) "Seismic gaps and likelihood of occurrence of larger earthquake in Northeast India", Current Science, 64, Issues 11&12, pp. 885-888.

Khattri, K.N. (1999) "Probabilities of occurrence of great earthquakes in the Himalayas". Earth & Planetary Sciences, 108, pp 87-92.

Khattri, K.N. and Tyagi, A.K. (1993) "Seismicity patterns in the Himalayan plate boundary and identification of the areas of high seismic potential". Tectonophysics, 96, pp 281-297.

Khattri, K.N. and Weiss, M. (1978) "Precursory variation of seismic rate in Assam Area, India". Geology, 6, pp. 685-688.

Molnar, P. and Tapponnier, P. (1977) "Relation of the tectonics of Eastern China to the India-Eurasia collision: application of slip-line field theory to large-scale control tectonics". Geology, 5, pp. 212-216.

Nandy, D.R. (2001) "Geodynamics of Northeastern India and the adjoining region", ABC Publications, Calcutta, p. 209.

Nath, S.K., Sengupta, P., Sengupta, S. and Chakrabarti, A., (2000) "Site response estimation using strong motion network: A step towards microzonation of the Sikkim Himalayas". Current Science, 9, pp. 1316-1326.

Nath, S.K., Singh, K.K., Thingbaijam and Raj, A. (2008) "Earthquake hazard in Northeast India – A seismic microzonation approach with typical case studies from Sikkim Himalaya and Guwahati city". J. Earth Syst. Sci. 117, S2, November 2008, pp. 809-831.

Gupta, H.K. (1993) "Patterns preceding major earthquakes in northeast India". Current Science, 64, Issues 11&12, pp. 889-893.

Ni, J. and York, J.E. (1978) "Late Cenozoic tectonics of the Tibetan plateau". Jour. Geophys. Res., 83, pp. 5377-5384.

Panthi, A., Shankar, D., Singh, H. N., Kumar, A., and Paudyal, H. (2011) "Time-predictable model applicability for earthquake occurrence in Northeast India and vicinity". Jour. Nat. Hazards Earth Syst. Sci., 11, pp 993–1002.

Rai, Durgesh C; Mondal, Goutam; Singhal, Vaibhav; Parool, Neha and Pradhan, Tripti, (2011) "A Brief Report On Impacts and Implications of 2011 Sikkim Earthquake", NICEE, IIT Kanpur.

Rai, Durgesh C; Mondal, Goutam; Singhal, Vaibhav; Parool, Neha and Pradhan, Tripti; (2011) "Sikkim Earthquake – Effects on built environment and a perspective on growing seismic risk", NICEE, IIT Kanpur.

Rajendran, Kusala, Rajendran, C. P. Thulasiraman, N., Andrews, Ronia and Sherpa, Nima (2011) "The 18 September 2011, North Sikkim earthquake". Current Science, 101, Issue. 11, pp 1475-1489.

Rao, N. P., Kumar, P., Kalpna, Tsukuda, T. and Ramesh, D. S., (2006) "The devastating Muzaffarabad earthquake of 8 October 2005: New insights into Himalayan seismicity and tectonics". Gondwana Res., 9, pp 365-378.

Sarmah, S.K. (1999) "The probability of occurrence of a high magnitude earthquake in Northeast". India. Jour. of Geophysics, 20, Issue 3, pp 129-135.

Shanker, D., and Singh, V. P. (1996) "Regional Time- and Magnitude-predictable seismicity model for north-east India and vicinity", Acta Geod. Geoph. Hung., 31, Issues 1-2, pp 181-190.

Sukhija, D.S., Rao, M.N., Reddy, D.V., Nagabhusanam, P., Hussain, S., Chadha, R.K. and Gupta, H.K. (1997) "Palaeoseismic investigations in the Shillong Plateau, northeast India: evidence of prehistoric major seismic events", Indo-US workshop on Palaeoseismicity, March 26-28, WIGH, Abs. Vol., p. 70.

Tiwari, R.P. (1999) "Disaster management- the utmost need of Mizoram", Proc. Sem. on Sci. & Tech. in Mizoram for 21st Century organised by Mizoram Science Society, Aizawl, pp 139-146.

Tiwari, R.P. (2000) "Earthquake hazards and mitigation in India with special reference to Northeastern India". ENVIS Bulletin, 8, Issue 2, pp 15-22.

Development of IWHR Centrifuge Controlled Tools System

Y.J. Hou, Y.F. Wen, X.D. Zhang, Q.L. Sun, C. Gauffre[1], P. Sabard[1] and D. Rames[1]

Department of Geotechnical Engineering, China Institute of Water Resources and Hydropower Research (IWHR), Beijing, China
[1] *Actidyn Systèmes SA, Elancourt, France*
E-mail: Houyj@iwhr.com

ABSTRACT

This paper presents the main features of the Controlled Tooling System (CTS) in China IWHR, set to operate on its 450g-ton geotechnical centrifuge and currently under commissioning. On top of more conventional tools usually equipping centrifuge robots in its category, IWHR installed an in-flight sand-raining device designed to simulate embankment construction. The IWHR CTS is a four-degrees of freedom robot capable of operating under centrifuge accelerations of up to 120g atop a large model container supporting the CTS frame. The tool magazine can host up to three different automatically recognized interchangeable tools. Finally, a water re-circulation system can also be mounted in the centrifuge basket to simulate seepage or the overtopping failure of earth dams.

1. INTRODUCTION

Geotechnical centrifuges provide an effective platform for the modelling of soil foundations and their behaviour under different conditions. Together with the development of the centrifuge, the Controlled Tooling System (CTS), or centrifuge robot, is recognized as a necessary device to allow in-flight manipulation of tools and model testing or modification during centrifuge rotation. The China Institute of Water Resources and Hydropower Research (IWHR) also installed a CTS recently on its 450 g-ton Centrifuge. The centrifuge is 5.03m in radius at the basket platform and generates a maximum acceleration of 300g[Y. Du et al 1994]. Several important high dams have been modelled with this big geotechnical centrifuge over the past decade.

The first centrifuge robot was installed by ACTIDYN in 1980 at Laboratoire Central des Ponts et Chaussées (LCPC), in Nantes, [Derks et al, 1998]. The robot can be operated in a centrifuge basket up to about 100g and can move tools in an (X, Y, Z) Cartesian referential, as well generate a rotation movement around the axis orthogonal to the basket platform (Z axis). It can be used for the simulation of pile driving, excavation and refill. Some testing tools such as a cone penetrometer (CPT) or shear vane can be adopted for dynamic testing of foundation models.

HKUST developed its own four-axis robotic manipulator to serve as a platform for the in-flight simulation of geotechnical activities such as excavation, loading, CPT, and surface scanning [Ng et al, 2002]. The structure of the robotic manipulator consists of linear and rotary driving mechanisms, a tool changer with multiple tool adapters, and a motion controller. The total weight of the robot is about 296kg and can be operated up to a maximum centrifuge acceleration of 50g.

This 4-axis robotic manipulator offers a powerful tool for the simulation of a variety of complex tasks at an elevated g field. The tasks include excavation and/or fill construction, application of pre-defined horizontal, vertical and torsional loads onto the soil, insertion of measurement probes into the soil model, soil characterization using a CPT at pre-defined locations in the soil model, and carriage of non-contacting displacement sensors for automated scanning measurement of topography of the soil surface. The robotic manipulator has been successfully used on the HKUST centrifuge to conduct laterally loaded and torsionally loaded single pile in sand [Kong & Zhang, 2006]. Tests in both loose and dense sands were carried out.

The ACTIDYN in-flight four-axis robot at RPI is designed to operate at accelerations up to 100g [Ubilla et al, 2006]. It is supported inside the centrifuge basket directly above the model and remains in place independently from the model container with a maximum stroke of 0.8m in the X direction, 0.8m in the Y direction and 0.5m in the Z direction orthogonally to the basket platform. The loading capacity is 1kN in both X and Y directions, and 5kN long the Z axis. The tool holder can generate rotations around the Z axis with a maximum amplitude of 270 degrees. The robot is designed to conduct in-flight model constructions, ground excavations, simulations of contaminants penetration, pile driving, static cone penetrations and vane in situ testing.

2. IWHR CONTROLLED TOOLING SYSTEM

2.1 Specifications

The new CTS at IWHR incorporates all the features and performances of former centrifuge robots and was specifically designed to fit the IWHR centrifuge (see Figure1). Its main purpose will revolve around the simulation of dam or embankment construction and testing. The CTS is designed to operate at centrifuge accelerations from 10g to 120g, and to carry water or other fluids to the model via the tool head. Two other hydraulic circuits bring oil and air to the tool head for tool clamping and optimal and reliable in-flight manipulation of tools. The CTS weighs about 260kg, measures 1.5m in length, 1.0m in width and 0.9m in height, excluding the container on which it operates. The maximum operating pressure is 12 MPa for the water hydraulic network and 1 MPa for the oil. The main characteristics of the CTS at IWHR CTS are summarized in Table 1. The total displacement of the CTS is limited by the size of the existing swing basket and can be extended for larger centrifuge basket.

Table 1: Specifications of IWHR CTS

X axis		
Total displacement	0.7	m
Velocity max	0.07	m/s
Axis resolution repeatability	1×10^{-4}	m
Force	1	kN
Y axis		
Total displacement	0.45	m
Velocity max	0.07	m/s
Axis resolution repeatability	1×10^{-4}	m
Force	1	kN
Z axis		
Total displacement	0.4	m
Velocity max	0.04	m/s
Axis resolution repeatability	1×10^{-4}	m
Force	5	kN
Z axis Rotating		
Total displacement	± 145	°
Velocity max	5	°/s
Torque	1	Nm

The IWHR CTS consists of the control system, hydraulic station, data acquisition system and the robot body and tools. On board cameras and lights are also installed. The CTS frame consists of 2 main beams, which are supported from the platform lateral supports on top of the container. All other mechanical elements, used for translation along the X, Y, Z axis and the rotation along z are assembled to these beams. The X and Y translation mechanisms are similar in nature. Each of them consists of two linear guiding rails, four sliding skids and a dual set of ball screw nut assembly, powered by an AC brushless motor through a set of timing belt and pulleys. The output signal, of directly driven resolvers attached to the drive motor is used to control the translation axis position. The left right motion synchronization results from the use of synchronized timing belt. Failsafe electrical brakes lock the moving parts in position when the system is out of control or the motors are stopped.

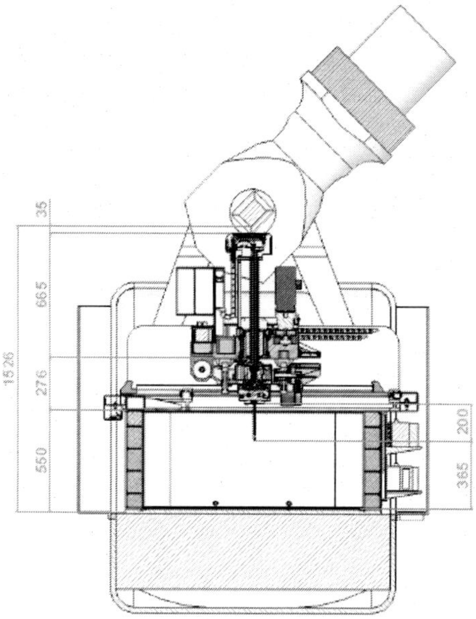

Side view

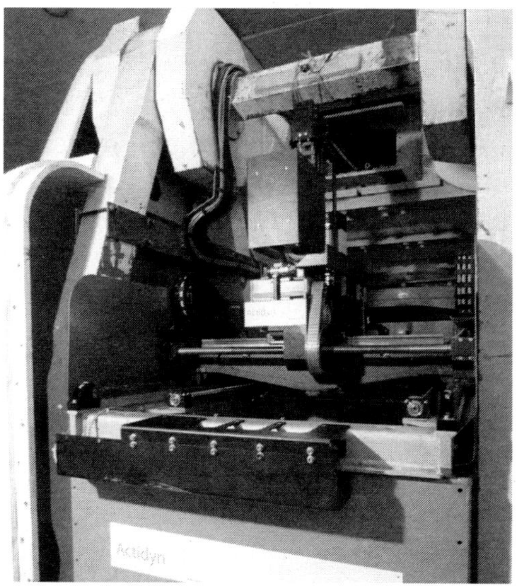

On board CTS

FIGURE 1: IWHR CTS installation

2.2 Sand Rainer and Testing Tools

The rain sand device consists of a sand reservoir and a sand rainer as shown in Figure 2 (a) and (b). It is designed to permit sand embankment construction which mechanical characteristics can be automatically controlled with good reproducibility. The sand rainer is attached to standard tool head, and it makes use of the electrical connector for controls and sensing functions. The sand rainer is filled in flight from a sand reservoir that may contain up to 20 kg of sand. The volume of the sand rainer is $0.001m^3$. It is 0.15m in width and 0.38 in total height. To satisfy the simulation of construction phases, the height of the rain fall, the speed of displacement along the X and the Y are controllable as well as the hopper aperture opening and closure.

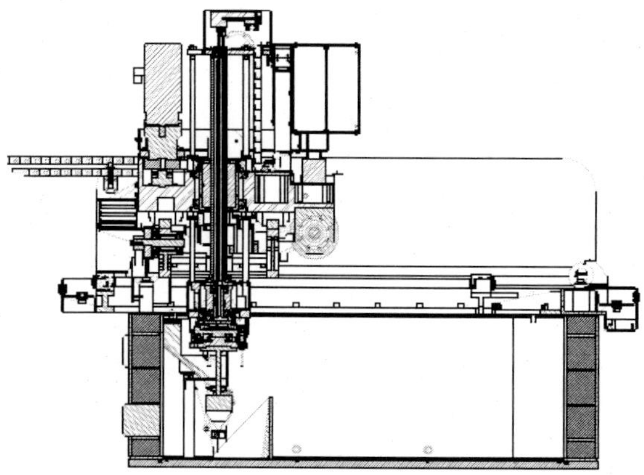

(a) Position of sand reservoir

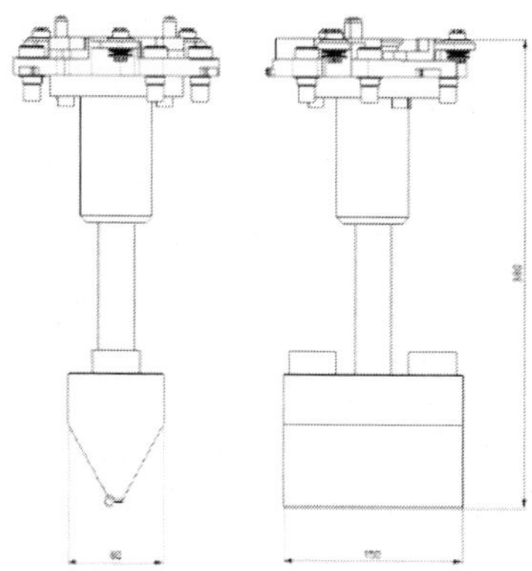

(b) Sand rainer

(c) Photo of sand rainer

FIGURE 2: Rain sand device

The sand rainer can be re-stored in the tool magazine and replaced by the cone penetrometer (Figure 3 (a)) or shear vane (Fig.3 (b)) and then the tools can be switched again to simulate a coordinated soil embankment construction. The tool magazine accessible by remote control during the centrifuge operation can store three tools. Tools can be stored and retrieved in any order and at any time during the centrifuge rotation and then positioned in the CTS operating volume with a repeatability of 0.1 mm.

The cone penetrometer (CPT) represented in the Figure 3, includes a penetrometer tip, a force sensor for tip resistance measurement, a rod, and the force signal conditioner. The tool is attached to standard head, and it makes use of the electrical connector for controls and sensing functions. The rod is driven into the soil by the translation motion of the Z axis of the CTS. The stroke is controllable in both force and displacement.

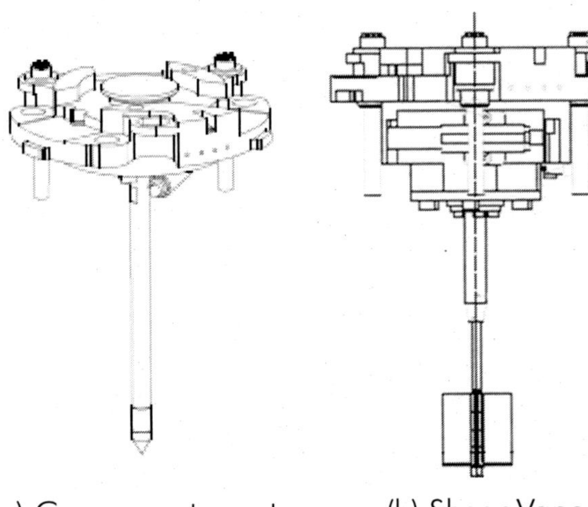

(a) Cone penetrometer (b) Shear Vane

FIGURE 3: Testing tools

2.3 Remote Control System and Software

The monitoring and control system is built around an industrial grade programmable controller. The system permits control of the CTS from a PC located either in the centrifuge control room or at the close vicinity of the CTS on centrifuge through Ethernet or fiber optical rings. The on board digital controller executes all control and surveillance functions while executing orders received via the Ethernet network. A personal computer is used as a human and network interface. Individual industrial grade solid state amplifiers servo loop controllers deliver the power command to the torque motors.

System communication is made through a dedicated Ethernet network that connects hardware elements located in the control room, on the centrifuge arm, in the central cabinet near axis and in the centrifuge basket. The hardware in control room consists of the operator PC, which is directly connected to the centrifuge local Ethernet network. The intelligent portion of the control system is located close to the central section of the centrifuge arm in a reinforced cabinet. The intelligent device controls the loops, limits the axis dynamics and prevents local collision. This unit includes hardware and software, fixed and programmable limits which fully protects the equipment's under hardware failure or operator mistakes. The communication with the control station is made through an Ethernet hub connected by optical fibers through an optical rotary joint. It is important to provide ventilation to the driver units attached to the container.

The display and communication panels are developed using Visual Basic software that communicates directly with the CTS controller (Figure 4). The user-friendly interface also display of all axis position and motion parameters, rate and acceleration as well as a 3D CTS representation. The PC software is also used to make trajectory estimates to avoid collisions with the centrifuge environment, the soil container and the model. It is also convenient to operate the CTS at view to verify its operational programming prior to a new test.

There are two operating modes available, the manual mode and automatic mode. The operator can command each axis by manual entry of discrete commands while the CTS is under visual observation. And in automatic mode, the CTS execute a test sequence created by the application software. The sequence may consist of a chain of repeated tasks such as automatic entry of origin kiting retrieving and sorting tools in and out of the storage compartment, operation of the pincer tool for opening, closing and transporting.

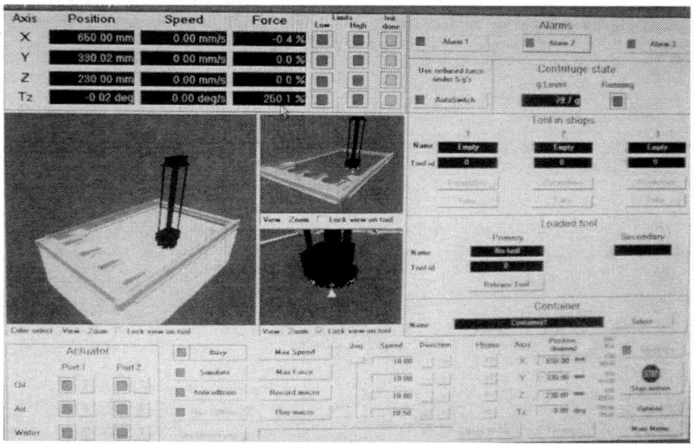

FIGURE 4: CTS operator control panel

The control software has the functions of axis position command generator, position limits survey and override protection, real time computation and security check. The CTS is due to operate in a geometrically limited environment in manual and in automatic mode. In both cases the control software takes into account the multi-axis interactive displacement limit constraints on all three axis. In the automatic mode of operation the system defines the non-optimal anti-collision trajectory for a given displacement. In manual mode it may let the operator select a preferred trajectory.

The control software is also designed to adapt to alteration of the available moving space geometry, within the firmware fixed limits, either in automatic or semi-automatic mode. Spatial displacement limitation may result from construction of models in the container or modification of such models by the ambient condition (slope failure or large deformation). Nonetheless the CTS is also used to modify the model geometry and thus is capable of over-ridding limits in pre-established area.

During the CTS operation all CTS parameters are displayed on the computer screen, as well as a three D representation of the stored model and the CTS head. The display can also accommodate a video image from an embarked video camera. The anti-collision system uses 3D representation files of the CTS, tools, container and model used during a test. These files can be generated with mechanical design software like Solidworks. To generate these files, all the 3D representations have to be referenced to the same zero point which is the point of intersection between the axis of rotation of the Z axis and the bearing surface of a tool and the tool holder.

2.4 Water Re-circulation System

The water re-circulation system in centrifuge model test is developed for the simulation of reservoir water level variation both upstream and downstream of model dam or embankment under seepage or slope failure. Water can be supplied through the hydraulic rotary joint and stored in the upper reservoir. During the model test, water may flow into the lower reservoir after seepage through model dam, and then pumped into the upper reservoir again after filtering. The water level variation can be detected by water pressure sensors and the pump can be remotely controlled automatically or manually.

The container with re-circulation pump mainly consists of a water pump for which the electrical motor is powered using a Gefran ADV200 power drive. The power drive is useful to select the operating speed of the motor which define the instantaneous flow of the water pump. The pump is a BC 25-250/110 type rated for a nominal speed of 2900 rpm

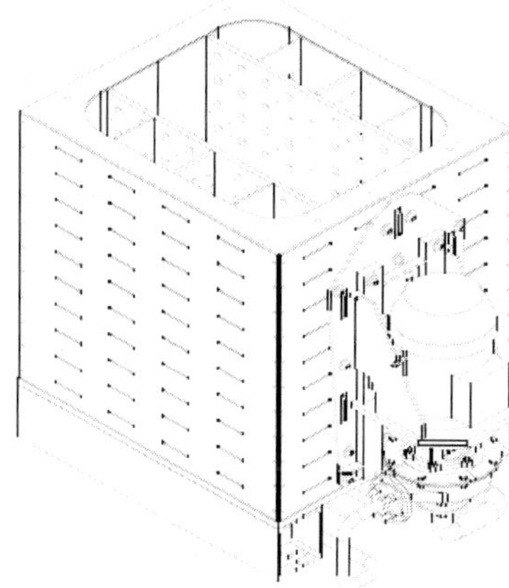

FIGURE 5: The container with re-circulation pump

The lower reservoirs are connected with a pipe at bottom. When the water from upper reservoir slowly flows through the earth dam and into the lower reservoir and reach a certain water level, the re-circulation pump will be triggered to operate and lift the water into the upper reservoir. The maximum water levels variation between upstream and downstream of the model reaches 0.5m under centrifuge acceleration of 100g. The inner size of the container for model construction is 1.0m × 0.4m × 1.0m (L × W × H).

3. SUMMARY

The controlled tools system is installed for China IWHR 450g-ton geotechnical centrifuge and its main features are described in this paper. The IWHR CTS has an in-flight rain sand device which enable the possibility of embankment construction simulation in centrifuge, apart from other functions such pile driving, excavation, refill and dynamic testing. The CTS can be operated under centrifuge acceleration of 120g with a large model container to support the CTS frame. There are three tool magazines for holding different tools for construction simulation or model tests. A water re-circulation system is also developed for the simulation of water level changes in reservoir, which is useful in analysing the seepage or overtopping failure of earth dams.

ACKNOWLEDGEMENTS

The project is founded by China Ministry of Water Resources with the approval document SGJ[2009]17 and supported by National Natural Science Fund with project No. 51078352. The authors wish to acknowledge the assistance of all laboratory technicians and the effort of Actidyn engineers.

REFERENCES

Derks, F., Merliot, E. Garnior, J and Cottineau, L.M., (1998). "On board remote controlled centrifuge robot", Centrifuge 98, Tokyo, Balkema, pp. 97-102.

Du, Y. and Zhu, S. (1994). "LXJ-4-450 Geotechnical centrifuge in Beijing", Centrifuge 94, Singapore, Balkema, pp. 35-40.

Kong, L.G. and Zhang, L.M. (2006). "Use of a robotic manipulator to conduct pile tests in centrifuge", International conference on Physical Modelling in Geotechnics, Balkema, pp. 199-204.

Ng C.W.W., Van Laak P.A. and Zhang, L.M. (2002). "Development of a four-axis robotic manipulator for centrifuge modeling at HKUST", Proc. Int. Conf. on Physical Modelling in Geotechnics, Newfoundland, Canada, pp. 71-76.

Ubilla, J., Abdoum, T. and Zimmie, T. (2006). "Application of in-flight robot in centrifuge modeling of laterally loaded", International conference on Physical Modelling in Geotechnics, Balkema, pp. 259-264.

Analytical Model for Stress-Strain Response of Plastic Waste Mixed Soil

Sandeep K. Chouksey and G.L. Sivakumar Babu

Department of Civil Engineering, Indian Institute of Science, Bangalore

E-mail: sandeepkc@civil.iisc.ernet.in; gls@civil.iisc.ernet.in

ABSTRACT

Recycling plastic water bottles has become one of the major challenges world wide. The present study provides an approach for the use plastic waste as reinforcement material in soil, which can be used for ground improvement, sub-bases and subgrade preparation in road construction. The experimental results are presented in the form of stress-strain-pore water pressure response and compression paths. Based on experimental test results, it is observed that the strength of soil is improved and compressibility reduced significantly with addition of a small percentage of plastic waste to the soil. In this paper, an analytical model is proposed to evaluate the response of plastic waste mixed soil. It is noted that model captures the stress-strain and pore water pressure response of all percentages of plastic waste adequately. The paper also provides a comparative study of failure stress obtained from different published models and proposed model which are compared with experimental results. The improvement in strength due to inclusion of plastic waste can be advantageously used in ground improvement projects.

1. INTRODUCTION

According to the international bottled water association, sales of bottled water have increased by 500 percent over the last decade: 1.5 million tons of plastic are used to bottle water every year, unfortunately the recycling process is messy and inefficient. Plastic bottle recycling has not kept pace with the continuously increase in virgin resin polyethylene terephthalate (PET) sales, particularly for PET bottles. It's interesting to note that the rate of consumption of plastic water reaches to 800 bottles per second and on the other only 100 bottles are recycled. Fig. 1(a) shows a typical group of the water bottles available in market used for the drinking purpose. Fig. 1(b) shows a typical collection of plastic bottles present in garbage.

This study presents simple way of recycling plastic water bottles in the field of civil engineering as reinforcing material. Reinforced soil construction is an effective and reliable technique for improving the strength and stability of soils. The technique is used in a variety of applications, ranging from retaining structures and embankments to subgrade stabilization beneath footings and pavements. Several researchers in literature presented the technique of using discrete fibers to enhance the strength of soil. Most of them used different types of fibers as reinforcing materials, such as natural fibers, glass fibers, plastic fibers polypropylene and polyester fibers. Experimental results reported by various researchers (Gray and Ohashi, 1983; Gray and A1-Refeai, 1986; Gray and Maher, 1989; Maher and Gray, 1990; AI-Refeai, 1991, Babu and Vasudevan 2008 a, b; Babu and Chouksey 2010) showed that the fiber reinforced soil is a potential composite material which can be advantageously employed in improving the structural behavior of soils. It is also noted that fiber reinforcement causes significant improvement in shear strength of soil showing higher post-peak strength compared to unreinforced soil.

Ranjan et al. (1996) conducted a series of triaxial compression tests on soil reinforced with discrete, randomly distributed fibers (both synthetic and natural) and influence of fiber characteristics (i.e. weight fraction, aspect ratio, and surface friction), soil characteristics and confining stress on shear strength of reinforced soil were analyzed and presented a mathematical model based on regression analysis of test results. Zornberg (2002) proposed a discrete approach to characterize the contribution of randomly distributed fibers to the strength improvement of soils. In the discrete approach, fiber reinforced soil is characterized as a two component material consisting of soil and fiber material. The methodology treats the fibers as discrete elements that contribute to stability by mobilizing tensile stresses along the shear plane. Michalowski and Cermak (2003) proposed a model for prediction of the failure stress in triaxial compression. They proposed a modified homogenization technique for predicting the shear parameters of fiber reinforced soil.

Fig.1(a) Fig.1(b)

FIGURE 1: Samples of water bottles

It is noted that, in the literature very few studies are available on the use of plastic mixed soil. The possible advantages of using the plastic wastes are that the plastic waste can be consumed in useful civil engineering applications. The plastic waste results in improvement of soil response in the case of roads and embankments, soil being a natural resource, the quantity of soil can be reduced. In this study, a series of consolidated undrained (CU) and one dimensional compression tests at different confining pressures were performed to determine stress-strain-pore water pressure and compression behaviour of plastic waste mixed soil and observe the influence of plastic waste on shear strength of soil with various percentages.

2. MATERIALS AND TESTING

The basic properties of soil are presented in Table 1. The soil had liquid limit of 39%, plasticity index of 26% and tests were conducted at initial water content of 17.8% and bulk density of 19.9 kN/m^3. UCC, CU and one dimensional compression tests are performed using red soil with dry unit weight of 16.9kN/m^3. Plastic water bottle wastes in the form of chips have been used as reinforcing material in the present study. The size of plastic chip used is 12mm long and 4mm in width. Plastic waste at different percentages (0 %, 0.5%, 0.75% and 1.0% by dry weight of soil) is mixed with soil and tests were conducted. To avoid segregation during sample preparation, plastic waste chips were distributed as evenly and randomly as possible throughout the soil.

Table 1: Properties of Red Soil Used in the Present Study

Properties	*Values*
Liquid limit (%)	39
Plastic limit (%)	26
Shrinkage limit (%)	20
Specific gravity (G)	2.65
Optimum moisture content (%)	17.8
Dry unit weight (kN/m^3)	16.9
Silt + clay size (%)	10

3. PROPOSED MODEL

Babu et al. (2010a, b) proposed generalized constitutive model based on critical state soil mechanics framework for the prediction of stress-strain-pore water pressure response and long term settlement for municipal solid waste. The model accounts volumetric strain due to elastic, plastic, creep and biodegradation for the predictions. Babu and Chouksey (2010) proposed a constitutive model for the prediction of stress-strain and pore water pressure response based on critical state soil mechanics framework. The model accounts fiber effect present in soil which is used for the prediction of the stress-strain and pore water pressure response of fiber reinforced soil. Based on the model proposed by Babu and

Chouksey (2010), stress ratio for unreinforced soil is expressed as:

$$\eta = q/p' \qquad ...(1)$$

where, $q = (\sigma'_1 - \sigma'_3)$ is the deviatoric stress and $p = (\sigma'_1 + 2\sigma'_3)/3$ mean effective stress.

It is observed from experiments that deviatoric stress and stress ratio are functions of plastic waste present in soil. A typical observations of stress ratio (η) versus strain (in %) presents in Fig.2 for plastic waste mixed soil at confining pressure of100 kPa.

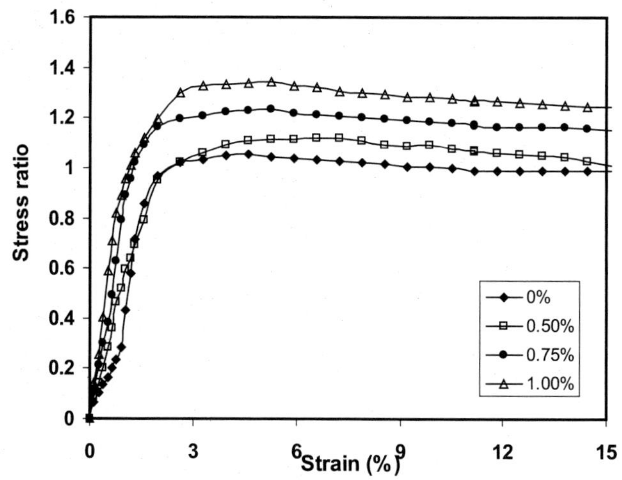

FIGURE 2: Stress ratio vs. strain response at confining pressure of (100 kPa)

The experimental observations of stress ratio with strain show that up to a certain strain level, stress ratio increases and becomes constant. Hence, to model the behavior of plastic waste mixed soil, it can be assumed that the stress ratio increases and subsequently becomes constant with increase in strain. This can be expressed as an exponential form as a function of plastic waste as follows:

$$\eta = \frac{q}{p'}\left(2 - e^{-\mu\chi}\right) \qquad ...(2)$$

where, χ is percentage of plastic waste present in soil (χ = 0.0, 0.50, 1.0 and 2.0) and μ is material constant.

$$\eta p' = q\left(2 - e^{-\mu\chi}\right) \qquad ...(3)$$

Therefore, taking differential both sides,

$$p'd\eta + \eta dp' = 2dq - \left(e^{-\mu\chi}dq - q\mu e^{-\mu\chi}d\chi\right) \qquad ...(4)$$

$$p'd\eta + \eta dp' = \left(2 - e^{-\mu\chi}\right)dq + \left(q\mu e^{-\mu\chi}d\chi\right) \qquad ...(5)$$

Let us assume that the slope of the yield curve at any point (p', q) in Fig. 10 be ψ. Since q decreases with p', the sign of ψ is negative.

$$dq = -\psi dp' \qquad ...(6)$$

Substituting value of dq in equation (5)

$$p'd\eta + \eta dp' = \left(2 - e^{-\mu\chi}\right) - \psi dp' + \left(q\mu e^{-\mu\chi}d\chi\right) \qquad ...(7)$$

$$p'd\eta + \left[\eta + \left(2 - e^{-\mu\chi}\right)\psi\right]dp' = \left(q\mu e^{-\mu\chi}d\chi\right) \qquad ...(8)$$

By rearranging Eq. (8) can be written as:

$$\frac{d\eta}{\left[\eta + \left(2 - e^{-\mu\chi}\right)\psi\right]} + \frac{dp'}{p'} = \frac{\left(q\mu e^{-\mu\chi}d\chi\right)}{\left[\eta + \left(2 - e^{-\mu\chi}\right)\psi\right]p'} \qquad ...(9)$$

The Eq. (9) defines a yield locus. Since for this model, the successive yield loci are geometrically similar, ψ is a function of η (stress ratio) only. Therefore, any yield curve passing through a known point can be obtained by integrating Eq. (9).

$$\int_0^{\eta} \frac{d\eta}{\left[\eta + \left(2 - e^{-\mu\chi}\right)\psi\right]} + \int_{p'_0}^{p'} \frac{dp'}{p'} = \int_0^{\chi} \frac{\left(q\mu e^{-\mu\chi}d\chi\right)}{\left[\eta + \left(2 - e^{-\mu\chi}\right)\psi\right]p'} \qquad (10)$$

Here in the equation (10) ψ is function of stress ratio. Now, our objective is to find the ψ in terms of stress ratio. The value of ψ is obtained by energy dissipated equation (Wood, 1990)

$$\psi = \frac{M^2 - \eta^2}{2\eta} = \frac{\left[(p'M)^2 - q^2\right]}{2p'q}$$

Multiplying numerator and denominator by $(2 - e^{-\mu\chi})^2$

$$\psi = \frac{\left[(p'M)^2 - q^2\right]}{2p'q}\frac{\left(2 - e^{-\mu\chi}\right)^2}{\left(2 - e^{-\mu\chi}\right)^2} \qquad (11)$$

On simplification of Eq. (11)

$$\psi = \frac{\left[M^2\left(2 - e^{-\mu\chi}\right)^2 - \eta^2\right]}{2\left(2 - e^{-\mu\chi}\right)\eta} \qquad (12)$$

Substituting value of ψ in Eq. (10)

$$\int_0^{\eta} \frac{d\eta}{\left[\eta + \left(2 - e^{-\mu\chi}\right)\dfrac{\left[M^2\left(2 - e^{-\mu\chi}\right)^2 - \eta^2\right]}{2\left(2 - e^{-\mu\chi}\right)\eta}\right]} + \int_{p'_0}^{p'} \frac{dp'}{p'} = \int_0^{\chi} \frac{\left(q\mu e^{-\chi}d\chi\right)}{\left[\eta + \left(2 - e^{-\mu\chi}\right)\dfrac{\left[M^2\left(2 - e^{-\mu\chi}\right)^2 - \eta^2\right]}{2\left(2 - e^{-\mu\chi}\right)\eta}\right]p'} \qquad ...(13)$$

On simplification of Eq. (13)

$$\ln\left\{\left(1 + \frac{q^2}{M^2 p'^2}\right)\frac{p'}{p'_0}\right\} = \frac{2q^2}{\left[M^2 p'^2 + q^2\right]}\ln\left(2 - e^{-\mu\chi}\right) \qquad (14)$$

On further simplification of Eq. (14) with expanding exponential series and assuming higher terms are neglected, final expression turns to:

$$q = Mp'\sqrt{\left\{\frac{p'_0}{p'}\exp\left[2\ln\left(2-e^{-\mu\chi}\right)\right]-1-\left[2\ln\left(2-e^{-\mu\chi}\right)\right]\right\}} \quad (15)$$

Thus, Eq. (15) represents the equation for deviatoric stress which incorporates behavior of plastic waste under different loading conditions (Chouksey, 2009). When the plastic wastes are ignored, it reduces to standard form of the modified cam clay model, i.e. when plastic waste $\chi = 0$, above equation changes to:

$$q = Mp'\sqrt{\left(\frac{p'_0}{p'}-1\right)} \quad ...(16)$$

4. RESULTS AND DISCUSSION

4.1 Model Parameters

The proposed model requires the same parameters as the modified cam clay model. These parameters are frictional constant $M = 6 \sin \phi'/(3 - \sin \phi')$, compression index ($\lambda$), recompression or swelling index (κ), p' is mean effective stress $(\sigma'_1 + 2\sigma'_3)/3$ and p'_0 is a pre-consolidation pressure. All these model parameters are obtained from standard triaxial and one dimensional compression tests. The compression and recompression indices (λ and κ) are obtained from the compression test results shown in Fig. 8. The slope of loading path in e-log p curve is given by λ and the slope of unloading path is given by κ. The summary of model parameters used for the validation of proposed model is presented in Table 2 for the plain and plastic waste mixed soils respectively. The average value of material constant (μ) used for prediction of stress-stain and pore water pressure behavior is taken as 0.52 for analysis in the present study. It can be noted from the results that, the compression parameters (λ and κ) decrease and frictional constant (M) parameter increases as percentage of plastic waste increases. The experimental results clearly point out that the plastic waste mixed soil improves stress-strain behavior. The response of the plastic waste mixed soil obtained from experiments is discussed in relation to the results obtained from analytical model in the following sections.

Table 2: Summary of Proposed Model Parameters Plastic Waste Mixed Soil

S. No.	F.c %	ϕ'	λ	κ	M
1	0.00	23.20	0.150	0.014	0.91
2	0.50	27.30	0.131	0.013	0.95
3	0.75	33.40	0.120	0.011	0.99
4	1.00	36.50	0.092	0.008	1.06

4.2 Model Predictions

To validate the proposed analytical model experimental results are compared with predicted results. In the earlier section, a generalized constitutive model is proposed which accounts

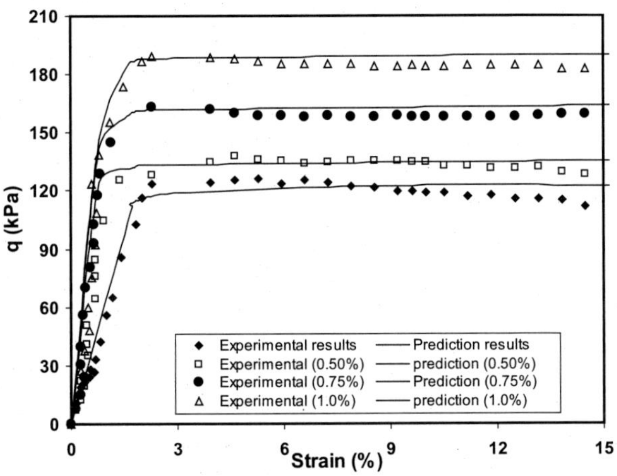

FIGURE 3(a): Comparison of experimental and prediction results for stress strain response (at confining pressure of 50 kPa)

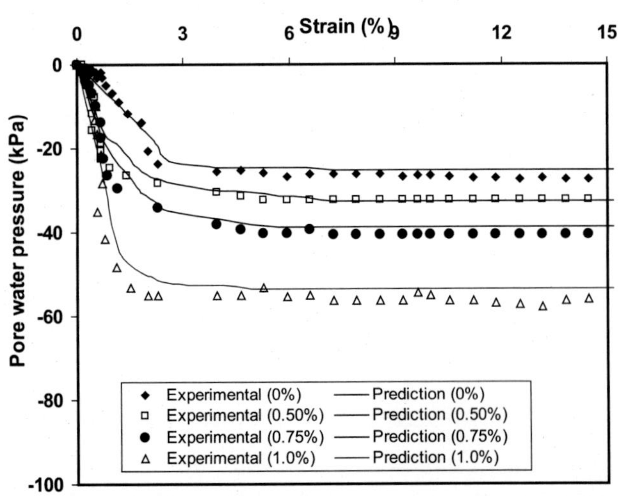

FIGURE 3(b): Comparison of experimental and prediction results for pore water pressure (at confining pressure of 50 kPa)

the effect of plastic waste present in soil, based on critical state soil mechanics framework. In the following sections, the stress-strain and pore water pressure response of plain and plastic waste mixed soil at different confining pressures are examined. Figs. 3(a), 3(b), 4(a), 4(b), 5(a) and 5(b) show the comparison of stress-strain and pore water pressure response for various percentages of plastic waste at confining pressures of 50, 100 and 150 kPa. Predictions using from the proposed model using (Eq. (16) at $\chi = 0$ for plain soil and $\chi = 0.50$, 1.0 and 2.0 for plastic waste mixed soil) with the experimental results are also shown. It can be seen that the proposed model gives reasonably good agreement with experimental results. It is clear from the results presented that the stress-strain and pore water pressure response of the plastic waste mixed soil can be well represented using the proposed model for different confining pressures. This agreement is attributed to the main feature of the proposed constitutive model that the critical state soil mechanics framework has been extended to include the effect of plastic waste. The proposed model requires material constant (μ). For the best fitting of experimental results with proposed model, material constant (μ) is taken as 0.52.

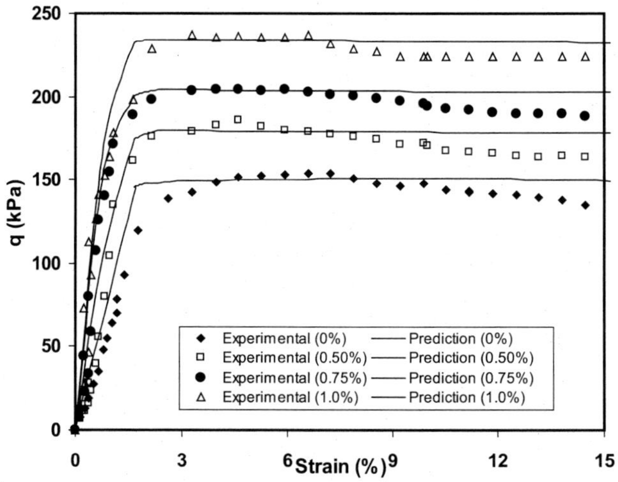

FIGURE 4(a): Comparison of experimental and prediction results for stress strain response (at confining pressure of 100 kPa)

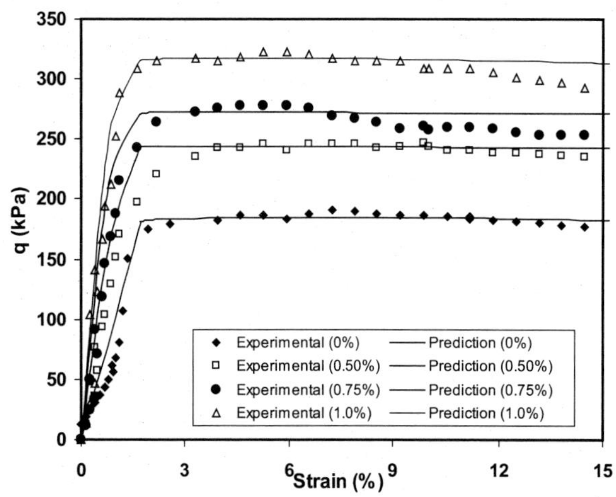

FIGURE 5(a): Comparison of experimental and prediction results for stress strain response (at confining pressure of 150 kPa)

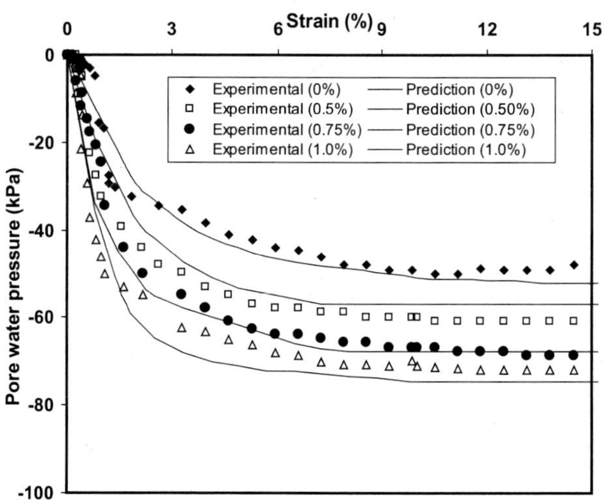

FIGURE 4(b): Comparison of experimental and prediction results for pore water pressure (at confining pressure of 100 kPa)

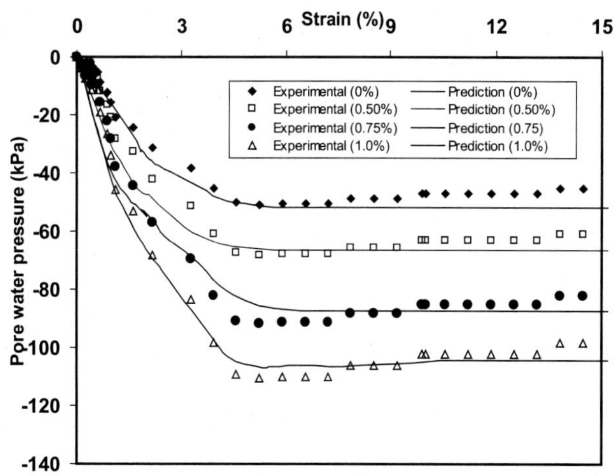

FIGURE 5(b): Comparison of experimental and prediction results for pore water pressure (at confining pressure of 150 kPa)

4.3 Comparative Study of Different Models

Ranjan et al. (1996) conducted a series of triaxial compression tests on soil reinforced with discrete, randomly distributed fibers (both synthetic and natural) and influence of fiber characteristics (i.e., weight fraction, aspect ratio, and surface friction), soil characteristics and confining stress on shear strength of reinforced soil were analyzed and presented as a mathematical model based on regression analysis of test results. Since the failure envelopes of fiber–reinforced soils were curvilinear with a transition at certain confining stress, termed as critical confining stress (σ_{crit}), when the pull out failure of the fiber governs the response and is given by (for $\sigma_3 \le \sigma_{cr}$),

$$\log(\sigma_{1f}) = 1.09 + 0.4\ \log(f_c)\ + 0.28\ \log(\frac{l}{d}) + 0.27\log(f^*)\ + 1.1\ \log(f)\ + 0.68\ \log\ (\sigma_3) \quad (17)$$

where, σ_{1f} is shear strength of fiber-reinforced soil, f_c is fiber content, l/d is aspect ratio, $f = (c/\sigma_N + \tan\phi)$, σ_3 is minor principal stress or cell pressure, $f^* = (c_a/\sigma_N + \tan\delta)$ is skin friction coefficient, c_a is adhesion intercept and σ_N is vertical stress and δ is angle of skin friction.

Zornberg (2002) proposed a discrete approach to characterize the contribution of randomly distributed fibers to the strength improvement of soils. Fiber reinforced soil fails by pulling out the fibers if the average normal stress acting on the fibers $\sigma_{n,\ avg}$ is less than the critical normal stress or other wise tension failure may occur. The critical normal stress, $\sigma_{n,\ crit}$ which defines the change in governing failure mode, is the normal stress at which failure occurs simultaneously by pull out and tensile breakage of the fibers.

$$\sigma_{n,crit} = \frac{\sigma_{f,ult} - \eta c_{i.c} . c}{\eta . c_{i.\varphi} . \tan\varphi} \quad (18a)$$

The relationships for the prediction of shear strength parameters are given by the following equations:

The equivalent shear strength of the fiber reinforced soil when failure when failure is governed by pull-out ($\sigma_{n,\ avg} < \sigma_{n,\ crit}$) can be written as

$$S_{eq.p} = S + t_p \quad (18b)$$

where, S = shear strength of unreinforced soil $(s = c + \sigma_n \tan\phi)$ and t_p is the fiber- induced distributed tension when failure is governed by pull out of individual fibers, i.e.

$$S_{eq.p} = (c + \sigma_n \tan\phi) + \alpha\chi\eta\left(c_{i,c.}c + c_{i,\phi} \tan\phi . \sigma_{n,ave}\right) \quad (18c)$$

$\sigma_{n,ave}$ = average normal stress acting on the fibers assumed to be equal to normal stress σ_n acting on the failure plane, i e.

$$S_{eq.p} = c\left[1 + \alpha\chi\eta c_{i,c}\right] + \left[1 + \alpha\chi\eta c_{i,\phi}\right]\sigma_n \tan\phi \quad (18d)$$

i.e.

$$S_{eq.p} = c_{eq.p} + (\tan\phi)_{eq.p}\,\sigma_n \quad (18e)$$

$c_{eq.p}$ = cohesive component of equivalent shear strength when $\sigma_n < \sigma_{n,crit}$

$(\tan\phi)_{eq.p}$ = frictional component of equivalent shear strength when $\sigma_n < \sigma_{n,crit}$

Equivalent cohesion,

$$c_{eqp} = c\left[1 + \alpha\chi\eta c_{ic}\right] \quad (18f)$$

Equivalent friction,

$$(\tan\phi)_{eqp} = \tan\phi\left[1 + \alpha\chi\eta c_{i\phi}\right] \quad (18g)$$

The term α is an empirical coefficient, introduced to take into account the effect of fiber orientation. For randomly oriented fibers α = 1.0.

Volumetric fiber content,

$$\chi = \frac{V_f}{V} = \frac{\chi_w \gamma_d}{(1 + \chi_w) G_f \gamma_w} \quad (18h)$$

Aspect ratio of individual fibers,

$$\eta = f_l / f_d \quad (18i)$$

Gravimetric fiber content,

$$\chi_w = f_w / W_s \quad (18j)$$

where, V_f is volume of fibers; V is total volume of fiber-reinforced soil; G_f is specific gravity of fibers; where f_w is weight of fiber; W_s is dry wt of soil; γ_d is dry unit weight of fiber reinforced soil composite; γ_w is unit weight of water, f_l is length of fiber; and f_d is equivalent diameter of a single fiber. Terms c_{ic} and $c_{i\phi}$ are interaction coefficients for the cohesive and frictional components of the interface shear strength respectively and usually taken as equal to 0.80.

$\tan \phi_{eqp}$ = frictional component of equivalent shear strength when $\sigma_n < \sigma_{n,crit}$

Equivalent cohesion,

$$c_{eqp} = c[1+\alpha\chi\eta c_{ic}] \quad (18\text{ k})$$

Equivalent friction,

$$(\tan\phi)_{eqp} = \tan\phi[1+\alpha\chi\eta c_{i\phi}] \quad (18\text{ l})$$

The term α is an empirical coefficient, introduced to take into account the effect of fiber orientation. For randomly oriented fibers α = 1.0.

Volumetric fiber content,

$$\chi = \frac{V_f}{V} = \frac{\chi_w \gamma_d}{(1+\chi_w)G_f \gamma_w} \quad (19a)$$

Aspect ratio of individual fibers,

$$\eta = f_l / f_d \quad (19b)$$

Gravimetric fiber content,

$$\chi_w = f_w / W_s \quad (20)$$

where, V_f is volume of fibers, V is total volume of fiber – reinforced soil, G_f is specific gravity of fibers, f_w is weight of fiber, W_s is dry wt of soil, γ_d is dry unit weight of fiber reinforced soil composite, γ_w is unit weight of water, f_l is length of fiber; and f_d is equivalent diameter of a single fiber.

For a confining pressure σ_3, failure stress (σ_{1f}) is given by $\sigma_{1f} = \sigma_3 \tan^2(45+\phi/2)$ and deviatoric stress (σ_d) is given by $\sigma_d = (\sigma_{1f} - \sigma_3)$.

Michalowski and Cermak (2003) conducted tri-axial compression tests on fiber-reinforced sand and proposed a model for prediction of the failure stress in triaxial compression. They proposed a modified homogenization technique for predicting the shear parameters of fiber reinforced soil. The concept of macroscopic internal frictional angle was introduced to describe the failure criteria of fiber-reinforced sand. This angle includes the influence of both the sand and fibers. Equivalent (or macroscopic) internal friction angle, which is given by,

$$\phi = 2\tan^{-1}\left(\sqrt{\frac{\rho\eta M \tan\phi_w + 6K_p}{6-\rho\eta M \tan\phi_w}}\right) - \frac{\pi}{2} \quad (21)$$

where, $M = K_p \sin \theta_0$; $K_p = \tan^2 (45 + \phi/2)$; $\theta_o = \tan^{-1}\left(\sqrt{(K_p/2)}\right)$; ϕ is sand/fiber interface friction angle and ρ is volumetric fiber content. For a confining pressure σ_3, failure stress (σ_{1f}) is given by $\sigma_{1f} = \sigma_3 \tan^2(45+\phi/2)$ and deviatoric stress (σ_d) is given by $\sigma_d = (\sigma_{1f} - \sigma_3)$

In the following sections, the experimental and predicted results of the failure stress (σ_{1f}) obtained from the present study are compared with the values obtained from Ranjan et al. (1996), Zornberg (2002) and Michalowski and Cenmark (2003). Table 3, 4 and 5 present the model parameters used for the calculation of failure stress for the published models. Table 6 shows the comparison of failure stress obtained from experiments and published models. It is observed that the failure stress calculated from experiments are in agreement with the values obtained from the model of Ranjan et al. (1996) and proposed model in the present study. However there is a divergence when the values are compared with the values obtained using the other two methods.

Table 3: Model Parameters for the Ranjan et al. (1996)

w_f	l	d	l/d	f*	σ_3
0	0	0	0	0.22	50
0.5	12	3	4	0.22	50
0.75	12	3	4	0.22	50
1	12	3	4	0.22	50
0	0	0	0	0.22	100
0.5	12	3	4	0.22	100
0.75	12	3	4	0.22	100
1	12	3	4	0.22	100
0	0	0	0	0.22	150
0.5	12	3	4	0.22	150
0.75	12	3	4	0.22	150
1	12	3	4	0.22	150

Table 4: Model Parameters for the Zornberg (2002)

$100\,X_w$	X_w	X	η	c	c_{ic}	c_{eqp}
0	0	0	0	40.19	0.8	40.19
0.5	0.005	0.00764	4	33.59	0.8	34
0.75	0.0075	0.01144	4	35.94	0.8	37
1	0.01	0.01521	4	47.51	0.8	50

Table 5: Model Parameters for the Michalowski and Cenmark (2003)

$Fc\%$	f'	$tan\ \phi_w$	η	$tan\ \alpha$	K_p	θ_0	M	ϕ'
0	23.2	0.25	0	56.6	2.29	46.90	1.67	23.20
0.5	27.3	0.30	4	58.6	2.68	48.99	2.02	27.39
0.75	33.4	0.39	4	61.7	3.44	52.64	2.73	33.54
1	36.5	0.44	4	63.2	3.92	54.4	3.19	36.67

5. CONCLUSIONS

In this paper, an approach to model stress-strain behavior of plastic waste mixed soil is presented. To investigate the effects of plastic waste mixed in strength of soil, series of triaxial compression and one dimensional consolidation tests have been performed with various percentages of plastic waste in soil at different confining pressures. The experimental results are presented in the form of stress-strain-pore water pressure response and compression paths. The conclusions that emerge from the present study are as follows:

1. The experimental results show that there is a good improvement in the strength of soil with inclusion of plastic waste. This increase in strength of soil is due to increase in friction between soil and plastic waste as results development of tensile stress in the plastic waste.
2. Based on experimental observations, a simple generalized constitutive model is proposed, which is extension of critical state soil mechanics framework. The experimental results are compared with analytical model, and the agreement is satisfactory.
3. The model parameters used for the proposed model are calculated from simple standard triaxial shear tests and one dimensional compression tests.
4. The usefulness of the model lies in capturing the complete stress strain curve of the plastic waste mixed soil where as the other models can give only failure stress.

Table 6: Values of Failure Stresses Obtained from Different Approaches

Fiber (%)	σ_3	Exp	Gopal Ranjan et al (1996)	Michalwoski et al. (2003)	Zornberg (2002)	Model Prediction
0	50	124	108.6	65.0	186.5	121
0.5	50	134	127.2	84.9	201.5	133
0.75	50	157	158.4	123.0	270.0	162
1	50	185	183.8	147.7	361.0	188
0	100	154	154.1	130.1	251.2	150
0.5	100	178	183.9	169.8	289.0	179
0.75	100	203	233.9	246.1	399.0	203
1	100	236	274.6	295.5	518.8	233
0	150	190	184.8	195.1	316.0	184
0.5	150	246	224.1	254.7	373.7	243
0.75	150	275	289.9	369.1	523.1	272
1	150	319	343.6	443.3	667.2	316

REFERENCES

Al-Refeai, T. (1991) "Behaviour of granular soils reinforced with discrete randomly oriented inclusions". Journal of Geotextile and Geomembrane, 10, Issue 4, pp 319-333.

Gray, D. H., and Maher, M. H. (1989) "Admixture stabilization of sand with discrete, randomly distributed fibers", Proc. XIIth Int. Conf. on SMFE, Rio de Janeiro, Brazil, pp 1363-6

Gray, D. H., and Al-Refeai, T. (1986) "Behavior of fabric versus fiber reinforced soil". Journal of Geotechnical Engineering, ASCE, 112, Issue8, pp 804-820.

Gray, D. H., and Ohashi, H. (1983) "Mechanics of fiber reinforcement in sand". Journal of Geotechnical Engineering, ASCE, 109, Issue 3, pp 335-353.

Maher, M. H., and Gray, D. H. (1990)"Static response of sand reinforced with randomly distributed fibers". Journal of Geotechnical Engineering, ASCE, 116, Issue 11, pp 1661-1677.

Michalowski, R. L., and Cermak, J. (2003) "Triaxial compression on sand reinforced with fibers". Journal of Geotechnical and Geoenvironmental Engineering, ASCE 129, Issue 2, pp 125-136.

Ranjan, G., Vasan, R. M., and Charan, H. D. (1996) "Probabilistic analysis of randomly distributed fiber reinforced soil". Journal of Geotechnical Engineering, ASCE, 122, Issue 6, pp 419-426.

Sivakumar Babu, G. L., Reddy, K. R., and Chouskey, S. K. (2010a) "Constitutive Model for Municipal Solid Waste Incorporating Mechanical Creep and Biodegradation-Induced Compression". Waste Management Journal, 30, Issue 1, pp 11-22.

Sivakumar Babu, G. L., Reddy, K. R., Chouskey, S. K.,

and Kulkarni, H. (2010b) “Prediction of Long-term Municipal Solid Waste Landfill Settlement Using Constitutive Model”. Practice Periodical of Hazardous, Toxic, and Radioactive Waste Management, ASCE, 14, Issue 2, pp 139-150.

Sivakumar Babu, G. L., and Chouksey, S. K. (2010) “Model Analysis of fiber reinforced clayey soil”. Geomechanics and Geoengineering Journal, Taylor & Francis, 5, Issue 4, pp 277-285.

Sivakumar Babu, G. L., and Vasudevan, A. K. (2008a) “Strength and stiffness response of coir reinforced tropical soils”. Journal Materials in Civil Engineering, ASCE, 20, Issue 9, pp 571-577.

Sivakumar Babu, G .L., and Vasudevan, A. K. (2008b) “Seepage velocity and piping resistance of coir fiber mixed soils”. Journal of Irrigation and Drainage Engineering, ASCE, 134, Issue 4, pp 495- 492.

Wood, D. M., (1990). Soil behavior and critical state soil mechanics. Cambridge University Press, UK.

Zornberg, J. G. (2002) “Discrete frame work for limit equilibrium analysis of fiber-reinforced soil”. Geotechnique, 52, Issue 8, pp 593-604.

Rainfall-Induced Slope Instability – Implications for Global Climate Change

H. Ling and H.I. Ling[1]

Columbia College, Columbia University, New York, NY 10027, USA
[1] Department of Civil Engineering & Engineering Mechanics, Columbia University, New York, NY 10027, USA
E-mail: Ling@civil.columbia.edu

ABSTRACT

Global warming and associated disasters have received wide social and political attention. This study focused on landslides resulting from the precipitation brought by hurricanes that may grow in intensity in the near future. A series of centrifuge simulations on a rainfall-induced landslide that occurred in Japan during 2005 Typhoon Nabi was performed. The procedures of the simulation on the slope (height=27 m, inclination=29 degrees) were illustrated. The precipitation was applied in increments to the slope surface until it exceeded that of the field measurement. The instability was examined using an infinite slope analysis and the mechanism of rainfall-induced failure was discussed. The study showed that incremental rainfall of less than 200 mm led to local failure, whereas that exceeding 200 mm resulted in global slope failure. A loss of apparent cohesion in the soil as well as an increase in pore pressure due to infiltration were responsible for the slope instability. Wide-spreading landslides, mudflows and debris flows are expected under an environment affected by global warming.

1. INTRODUCTION

It is evident from statistical records that the earth's climate is facing a gradual increase in temperature. Several studies, such as those by the National Aeronautics and Space Administration (NASA), predicted an increase in ocean surface temperature of approximately 2 degrees Celsius by the end of century. The warming temperature is expected to melt icebergs and glaciers and thus raise the sea level. Many coastal cities and low-lying areas will be affected directly by such a sea level rise. Global warming may also affect the weather in several ways. One of the prime concerns is about the occurrence of hurricanes and typhoons (e.g., Landsea, 2005; Webster et al., 2005; Sauder and Lea, 2008; Walsh et al., 2009). The high wind speed may cause direct damages and surges at the coastal area, and hurricanes may also bring heavy rainfall which will lead to flooding.

Precipitation from rainfall or melting ice may lead to wide-spreading landslides. Many slopes have failed under heavy rainfall due to tropical depressions, tropical storms and hurricanes. Hundreds of lives and pieces of property are lost every year due to landslides (e.g., Schuster and Highland, 2001; Sidle and Ochiai, 2006). Landslides may cut off highway infrastructure, utilities, food supply, and communication networks for an extended period of time. Residences may be destroyed and inhabitants located at the landslides areas may be killed. Most recent landslide disasters include the 2009 Typhoon Morakot which buried entire Hsao-Lin Village in Taiwan and caused approximately 500 deaths, and the heavy rain of August 2010 that killed more than 14000 people in Zhouqu County in the Gansu Province, China.

Current classification of hurricanes under the Saffir-Simpson scale is solely based on the wind speed (categories and wind speed in km/h: 1=119-153, 2=154-177, 3=178-209; 4= 210-249, 5> 250) whereas rainfall characteristics, such as intensity and accumulation, are not considered. Through the information on the typhoons that brought considerable damages to Japan since 1989 (see

Fig. 1), there was a large variation in intensities and durations; the intensities varied between 40 and 150 mm/h (mean= 88 mm/h), whereas the total accumulations were between 120 and 1800 mm (mean= 645 mm). Recent numerical predictions by National Oceanic and Atmospheric Administration (NOAA) and Japan Meteorological Agency (JMA) have indicated a possible increase in the frequency of occurrences and the intensities of typhoons, such as the super-typhoon that has a wind speed exceeding 240 km/h, because of the increase in the ocean surface temperature. The precipitation is expected to increase by approximately 20 percent within 100 km of the storm center (e.g., Knutson and Tuleya, 2004; Oouchi et al., 2006).

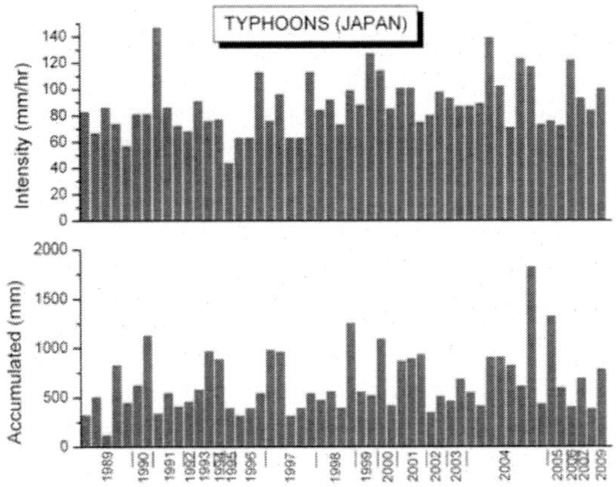

FIGURE 1: Rainfall characteristics of destructive typhoons in Japan since 1989 (compiled after data obtained from Japan Meteorological Agency)

This study presents a centrifugal simulation of rainfall-induced landslide due to typhoon. A centrifuge modeling technique was used. While previous studies (Ling et al., 2009; Ling et al., 2010) were on hypothetical slopes, this study attempted to simulate a real slope failure. The procedures of simulation and model construction are illustrated. Several different incremental rainfall accumulations were used in the testing. The instability was examined using infinite slope analysis and the mechanism of rainfall-induced failure was given.

2. TYPHOON NABI AND LANDSLIDE

There are very limited case histories of rainfall-induced landslides where the soil properties and rainfall characteristics are well known, allowing for validation of physical or numerical modeling technique. Typhoon Nabi (Typhoon #14), a super-typhoon, passed through the western part of Japan during 6-7 September, 2005. The maximum wind speed was 175 km/h with the central pressure depression of 925 hPa. Typhoon Nabi caused 21 deaths and 143 injuries in Japan. In addition, over 10,000 homes and 88 roads were damaged. There was a total of 168 landslides in the affected area (Sahera et al., 2005).

This study was related to a landslide along the Sangyo Expressway in the Yamaguchi Prefecture, located between the Kuga and Iwakuni interchanges. A slope, 27 m high and of inclination 29 degrees, failed 7 hours after the peak rainfall. The rainfall records at the nearby Kuga and Iwakuni stations had a total accumulation of 437 mm 341 mm, respectively, and the peak rainfall intensities varied between 40-50 mm/h . Note that the site was not the area that received the highest rainfall intensity nor the total accumulation during typhoon Nabi. The landslide buried several houses at the foot of the mountain and killed 3 residents. The traffic was interrupted for more than 3 months.

The slope of interest had a layer of cover soil underlaid by weathered rocks. Murata et al. (Murata et al., 2009) presented a series of results on the soil investigation conducted after the landslide. There was a significant variation in the grain size distribution among the samples retrieved from the site. The fines content was between 10 to 40 percent by weight.

3. CENTRIFUGE MODELING PROCEDURES

Considering the capacity of the centrifuge facilities, the model was prepared for testing at 100-g; thus the height of the slope model was 27 cm. The model was housed in a rigid aluminum box with inner dimensions of 94 cm × 39.5 cm × 38 cm (length × width × height). Figure 2 shows the configurations of the model. In constructing the model, the slope geometry was first traced onto the walls of the box. The bottom weathered rock was simulated using concrete cement. A sand-clay mixture (Nevada sand and kaolinite) was prepared with a fines content of 30 percent to simulate the cover soil. A series of compaction tests were conducted on the sand-clay mixtures that yielded a maximum dry density of 1.94 g/cm^3 at a water content of 8.9 percents (total density was 2.2 g/cm^3).

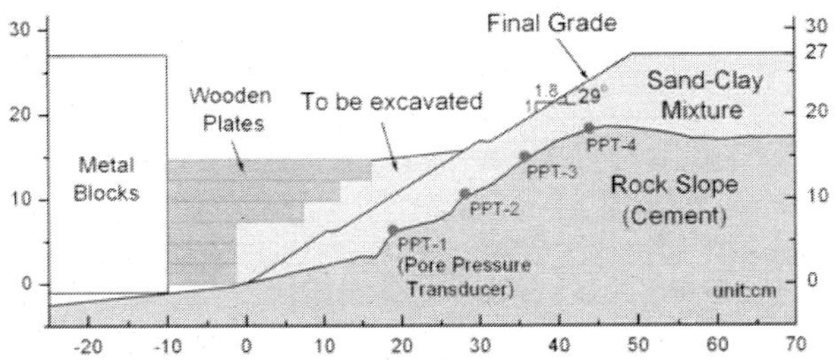

FIGURE 2: Configuration and preparation of slope model used in the centrifuge testing

The cover soil was compacted in 11 horizontal layers at maximum density. The required volume of soil with the optimum water content was used to compact each soil layer based on the desired density. Wooden plates were placed in the front end of the slope in order to minimize the volume of soil used in construction. Four pore pressure transducers were embedded at different locations in the slope in order to observe the change in pore pressure during rainfall infiltration. After attaining the full height, the wooden plates were removed and the slope was cut to its final grade of 1(V):1.8(H) or 29 degrees. The finished slope was covered with plastic sheets and left overnight before rainfall testing was conducted.

A simplistic approach was adopted where the precipitation was introduced to the slope by spraying water prior to centrifugal acceleration. Note that in Ling et al. (2009, 2010), the rainfall was introduced during centrifuge flight using spraying nozzles. For each test, different values of precipitation ranging from 50 mm to 200 mm were applied in two or more increments (Table 1). In this model, based on the total area of the slope surface, 240 cm^3 of water was equivalent to 100 mm of rainfall in the field. The centrifuge took about 5 minutes to attain the targeted 100g, after which it was kept for another 5 minutes before the centrifuge was brought to a stop in 5 minutes. Then, the next increment of precipitation was applied to the slope and the process was repeated until the total accumulation exceeded that of the field measurement.

In order to prevent the drying up of the slope surface by wind currents during centrifuge spinning, the box containing the slope was kept air tight with a cover on. Visual observation of the slope deformation and failure was done through a closed circuit camera. The response of the pore pressure transducers was also monitored through a wireless data acquisition system.

4. TEST RESULTS

As seen in Table 1, for the tests where precipitation was less than 150 mm in each increment (Tests 1, 2 and 3), only local failure of the slope was observed. That is, no global failure was observed even when the total accumulation exceeded that of the field measurements of 500 mm. The local failure typically occurred at the toe of the slope. Cracks are formed at the crest. When the increment was increased to 200 mm (Tests 4 and 5), local failure occurred in the first increment followed by a global failure in the second increment. Typical views of the slopes at failure, indicating both local and global failures, are shown in Fig. 3.

Table 1: Rainfall Testing Program and Results

Test #	*Rainfall Increment (mm)*	*Total Increments*	*Accumulated Rainfall (mm)*	*Failure Increment #*	
				Local	*Global*
1	50 [120 cm^3]	11	550	4	No
2	100 [240 cm^3]	7	700	2	No
3	150 [360 cm^3]	4	600	2	No
4	200 [480 cm^3]	2	400	1	2
5	200 [480 cm^3]	3*	500	1	2

*Third increment = 100 mm

Figure 4 shows a typical response of the pore pressure transducers for Test 5 recorded from the time of centrifuge spinning up. The pore pressure started to increase as the centrifuge was spun up.

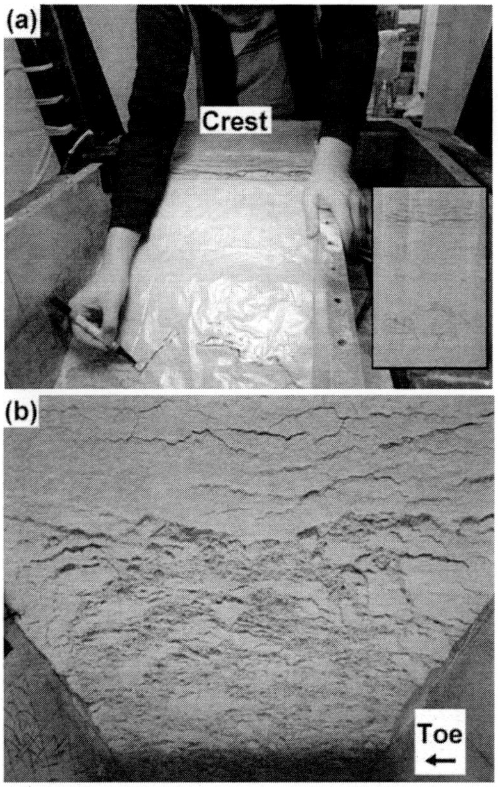

FIGURE 3: View of the slope at failure (Test 5): (a) local failure at the toe and cracks at the crest after first increment, (b) Global failure as viewed from the crest

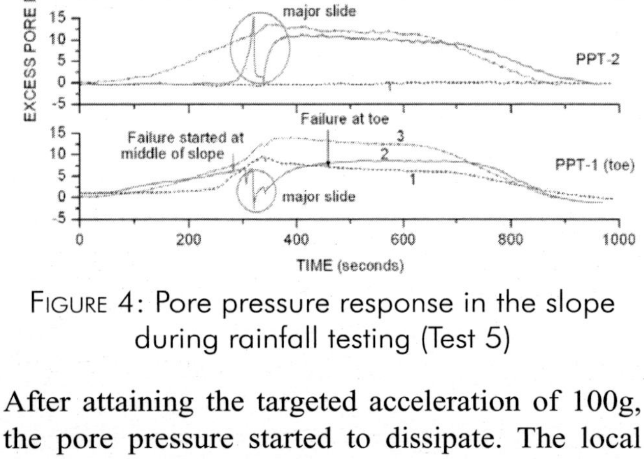

FIGURE 4: Pore pressure response in the slope during rainfall testing (Test 5)

After attaining the targeted acceleration of 100g, the pore pressure started to dissipate. The local or global failure occurred right after the slope attained peak acceleration. The two transducers located at the upper portion of slope did not record significant change in pore pressure. It was due to the fact that the moisture did not infiltrate to a depth where the transducers were located. Rather, the flow occurred in the lower portions of the slope as indicated by the response of the transducers. At the instant of failure, there was a sharp drop in pore pressure, after which the reading increased again.

5. MECHANISM OF INSTABILITY

In previous studies (Ling et al., 2009) it was shown that the modified Coulomb failure criteria is applicable to partially saturated soil:

$$\tau_f = c + \sigma_n' \tan\phi \qquad (1)$$

where τ_f, σ_n', c, and ϕ are the shear strength, effective normal stress, "apparent" cohesion, and angle of internal friction, respectively. The partially saturated soil possesses a small value of apparent cohesion due to the capillary action between the soil particles. Upon saturation, the apparent cohesion diminishes. Coupled with an increase in pore pressure, u, the effective stress, which is the difference between the total normal stress and pore pressure, is reduced ($\sigma_n' = \sigma_n - u$). Thus, the shear strength of the soil as a whole becomes smaller than that of its unsaturated state. This mechanism of failure was associated with initially unstable steep slopes of 60 degrees or greater, whereas this study dealt with rather gentle slopes of 29 degrees.

In order to validate this concept further, a series of direct shear tests was conducted on the sand-clay mixture using different percentages of moisture contents: 10%, 13%, and 14.4%. These values corresponded to a degree of saturation of 53.8% (slope as constructed), 72.5% (after 200 mm of precipitation), and 79.9% (after 400 mm of precipitation). The soil specimens were sheared rapidly at a displacement rate of 2 cm in 3 minutes. The results presented in Fig. 5 show that the soil at 10% moisture possessed an apparent cohesion of 25.6 kPa. The apparent cohesion reduced with an increase in moisture content, and it almost vanished when the moisture content reached 14.4% or 80% degree of saturation. The variation in the angle of friction with moisture was rather small.

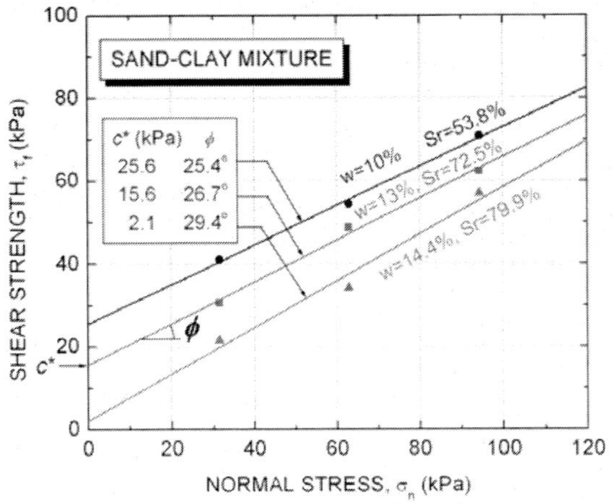

FIGURE 5: Effects of moisture content on the shear strength

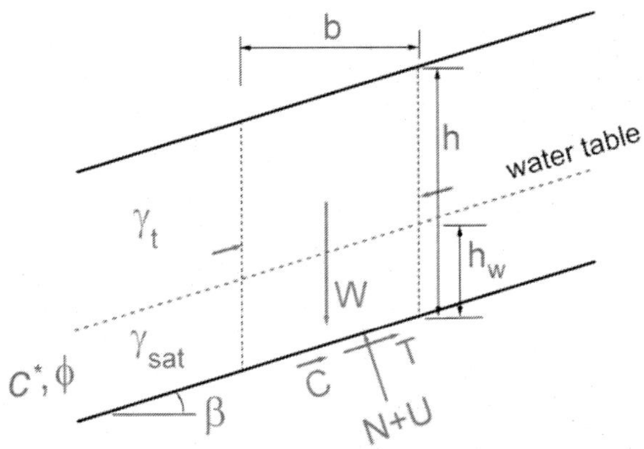

FIGURE 6: Infinite slope stability in the presence of water table (Ling and Ling, 2012)

An analysis was conducted to examine the stability of the slope using the test results obtained from the direct shear and centrifuge testings. The analysis considered an infinite slope (Fig. 6), since the slope was much longer than its thickness. The general equation to evaluate the factor of safety for the slope in the presence of apparent cohesion and water table is derived and given below in convenient form, where the first part is attributed to the apparent cohesion *c* and the second part due to the angle of friction ϕ (derivation of equation is found in Ling and Ling, 2012):

$$F_s = \frac{2c}{\{\gamma_t(1-\alpha)+\gamma_{sat}\alpha\}h\sin 2\beta} + \frac{\gamma_t(1-\alpha)+(\gamma_{sat}-\gamma_w)\alpha}{\gamma_t(1-\alpha)+\gamma_{sat}\alpha}\frac{\tan\phi}{\tan\beta} \quad (2)$$

where γ_t, γ_{sat}, γ_w, β, h and h_w are the total unit weight, saturated unit weights of soil, unit weight of water, slope angle, thickness of soil layer and depth of water, respectively ($\alpha = h_w/h$). Based on the measured pore pressure in the centrifuge test, an equivalent height of the water table h_w was obtained. The average value of the angle of internal friction of 26.2° was used. The values of these parameters are summarized in Fig. 7.

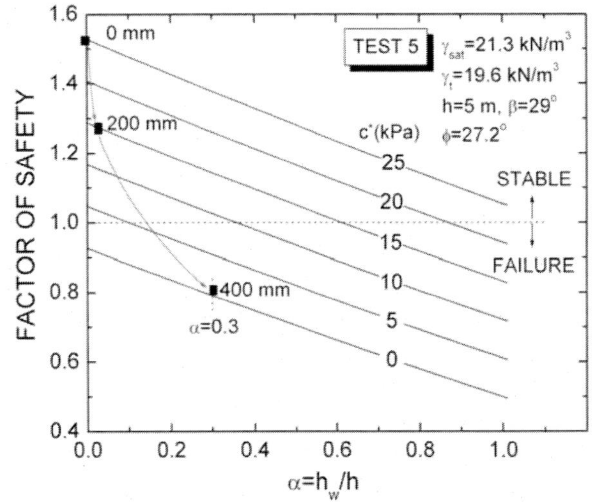

FIGURE 7: Effects of water table and apparent cohesion on slope stability

As can be seen in Fig. 7, a reduction in apparent cohesion and an increase in water table led to a reduction in the factor of safety. The factor of safety of the slope without rainfall was 1.52, indicating a stable slope. It reduced to 1.28 at 200 mm of precipitation, but dropped to 0.79 after 400 mm of precipitation that indicated an unstable slope. Thus, a reduction in apparent cohesion coupled by an increase in pore pressure due to the rising water table contributed to the failure of this particular slope.

6. CONCLUSIONS

The results of simulation showed that centrifuge modeling technique can be used to simulate full-scale slope failure induced by rainfall. An increased saturation due to infiltration led to a reduction in apparent cohesion and effective stress, thus a loss of shear strength. Relatively gentile slopes may become unstable in the events of heavy rainfall. Thus, under increasing amounts of precipitation during hurricanes in an environment affected by global warming, more landslides are expected.

ACKNOWLEDGEMENTS

The authors appreciated the communications with Dr. Yukio Nakata of Yamaguchi University, Japan. The assistance of Dr. Liming Li is also very much appreciated. The technical information on the Japanese typhoons was obtained through the JMA website.

REFERENCES

Knutson, T.R. and Tuleya, R.E., (2004). "Impact of CO2-induced warming on simulated hurricanes intensity and precipitation: sensitivity to the choice of climate model and convective parameterization." Journal of Climate, 17, Issue 8, pp 3477-3495.

Landsea, C.W., (2005). "Hurricanes and global warming". Nature 438, E11-E13.

Ling, H.I., Wu, M-H., Leshchinsky, D., and Leshchinsky, B., (2009). "Centrifuge modeling of slope instability". Journal of Geotechnical and Geoenvironmental Engineering, ASCE, 135, Issue 6, pp. 758-767.

Ling, H., Ling, H.I., Li, L., and Kawabata, T., (2010). "Centrifuge modeling of slope failures induced by rainfall". Proceedings of 7th International Conference on Physical Modelling in Geotechnics, London, pp 1131-1136.

Ling, H. and Ling, H. I., (2012). "Centrifuge model simulations of rainfall-induced slope instability." Journal of Geotechnical and Geoenvironmental Engineering, 138, Issue 9, pp1151-1157.

Murata, H., Takekuni, K., and Nakata, Y., (2009). "Slope failure of embankment in Sanyo Expressway due to passage of Typhoon No. 14 in 2005". Soils and Foundations, 49, Issue5, pp797-806.

Oouchi, K., Yoshimura, J., Yoshimura, H., Mizuta, R., Kusunoki, S., and Noda, A., (2006). "Tropical cyclone climatology in a global-warming climate as simulated in a 20 km-mesh global atmospheric model: frequency and wind intensity analyses." Journal of the Meteorological Society of Japan, 84, Issue2, pp259-276.

Saunders, M. A. and Lea, A. S., (2008). "Large contribution of sea surface warming to recent increase in Atlantic hurricane activity". Nature, 451, Issue 31,pp557-560.

Schuster, R.L. and Highland, L.M., (2001). Socio-economic and Environmental Impacts of Landslides in the Western Hemisphere: US Geological Survey Open-File Report 01-0276.

Sehara, Y., Suzuki, M., Yamamoto, T., Terayama, T., Tomokiyo, T., and Koichi Y., (2006). "Slope disasters caused by Typhoon No. 14 of 2005 in Yamaguchi Prefecture". Soils and Foundations, 46, Issue 6, pp817-830.

Sidle, R. C. and Ochiai, H., (2006). Landslides: Processes, Prediction, and land Use. American Geophysical Union, Washington, DC.

Walsh, K., Karoly, D., and Nicholls, N., (2009). "Detection and attribution of climate change effects on tropical cyclones." Hurricanes and Climate Change, Elsner J. B. and Jagger, T. H. (eds),pp1-20.

Webster, P. J., Holland, G. J., Curry, J. A., and Chang, H.-R., (2005). "Changes in tropical Cyclone number, duration, and intensity in a warming environment". Science 309, pp1844-1846.

Centrifuge Modelling of the Upheaval Buckling of Pipelines in Sand

S.K. Haigh and J.Wang

Department of Engineering, University of Cambridge, Cambridge, UK
E-mail: skh20@cam.ac.uk

ABSTRACT

Upheaval buckling owing to thermal expansion of buried pipelines can be a cause of large financial losses and great environmental damage. The reliable design of pipelines to resist upheaval buckling relies on the prediction of both the maximum uplift resistance that can be provided by the overlying soil and of the pipe movement that is required to mobilise this resistance. In this paper, full-scale 1g and reduced scale centrifuge models of buried pipeline segments were subjected to uplift in order to fully characterise the uplift response. It was shown that while current design methods accurately reflect the peak resistance that might be experienced by the pipeline, the pipeline movement required to mobilise this resistance is significantly greater than would be expected based on current design methods. This may potentially lead to unconservative designs.

1. INTRODUCTION

Over the past two decades, increasing oil prices have allowed the exploitation of oil and gas deposits in deeper water that were previously uneconomic. The great distance of many of these deposits from the coast has dramatically increased the length of pipeline required to deliver the products of these oilfields to onshore refining facilities.

Pipelines are often laid by the reel-lay technique, which requires thick-walled triple-layer polypropylene coated steel pipes in order to prevent local buckling during the lay process. This process is carried out at the ambient water temperature of around 4°C. During operation, crude oil is transported at approximately 140°C, resulting in expansive thermal strains of around 0.15% being experienced by the pipeline. These thermal strains must either result in longitudinal expansion of the pipeline, or induced compressive mechanical stresses within the pipe. If insufficient restraint is provided to the pipeline, this will result in buckling of the pipeline.

Surface laid pipelines can be induced to tolerate this thermal expansion by inducing long-wavelength lateral buckles into the pipeline. Offshore pipelines are, however, often buried, especially in shallow water, in order to provide thermal insulation and minimise the energy input needed to assure the smooth flow of oil within the pipeline, and to protect the pipeline from damage due to anchors and fishing activities. The preferential mode of buckling for buried pipelines is in the vertical direction, termed "upheaval buckling". The wavelength of these buckles is often very short, inducing high curvatures to the pipeline and potentially causing structural failure of the pipe. Additional problems may occur in shallow water with the buckled pipe potentially being snagged by anchors and pulled further out of the ground surface. Figure 1 shows an example of the extreme movements that can occur during the upheaval buckling process.

Upheaval buckling is resisted by the self-weight and strength of the overlying soil and design techniques are available to estimate the contribution from each of these resistances. As buckling is a stiffness-dependent phenomenon, the stiffness exhibited by the overlying soil in resisting pipe movement is also of importance. This paper will investigate the adequacy of existing design methods using both full-scale physical modelling and reduced-scale centrifuge modelling.

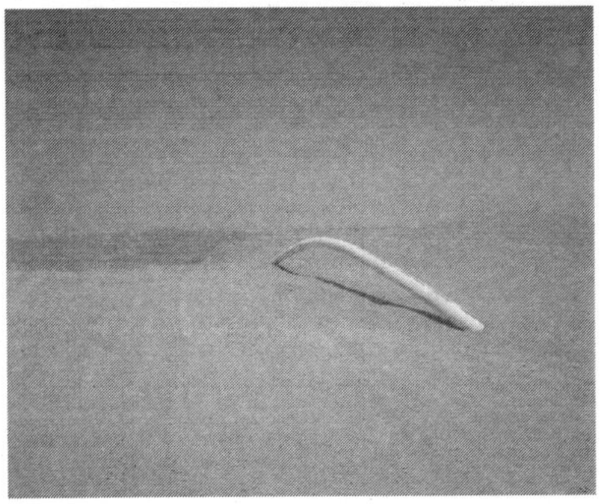

FIGURE 1: Upheaval buckling of an onshore pipeline

2. CURRENT DESIGN TECHNIQUES

Current pipeline designs against upheaval buckling largely follow the codal provisions outlined by DNV-RP-F110 (DNV, 2007). This suggests that the resistance of cohesionless soils to uplift can be captured using a tri-linear model with increasing capacity up to a maximum value, as shown in Figure 2. In order to use this tri-linear model, estimates must be made of the maximum uplift resistance R_{max} and the mobilization distance δ_f.

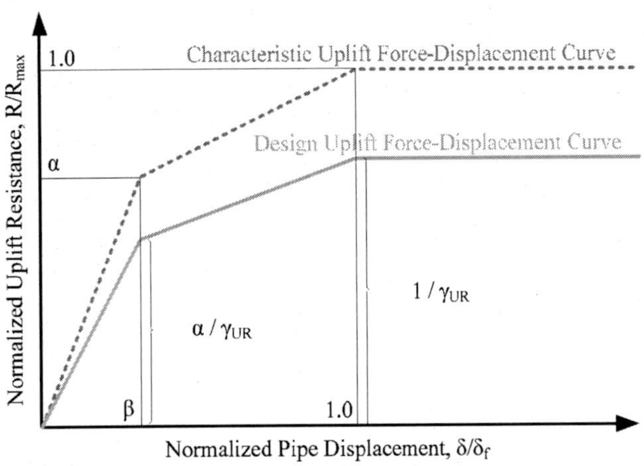

FIGURE 2: Force-displacement relationship from DNV-RP-F110

2.1 Maximum Uplift Resistance

Uplift resistance can be estimated using the simplified vertical slip surface model introduced by Trautman et al. (1985), presented in Figure 3.

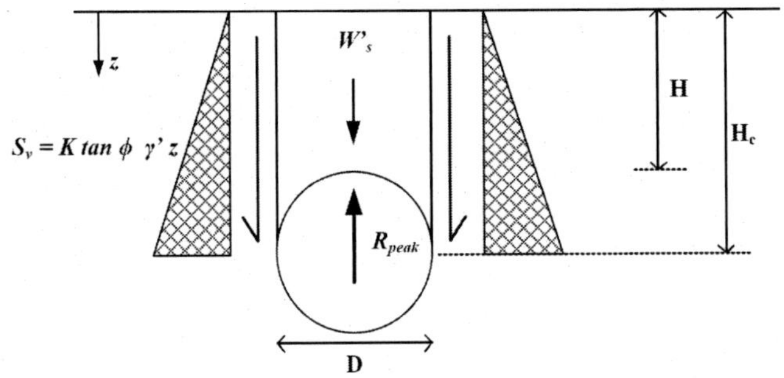

FIGURE 3: Forces resisting uplift

The model assumes uplift resistance to be provided by the sum of the buoyant unit weight of the backfill material in a vertical column above the pipe and the frictional resistance acting on the sides of this column. The frictional resistance τ is related depth by:

$$\tau = (K \tan\phi)\gamma' z = f_p \gamma' z \qquad (1)$$

where K is the lateral earth pressure coefficient, γ' is the buoyant unit weight of the backfill soil, z is the depth below the ground surface and ϕ is the soil's internal friction angle.

Integrating this resistance from the ground surface to the pipe centre results in the maximum available resistance R_{max} for a pipe of diameter D and length L buried to a depth H being given by:

$$\frac{R_{max}}{HDL\gamma'} = 1 + \left(\frac{1}{2} - \frac{\pi}{8}\right)\frac{D}{H} + f_p\left(\frac{D}{H}\left[\frac{H}{D} + \frac{1}{2}\right]^2\right) \qquad (2)$$

The code suggests values for the uplift resistance parameter f_p, as shown in Table 1.

Table 1: Soil parameters from DNV-RP-F110 (DNV, 1997)

Soil	*Parameter*	*Value*
Loose Sand	f_p	0.1 – 0.3
	δ_f	0.5% – 0.8% H
	α	0.75 – 0.85
	β	0.2
Medium-Dense Sand	f_p	0.4 – 0.6
	δ_f	0.5% – 0.8% H
	α	0.65 – 0.75
	β	0.2
Rock	f_p	0.5 – 0.8
	δ_f	20 – 30 mm
	α	0.35 D
	β	0.2

It should be noted that the uplift resistance parameters quoted are only stated to be valid for H/D ratios greater than 2 (3.5 for loose sands)

2.2 Mobilisation Distance

For sandy soils, the mobilization distance defined by DNV-RP-F110 is 0.5-0.8% of the cover height *H*, as shown in Table 1. This mobilization distance appears to come from the data of Trautmann et al. (1985) who presented the results of pull-out tests on pipe segments buried in sandy soils.

2.3 Required Uplift Resistance

Pipeline stability can be assessed based on the work of Palmer et al. (1990). In this work, the stability of a pipeline segment with an initial curved shape can be assessed based on the equation:

$$\Phi_w = A\left(\frac{\pi}{\Phi_L}\right)^2 - B\left(\frac{\pi}{\Phi_L}\right)^4 \quad (3)$$

where,

$$\Phi_w = R\frac{EI}{\Delta P^2} \quad (4)$$

$$\Phi_L = L\left(\frac{P}{EI}\right)^4 \quad (5)$$

where *R* is the required soil uplift resistance per unit length along the pipeline; *EI* is the flexural rigidity of the pipeline; *P* is the thermally-generated axial compressive force during equilibrium; *Δ* is the maximum height of the imperfection; *L* is half the total imperfection length; and *A* and *B* are constants to be determined numerically.

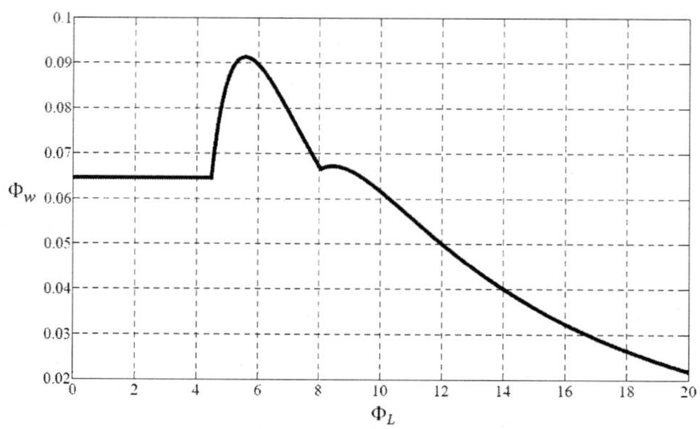

FIGURE 4: Non-dimensional required uplift resistance

Based on numerical calculations, Palmer et al. (1990) showed the relationship between dimensionless imperfection length Φ_L and dimensionless uplift resistance Φ_w to be as shown in Figure 4.

3. PHYSICAL MODELLING

Physical modeling was carried out both at prototype scale at 1g and at small-scale in a minidrum centrifuge in order to investigate both the peak uplift resistance and the mobilization distance required to achieve it for pipelines uplifting in saturated sand.

3.1 Centrifuge Modelling

Centrifuge modelling was carried out using the 1m diameter Mark II minidrum centrifuge at the Schofield Centre, University of Cambridge. This centrifuge is described in detail by Barker (1998). The centrifuge has been in operation since 1994 and has been heavily used for pipeline uplift studies. The centrifuge can reach 636g at a maximum speed of 1067 rpm, with an onboard PC providing 16 channels of data acquisition at up to 10 kHz. Figure 5 shows a photograph of the centrifuge in operation.

FIGURE 5: Minidrum centrifuge in operation

The centrifuge is equipped with a single degree of freedom actuator, as shown in Figure 6, which allows a buried pipeline segment to be pulled out of the soil at constant velocity. Whilst the drum centrifuge is designed to have a soil sample spread around the entire circumference of the channel, for this experimental programme, two model containers were placed within the ring channel, one containing the buried pipe, and one diametrically opposite as a counterweight. A photograph of the model container and pipe prior to sand pouring is shown in Figure 7. Sand was pluviated into the model container at a relative density of approximately 30% using a computerised sand pourer, (Madabhushi et al. 2006)

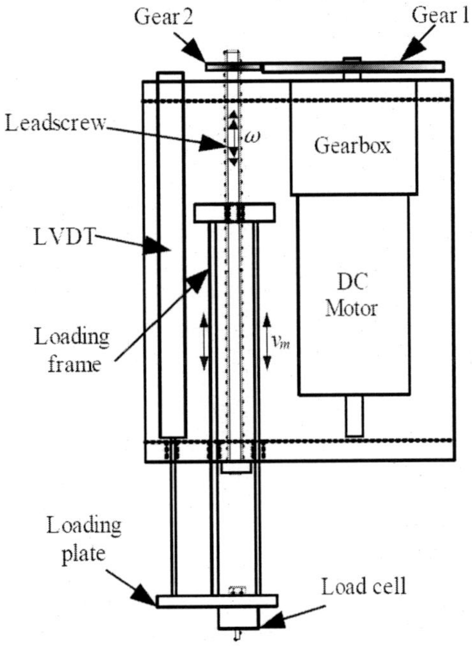

FIGURE 6: Minidrum actuator

FIGURE 7: Minidrum model container and pipe

3.2 Full-Scale Modelling

Large scale modelling was carried out in the test tank shown in Figure 8. This has a large Perspex window allowing accurate visualisation of failure mechanisms analysed using PIV. The tank has dimensions of 1000 x 850 x 76 mm (W x H x D). An actuator mounted above the tank allowed pipe sections with PTFE end caps to be pulled upwards through the backfill soil at a constant rate.

FIGURE 8: 1g test apparatus

4. TEST PROGRAMME

A comprehensive series of pipeline uplift resistance tests in sandy and rocky backfills have been carried out. The series can be subdivided into the following categories:

- Plane-strain testing in loose, fully saturated Fraction E sand using a 100-mm-diameter model pipe at 1g
- Plane-strain testing in loose, fully saturated Fraction E sand using a 258-mm-diameter model pipe at 1g
- Plane-strain testing in loose, fully saturated Fraction E sand using a 258-mm-diameter prototype pipe at multiple g-levels for H/D=1
- Plane-strain testing in loose, fully saturated Fraction E sand using a 258-mm-diameter prototype pipe at 30g

The plane strain testing set-up and procedures for uplift in sandy backfills are described in detail by (Wang et al. 2010). A schematic of the test

Table 2: Test Series Parameters (prototype scale)

Test No.	*Category*	*g-level*	*Backfill*	D_{50} *(mm)*	γ_{dry} *(kN/m³)*	γ_{sat} *(kN/m³)*	*D (mm)*	*H (mm)*
1								10
2								30
3								40
4	I						100	50
5								100
6		1g						200
7								350
8								110
9	II						258	129
10								258
11		10.2g	Saturated Fraction E sand	0.28	14.1 (± 1.0%)	18.5 (± 1.0%)		
12		13.4g						
13		20.4g						
14	III	27.4g					258	258
15		30g						
16		41g						
17		86g						
18								0
19								65
20	IV	30g					258	129
21								516
22								1032

tank is shown in Figure 4. The pipeline uplift was displacement controlled at a rate of 0.005 mm/sec. A pore pressure transducer (PPT) was embedded in the bottom of the model pipes. During all tests in categories I and II, no induced negative pore pressures were observed, i.e. drained conditions were achieved. The Fraction E sand had a relative density $I_D = 35\%$ and a saturated unit weight γ_{sat} of 18.5 kN/m³. The test parameters are summarized in Table 2.

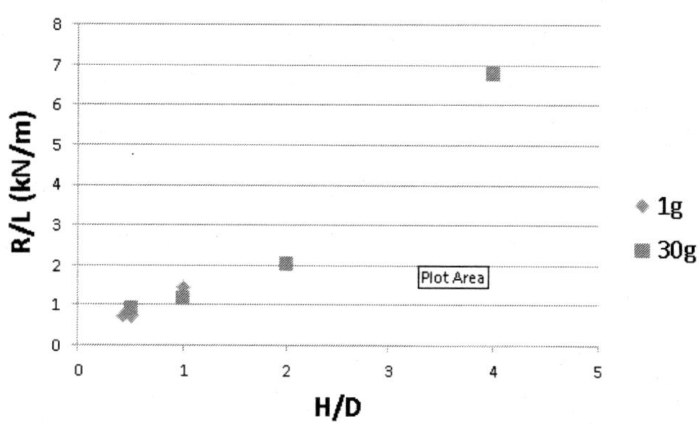

FIGURE 9: Maximum uplift resistance for 1g and 30g tests with 258mm pipe

5. RESULTS

5.1 Peak Uplift Resistance

The peak recorded uplift resistances for tests in categories II and IV are shown in Figure 9. It can be seen that while the range of H/D ratios in the two series are not identical, limitations in the equipment dictating a maximum H/D value of 1 for the 1g tests, the test series are in good agreement. Interpreting the data though the design framework given by equation 2, results in values of f_p for the 1g testing of 0.56 and for the 30g testing of 0.68. Both of these values are high compared to the range for loose sand suggested by DNV-RP-F110, being at the upper end of the range of values suggested for medium-dense sand.

Another notable feature of the data is that there is no significant reduction in f_p value at low *H/D* ratios. As the DNV recommendations are only stated as being valid for *H/D* values greater than 3.5, conventional practice is to ignore shear resistance for pipelines with H/D less than one, relying solely on the self-weight of the overburden soil. The data presented here and further analysed by Wang et al. (2010) suggests that no lower cutoff on *H/D* exists in the validity of equation (2).

The theoretical basis for concerns over the use of frictional strength at very low *H/D* ratios arises from the transition from continuum mechanics when there are many particles above the pipe to discrete mechanics when only a relatively small number of particles are interacting to provide shear resistance. Whilst in the experiments presented here the number of particles remains significant, (a minimum of around 100 particles), further tests on pipelines buried in gravels presented by Wang et al. (2010) suggest that if a lower cutoff value exists, this is small enough that most conceivable design scenarios would exceed this cover.

5.2 Mobilisation Distance

Whilst the uplift resistance values obtained agree closely between full-scale and centrifuge modelling, and correlate reasonably closely to those predicted using the current design approach, the mobilisation distances measured are more problematic.

Figure 10 shows the model scale mobilisation distances measured for tests in category III together with the equivalent full-scale test. It can be seen that while mobilisation distance decreases with g-level, it does so less rapidly than would be predicted using the standard linear scaling law for distance in centrifuge model testing. A comparison of the ratio of mobilisation distance to that measured at 1g (*M*) is shown in Figure 11.

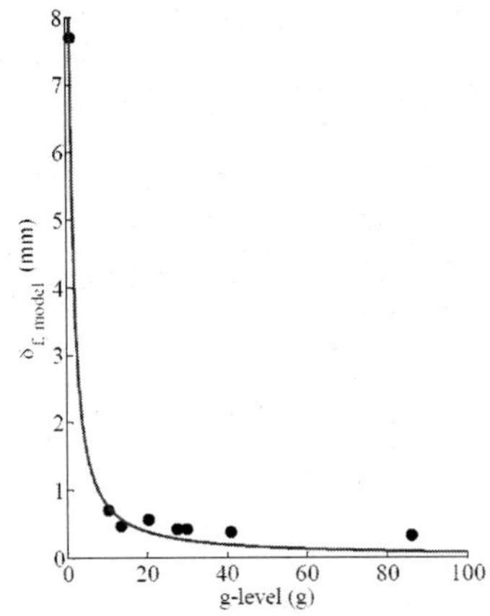

FIGURE 10: Mobilisation distance at varying g-level

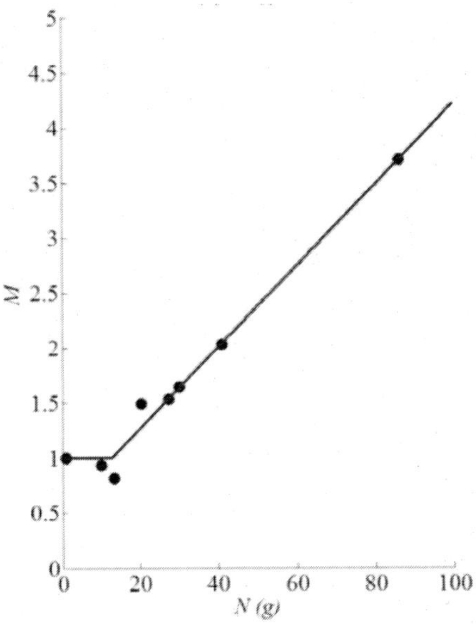

FIGURE 11: Amplification of mobilisation distance in centrifuge models

It can be seen that all centrifuge tests carried out at g-levels greater than 20g show exaggerated mobilisation distances relative to the 1g model. This is potentially a function of the soil grain size effect on shear band formation identified by Palmer et al (2003), though the conclusion of that paper that identical mobilisation distances would be observed in centrifuge model and prototype appears erroneous.

Further quantification of the effects of the ratio of cover height to grain size is currently being attempted in order to further investigate the appropriate scaling law for mobilisation distance and whether any upper-limit should be placed on the appropriate g-level for this class of centrifuge testing.

The mobilisation distances predicted by both the 1g and centrifuge models significantly exceed those predicted by the design code. 0.8% of the cover height in this case is approximately 2mm, a factor 4 smaller than that from the 1g model. This tendency for mobilisation distances to significantly exceed those proposed by DNV (2007) is further explored by Wang et al. (2012).

5.3 Deformation Mechanism

The transparent windows present in both the 1g and centrifuge test apparatus allow the observation of failure mechanisms to be carried out using PIV techniques. Figure 12 shows the deformation mechanisms measured for 1g tests with *H/D* ratios of 3.5 and 0.5 using a 100mm diameter pipe at peak uplift resistance. It can be seen that the failure mechanism at depth closely resembles that used in the design approach presented in figure 3, a vertical chimney of soil above the pipe being lifted with vertical slip planes against the remainder of the soil. At shallow depths, however, dilation of the soil results in a lateral expansion of the failure mechanism, resulting in an increased width of soil block being lifted.

The vertical slip surface model thus seems an appropriate model to use to estimate the uplift resistance of pipelines in sand, matching closely the mechanics of the physical problem.

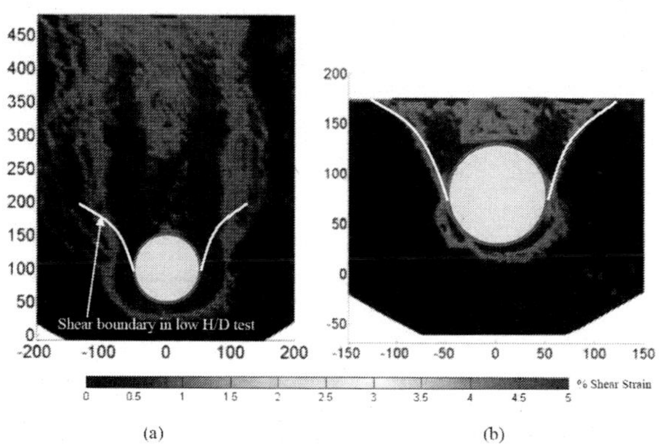

FIGURE 12: Shear strain distributions for high and low *H/D*

6. CONCLUSIONS

1g and centrifuge modelling play a valuable role in supporting the design process for critical infrastructure including offshore pipelines, where the financial and environmental costs of failure are large. Whilst physical modelling can offer unique insights into the behaviour of these complex systems and be used to both derive design parameters for particular projects and to validate design methodologies, it should always be used with caution.

In this paper it has been shown that for the particular case of the uplift resistance of buried pipelines in loose sands, centrifuge modelling can overestimate the mobilisation distance that would be expected during uplift. Whilst in this case that would result in an overly conservative design, rather than an unsafe one, this may not always be the case for different projects.

Modelling of models is a vital tool in assessing the accuracy of scale-modelling techniques, and should always be attempted when there is concern over scaling limitations such as shear band formation in granular materials.

REFERENCES

Barker, H. R., (1998). Physical modelling of construction processes in the minidrum centrifuge. *PhD thesis*, University of Cambridge

DNV (2007). DNV-RP-F110 Global buckling of submarine pipelines –structural design due to high temperature / high pressure. Det Norske Veritas, Norway.

Madabhushi, S. P. G., Haigh, S. K. and Houghton, N. E., (2006). "A new automatic sand pourer for model preparation at University of Cambridge". 6th International Conference on Physical Modelling in Geotechnics pp.217-222

Palmer, A. C., Ellinas, C. P., Richards, D. M. & Gujit, J., (1990). "Design of submarine pipelines against upheaval buckling". Offshore Technology Conference pp.551-560

Palmer, A. C., White, D. J., Baumgard, A. J., Bolton, M. D., Barefoot, A. J., Finch, M., Powell, T., Faranski, A. S. and Baldry, J. A. S., (2003). "Uplift resistance of buried submarine pipelines: comparison between centrifuge modelling and full-scale tests". Géotechnique, 53, Issue 10, pp 877-883

Trautmann, C. H., O'Rourke, T. D. and Kulhawy, F. H., (1985). "Uplift force-displacement response of buried pipe". ASCE Journal of Geotechnical Engineering, 111, Issue 9, pp 1061-1076.

Wang, J., Ahmed, R. and Haigh, S. K., (2010). "Uplift resistance of buried pipelines at low cover-diameter ratios". Offshore Technology Conference. Paper OTC 20912

Wang, J., Haigh, S. K., Forrest, G. and Thusyanthan, I., (2012). "Mobilization distance for upheaval buckling of pipelines". ASCE Journal of Pipeline Systems Engineering and Practice doi:10.1061/(ASCE)PS.1949-1204.0000099

Simulation of Soil Movement in Geotechnical Centrifuge Testing – Deep Excavations, Tunnelling, Deposit

D. König

Institute for Foundation Engineering, Soil and Rock Mechanics, Ruhr-Universität Bochum, Bochum, Germany
E-mail: diethard.koenig@rub.de

ABSTRACT

Foundation engineering works are mostly related to movement of soil masses. Excavation of construction pits, progress of tunnels, construction of embankments and dams, all this activities impose the take of, transport and deposit of more or less huge amount of soil masses. These processes of soil movement lead to change of stress, activation of shear resistance and deformation in the remaining or underlying soil. To control this effects special works like installation of retaining structures, soil stabilisation by nailing or injection, improvement of drainage conditions or others are carried out. These measures have to be designed taking into account the interaction with the surrounding soil and the construction process. Up to now not all of the relations between the different construction elements and soil reactions are fully understood. Independent on the progress in numerical modelling physical modelling is helpful to discover and analyse such interactive reactions. For this it is necessary to simulate soil movement in physical modelling. Especially in centrifuge modelling this is a challenge. Within this paper an overview is given of different methods to simulate soil movements in the geotechnical centrifuge and three examples on projects carried out in the Bochum Geotechnical Centrifuge ZI are described more in detail. The main topics are the excavation of construction pits, the excavation of tunnels and the deposit of soil masses.

1. INTRODUCTION

Prediction of ground movements arising from foundation works or tunnelling becomes more and more important. Density of infrastructure is increasing especially in urban areas and the sensitivity of historic or sophisticated structures gain in the focus of authorities and public interest. The complex behaviour of the soil and the multiple interactions between soil, construction procedures and structures force to improve numerical methods. Constitutive models started to take into account stress path depending soil behaviour, small strain stiffness and time depending effects. Computer capacity allows to model complex three dimensional situations including a high degree of detailing. Nevertheless to decide on the correct boundary conditions, to choose an appropriate idealisation, which is still necessary, and to assess the different parameters is still a challenge. Due to this a calibration and validation of our models is essential, which can be realised by comparing the numerical results with field measurements. Otherwise physical modelling give the chance to compare numerical data with physical measurements on a prototype implying all selected characteristics of the real field situation under well known boundary conditions, e.g. by centrifuge modelling. This promises the sufficient techniques to display the major aspects of the field situation especially in respect to depict relevant stress paths.

One important aspect considering this background is the simulation of soil excavation and soil deposit procedures also in centrifuge modelling under high acceleration levels, which have been developed since many decades. In this paper three processes are selected to give a short overview on the development of the model technique and for each process on example for a test design is described. These test designs have been developed at Ruhr-Universität Bochum and are used in the Bochum Geotechnical Centrifuge ZI, which characteristics are described by Jessberger & Güttler (1988).

2. CONSTRUCTION PITS

2.1 Methods

Different methods have been used to simulate an excavation in centrifuge modelling. Lydon & Pearson (1984) modelled the behaviour of sheet pile walls in sand. The sheet pile model was located in the final position in the strong box and the soil has been placed around. The strong box was exposed to the selected g-level. After reading the data of the different transducers the centrifuge was stopped and an excavation step performed at 1 g. This procedure has been repeated several times. Using this method a sequence of loading and unloading procedures is applied to the soil and the structure, which do not represent field conditions.

Bolton & al. (1988) simulated the excavation of a retaining wall in overconsolidated clay by replacing the soil in the excavation zone by a rubber bag filled with a zinc chloride solution and draining the fluid after a reconsolidation period. The zinc chloride solution was mixed to the same unit weight like the clay and therefore an initial effective earth pressure coefficient of K0 = 1 was applied, which may correspond to earth pressure conditions in overconsolidated clays. A fluid support has been chosen also by Toyosawa at al. (1994) (zinc chloride), Lade & al. (1981) (paraffin oil) and Schürmann (1994) (water), which allow to simulate smaller earth pressure coefficients, or Zheng et al.(2010).

After placing the model soil in the strong box Azevdo (1983, 1988) removed the wet sand within the zone which should be later excavated. Then a bag of fabric was placed all along the contour of the later excavation at 1g and the removed soil was filled inside the bag again. After exposing the model to higher g-level the excavation was simulated by lifting up the bag with the soil inside using an electric motor. The excavation is simulated in one step, not as a continuous procedure. A similar system has been developed by Allersma (1998) using also a fabric which is placed by loops in different layers and which allows a stepwise excavation. The influence of the reinforcement on the soil behavior by the fabric is unknown using such techniques.

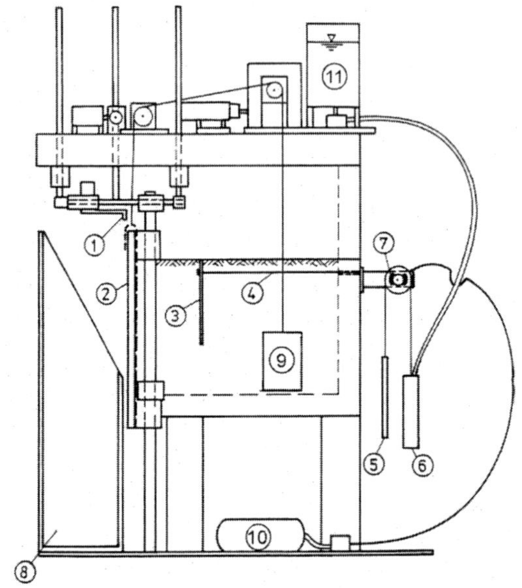

FIGURE 1: Model set up (explanation see text)

In 1994 Kimura et al. presented an in-flight excavator used in soft clay. The excavator removes the soil by scraping layer for layer (5 to 10 mm thick). This scraping technique has been modified by Loh at al. (1998) to simulate a three dimensional excavation. Such a procedure could also be performed with an on-board robot as presented by Derkx at al (1998) and Gaudin et al. (2002). A 2D servo-actuator has been developed at University of Cambridge (Haigh et al. 2010) and used for the simulation of the excavation by scraping in soft clay in front of a retaining wall, which is supported by props (Lam et al. 2010). The scraping technique does not allow a lift up of the soil which has to be excavated. Due to this the soil has to be moved throughout a barrier. The top edge of the barrier has to be on a lower level compared to the soil moved by the scraper. The barrier has to be removed step by step or has to be lowered continuously with the excavation. This can be realised by removing a segmented wall step by step or by a moving wall system. This method is up to now the most realistic way to simulate field conditions and has been chosen for a series of tests in the Bochum Geotechnical Centrifuge ZI for studying the earth pressure distribution acting on an anchored sheet pile wall (König 2002).

2.2 Bochum In-flight Excavator for Deep Excavations

2.2.1 Design

The model set up is shown in figure 1. The model is built up within a strong box with internal dimensions of 630 mm in length, 360 mm in width and 700 mm in height. The soil sample is placed inside the box including the model wall (3), which could be supported by one or two anchor levels. The position of the surface of the soil inside of the strong box depends on the length of the model wall, the number and position of anchors. At each anchor level a waling supported the model wall made of a steel bar with a cross section of about 4 to 8 mm.

Two steel rods (4) with a diameter of 2 mm are screwed into the waling and are passing the back wall of the strong box (figure 2). Each of them is connected first to a load cell (a) and then to a pulley (7) which is fixed on an axle (b) mounted at the back of the box. A gear weal (c) is also fixed on this axle and could be loaded by a chain with a counterweight (5) and a fluid container (6) at it ends. The fluid container can be filled by open the valve between the container and the water reservoir (11) on top of the box. Additionally the pulley can be blocked by a break (d) activated by air pressure (10). In that case no more displacement of the anchors can occur.

The front wall (2) is mounted movable hanging on two chains balanced by counterweights (9). The position is controlled by an electric chain drive placed on the top of the strong box. The wall is sealed against the box by a rubber membrane. Sliding occurs during lowering of the front wall between the front wall and plastic stripes glued in a horizontal position one above the other on the rubber membrane.

On top of the strong box the excavation mechanism is placed which is designed similar to the device presented by Kimura at al. (1993, 1994). The scraper (1) could be moved back and forth as well as up and down controlled by electric motors. In front of the strong box a second box is placed to collect the excavated soil.

The tests described in this paper have been performed with a fine grained sand used by Schürmann & Jessberger (1994). After fixing the strong wall in the initial position the side walls and the back wall of the strong box have been lubricated by a fine grease and covered by a plastic folio to reduce wall friction as fare as possible. The sand is placed by pluviation with a density of 1.68 g/cm³, which is close to 100 % relative density, up to the level of the toe of the model wall. The model wall with the waling is placed by a temporary support in the final position as well as the LVDT detecting the displacements at the toe of the wall.

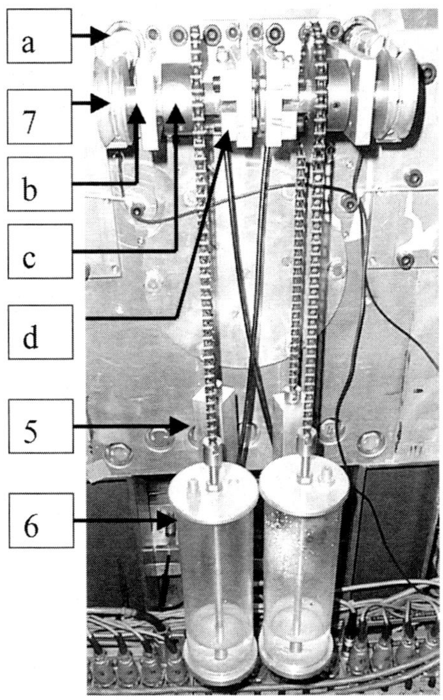

FIGURE 2: Fixation of anchors at the back wall of the strong box

The pluviation of the sand continues on both sides of the wall until an anchor level is reached. Then the anchor rods are placed within small tubes to reduce friction and connected to the waling as well as to the load cells and pulleys. In that moment the anchors are slackly embedded in the sand only stressed by a small pre-load due to a little bit of water inside the fluid containers

(Figure 1, No. 11). The LVDT's to measure the horizontal displacements at anchor level are placed at the same time. The pluviation is continued until the final soil surface is reached. Now the excavator, the chain drive for the front wall and the water reservoir for loading the anchors are mounted on the top of the box.

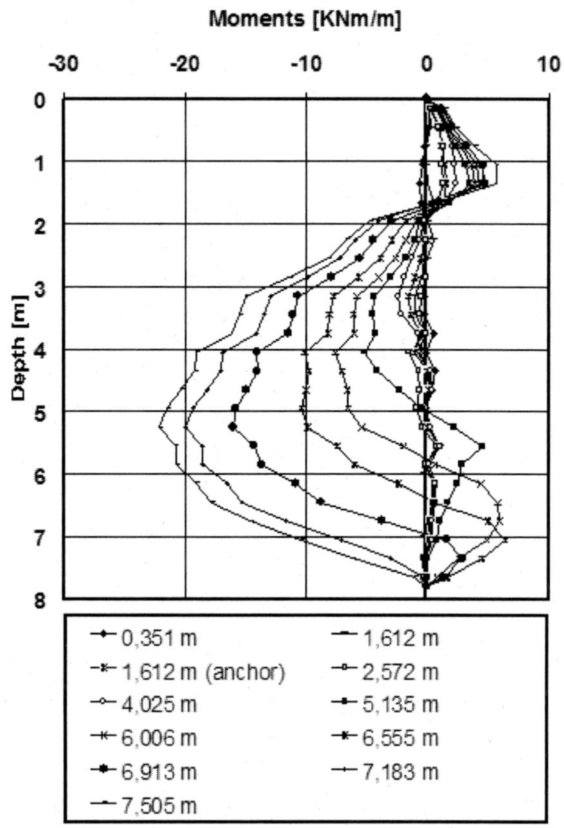

FIGURE 3: Bending moment distributions measured at different excavation depths

2.2.2 *Test Procedure*

The tests have been performed at an acceleration level of n = 30. After reaching this acceleration level the excavation starts with a short delay. First the front wall is lowered by about 3 cm. This leads to a failure in the sand close to the top of the front wall and some sand falls into the soil collecting box. A slope remains above the top of the front wall. This area is outside the active shear zone in front of the model wall so it is assumed that there is no influence on the model wall behaviour.

Now the scraper is moved close to the model wall and is pushed for a few millimetres (model dimension) into the sand. A layer of sand is removed and is slipped over the top of the front wall. The sand falls into the soil collection box. The scraper moves up and is driven back to the model wall to repeat the procedure. In front of the model wall a small slope remains, which height depends on the amount of each excavation step and the distance between scraper and model wall during penetration of the scraper into the soil. In the reported tests the height of this slope was about 2 cm (model dimensions).

After a series of excavation steps the front wall is lowered again. Reaching an anchor level the excavation is continued up to about 50 cm in prototype dimensions below the waling. Now the valve between the water reservoir (Figure 1, No 11) and the fluid container (6) is opened and the anchor, which has been loaded only by a small pre-load up to now, is loaded up to a defined force. Then the pulleys (7), which are connected with the anchors, are fixed by the breaks mentioned above

The excavation is continued up to the selected excavation level or up to the moment where large deformations of the model wall indicate failure of the soil in front of the model wall.

2.2.3 *Test Results*

Figure 3 shows the bending moment distributions measured at different excavation depths. At the beginning only very small bending moments are measured. The distribution is randomly. Also with the first excavation steps the load on the wall is small. Typical positive moments are observed at anchor level after tightening of the anchors. With further excavation the moments at anchor level increased as well as the negative moments in the field. The rate of increase of bending moment at anchor level becomes smaller compared to the rate of increase of the bending moment in the field after five meters excavation. This may be due to the limited bearing capacity of the soil above the anchor, where the wall moves in direction of the soil and passive earth pressure conditions develops.

The change of the interaction between soil and sheet pile wall in the lower part of the wall is obvious. The positive moments indicate a fixation of the wall in the soil. This moments disappear with continues excavation.

Processed test data are presented in Figure 9. The measured bending moments are approximated by spline functions as explained by Schürmann (1994). Due to the singularity at the anchor level two splines are used, one for the upper and one for the lower part. For this procedure it is necessary to introduce different boundary conditions. Most of them are well known by the measurements, some of them have to be estimated from the overall data. For example the exact bending moment at anchor level is not measured, so the peak which developed in theory at that level is not detected. This value is estimated from the tangents of the measured bending moment distributions. The total amount of the anchor force has been measured, which is equivalent to the difference of the shear forces just above and below the anchor level. But the values of the shear forces are not known and therefore are estimated.

Other tests have been performed to investigate the effect of wall plasticity on the response of varying wall system geometries (Bourne-Webb et al. 2011). In that case the excavation technique has been used successfully, but another challenge arose. In reality before formation of a plastic hinge within the sheet pile almost the full elastic wall bending stiffness would be available. The wall model was made from an aluminium plate with a weak section in the depth, where the plastic hinge formation has been expected. Due to this, large deformations corresponding to a strong bend of the wall within the weak section occurred with increasing excavation depth. These deformations caused extreme earth pressure reductions due to arching effects. This effect differs from the behaviour of a real sheet pile profile where elastic behaviour of the wall with constant stiffness dominates the deformation of the wall at the beginning of the excavation. At high excavation depth the formation of the plastic hinge would lead to a reduction of stiffness and a sudden increase of wall deformation. To reproduce this behaviour a more detailed wall model has to be developed showing a moment – curvature characteristic according to that of a real sheet pile profile.

3. TUNNELING

3.1 Methods

During the 1970`s and the 1980`s a model technique has been developed at the University of Cambridge (Potts 1976, Mair et al. 1984) in which air pressure is applied to control support conditions. This technique has been used later by e.g. Chambon et al. (1991) and König et al. (1991). The soil mass, which has to be removed

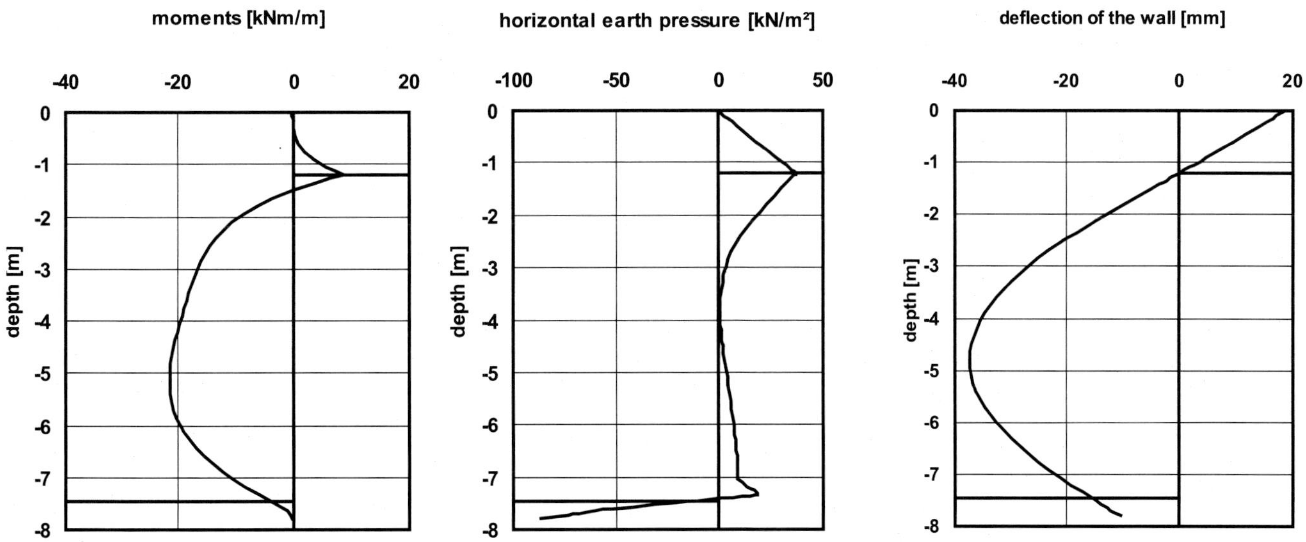

FIGURE 4: Fitted bending moments, earth pressure distribution, deflection of the wall examined for a test with density $\rho = 1.69$ g/cm^3, anchor force A = 56.7 KN/m, wall stiffness EI = 3100 KNm^2/m

during the excavation, is represented by a rubber membrane pressurized with air pressure. The shape of the rubber membrane is identical with the contour of the excavation. The equilibrium is kept between internal air pressure and external earth pressure during model preparation and increasing g-level. Initial stress conditions can be observed in the model after reaching the selected g-level. The excavation is simulated by decreasing the internal air pressure.

To keep equilibrium between internal air pressure and the earth pressure it would be necessary to apply internal support according to the theoretical earth pressure at state of rest. This support can not be achieved by the air pressure technique, because the air pressure remains constant along the contour of the excavation whereas the earth pressure varies with the change of orientation from vertical to horizontal and due to increase of vertical stress level with depth.

To overcome this problem Sharma et al. (2001) used a support by a rigid Styrofoam block instead of air pressure. The Styrofoam block covered by a rubber bag is dissolved by a solvent injected by pipes to reduce the support of the excavation zone. Yeo et al. 2010 discussed these two methods. Corresponding to the simulation technique for construction pits developed by Bolton et al (1988) they used a Zinc chloride solution for stabilizing the excavation zone before modelling the excavation process by releasing the fluid pressure. These techniques are mainly used for analysing failure mechanisms and limit state conditions.

Other techniques have been presented to study single processes during tunnelling by Bezuijen & van der Schrier (1994) and Yoshimura et al. (1994). Bezuijen & van der Schrier (1994) focused on the interaction between shield tunnelling process and an existing pile foundation. The influence of the gap between tunnel lining and soil behind the shield tail on the displacement field in the soil and the pile foundation was modelled by using a two dimensional tunnel model of which the diameter can be reduced. In this case only the part of the process was modelled which seems to be most important for the investigated phenomenon. The complete tunnelling process was not modelled,. Another solution for this task has been demonstrated by Farrell & Mair (2010). They modelled the volume loss taken place during a tunnelling process in a two dimensional model (plain strain conditions) by withdrawing a fluid out of an fluid filled annulus between the model tunnel lining and outer latex rubber lining.

Yoshimura et al. (1994) have investigated the influence of a shield tunnelling process on an existing tunnel. The existing tunnel is located parallel to a tunnel which is under construction. The effect of the shield tunnelling process on the stress strain conditions in the soil and the existing tunnel have been simulated by applying support pressure of varying magnitude on the face plate of a model shield. In this case an excavation process was not modelled. To study the failure mechanism in front of a collapsing tunnel face and to determine the minimum support pressure for stabilizing a tunnel face different authors used a rigid support of the tunnel face. This support is moved backwards to simulate the decrease of support pressure (Walter et al. 2010) or forward to activate passive earth pressure conditions (Wong et al.2010). In all these cases an in-flight excavation process was not realised.

In 1994 Nomoto et al. presented a miniature shield machine, which allowed the in-flight excavation of soil and also the in-flight installation of the lining. The soil is scraped by a cutting well and transported by a screw conveyor to the outside. The cutting head is pushed forward into the soil followed by a tube, which diameter is slightly smaller than the one of the cutter. This tube simulates the lining.

3.2 Bochum In-flight Excavator for Tunnelling

3.2.1 Design

The Bochum in-flight excavator (König 1998) is designed for modelling one step of a tunnel excavation in a silty sand with low water content and to analyze stress redistributions around the excavation area at the tunnel face initiated by the excavation process. An overview of the model with the in-flight excavator is given in Fig. 5. The entire model is placed on a bottom plate. A strong box is housing the soil mass and supporting a hydraulic pressure jack as well as a carriage. The carriage can be moved horizontally by the hydraulic jack about ± 4 cm. This is equivalent to a maximum driving displacement of the in-flight excavator of 8 cm at model scale.

The tunnel lining is placed inside the soil. The lining is supported in the front wall of the strong box. This support allows a vertical displacement, but impedes a tipping of the lining. The in-flight excavator is located inside of the tunnel lining. The cutting head is placed in start position inside the end of the lining in front of the tunnel face. The drive shaft of the excavator is connected by a cardan shaft with an electric motor. The cardan shaft allows a length compensation when the

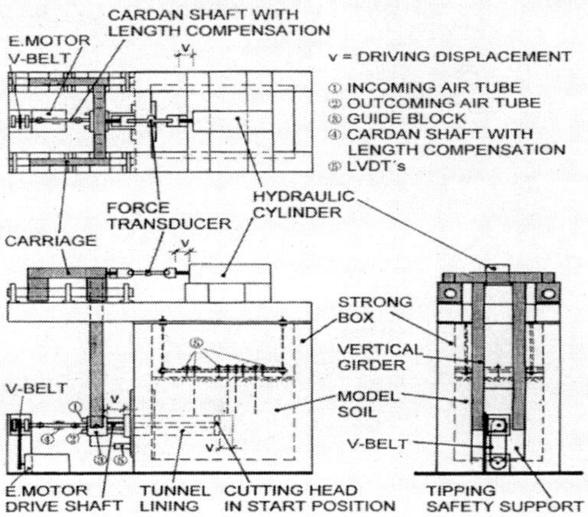

FIGURE 5: Setup of the model with in-flight excavator (a) top view, (b) side view, (c) front view

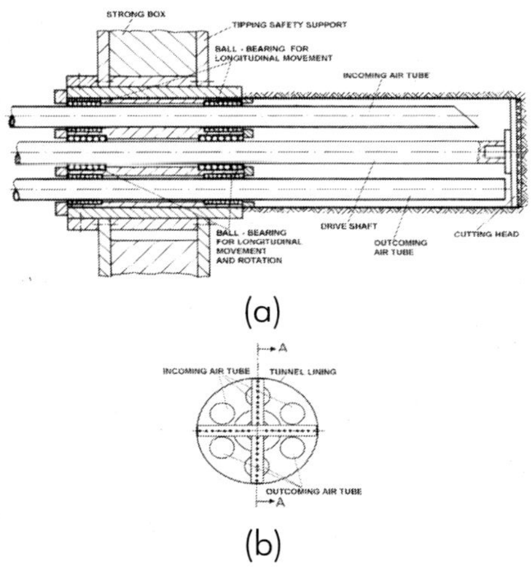

FIGURE 6: In-flight excavator (a) section A-A, (b) front view

excavator is moved. The carriage is connected with the in-flight excavator by vertical girders and a guide block.

The in-flight excavator is presented in figure 6 in detail. The excavator is supported in the tipping safety support. The excavator consists of the drive shaft with the cutting head and of six tubes. The six tubes and the drive shaft can be moved in longitudinal direction. The cut soil is transported to the outside of the strong box through three of the tubes by suction. The other three tubes allow air to flow behind the tunnel face and the cutting head to create an air flow in order to grasp all particles.

During the excavation process surface settlement (Figure 6) are measured as well as vertical displacements close to the excavation by LVDT. Additional LVDT embedded in the soil are monitoring the horizontal displacements close to the excavation in the depth of the springline of the tunnel. Tangential strains of the tunnel lining are detected in crown, springline and invert at different distance from the end of the tunnel lining (4 measurement sections in 2.5, 10, 20 and 40 mm distance) by strain gauges. At least the force is measured by a force transducer necessary to move the cutting head forward.

3.2.2 Test Procedure

The tipping safety support is installed in the strong box. The hole in tipping safety support system is closed. Now the strong box is filled with the model soil up to a given height. In this case it is either a mixture of fine sand (90 %), fine quartz (5%) and kaolin (5 %) with a water content of 2 % (model soil 2) or a mixture of fine sand (95 %), fine quartz (2.5%) and kaolin (2.5 %) with a water content of 1 % (model soil 3). The model soil is compacted to a given density. The strong box is placed in the swinging basket of the centrifuge and accelerated up to the selected g-level. After consolidation the machine is stopped and the model is removed from the basket.

Now the hole in the tipping safety support is opened and by the use of the in-flight excavator a cylindrical cavity is excavated which diameter is slightly less than the outer diameter of the tunnel lining. The tunnel lining is inserted into this cavity and fixed in the tipping safety support. The soil cut during the placement of the lining is sucked out. Now the in-flight excavator is placed inside the lining. The cutting head is located inside the end of the lining directly in front of the tunnel face.

After installation of all the equipment described above and also of the measurement devices the complete model is placed in the swinging basked of the centrifuge and is accelerated to the selected g-level. Initial stress conditions can be observed after a short reconsolidation time. Then the excavation starts.

The cutting head is rotated by the electric motor. At same time the cutting head is pushed forward into the soil by the hydraulic cylinder. The force of the cylinder is transferred by the carriage and the vertical girders to the guide block and from there to the drive shaft and the cutting head. The hydraulic cylinder is operated with constant displacement rate. The released soil is sucked out through the three lower tubes.

3.2.3 Test Results

From the readings of the strain gauges the change in normal force in the tunnel lining due to one excavation step is calculated. The changes of normal forces determined in one measurement section at one distance from the end of the lining were nearly the same in crown, springline and invert and so the average value has been calculated. This change in normal force in the tunnel lining ΔN has been normalized by the unit weight of the soil γ and the square of the diameter of the tunnel D and plotted versus the normalized distance from the end of the lining a/D in Figure 7.

Increased normal forces are detected at the end of the lining. This increase may be caused by a longitudinal arching, which is mobilised due to movements into the excavation and which loads the stable soil in front of the tunnel face and the tunnel lining. This longitudinal arch is acting simultaneous to a circumference arching effect.

The intensity of the longitudinal arching effect is increasing with excavation length L/D, with increase of cover in the range of a shallow tunnel (C/D = 1) and a deep tunnel (L/D = 4) and with decreasing soil stiffness (model soil 3 is stiffer then model soil 2) for the tested boundary conditions.

Figures 8 shows the excavation area for two tests after the tests have been finished and the models have been removed from the swinging basket and the soil mass have been cut. In the test with model soil 2 and a cover of C/D = 1 the excavation was not stable and collapse occurred. In case of model soil 3 and C/D = 4 the excavated area was stable.

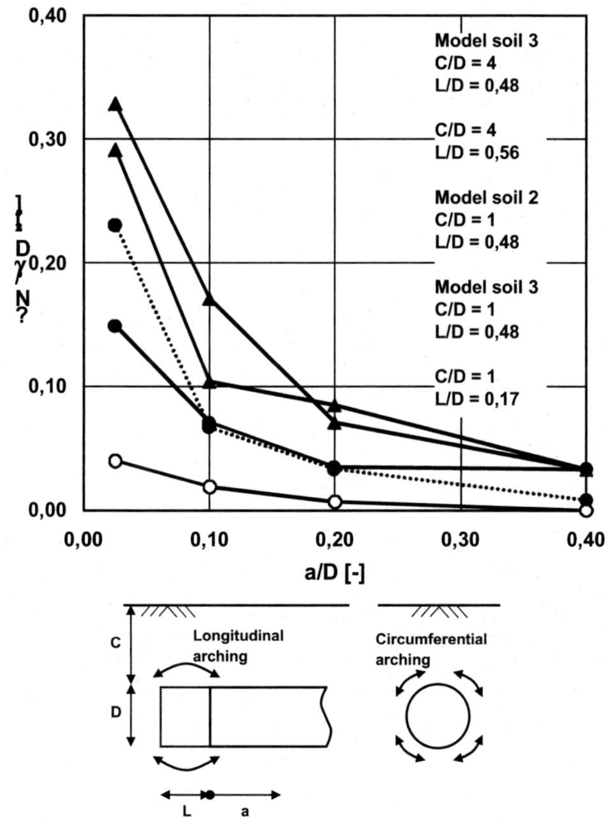

Figure 7 Normalised change in lining normal force due to one excavation step $\Delta N/\gamma D^2$ versus normalized distance from the end of the lining a/D (γ unit weight of soil, D diameter of tunnel, C soil cover from tunnel crest, a distance from end of tunnel lining, L excavation length)

4. DEPOSITS

4.1 Methods

To simulate the construction of an embankment

in a geotechnical centrifuge under increased gravity two methods are applied in general, fixed and movable hoppers. Davies & Parry (1985) described a hopper consisting of a number of cells with an outlet at the bottom which can be opened or closed by a two-way solenoid valve system. The hopper is installed on top of the strong box and each cell is filled with sand before the model is accelerated to the selected g-level. For constructing the embankment in-flight, the cells are opened and a layer of the embankment is poured. The geometry and thickness of the layer depend on the position of the cells, the amount of sand within each cell and on the opening time. By filling only a part of the cells of the hopper the geometry of the embankment can be adapted. For defining which cells should be filled to reach a certain geometry of the embankment the Coriolis effect has to be taken into account. This effect occurs when particles, are released from a part of the rotating model, e.g. from a hopper. These particles move independent from the rotating system and follow their own trajectories, which are different from the movement of the model (Taylor 1995). The effect and its consequences can be predicted. Other hopper systems based on this concept have been developed, e.g. by Allard et al. (1994).

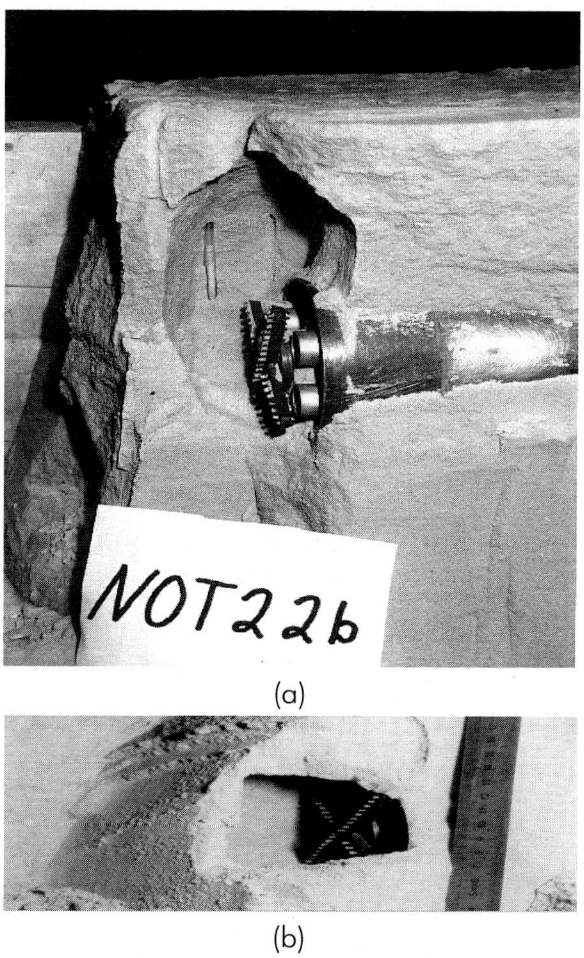

(a)

(b)

FIGURE 8: View of the unsupported cavity (excavation area) after the test a) in model soil 2 with a cover of C/D = 1 (collapse) and b) in model soil 3 with a cover of C/D = 4 (stable)

Allersma (1994) gave details on a hopper system working with a movable funnel. Before accelerating the centrifuge the funnel is filled with sand. At increased g-level the outlet is opened and sand starts to sprinkle on the model surface. Simultaneously the hopper is moved. The geometry of the embankment depends on the amount of sand flowing out of the funnel and on the speed of the hopper. Also in this case the Coriolis effect has to be taken into account for predicting the final geometry of the embankment. The device used for the investigations performed in Bochum follows the principle of a moving hopper system.

4.2 Adaptive Foundation System for an Embankment on Soft Soil

The placement of embankments on very soft soil requires foundation works. Within a research project a new innovative adaptive foundation system for such situations is analysed by means of numerical calculations and centrifuge model tests in cooperation between HUESKER Synthetic GmbH and Ruhr-Universität Bochum.

The adaptive foundation (Figure 9) system consists of two vertical and parallel walls (e.g. sheet pile walls) which are introduced at a certain distance between each other into the soft soil and connected to each other by a tension membrane (e.g. geotextile) at the existing ground level. The vertical walls may end within the soft soil layer or reach further down into a firm layer. The soft soil beneath the embankment is therefore confined by the vertical and horizontal elements. The embankment is constructed above the tension membrane. The load from the embankment onto the soft soil generates a horizontal pressure onto the vertical walls which provokes outward movements. These movements are restricted by the tension membrane. At the same time an additional tension force is mobilized within the membrane due to settlements beneath the embankment. This additional tension force may lead to a further restriction of the outward movements. The foundation system ensures the global stability of the embankment (e.g. bearing failure and extrusion) and prevents or reduces the system deformations.

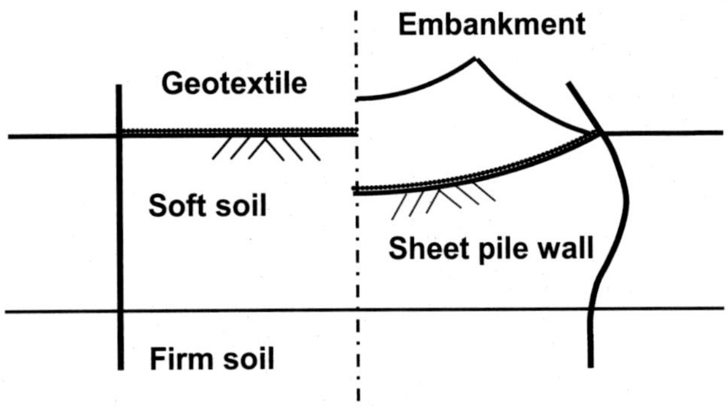

FIGURE 9: Adaptive foundation system for an embankment on soft soil, (a) undeformed system before placing of the embankment, (b) deformed system some time after placing the embankment

The stress and strain of the different system components, vertical walls, tension membrane and soft soil, are strongly influenced by their interaction. Due to consolidation processes in the soft soil these interactions are time dependent. So the stiffness of the soil as well as the total stress on the walls are changing with the consolidation from undrained condition at the beginning of the embankment construction to drained conditions in the final state. The system behaviour strongly depends on the distance between the vertical walls, their length and degree of fixation. Furthermore it depends on the thickness, shear strength and stiffness of the soft soil layer as well as the bending stiffness of the vertical walls and tension stiffness of the membrane and the relation of the latter both between each other. Within a series of centrifuge model tests some principle configurations of the systems should be analysed before starting a systematic investigation by numerical simulations. In the centrifuge model tests only the half of the system is modelled to take advantage of its symmetry. Therefore a hopper system has been developed to build up a half embankment under increased acceleration in three layers. This system will be described.

4.3 Bochum Hopper System and Test Set Up

4.3.1 Design of the Hopper System

The hopper (Figure 10) was developed to allow the staged construction of the embankment in-flight and is mounted on top of the strong box. The device consists of a funnel (1) with a variable opening slit and a storage drum (2). The funnel is driven by a threaded rod (3) which is attached to a motor (4) and can move horizontally forwards and backwards along two steel bars (5). The speed of the funnel is fully adjustable. The sand can flow out from the funnel through a slit which opening width was calibrated in preliminary tests to allow for a well defined and regular sand flow.

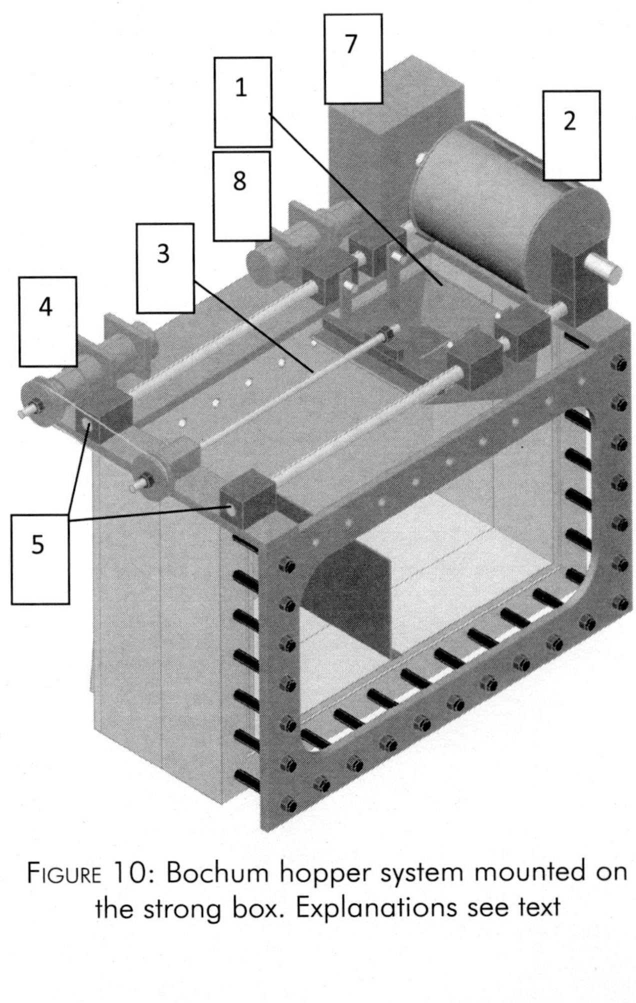

FIGURE 10: Bochum hopper system mounted on the strong box. Explanations see text

In start position as shown in Figure 10 this slit is blocked.

Once the funnel is empty it can be refilled in-flight from a storage drum (2). Therefore the storage drum can be rotated by an electric motor (6) and a gear system (7). The sand volume which is poured into the funnel through a 5 cm wide opening slit can be controlled by the rotation angle of the drum.

4.3.2 Test Set Up

A detailed description of the test set up is given by Detert et al. (2012). Two similar strong boxes are used for the test series. The inner dimensions of the strong boxes are 90 cm width, 36 cm depth and 60 cm height. One side wall is made out of acrylic glass to observe the system during the test. The thickness of the soft soil layer and the height of the embankment are 20 cm. The embankment width is 40 cm and the crest width about 5 cm. The soft soil layer is consolidated out of a Kaolin slurry. A 20 mm sand drainage layer is placed beneath the slurry. A geotextile is located in between the sand and the slurry as a separation and filtration layer. As only half of the system is modelled only one sheet pile wall model made of an aluminium plate is used. In the axis of symmetry at the side wall of the strong box the horizontal displacement of the geotextile should be zero whereas vertical movements should be allowed. This is realised by a special bearing element. The geotextile is fixed on one side to the model wall and at the other side to this special bearing element. Down-scaled geogrids according to Springman et al. (1992) are used for modelling the tension membrane.

4.3.3 Instrumentation

During the centrifuge tests total vertical pressure and pore water pressures are measured at various positions. The aluminium wall is instrumented with strain gauges at the centre line. At the connection between aluminium wall and geogrid two load cells are installed to measure the connection forces. Two displacement transducers are fixed to the top end of the aluminium wall to measure the horizontal displacements. A draw-wire sensor is attached to the geogrid bearing device at the axis of symmetry to detect vertical displacements. Two more displacement transducers measure the settlement of the Kaolin surface during the entire test.

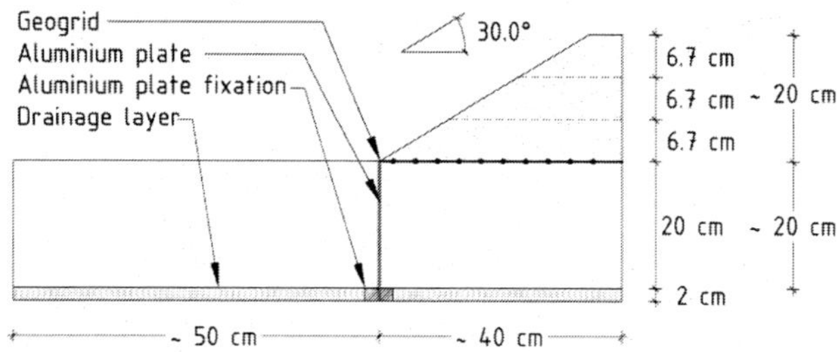

FIGURE 11: Centrifuge test set-up geometry

4.3.4 Test Procedure

First the model wall is placed in the strong box in its final position and also the bearing element with the geotextile is installed, but the geotextile is not connected to the model wall at this time. A slurry prepared from Kaolin and water (w = 100 %) is filled in the strong box and is consolidated in the centrifuge at 50 g preparing a normally consolidated soft soil (OCR ~ 1). Since the consolidation takes a long time, two strong boxes are prepared and consolidated simultaneously in the two baskets of the beam centrifuge. After consolidation of the slurries, the centrifuge is stopped and one strong box is replaced by counterweights for the next test phase.

Excess Kaolin is extracted from the surface of the clay layer so the total height of the soft soil layer becomes 20 cm. The geogrid including the load cells is now connected to the aluminium plate and the hopper mechanism is mounted. After spinning up the model to 50 g and a reconsolidation phase of about 1 to 2 hour the embankment is constructed in three stages. In each stage a layer thickness of 1/3 of the final height is poured. A consolidation phase follows after each construction step.

4.4 Test Results

Figure 12 shows the sequence of embankment construction at an acceleration level of 50 g. The toe of the embankment is in direct contact with the upper edge of the model wall, which

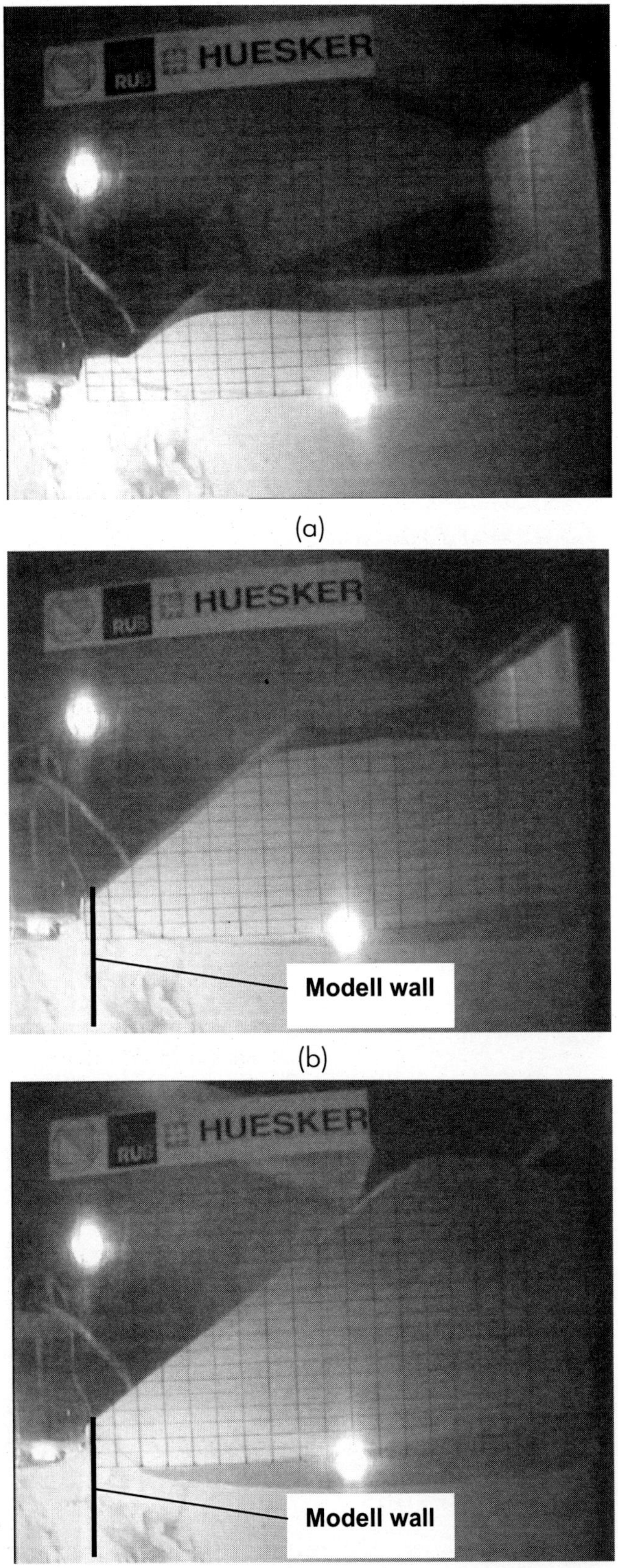

(a)

(b)

(c)

FIGURE 12: Construction of the embankment under increase acceleration in three stages

exceeds the surface of the soft clay. Besides the model wall the clay is slightly coloured close to the contact surface with the Perspex glas. This is for getting a better contrast for evaluating the displacement field around the wall by digital image correlation technique from additional camera sequences. The lighting of the front side of the strong box is optimized for the additional digital cameras observing the soil mass beside the model walls.

A slightly deformed clay surface is visible at the first construction step (Figure 12 a). With increasing time and increasing height of the embankment (Figures 12 b and c) settlements of the clay surface are observed, which show a maximum in the middle of the system (axis of symmetry, side wall of strong box). Compared to a conventional basal reinforced embankment, where significant settlements can be also observed under the toe of the embankment, in this case the settlement contour vanish towards the model wall. This shows the influence of the combined system of the walls connected with the tension membrane on the deformation behaviour.

The readings of the total stress transducer 7, located beside of the embankment, and of total stress transducer 3, located below the embankment, at the bottom of the clay layer are presented in Figure 13. The measured total stress is plotted versus time. The origin of the time axis corresponds to the start of the centrifuge.

First the reconsolidation of the clay has taken place before the embankment is constructed in the three stages. In addition the total stress values are marked, which have been predicted analytically from layer thickness, soil density and acceleration.

Pressure cell 7 outside of the embankment detects constant total stress close to the predicted value. Pressure cell 3 below the embankment shows at the beginning similar total stress to cell 7. With construction of the embankment total stress is increasing. After the first construction step the measured total stress matches well the predicted value. With further construction steps an increasing discrepancy between measured and predicted total stress is observed. This may be caused by the contribution of the tension membrane to the load transfer of the self weight of the embankment. The tension membrane transfers a part of the load to the vertical structure elements.

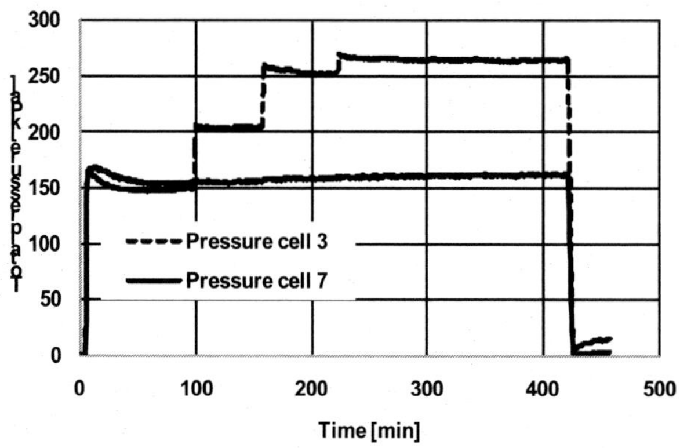

FIGURE 13: Readings of total pressure cells during reconsolidation and embankment construction versus time (model dimensions) and predicted total stress values for each construction stage (triangles)

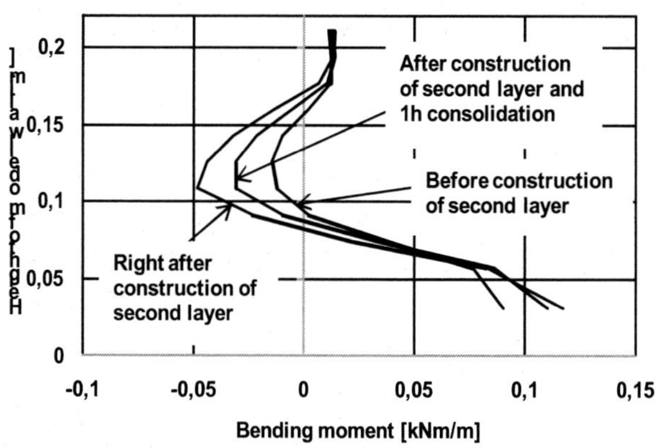

FIGURE 14: Bending moments distribution of the model wall before and right after construction of second embankment layer as well as after 1 h consolidation (model dimensions)

The bending moment distribution of the model wall measured before and right after construction of second embankment layer as well as after 1 h consolidation is shown in Figure 14. The distribution is typical for a wall embedded at the toe and supported at the top. Due to the construction of the new layer of the embankment the bending moments increase. With ongoing consolidation the bending moments decreases significantly. This decrease corresponds to excess pore pressure dissipation, increase of effective stress and therefore reduction of earth pressure coefficient. The vertical structure elements in combination with the tension membrane restrict the horizontal deformation of the soil block below the embankment and lead to a stress concentration with corresponding consolidation process. Due to this the soil below the embankment gains shear strength and stiffness.

5. CONCLUSION

An overview is given on concepts and techniques to realize soil movements in centrifuge model testing under increased gravity related to the simulation of deep excavations, tunnelling and embankment construction. It is shown that also under increased acceleration complex excavation or deposit procedures can be simulated. Modern control and measurement systems will broaden the technical capabilities and will expedite the developments.

The various methods show different levels of complexibility to display the real processes more or less detailed. To support an excavation area by a fluid or air pressure at 1g acceleration and during spinning up the centrifuge until the selected acceleration level is reached and to model the excavation by the release of the pressure is an approved method and can be conducted with relatively low technical effort. A corresponding test design can be developed and prepared within a short time. On the other side this leads to different stress paths in the soil compared to the real situation. The differences in stress path can be reduced by increasing the degree of detailing of the excavation equipment. The development of such systems is more costly and time consuming. Robot systems acting in 2D or 3D space allow a manifold adaption and more economic test designs although complex construction procedures are simulated.

The choice on an adequate technique depends on the aim of the tests. Whereas failure mechanisms could be sufficiently triggered and visualized with simplified techniques to capture changes in stress and strain for small deformations e.g. in the state of serviceability the modelling of correct stress paths requires more complex techniques. But in all cases it is important to figure out the processes dominating the mechanisms and effects which want to be discovered for the field situation. These processes should be modelled carefully under idealized but well defined boundary conditions.

REFERENCES

Allard, M. A., Hjortnaers-Pedersen, A. G. I. and Out, J. G., (1994). "Performance of a sand hopper". Proceedings of Int. Conf. Centrifuge94, Singapore, Balkema, pp 83-88.

Allersma, H. G. B., (1998). "Development of cheap equipment for small centrifuges". Proceedings of Int. Conf. Centrifuge98, Tokyo, Balkema, pp 61-67.

Allersma, H. G. B., (1994). "The university of delft geotechnical centrifuge". Proceedings of Int. Conf. Centrifuge94, Singapore, Balkema, pp 47-52.

Azevedo, R., (1983). "Centrifugal and analytical modelling of excavations in sand". Phd University of Colorado.

Azevedo, R.F. (1988). "In-flight centrifuge excavation tests in sand". Proceedings of Int. Conf. Centrifuge88, Paris, Balkema, pp 119-124.

Bezuijen, A, Schrier, J van der (1994). "The influence of a bored tunnel on pile foundations". Int. Conf. Centrifuge94, Singapore, Balkema, pp 681-692.

Bourne-Webb, P. J., Potts, D., König, D. and Rowbottom, D., (2011) "Analysis of model sheet pile walls with plastic hinges", Geotechnique, 61, Issue 6, pp 487 –499.

Bolton, M.D., Powerie, W. and Stewart, D.I. (1988). "The design of deep in-situ walls". Proceedings of Int. Conf. Centrifuge88, Paris, Balkema, pp 405-414.

Chambon, P., Corté, J. F., Garnier, J. and König, D. (1991). "Face stability of shallow tunnels in granular soils". Proceedings of Int. Conf. Centrifuge 94, Boulder, Colorado, Balkema, pp 99 – 106.

Davies, M. C. R., Parry, R.H.G., (1985). "Centrifuge modeling of embankments on clay foundations". Soils and Foundations, 25, Issue 4, pp 19-36.

Derkx, F., Merliot, E., Garnier, J. and Cottineau, L. M., (1998). "On board remote-controlled centrifuge robot". Proceedings of Int. Conf. Centrifuge98, Tokyo, Balkema, pp 97-103.

Detert, O., König, D., Schanz, T., (2012). "Centrifuge modeling of an adaptive foundation system for embankments on soft soils". Geotechniek, 10/2012, pp 46-49.

Furrel, R. P., Mair, R. J., (2010). "Centrifuge modeling of the response of buildings to tunneling". Proceedingsof International Conference on Physical Modelling in Geotechnics, Zürich,Switzerland, Balkema, pp 549-554.

Haigh, S. K., Houghton, N. E., Lam, S. Y., Li, Z. and Wallbridge, P. J., (2010). "Development of a 2D servo-actuator for novel centrifuge modeling". Proceedings of International conference on Physical Modelling in Geotechnics, Zürich Switzerland, Balkema, pp 239-244.

Gaudin, C., Garnier, J., Gaudicheau and Rault, G., (2002). "Use of a robot for inflight excavation in front of a embedded wall". Proceedings of International Conference on Physical Modelling in Geotechnics, Canada, pp77-82.

Jessberger, H. L. and Güttler, U., (1988). "Bochum geotechnical centrifuge". Proceedings of Int. Conf. Centrifuge88, Paris, Balkema, pp 37-44.

Kimura, T., Takemura, J., Hiro-oka, A., Okamura, M. and Park, J., (1994). "Excavation on soft clay using in-flight excavator". Proceedings of Int. Conf. Centrifuge94, Singapore, Balkema, pp 649-654.

Kimura, T., Takemura, J., Hiro-oka, Suemasa, N. and Kohda, M., (1993). "Stability of unsupported and supported vertical cuts in soft clay". 11th Southeast Asian Geotech. Conf., Balkema, pp 61-70.

König, D., (2002). "Modeling of deep excavations". Proceedings of International Conference on Physical Modelling in Geotechnics, Canada, pp 83-88.

König, D., (1998). "An inflight excavator to model a tunneling process". Proceedings of Int. Conf. Centrifuge 98, Tokyo, Balkema, pp 707 – 712.

König, D. Güttler, U. and Jessberger, H. L., (1991) "Stress redistributions during tunnel and shaft constructions". Proceedings of Int. Conf. Centrifuge91, Boulder, Colorado, pp 129 – 135.

Lade, P. V., Jessberger, H. L., Makowski, E. and Jordan, P, (1981) "Modelling of deep shafts on centrifuge tests". Proc. 10th ICMFE, Vol. 1, pp 683-691.

Lam, S. Y., Elshafie, M. Z. E. B., Haigh, S. K. and Bolton, M. D., (2010). "Centrifuge modeling of deep excavation in soft clay",Proceedings of International Conference on Physical modelling in Geotechnics Zürich Switzerland, Balkema, pp 475-480.

Loh, C. K., Tan, T. S. and Lee, F. H., (1998). "Three-dimensional excavation tests". Proceedings of Int. Conf. Centrifuge 98, Tokyo, Balkema, pp 650 – 655.

Lyndon, W. and Pearson, R. A., (1984). "Pressure distribution on a rigid retaining wall in cohesionless material", Proc. of the Symp. on the application of centrifuge modeling to geotechnical design, Manchester.

Mair, R. J., Rain, P., Schofield, A. N. and Taylor, R. N., (1984). "Application of centrifuge modeling to the design of tunnels and excavations in soft clay". Proc. Of Int. Symp. on the application of centrifuge modeling to geotechnical design, Manchester, pp 356-366.

Nomoto, K., Mito, K., Imamura, S., Ueno, K. and Kusakabe, O., (1994). "A miniature shield tunneling machine for a centrifuge". Proceedings of Int. Conf. Centrifuge94, Singapore, Balkema, pp 699-704.

Potts, D. M., (1976). "Behavior of lined and unlined tunnel in sand" PhD Thesis, Cambridge University, Churchill College.

Schürmann, A. and Jessberger, H. L., (1994). "Earth pressure distribution on sheet pile walls". Proceedings of Int. Conf. Centrifuge94, Singapore, Balkema, pp 95-100.

Springman, S., Bolton, M., Sharma, J. and Balachandran, S., (1992) "Modelling and instrumentation of a geotextile in the geotechnical centrifuge". Int. Symp. On Earth Reinforcement Practice, Balkema, Rotterdam, pp 167–172.

Shama, J. S., Bolton, M. D. and Boyle, R. E., (2001). "A new technique for simulation of tunnel excavation in a centrifuge". ASTM Geotechnical testing Journal, 24 , Issue 4, pp 343-349.

Taylor, R. N., (1995). "Geotechnical centrifuge technology" Chapmann & Hall, ISBN 0 7514 0032 7.

Toyosawa, Y, Horii, N., Tamate, S., Hanayasu, S. and Ampadu, S. K., (1994). "Deformation and failure characteristics of vertical cuts and excavations in clay". Proceedings of Int. Conf. Centrifuge94, Singapore, Balkema, pp 655-663.

Walter, H., Coccia, C. J., Wallen, R. B., Ko, H. Y. and McCartney, J.S., (2010). "Modelling of a tunnel lining deformation due to face instability". Proceedings of International Conference on Physical Modelling in Geotechnics, Balkema, pp 593-598.

Wong, K. S., Ng, C. W. W., Chen, Y. M. and Bian, X. C., (2010). "Centrifuge modeling of passive failure of tunnel face in saturated sand". Proceedings of International Conference on Physical Modelling in Geotechnics, Balkema, pp 599-604.

Yeo, C. H., Lee, F. H., Hegde, A. and Juneja, A., (2010). "Centrifuge modeling of a steel pipe umbrella arch for tunneling in clay", Proceedings of International Conference on Physical Modelling in Geotechnics, Balkema, pp 611-616.

Yoshimura, H., Miyabe, K. and Tohda, J., (1994) "Response of a tunnel lining due to an adjacent twin shield tunneling".Proceedings of Int. Conf. Centrifuge94, Balkema, pp 693-698.

Zheng, G., Wei, S. W., Peng, S. Y., Diao, Y. and Ng, C. W. W., (2010). "Centrifuge modeling of the influence of basement excavation on existing tunnels". Proceedings of International Conference on Physical Modelling in Geotechnics, Balkema, pp 523-527.

Modelling in a Geotechnical Drum Centrifuge

J. Laue, J.N.F. Gautray, W.F. Morales and A. Askarinejad

Institute for Geotechnical Engineering, ETH Zürich, Zürich, Switzerland
E-mail: jan.laue@igt.baug.ethz.ch; jean.gautray@igt.baug.ethz.ch; amin.askarinejad@igt.baug.ethz.ch; ferney.morales@igt.baug.ethz.ch

ABSTRACT

Physical Modelling under enhanced gravity in a geotechnical drum centrifuge offers different opportunities and faces to some extent different challenges than physical modelling in a geotechnical beam centrifuge, Following a short descriptions of the developments made at the Swiss Federal Institute of Technologies geotechnical drum centrifuge, which has been inaugurated in 2001, this opportunities and challenges will be discussed based on three examples of recent research work. This projects covers the modelling of the behaviour of river levees during flood events, the performance of natural slopes during heavy rainfall events and the modelling of stone columns as ground improvement measure for soft soil. Focus is given in this contribution on the modelling techniques.

1. INTRODUCTION

Physical modelling in a geotechnical drum centrifuge faces - to some extent - different challenges compared to modelling in a beam centrifuge. On the other hand there are advantages as size of the model is less limited in one of the dimensions, which offers flexibility in the size of a test setup. In addition, the access to the model during operation is easily made possible using a tool platform or tool table, which can be controlled independent of the soil model and can be stopped and accessed while the soil model is still spinning.

The geotechnical drum centrifuge, inaugurated in 2000, has an outside diameter of 2.2m with a depth of the channel of 30cm. Strongboxes can be placed on base plates with an even surface reducing model heights to a maximum of 20 cm. The setup of the centrifuge is shown in Figure 1.

The use of the centrifuge developed over the years with topics needing an increasing complexity of the test designs. After the challenges of preparing models in flight as well as model making in the laboratory floor with transport and inbuilt in the stationary centrifuge have been solved (comp. Laue et al. 2005, Weber et al. 2006) main focus of research projects have been laid to modelling of soil structure interaction problems of foundations (e.g. Nater, 2007; Arnold, 2011) and improvement of soft soil (e.g. Weber, 2008; Pooley et al. 2009) as well as modelling problems raised by mainly gravity driven natural hazards. The latter included modelling of wet- (Bowman et al. 2010) and dry debris flows (Imre, 2010) as well as simulation

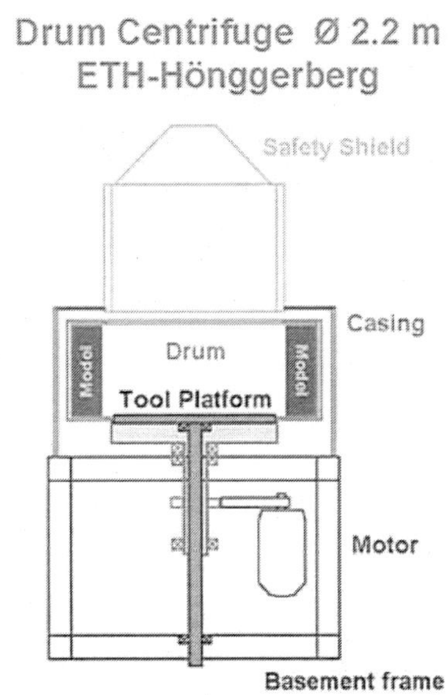

FIGURE 1: Setup of the geotechnical drum centrifuge at ETH Zürich (Springman et al. 2001)

of rock fall events (Chikatamarla, 2006). Measurement techniques developed from classical solutions towards the use of flexible pressure pads to measure stress distributions (Springman et al. 2002), use of optical fibres (Kapogianni et al. 2010) or TDRs. Observational tools allow the use of a fast camera (Chikatamarla et al. 2004) or a set of several cameras to evaluate the data in a geodetically fashion (e.g. Akca et al. 2010 used in Askarinejad, 2012). Data acquisition developed towards faster acquisition and data from the tool table can be transferred to the stationary world by use of a WLAN connection. In most cases a tool table with a fixed built-in actuator tower has been used to apply actions onto the soil model. The use of this actuator tower (Figure 2) allows 3 dimensional access to the model as the arms of the actuator can be moved in and out as well as up and down while movements of the toolplate relative to the channel enables movement in the third direction.

FIGURE 2: Top view on the actuator tower installed in the centrifuge. For safety reasons relative movements between channel and drum are prevented by the use of chains in this particular case

In this contribution the recent developments in test designs and modelling procedures are summarised based on recent contributions to the conference Eurofuge at Geodelft in April 2012 and ISGI in May 2012 (Morales et al, 2012 a,b; Askarinejad et al. 2012, Gautray et al. 2012).

2. MODELLING OF LEVEES

2.1 Motivation

Understanding the performance of levees during a flood is necessary when designing governmental prevention and assistance plans. This acquires more relevance as flood events occur more frequently, cost of hazards in terms of materials and life have increased in the last decades and land use proceeds further into former flood plans (e.g. Hoyois & Guha-Sapir 2003, Bezzola et al. 2008).

Complete modelling of river levees prior and during breaching should include a transient water level analysis reflecting the state of the levee prior to the overflow process. Springman et al. (2008) and Mayor et al. (2008) show how this has a strong influence on the unsaturated condition of the levee by conducting a field test on an existing dyke supported with numerical analyses. They showed for this specific case that the unsaturated seepage is an important factor in the analyses as the failure mechanism of levees is closely related to the change of matric suction of unsaturated soils.

Previous research projects under an increased gravity field were often limited by space in the available strongboxes or by the lack of an effective water control prior and during overflow. This caused models to be created with steep slopes or models that had to cut either the air or the water side slope, as was done by Cargill & Ko (1983), Kusakabe et al. (1988), Okumura et al. (1998) or Seo et al. (2006) among others.

Using the advantages of the geotechnical drum centrifuge an attempt has been made to overcome these restrictions. It is composed of a new strongbox, a new device to allow water levels to be controlled and a novel model construction procedure. The new system allows all possible water conditions in a levee to be studied in one model.

2.2 Challenges

2.2.1 Strong Box

The test setup of the model is shown in Figure 3. The form of the box is an annular sector of dimensions 1000x500x 300mm. The length (1000mm) is determined by the top and bottom plates). The height of the box and thus the width of the model (500mm) is given by the length the connection struts, which are designed so that the model can stand without both of the lateral walls. The width (300mm) is given by the length of the shortest dimension of the side walls so that the safety shield and thus the start and stop function of the tool plate can be used in further tests.

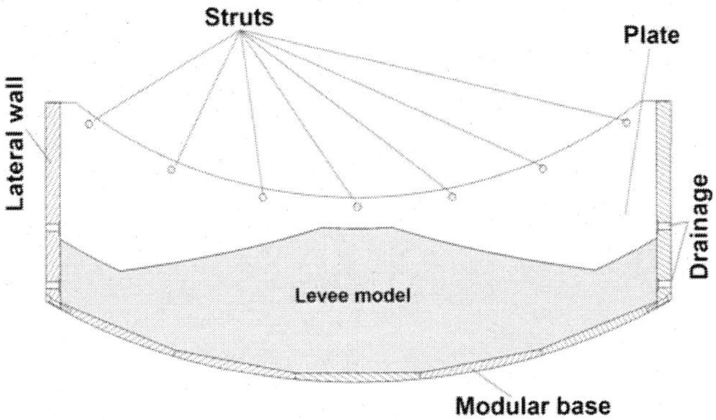

Figure 3 Scetch of the model to study river levve braches (Morales et al. 2012a)

This new box allows levee models to be built with different slope gradients and levee heights. For instance, having a 1:2.5 slope gradient, a model of 0.2m height (representing 20m at 100 g) can be built inside the box. Models can be built outside the centrifuge in the preparation laboratory. The material is subjected to unsaturated condition, in which suctions help the model to remain stable, so it can be brought into the drum. This issue will be discussed in the further sections.

The new strongbox (Figure 4) is composed of two plates (bottom and top) representing the side walls in the model perspective, two lateral walls, which are tilted against the g-field and a curved modular base and seven connecting struts. The box is fixed to the channel of the drum centrifuge by eight M12 screws.

The new strongbox has a similar weight than the usually used strong boxes model dimensions of 40*40*20cm or 40cm in diameter. It can hold about 6.3 and 4.9 times the volume of the cylindrical and rectangular boxes, respectively. This allows larger models to be created, which may be more representative of the physical problems to be analysed. The new box was also designed to allow large-size measuring sensors to be inserted as inlets for PPTs as well as for Time Domain Resistivity measurements to be incorporated.

As weight – for practical installing purposes and stability - are both concerns for the design of the strong box, the design has been improved, after conducting the basic design, using a product life cycle management system (Morales et al. 2012b). This system allows design and calculation of a structure based on a CAD framework till manufacturing in CEC devices.

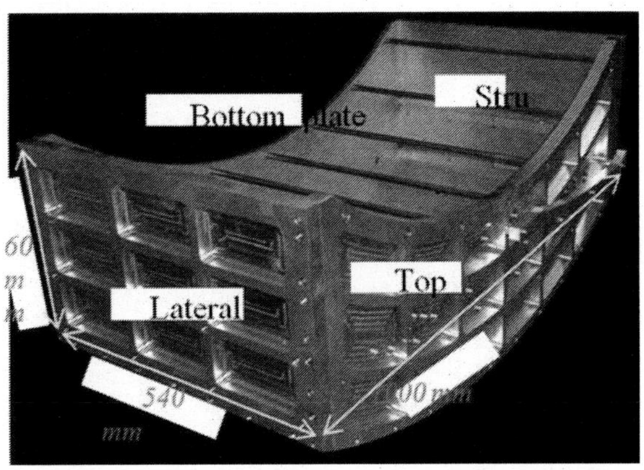

FIGURE 4: Picture of the model container to study river levee breaches including dimensions (Morales et al. 2012a)

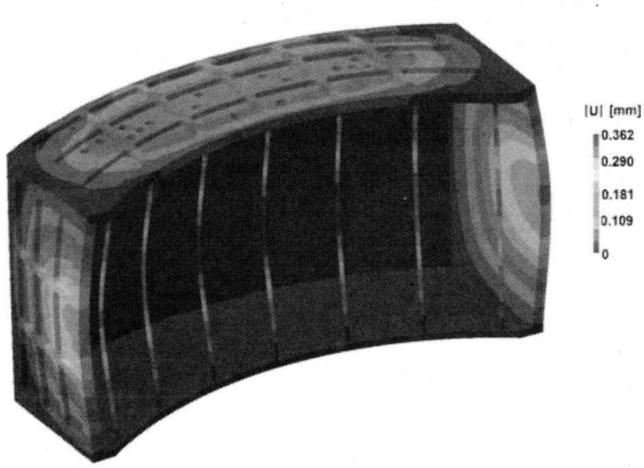

FIGURE 5: Deformations of the model container under full filling at 50g following a calculation using a PLM calculation (Dassault Systems, 2000, Morales et al. 2012b)

2.2.2 *Water Infill System*

Beneath the container allowing for a reasonable size of the model, the control and fast inlet of water faces the next challenge of modelling the water household of levees. Usually flood events in alpine regions occur relatively quick in hours and days in real scale. Nevertheless, this has to be scaled – given laminar flow conditions during phases without overflow and using water as fluid – by a factor of n^2 - in a model tested in a centrifuge. Thus a new system of incorporating water had to be implemented for this research project.

A system is designed to set the water levels and the transient cycles of raising and decreasing the water level followed by holding a constant water level during an overflow phase. A two-chamber box has been developed to provide a controlled water level to the system. A thin wall separates two chambers of this device, as shown in Figure 6 and 7. The device is connected to the arm of one of the two actuators placed on the tool plate of the drum centrifuge. The location of the box can be varied along a radius from the centre of the drum, as shown in Figure 8, so that different positions, hence water levels, can be set in-flight. The external dimensions of the device are 200x100x85mm with 10mm thick walls, enclosing a volume of 1.22 litres. The device has a maximum discharge rate of 1500ml/s connected by three tubes with the container of the model.

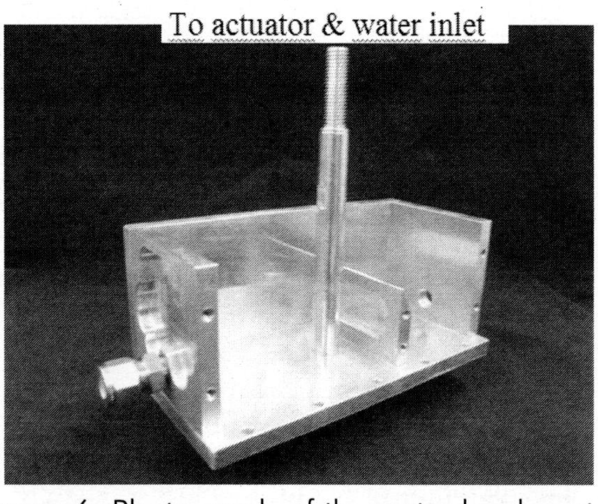

FIGURE 6: Photograph of the water level control device, (Morales et al. 2012a)

Water flows continuously through 3 pipes connected to an external water supply into the bigger chamber of the device. The outlet of this chamber is connected to the upper drainage line of the strongbox by a plastic hose of 10mm in diameter. When the water reaches the height of the dividing wall, it overflows into the second chamber, which lets the water flow out of the system, maintaining a fixed height given by the height of the dividing wall, as shown in Figure 8.

The water surface is curved due to the acceleration field. The water height in both the water-level device and the strongbox is the same, due to Archimedes' principle. The water level is measured on the water side of the levee by a PPT sensor.

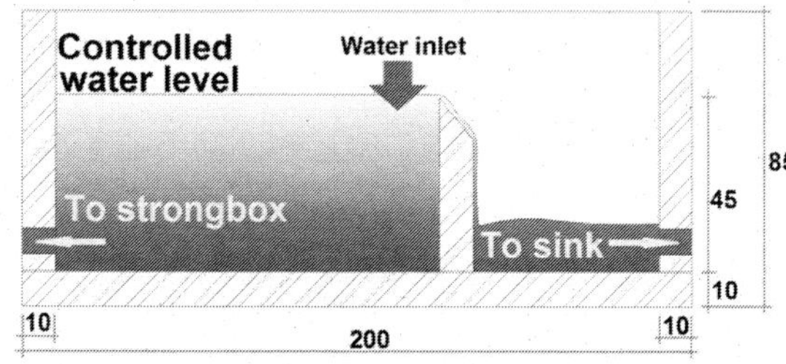

FIGURE 7: Sketch of the water level control device, Morales et al. (2012a)

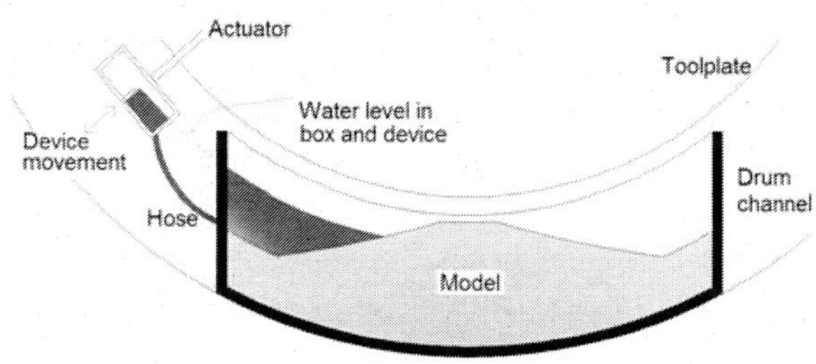

FIGURE 8: Principle sketch of water level control in river levee physical model tests using Archimedes principle (Morales et al. 2012a)

2.2.3 Model Making Procedures

A basic challenge of physical modeling using containers in a drum centrifuge is the undisturbed transport of the model prepared on the laboratory floor into the machine. During this procedure the model has to be saturated and dewatered to cause unsaturated conditions to allow enough suction to be developed to remain able while being tilted by 90°. This had been confirmed by modelling of the stability of the sample using both, limit equilibrium as well as finite element calculations (Roddemann, 2003). This calculation showed that deformations of 0.15mm on the top of the model could be expected (Morales et al. 2012). The estimated factor of safety for stability of the model with the c-phi reduction procedure has been 1.56. A two-dimensional analysis of the central section with the Morgersten-Price method leads a value of 3.84. Figure 9 shows the failure surface for that analysis. It is a wedge almost triangular, which had been observed already in previous tests on very low dense models (Laue et al., 2005). It contrasts with the displacements obtained with the c-phi reduction method. The difference in the failure mechanism might explain the difference in the estimated factors of safety.

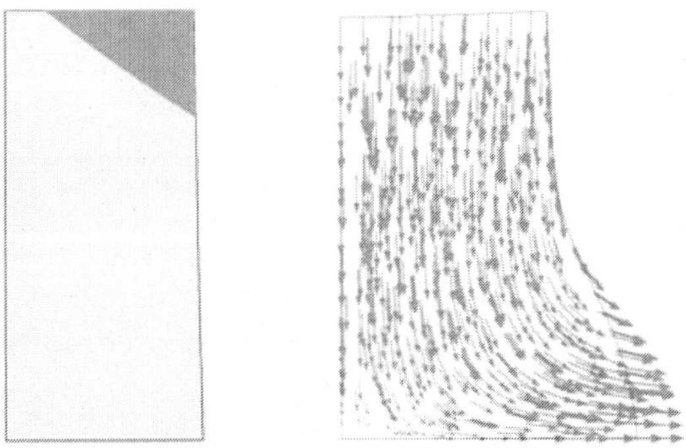

a) Limit equilibrium. b) C-phi reduction.

FIGURE 9: Failure mechanism of the centre of the model calculated with limit equilibrium method and numerical methods using a c –phi reduction (Morales et al. 2013)

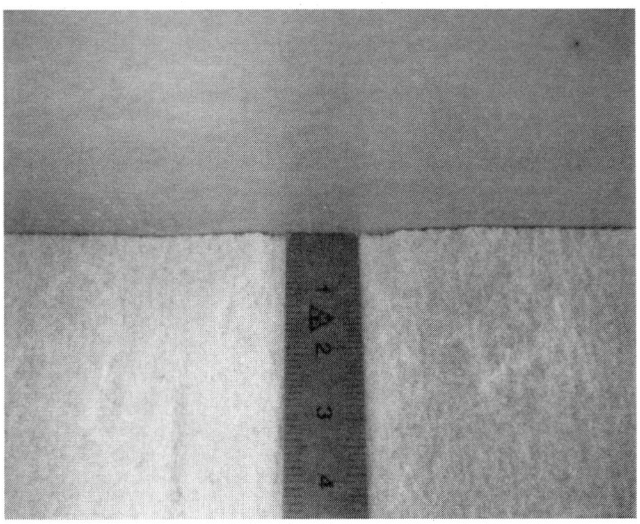

FIGURE 10: Gap occurring on the top of a sand sample, Morales et al., 2013

Installing the model into the centrifuge unfortunately caused a gap of of approximately 0.5mm to develop (Figure 10). This deformation, caused by vibrations during the installation procedures is supposed not to be a problem in most of the test setups but affects the performance of the dyke model, Thus the gap problem is overcome by taking two measures. First, a PTFE (generic term for Teflon®) film is stuck to the bottom of the drum channel, as well as to the bottom of the strongbox. A Teflon-Teflon contact has an approximate coefficient of friction of 0.04, reducing the required pushing force of sliding the model container into the drum channel.

The second measure consists of using a thin layer of a soil composed of a mix of the sand with an expansive material (bentonite MX-80) at the upper part of the model. This material does not prevent the gap, but once it starts being in contact with water the expansion of the material closes the gap. Figure 11 shows the results of the test. The content of bentonite changes the permeability of the material, as well as the time required for soaking the sample, hence the x-axes of the chart is a normalized time of the total time and the time required to soak the sample. The time to soak the sample is given beside each curve.

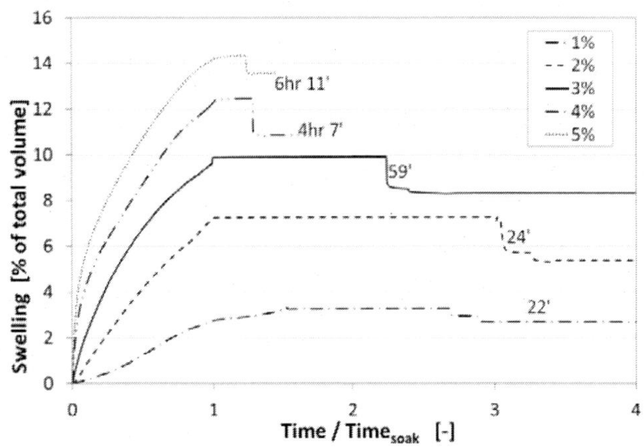

FIGURE 11: Results of swelling tests of bentonite – Perth sand mixtures, Morales et al., 2013

Figure 12 shows a model using the bentonite mixture in the upper section. No visible crack could be observed and no weak section in the model was found during testing.

2.3 Results and Outlook

Results of the on-going tests are promising. Various combinations of slope angles and stabilisation measured like walls and filters are studied using the physical models in this research work. As an example, a failure mechanism, which has recently been modelled using a cut of wall is given in Figure 13.

Figure 12 Dyke sample with a 2 cm layer of sand bentonite mixture placed into the channel of the drum showing now crack at the top, Morales et al., 2013

FIGURE 13: Visual of failure mechanism observed on a sample of a dyke with cut off walls after two flooding events without overflow.

3. MODELLING OF RAINFALL INDUCED LANDSLIDES UNDER CLIMATE CONTROLLED CONDITIONS

3.1 Motivation

Landslides induced by rainfall impose considerable damage to infrastructure and cause major casualties worldwide. Several attempts have been undertaken to investigate the triggering mechanisms of these frequent and often powerful natural hazards analytical approaches, numerical simulations and also physical modelling techniques. Physical modelling of landslides due to rainfall covers the wide range of full scale testing, small scale laboratory experiments under 1g conditions (a summary of the references can be found e.g. in Askarinejad et al. 2012) or in geotechnical centrifuges (Kimura et al., 1991, Take et al., 2004, Hudacsek et al., 2009, Ling et al., 2010 among others). Different aspects of the problem have been addressed but less emphasis has been put on the hydro-mechanical interactions between the bedrock and the soil slope.

Accordingly, a climate chamber has been designed and constructed for the use in the geotechnical drum centrifuge at the Swiss Federal Institute of Technology in Zurich (ETHZ) in order to simulate and replicate the boundary conditions of two landslide experiments, which have been conducted in October 2008 and March 2009 on a 38° steep slope on the east facing banks of the river Rhine in northern Switzerland near the village of Ruedlingen (Springman et al., 2010, Askarinejad et al., 2012b). In this experiment, an amount of about 1500 mm rainfall was sprinkled over the slope within 4.5 days during the first experiment with more concentration on the lower part where large fissures in the bedrock were detected. No failure was observed in the slope after this extreme event, although the whole profile of the soil cover had very low values of suction and even though piezometers showed slight increases (maximum

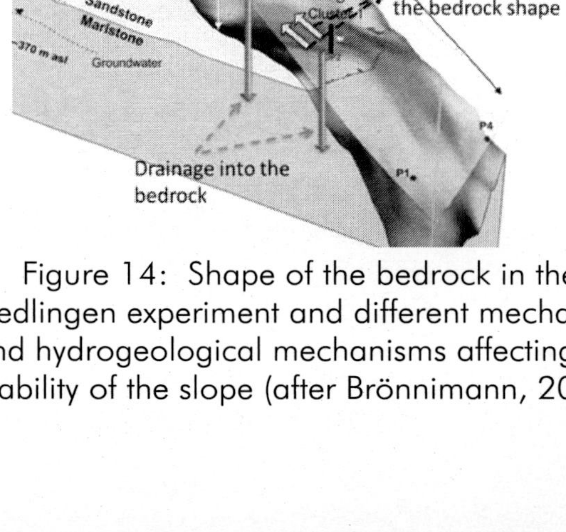

Figure 14: Shape of the bedrock in the Ruedlingen experiment and different mechanical and hydrogeological mechanisms affecting the stability of the slope (after Brönnimann, 2011)

of 170 mm) in the water table. The depth of the bedrock varied between 0.5 m to more than 5 m. A convex bulge was identified in the lower part of the slope, which was thought to have a potentially stabilising effect. The shape of the bedrock allocated is illustrated in Figure 14 based on the probing results and geological observations.

In March 2009, the rainfall was concentrated in the upper part of the slope where less root reinforcement and shallower bedrock were expected. In this second experiment, rainfall with a total amount of 225 mm within 15 hours was enough to trigger a 150 m^3 landslide. The piezometers located in the failed area showed significant rises (~1 m) in the groundwater table above the bedrock. These measurements can be attributed to the shallower and less fissured rock in the upper part of the slope. Moreover, based on the photographs taken during the failure, water was exfiltrating from the back of the failure wedge. Together with shallower bedrock and less vegetation effects, this exfiltrated water might have been a key factor in destabilising the slope.

In order to investigate the triggering mechanisms of landslides due to rainfall, a climate chamber had to be developed allowing for controlled conditions of rainfall intensity and duration, ambient relative humidity and drying phases caused by wind and temperature. As the tests to be conducted in the climate chamber aimed at studying the possible failure mechanisms proposed for the full scale landslide experiment, different shapes and hydraulic properties of the bedrock, in terms of drainage and exfiltration, had to be provided for the model as well. A three dimensional close range photogrammetric technique had been chosen to track the movements and monitor the volumetric changes of the ground during cycles of wetting and drying. In addition, a high speed camera could be used from the side of the climate chamber and the deformation vectors and strains could be elaborated using the PIV method.

3.2 Climate Chamber and Rain Simulator

The main goal of the climate chamber is to provide the possibility to investigate several triggering mechanisms of landslides due to rainfall and hydro-mechanical interactions with the bedrock. Therefore, a rain simulator has been designed that enables the user to apply different rain intensities on the slope with different durations according to the scaling laws. The size of the droplets and the *Coriolis Effect* (Schofield, 1980) had to be taken into consideration when planning application of the rainfall. The size of the rain drops in a prototype varies from 300 microns for fine drizzles to more than 2.6 mm in diameter for heavy rain and squalls, depending on the rain intensity (Tamate et al., 2010). The prototype rain intensities when performed at 50g in the tests of this study are more than 1 mm/hr and accordingly the model rain drops are scaled 50 times smaller than the real rain drops of an intense rainfall and they are 31.4 to 54.9 microns in diameter. The rainfall is applied on the slope by means of non-uniformly distributed nozzles, which have been extended below the model roof to a uniform height above the slope surface to reduce the Coriolis effect without washing fines

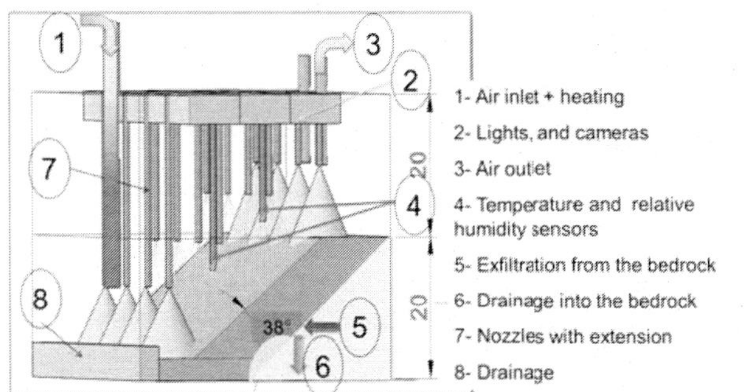

FIGURE 15: The conceptual schematic of the climate chamber with dimensions in centimetres (Askarinejad et al. 2012)

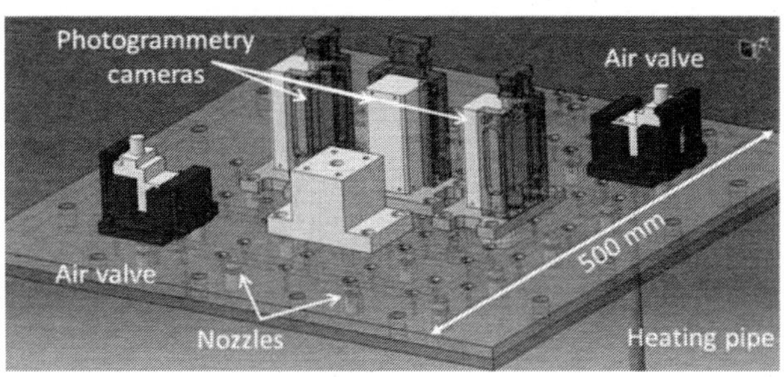

FIGURE 16: The schematic of the top plates of the climate chamber and location of the photogrammetry cameras. The nozzles are connected with two separate channels (Askarinejad et al. 2012)

out, which happens when the nozzles are placed to close to the surface. Drainage and exfiltration on the slope is provided by drains at a height of 1/3 of the slope.

In the drying phase of the tests, it is necessary to control the process of evaporation by means of air flow and increase in the ambient temperature. The air flow is provided by means of two air valves, one of which is installed in the direction of the centrifugal rotation and the other facing in the opposite direction. These two valves open and close together. The pipe connected to the inlet valve is covered by a high resistance wire coil to heat the air. The relative humidity and temperature of the chamber is measured at two points over the slope and these measurements, together with the suction build up in the soil, provide information to control the evaporation process.

The rain simulator of this climate chamber is composed of a central water tank (CWT), magnetic valves, pipes, water channels grooved into the top plates of the chamber, and the nozzles (Figures 16 and 17). The CWT is a cylindrical aluminium container that is installed in the middle of the tool table on a platform with a capacity of 24.6 litres. The water in the CWT is pressurised by a smaller air tank to 10 bar and regulated by a valve with an output pressure at 2.5 bars to ensure there is enough water pressure as the water reaches the nozzles. The water height in the CWT is monitored by a pressure difference sensor (PDS-Keller PD11), which is connected to the top and bottom of the CWT and measures the difference in the pressure at these two points.

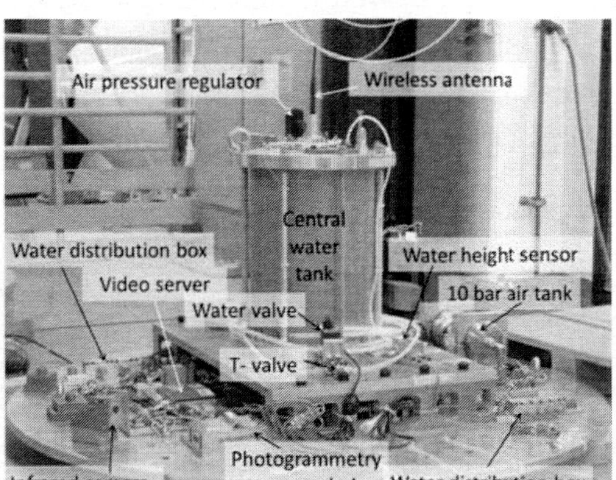

FIGURE 17: Water supply system to the climate chamber mounted on the tool table (Askarinejad et al. 2012)

The water flow from the CWT is controlled by means of two magnetic water valves. These valves are connected to two T-valves. High pressure pipes convey the water from the tool table to the climate chambers, which are installed completely diametrically opposite to each other in the centrifuge channel. The pipes are connected to the top plates of the climate chamber. The inner part of the upper plate is grooved to provide two systems of channels for the water flow. The inner channel is connected to 4 inner nozzles and the outer channel to 18 outer nozzles. The two plates are attached together by especial epoxy glue and several screws along the channels to make them completely water tight. This separate channel system provides the possibility of applying 3 different rain intensities during the tests. Technical

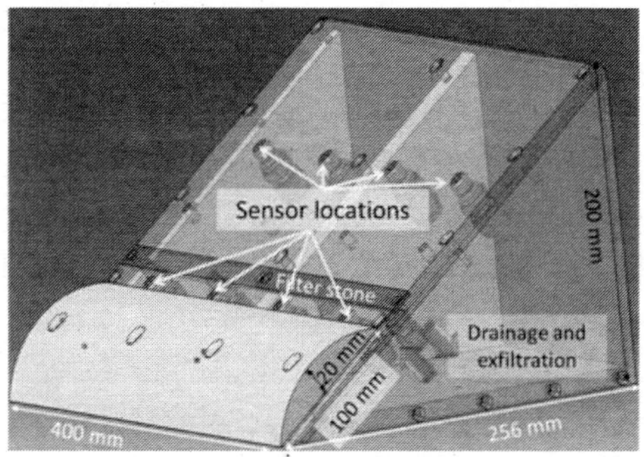

FIGURE 18: The bedrock, location of the sensors, and the hydraulic interaction between the soil and rigid bedrock. (Askarinejad et al. 2012)

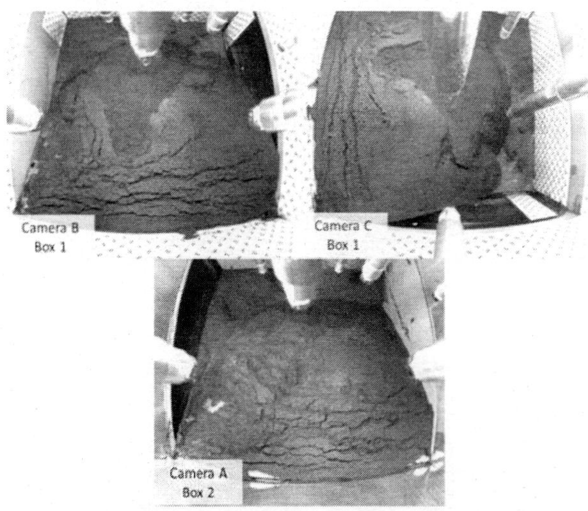

FIGURE 19: Pictures from different photogrammetry cameras from two parallel tests showing failure scarp and tension cracking.. (Askarinejad et al. 2012)

details of the nozzles used can be found in Askarinejad et al. (2012).

An aluminium triangular frame, with an inclination of 38°, is constructed to provide a rigid underlying layer for the soil slope. The top surface of this *bedrock* is roughened by gluing soil on it. A semi-circular element is also constructed to model the *buttress effect* (Doglioni et al., 2011) at the lower part of the slope (Figure 18). A shallow notch is grooved laterally on the top surface to install filter stones below, from which there are connections to 4 water pipes for drainage or exfiltration. The top surface of the *bedrock* has 8 places to install different sensors, such as *Druck PDCR81* or *Keller 2MIX* Pore Pressure Transducers (PPTs). The sensors can be located either at the interface between the bedrock and the slope, or inside the soil mass. Figure 19 shows an exemplary tests result of two parallel tests conducted with silt from the site of the field experiment.

4. MODELLING OF STONE COLUMNS

4.1 Motivation

Stone columns have proven to be an efficient ground improvement technique in soft soils. They stiffen the ground and provide drainage while changing the preferred drainage path from vertical to horizontal. However, smear and plastic annuli appear around these columns during installation. While plastic hardening supports an inclusion and promotes settlement control, the smear and plastic zones can also have a negative influence on the efficiency of the accelerated drainage and the bearing capacity. Following a successful research work on the behavior of stone columns loaded with an in flight built embankment (Weber, 2008) main focus of the research is given to the distribution and extent of the disturbed and plastic zones. This is accompanied by a detailed analysis of the pore pressures developing at different locations around the stone columns and separation of the influences of the various phases during the installation of the columns.

4.2 Installation of Stone Columns

A specially designed apparatus (Weber et al, 2005; Figure 20 and 21) was used to install the stone columns into a clay model. This stone column installation tool is connected to the actuator tower.

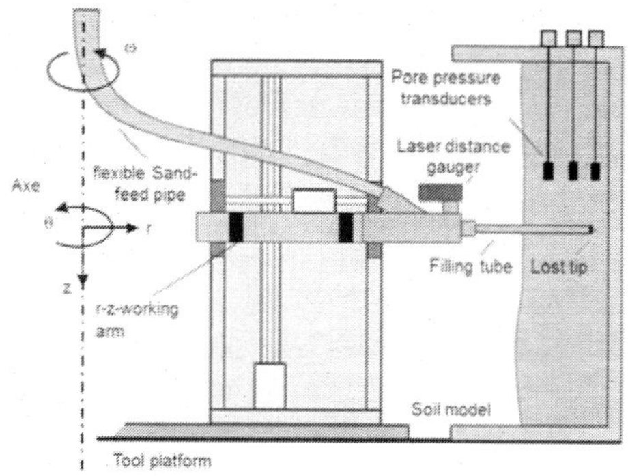

FIGURE 20: Sketch of the stone column installation tool (Weber et al., 2005)

FIGURE 21: Stone column installation tool (Weber et al. 2006)

FIGURE 22: Set up of the tool table with the stone column installation tool (Weber, 2008)

The installation tool has a diameter of 10mm and is driven at a constant speed into the model till the desired depth of the column is reached. A drawing pin is installed on the surface of the model in order to locate the stone column to be constructed and to block the extremity of the installation tool and so to prevent any plugging of the tip of the tool. Once the installation tool reaches the desired depth, quartz sand, used as column material, is fed from the top of the centrifuge from the stationary world through a flexible pipe to the tool (Green tube in Figure and 22). Installation can be conducted in various regimes. Usually a compaction regime was used withdrawing the installation tool for 15mm in model scale and pressing it back into the clay model for 10 mm (compare Fig. 23).

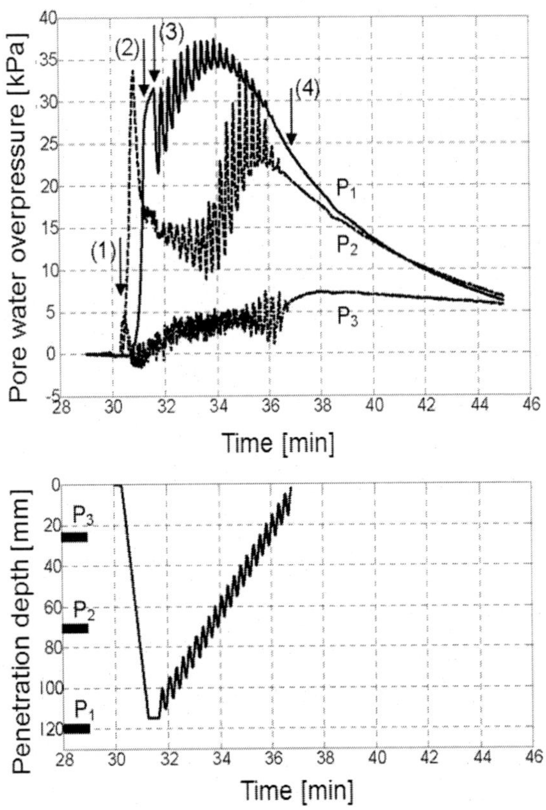

FIGURE 23: Installation regime (bottom) and recorded pore pressure distributions (top) during stone column installation. The arrow (1) indicates the penetration of the installation tool into the soil model. The arrow (2) corresponds to the point when the tip of the installation tool reaches the desired installation depth and when sand is fed into the feed pipe (cf. Figure). The compaction process starts by arrow (3) while arrow (4) marks the end of the building process. The distance of the PPTs to the axis of the column is 24mm (Weber et al. 2006)

Investigations on the behaviour of stone columns loaded with embankments built in flight have been conducted both on laboratory prepared slightly over consolidated samples and on almost normally consolidated samples (Fig. 24 and 25). The model soil used is a reconstituted silty clay from a construction site in the suburbs of Zurich (Birmensdorf). Its properties are summarized in Table 1. To build the embankment, lead shots with a filling density of about 6.72 g/cm^3 are used to overcome the limits in height of the centrifuge. The embankment was installed in a similar way than the stone columns (Figure 22).

Table 1: Properties of the Reconstituted Birmensdorf Clay (after Panduri, 2000, Küng, 2003)

USCS Classification	*CH*
Clay particle content from sedimentation analysis < 2μm [%]*	42
Liquid limit w_l [%]*	45-62 (av. 60)
Plastic limit w_p [%]*	18-26 (av. 21)
Plasticity index I_p [%]*	27-36 (av. 30)
Effective angle of friction φ_{cv} [°]	24.5
Effective cohesion c' [kN/m²]	0
Grain specific gravity G_s	2.75
Grain size d_{50} [μm]	4

* by weight

FIGURE 24: Test Set up for two tests on a laboratory prepared sample. Comparison of an unreinforced are (left) and an area reinforced with an area ratio of improvement of 10% (Weber, 2008)

FIGURE 25: Top view of two test sections using the whole channel for the soil model varying the area ratios of improvement between 5% and 10 (Weber, 2008)

4.3 Exemplary Test Results

Test results shown here are presented for one case using a laboratory prepared sample with an area ration of improvement of 10%. The sample was consolidated in the laboratory to a vertical stress of 100kPa leading to an over-consolidated behaviour (cf. Figure). The undrained shear strength of the soil sample has been derived using a T-bar. From the measured values, the parameters a = 0.24 and b = 0.93 could be derived for the equation 1.

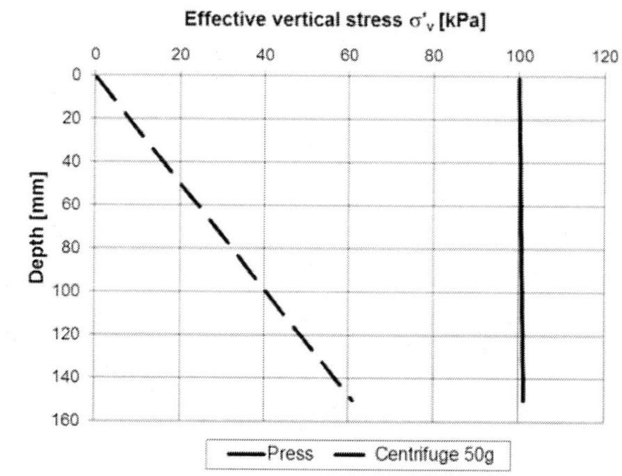

FIGURE 26: Comparison of the vertical stresses under the press and in the centrifuge (Weber, 2008)

$$s_u = a * OCR^{b} \quad (1)$$

The test has been conducted at 50 times gravity and the installation of the stone columns was performed to a depth of 75mm, 100mm or 120mm (given by the available length of the flexible installation pipe) with a compaction regime of 15/10mm.

After installation of the stone columns, the centrifuge had to be stopped and a model geotextile has been placed on the surface of the model to prevent punching of the lead balls. The installation sequence of the lead balls is given in Figure 28.

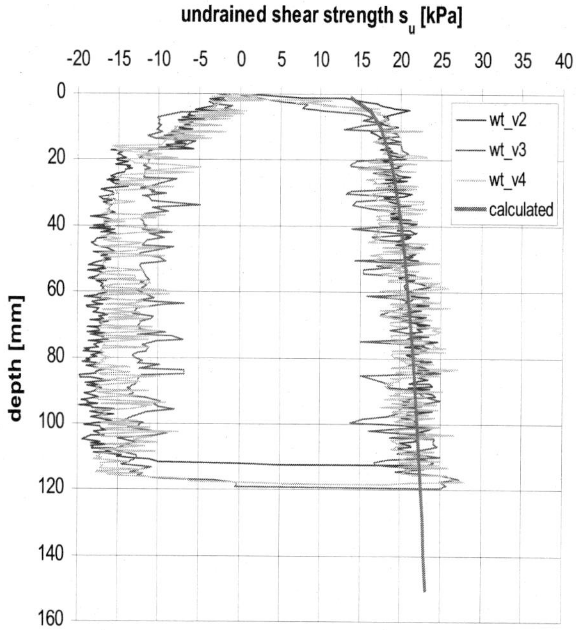

FIGURE 27: Results of T-Bar tests before installing stone columns (Weber, 2008)

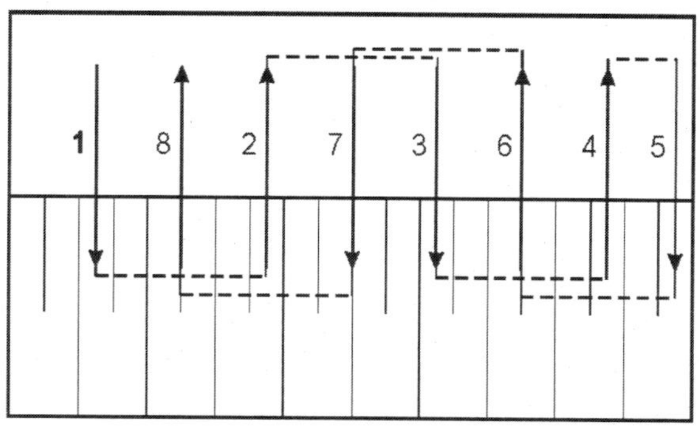

FIGURE 28: Building sequence of the embankment (Weber, 2008)

The embankment has been built on top of the clay surface using the stone column installation tool. The tool platform was first brought into a certain position, and the working arm was moved along a line, while the embankment material was continuously poured. Once the first line was built,

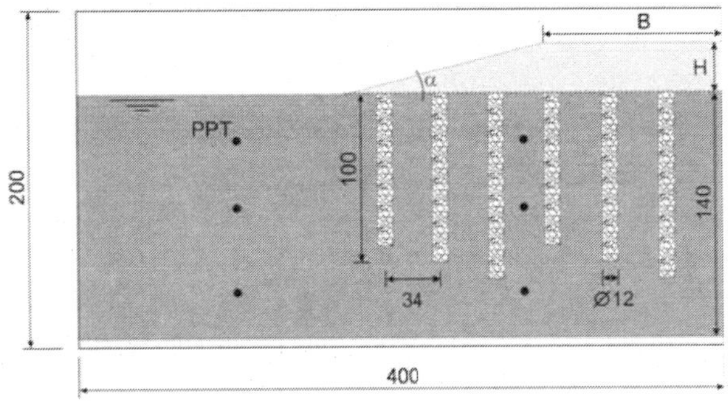

FIGURE 29: Cross-section of the model (Weber et al. 2006)

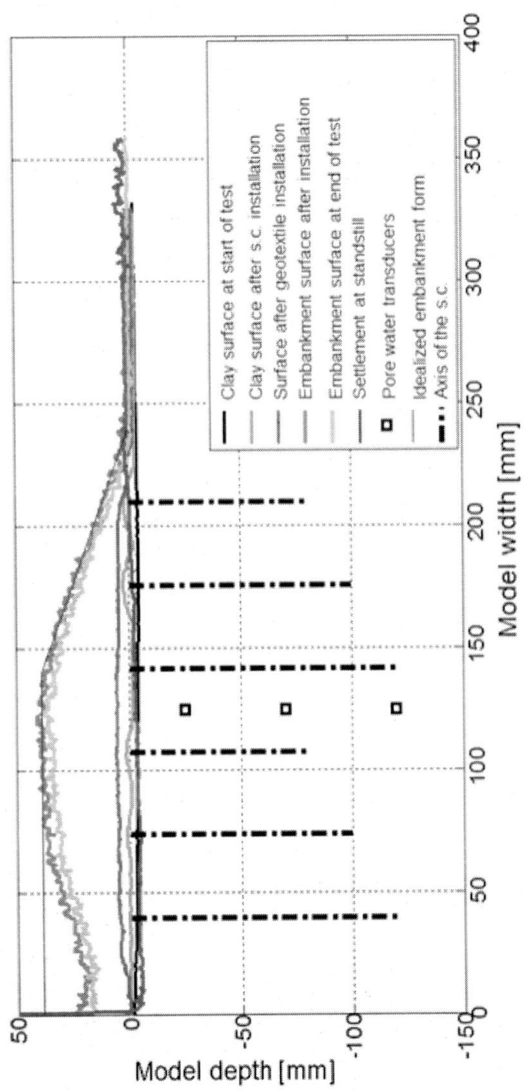

FIGURE 30: Deformations measured at various stages of the test (Weber, 2008)

the tool platform was moved and the process started again. The distance between rows 1 to 4 and 5 to 8 in Figure 28 was 100mm so that the distance between the rows was in the end 50mm. A change of the form of the embankment after construction was not possible. Figure 30 shows the measured deformations by means of laser scans (see also Figure 21). The embankment had a height (H) of mm and a slope angle (α) of 18°.

The comparison of the observed settlements in the two sectors (Figure) allows for a clear appraisal of the positive effect that stone columns have on the development of settlements (Figure 31).

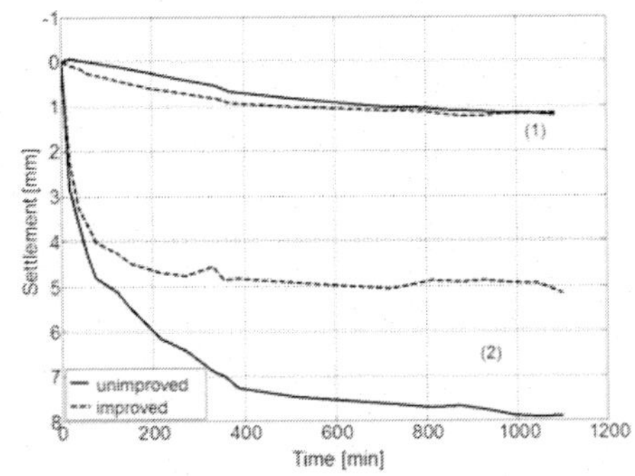

FIGURE 31: Settlements of the embankment toe (1) and crown (2) (Weber et al. 2006)

Figure 32 shows the distribution of the pore water pressure after the construction of the embankment in the two sectors. The depths are similar as those given in Figure 23. It can be seen that the dissipation of the pore water overpressures due to the embankment loading is significantly faster in the improved sector, which confirms the role that stone columns have for the acceleration of the consolidation process.

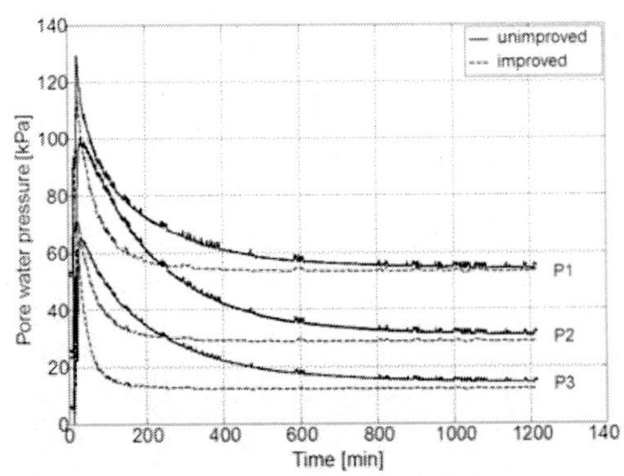

FIGURE 32: Distribution of the pore water pressure after embankment construction (Weber, 2008)

As opposed to the unimproved area, no parabolic form of the isochrones can be seen in the improved sector, but a parallel movement of the isochrones is noted, which indicates a horizontally-acting drainage (Figure 33).

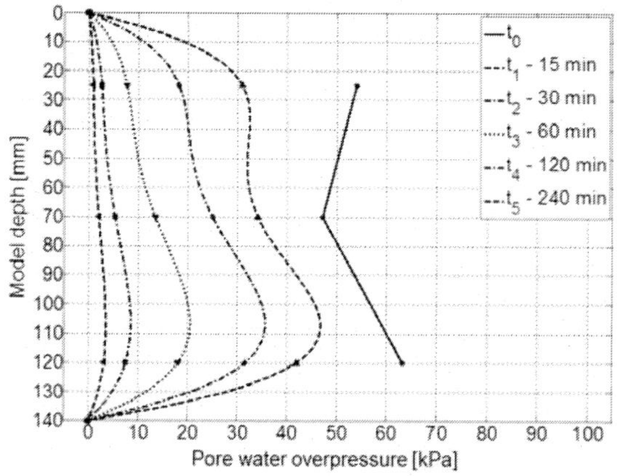

FIGURE 33: Distribution of the pore water overpressures over the model depth after embankment construction in the improved sector (Weber, 2008)

4.4 Outlook

In order to understand the changes in the soil due to installation of a stone column, samples were taken across one section and investigated using mercury intrusion porosimetry (MIP) and electron microscopes. In doing this, 4 areas next to the stone columns could be identified and are shown in Figure 34:

- a first zone in which stones and clay are mixed (zone 1), with a thickness of approximately 2mm, corresponding on the one hand to the cavity produced by the installation tool and on the other hand to the expansion of the stone column into the host soil.

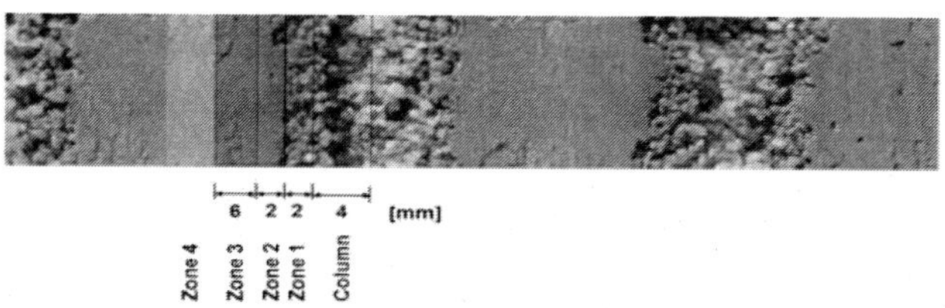

FIGURE 34: Spatial distribution of the installation effects around stone columns (Weber et al. 2010)

- a second zone in which the clay particles are reoriented in a semi-vertical direction (zone 2), with a thickness of approximately 2mm
- a third zone in which the host soil is compacted radially (zone 3), with a thickness of about 6mm.
- a fourth zone in which the horizontal stresses are moderately increased while no compaction can be measured (zone 4).

Further tests are on-going to study the distribution of this zones in more detail. At first columns are built in various step to subdivide the changes after each of the installation procedures (Gautray et al. 2012). Second, a system has been prepared and set up, which allows easier and faster progression in conducting the physical model tests, where consolidation in the big oedometer is always the limiting aspect. Multiple clay samples can be prepared in the laboratory at ETH Zurich with a diameter of 25 cm. These samples are then placed in the centre of a circular container with a diameter of 40cm and surrounded by dense sand. A column can be constructed in the centre of this sample and even loaded with a round foundation, so that, beneath sampling and studying the micro effects, also data is available representing ideal axisymmetric conditions to be modelled numerically.

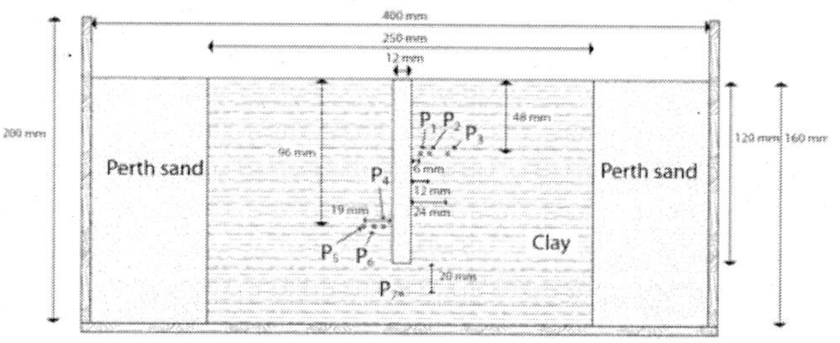

FIGURE 35: Modified test setup for studying stone columns (Gautray & Laue, 2013)

Latest improvement for the research of the installation effects of stone column is the development of a needle to investigate the technique. The electrical needle used in the geotechnical centrifuge of the ETH Zürich consists of a 1mm-diameter stainless steel cylinder in which an isolated cable is installed. The cable is let free of isolation at the tip of the needle and thus enables the measurement of the electrical impedance while driving the needle into the soil sample.

First results (Gautray & Laue, 2013) provide further insight into the pore distribution around a stone column and the variations with depth. Different scenarios tested in the centrifuge show that the pore volume indeed decreases with increasing depth and that the stone column, as well as loading histories, do have an influence on it.

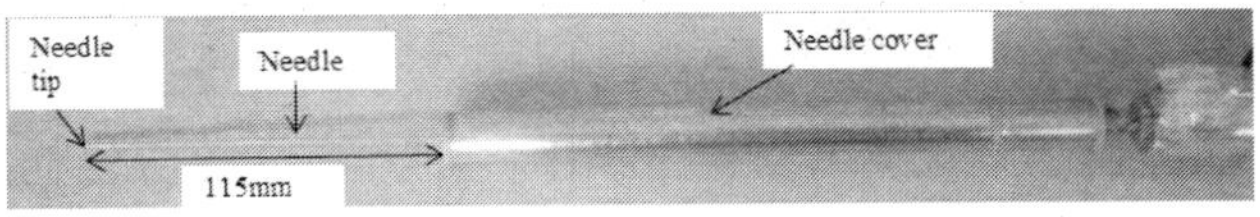

FIGURE 36: Electrical needle used in the centrifuge (Gautray & Laue, 2013)

5. SUMMARY

This contribution summarises the recent research efforts made at the institute of geotechnicak engineering at ETH Zurich at the professorship of geotechnics held by Prof. Sarah Springman. It focusses on modelling methods and techniques rather than on analysing and discussing the results of the various test set ups presented. As all of the work described above is ongoing, Final conclusions on the results have not been drawn yet.

ACKNOWLEDGEMENTS

The authors want to acknowledge the ongoing support from the Swiss National Science Foundation (SNF) and of the various fundings of the Swiss National Institute of Technology (ETH) without the development of the centrifuge facility would not have had reached the actual stage. In articular, the project on river levees has been sponsored by the competence centre of environment and sustainibility in the frame of the project APUNCHE. Work on unsaturated slopes has been part of the project TRAMM of the same competence centre. The inititaing part of the research work on ground improvement had been sponsored by SNF with the grants No. 200021-100362/1 and No. 200020-107483.

REFERENCES

Akca, D., Gruen, A., Askarinejad, A. and Springman, S. M., (2011). "Photogrammetric monitoring of an artificially generated landslide", Paper presented at the International Conference on Geo-information for Disaster Management (Gi4DM), Antalya, Turkey.

Arnold, A., (2011) Tragverhalten von nicht starren Flachfundationen unter Berücksichtigung der lokalen Steifigkeitsverhältnisse. Diss Nr. 19516; ETH Zürich.

Askarinejad, A., Laue, J., Zweidler, A., Iten, M., Bleiker, E., Buschor, H. & Springman, S.M. (2012). Physical modelling of rainfall induced landslides under controlled climatic conditions. 2nd Conference Eurofuge, Deltares, Delft.

Askarinejad, A., Bischof, P., Beck, A., Casini, F. and Springman, S. M. (2012b) "Rainfall induced instabilities in a silty sand slope: a case history in northern Switzerland". submitted to RIG (Italian Geotechnical Journal).

Bezzola, G. R., Hegg, C., Frank, F. and Koschni, A. (2008). Vom Hochwasser lernen - Standortbestimmung. TEC21, 39, pp 24-27.

Bowman, E.T., Imre, B., Laue, J. & Springman, S.M. (2010) "Experimental modelling of debris flow behaviour using a geotechnical centrifuge". Canadian Geotechnical Journal, Canadian Geotechnical Journal. 47(7): pp 742-762. doi:10.1139/T09-141.

Brönnimann, C. (2011) "Effects of Groundwater on Landslide Triggering", PhD thesis, EPFL.

Cargill, K. W. and Ko, H. Y. (1983) "Centrifugal modeling of transient water flow". Journal of Geotechnical Engineering, 109 (4), pp 536-555.

Chikatamarla, R. Laue, J. & Springman, S.M. (2004) "Rockfall impact on protection galleries", Second International Conference on Structural Engineering Mechanics and Computaion, Kapstadt, Zingoni ed., Balkema, Leiden, pp. 1139-1144.

Chikatamarla, R. (2005). Optimisation of Cushion Materials for Rockfall Protection Galleries Diss. Nr. 16315. ETH Zürich

Dassault Systèmes. (2000). CATIA V5 R19 User›s documentation.

Doglioni, A., Galeandro, A. and Simeone, V. (2011), A simple model for passive failure compression structure at the toe of landslide. in The Second World Landslide Forum, eds. pp 98-113.

Gautray, J., Laue, J. & Springman, S.M. "Identification of pore size distribution to identify plastic zones around stone columns", IS-Ground Improvement Brussels 2012. BGI, Brussels, pp 433-442.

Gautray, J.N.F. & Laue, J., (2013) "Electrical resistivity needle in soft clay under enhanced gravity", Paper in preparation for ASCE Geotechnical Testing Journal

Hoyois, P. and Guha-Sapir, D. (2003) "Three decades of floods in Europe: a preliminary analysis of EMDAT data", WHO collaborating Centre for Research on the Epidemiology of Disasters (CRED), Catholic University of Louvain, (Pub).

Hudacsek, P., Bransby, M. F., Hallett, P. D. and Bengough, A. G. (2009) "Centrifuge modelling of climatic effects on clay embankments. Engineering Sustainability". 162 (ES2), pp 91-100.

Imre, B. Micromechanical Analyses of Sturzstroms (Rock Avalanches) on Earth and Mars. Diss Nr. 18854; ETH Zürich.

Kapogianni, E., Laue, J. Sakellariou, M. & Springman, S.M. (2010) "The use of optical fibers in the geotechnical centrifuge", Proc. of the 7th International Conference on Physical Modelling in Geotechnics, Springman, Laue & Seward eds.), Taylor and Francis, pp 343-348.

Kimura, T., Takemura, J., Suemasa, N. and Hiro-Oka, A. (1991) "Failure of fills due to rainfall", Centrifuge 91, eds. H. Y. Ko & F. G. McLean, pp 509-516.

Küng, H., 2003. Undrainierte Scherfestigkeit an aufbereitetem Seebodenlehm, Diplomarbeit, Institut für Geotechnik, ETH Zürich.

Kusakabe, O., Hagiwara, T. and Kawashima, A. (1988) "Levee failures modelled in a drum centrifuge", Centrifuge 88: Proceedings of the International Conference on Geotechnical Centrifuge Modelling, J. F. Corté (eds.), pp 147-153.

Laue, J., Nater, P., Springman, S.M. & Grämiger, E. (2002) "Preparation of soil samples in drum centrifuges", Proceedings of the International Conference of Physical Modelling in Geotechnical Engineering, St. Johns, Phillips et al. eds., Balkema, Rotterdam, pp 143-148.

Laue, J., Springman, S.M., Nater, P., Graemiger, P. & Ducksch, A. (2005) "Creating seafloor conditions in geotechnical centrifuges", Proceedings of the XVI International Conference on Soil Mechanics and Geotechnical Engineering, Osaka, Millpress, Rotterdam, pp. 1741-1744.

Ling, H. I., Wu, M. H., Leshchinsky, D. and Leshchinsky, B. (2009) "Centrifuge Modeling of Slope Instability". Journal of Geotechnical and Geoenvironmental Engineering, ASCE, 135 (6), pp 758-767.

Mayor, P. A., Springman, S. M. and Teysseire, P. (2008) "In situ field experiment to apply variable high water levels to a river levee", Unsaturated Soils. Advances in Geo-Engineering : 1st European Conference on Unsaturated Soils, D. G. Toll, C. E. Augarde, D. Gallipoli & S. J. Wheeler (eds.), pp 947-952.

Morales, W.F., Laue, J. & Springman, S.M. (2012a) "Controlled water-level system for breach analysis of levees within an enhanced acceleration field", 2nd Conference Eurofuge, Deltares, Delft.

Morales, W.F., Laue, J. & Springman, S.M. (2012b) "Design and analysis procedure for centrifuge devices with a product lifecycle management (PLM) system", 2nd Conference Eurofuge, Deltares, Delft.

Morales, W.F., Laue, J. & Springman, S.M. (2013) "On the use of unsaturated properties of a sandy material for centrifuge model preparation", Paper accepted for the 1st Pan American Conference on unsaturated soils, Cartagena, Columbia.

Nater, P. (2007). Belastungs- und Verformungsverhalten von geschichteten Bodensystemen unter starren Kreisfundamenten. Veröffentlichungen des Instituts für Geotechnik, ETH Zürich, Band 226, VDF-Verlag, ETH Zürich.

Okumura, T., Narita, K. and Ohne, Y. (1998) "Failures of earth dam due to flooding", Centrifuge 98: Proceedings of the International Conference, IS-Tokyo ‹98, T. Kimura, O. Kusakabe & J. Takemura (eds.), pp 633-636.

Panduri, R, 2000, Versuchsschüttung Nordwestumfahrung Zürich: CPT, Laborversuche & numerische Modellierung, Institut für Geotechnik, ETH Zürich.

Pooley, E., Laue, J. & Springman, S.M. (2009) "Assessment of the use of dynamic compaction on double porosity clay landfill", 17th International Conference on Soil Mechanics and Geotechnical Engineering. Hamza et al. eds., IOS Press, pp 2252-2255.

Roddeman, D. 2003. Tochnog User's Manual–a free explicit/implicit FE program.

Schofield, A. N. (1980) "Cambridge geotechnical centrifuge operations". Géotechnique, 30 (3), pp 227-268.

Seo, M. W., Ha, I. S., Im, E. S., Kim, H. S., Kim, Y. S. and Kim, T. H. (2006) "Centrifuge test for evaluating effects of water level change on fill dam behavior", Physical Modelling in Geotechnics - 6th ICPMG ‹06. Proceedings of the International Conference, C. W. W. Ng, Y. H. Wang & L. M. Zhang (eds.), pp 431-434.

Springman, S.M., Laue, J., Boyle, R., White, J. & Zweidler, A., 2001, The ETH Zürich geotechnical drum centrifuge, Int. J. Phys. Modelling Geotech. 1(1), 59-70.

Springman, S.M., Nater, P., Chikatamarla, R. & Laue, J. (2002) "Use of flexible tactile pressure sensors in geotechnical centrifuges", Proceedings of the International Conference of Physical Modelling in Geotechnical Engineering, St. Johns, Phillips et al. eds., Rotterdam, Balkema, pp 113-118.

Springman, S. M., Mayor, P. A., Müller, P. and Teysseire, P. (2008) "The influence of river floods and rainfall on protection levees in the valais. in Long time effects and seepage behavior of dams", Proceeding of the 1st International Conference on long time effects and seepage behavior of dams (LTESBD08), Zhu, Liu & Qiang (eds.), pp 155-162.

Springman, S. M., Askarinejad, A., Kienzler, P., Casini, F., Sperl, M., Bleiker, E., Akca, D. and Gruen, A. (2010) "The Rüdlingen Monitoring and Landslide Experiment", D-Baug Highlights 2009, ETH Zürich, 26-31.

Take, W. A., Bolton, M. D., Wong, P. C. P. and Yeung, F. J. (2004) "Evaluation of landslide triggering mechanisms in model fill slopes", Landslides, 1, pp 173-184.

Tamate, S., Suemasa, N. and Katada, T. (2010) "Simulating shallow failure in slopes due to heavy precipitation", 7th International Conference of Physical Modelling in Geotechnics, eds. S. M. Springman, J. Laue & L. Seward, pp 1143-1149.

Weber, T.M., Laue, J. & Springman, S.M. (2005) "Modelling the inflight construction of sand compaction piles in the centrifuge", Proceedings of the XVI International Conference on Soil Mechanics and Geotechnical Engineering, Osaka, Millpress, Rotterdam, pp 1291-1294.

Weber, T. M., Laue, J. and Springman, S. M. (2006) "Centrifuge modelling of sand compaction piles in soft clay under embankment load", VI International Conference on Physical Modelling in Geotechnics, Hong Kong, Taylor & Francis, London 1, pp 603-308.

Weber, T. M. (2008) Modellierung der Baugrundverbesserung mit Schottersäulen, Veröffentlichungen des Institutes für Geotechnik, ETH Zürich, Band 232, VDF-Verlag, ETH Zürich.

Weber, T.M., Plötze, M., Laue, J., Peschke, G. & Springman, S.M. (2010) "Smear zone identification and soil properties around stone columns constructed in-flight in centrifuge model tests". Géotechnique, 60 (3), pp 197-206.

Centrifuge Model Studies on the Performance of Rail Embankments Constructed with Coal Ash as a Structural Fill Material

B.V.S. Viswanadham and V.K. Mathur[1]

Department of Civil Engineering, Indian Institute of Technology Bombay, Powai, Mumbai
[1] *Environment, Ash Management, NTPC Limited, NFL Premises, Sector-24, Noida*
E-mail: viswam@civil.iitb.ac.in; vkmathur@ntpceoc.co.in

ABSTRACT

The objective of the present study is to understand the stability and deformation behavior of rail embankments constructed with coal ash as a structural fill material through centrifuge model tests. Two types of railway embankments were modeled: 1) Clay confined coal ash embankment with 1V:2H slope of 11 m height and 2) Geogrid reinforced coal ash embankment with 2V:1H slope of 4 m height using a large beam centrifuge facility available at IIT Bombay. Considering the nature of railway embankments and possibility of building-up of ground water table due to rainfall, a seepage tank simulator was used to induce seepage into the rail embankment during centrifuge tests. After establishing steady-state seepage conditions, the crest of the embankment was subjected to an incremental static loading pressure distributed through a rigid strip footing up to 700 kPa or failure, whichever occurred first. Additionally, an option of provision of drainage at the mid-height of the railway embankment was also explored for clay-confined coal ash embankments. Based on the analysis and interpretation of centrifuge model test results, for a soil confined coal ash embankment with 1V:2H slope with a confinement layer of 1.5 m thickness in the top-half height and berms in the bottom-half zone was observed to sustain an ultimate load bearing pressure of 400 kPa. Geogrid reinforcement layers in a reinforced coal ash embankment with 2V:1H slope were observed to experience straining after applying a bearing pressure of 250 kPa in the top-half zone and indicating a need to provide high strength layers in top-half zone. Further, results of stability analysis of soil confined coal ash embankment models constructed with coal ash as a fill material were found to corroborate well with physically observed centrifuge test results.

1. INTRODUCTION

Coal ash is the by-product of coal based thermal power plants and its disposal is a major problem from an environmental point of view in many countries and also requires a lot of disposal areas. Coal ash is referred for ash accumulated in the ash pond dam sites. It is estimated that by the year 2012 coal ash generation in India will be reaching nearly 160 million tones. Coal ash has been successfully used as a structural fill material for constructing highway embankments in number of locations throughout the world. However, studies pertaining to the usage of coal ash as a structural fill material for constructing railway embankments are very much limited.

Many investigators like [Gray and Lin (1972); Digioia Jr. and Nuzzo (1972); Leonards and Balley (1982); Toth et al. (1988); Martin et al. (1991); Das et al. 2009] have analyzed coal ash generating from different power plants all over the world for different applications such as, compacted coal ash embankments, sub-base in road construction, filling low-lying areas, etc. The main emphasis is to identify suitable methodology for bulk consumption of the material like coal ash generated from the coal based thermal powers stations located all over the world. To minimize the potential for environmental problems with structural coal ash embankments, it is logical to use only coal ash that has been shown to have safe leaching characteristics and also to restrict the rate of water permeability of fills. This was envisaged through capillary cut-offs at the bottom of embankment and provision of confinement layer along side slopes. The major concerns with respect to the potential impact of a coal ash embankment on the local environment are wind erosion, surface-water erosion, dissolution in surface runoff and dissolution in rainfall percolating to ground water. To counter some of the above-mentioned factors, the coal ash is encapsulated with a relatively low permeable soil cover. The low-permeable soil cover generally consists of clayey silt or silty clay type soils. The soil cover impedes or controls dissolution of coal ash due to surface run-off and also acts as a confinement layer. With this sort of coal ash confining system, it is possible to retain strength properties of coal ash. According to Lewis (1976); Faber and DiGioia Jr. (1976); Toth et al. (1988); Martin et al. (1990), on an average a slope of 1V: 2H (26.5° with the horizontal surface) was adopted for constructing highway embankments with a confinement layer ranging from 450 mm to 2000 mm thickness is provided all around to prevent erosion and contain coal ash safely. It is also in practice to provide thin horizontal barriers made of impermeable soil to contain the coal ash in cells.

As these soil confined coal ash embankments are constructed on a soil having adequate bearing capacity, the following five modes of failures are envisaged. They are: 1) Failure of cover due to raising water pressure in coal ash, 2) Global failure surface extending from the edge of the loaded area, 3) Punching of loaded area in blanket layer causing differential movements at loading edges, and 4) Sliding movement of blanket portion above embankment level. Failure of cover soil can occur due to inadequate cover thickness. This can also occur due to seasonal drying and wetting of soil cover leading to cracking or opening of soil cover. Opening of soil cover supports eroding tendency of coal ash out of the bund. This can result in differential sinking of rail tracks with time. Figures 1-2 depict a typical observed feature in the field wherein soil confined coal ash embankment was constructed (NTPC report, 2008).

As can be noted from the above cited literature, coal ash has been used as a structural fill material in majority of approach roads/highway embankments. Literature pertaining to the utilization of coal ash as a structural fill material for railway embankment construction is very much limited. Sunaga and Sekine (1992) reported about utilization of potential of coal ash as a structural material for railway embankments/construction of raised intersections for rail metro in Japan. Apart from these, details pertaining to the behavior and application are very much scattered. Hence, this forms the relevance and current interest of the present study. In addition, to the best of authors knowledge, attempts towards understanding the response of rail embankments constructed with coal ash as a structural fill through centrifuge based small-scale physical modeling are limited. The main drawback of the reduced scale physical modeling at normal gravity is that stress levels are much smaller than that of the prototype. As the behavior of soils is mainly governed by confining stresses, the reduced scale models may not represent the true behavior. Identical stress field in both model and prototype can be achieved by adopting centrifuge modeling technique. In a centrifuge, the behavior of prototype structures can be studied as scaled-down models in a controlled environment, while preserving the stress states required developing appropriate soil properties. Therefore, centrifuge testing is an appropriate tool for studying the behavior of embankments constructed with coal ash as a structural fill material.

FIGURE 1: Status of a slope of Tamluk Digha railway embankment (South Eastern Railway, India) constructed with coal ash as a structural fill material in (NTPC report, 2007)

FIGURE 2: Close view of the slope along Tamluk Digha railway embankment (NTPC report, 2007)

FIGURE 3: View of a 4.5 m radius large beam centrifuge facility at IIT Bombay

In this paper, centrifuge test results of two soil confined coal ash embankments of 11 m height were compared. In addition, centrifuge test results of a 4 m high 2V:1H slope geogrid reinforced slope were discussed. Considering the nature of railway embankments and possibility of building-up of ground water table due to rainfall, a seepage tank simulator was used to induce seepage. After establishing steady-state seepage conditions, the crest of the embankment was subjected to an incremental static loading pressure distributed through a rigid strip footing up to 700 kPa or failure, whichever occurred first. Additionally, significance of confining layer and an option of provision of drain vent at the mid-height of a clay-confined coal ash embankment was also explored.

2. EXPERIMENTAL STUDIES IN A CENTRIFUGE

The 4.5m radius large beam centrifuge facility available at Indian Institute of Technology Bombay was used for model tests presented in this paper. The centrifuge capacity is 2500 g-kN with a maximum payload of 25 kN at 100g and at higher acceleration of 200g the allowable payload is 6.25 kN. Specifications in detail are discussed by Chandrasekaran (2001) and Viswanadham et al. (2009). Figure 3 depicts a large beam centrifuge facility used in the present study.

2.1 Scaling Considerations

Conventionally, it is assumed that the soil in the centrifuge model and prototype are identical. In a centrifuge, stress similarity is achieved by accelerating a model of scale 1/N to N times the earth's gravity (N = Scale factor or g-level). Scaling considerations and errors due to centrifuge modeling are discussed in detail by Schofield (1980). The soil particles in a centrifuge model can not be scaled down to scale the soil particles of prototype, while other model dimensions can be modeled down and this effect is called as grain size effect. However, with respect to centrifuge model tests on rail embankments with coal ash as a structural fill, this aspect is satisfied because of similarity of grain size distribution of coal ash in the model and in the field. In the present study, scale factors for seepage velocity $(v_s)_m = N(v_s)_p$, seepage time $t_m = t_p/N^2$ and pore water pressure $u_m = u_p$ were used (where suffix m = centrifuge model; p = prototype).

2.2 Model Materials

2.2.1 Coal Ash and Local Soil

Coal ash was collected from Kahalgaon Super thermal power plant in Bihar state of India. The grain size distribution of the coal ash showed that it contains 64% sand, and 36% silt size particles. It is classified as an ASTM class F coal ash with a calcium content of 1.94% (Sridharan et al.

2000). The effective particle size d_{10} is 0.02 mm and is classified as SM according to Unified Soil Classification system. The maximum dry unit weight $\gamma_{d,max}$ and optimum moisture content OMC of the coal ash are 13.2 kN/m^3 and 22% (standard Proctor compaction). Saturated direct shear box tests on coal ash moist-compacted at its $\gamma_{d,max}$ and OMC gave an internal friction angle ϕ of 34° and cohesion as zero. The coefficient of permeability k of coal ash moist-compacted at $\gamma_{d,max}$ and OMC is 2.58×10^{-6} m/s. In order to construct base soil, side cover and top cover soil, locally available soil at Kahalgaon Super thermal plant site was used. The grain size distribution of the Kahalgaon soil contains 20% sand, 57% silt, and 23% clay size particles. Liquid limit and plasticity index are 46% and 26% respectively. It is classified as CL according to Unified Soil classification system. The maximum dry unit weight $\gamma_{d,max}$ and optimum moisture content OMC of the local soil are 17.4 kN/m^3 and 16% (standard Proctor compaction). The cohesion and angle of internal friction ϕ of the local soil moist-compacted at $\gamma_{d,max}$ and OMC are 50 kN/m^2 and 16° (Direct shear box test). The coefficient of permeability k of local soil moist-compacted at $\gamma_{d,max}$ and OMC is 2.5×10^{-8} m/s.

2.2.2 Geogrid

A major difficulty encountered in model studies involving geogrid materials is selection, modelling and instrumentation of ideal materials. Contrary to soils, the similitude condition does not allow the use of identical materials in model and prototype studies. The reinforcement elements are mainly planar layers like uniaxial geogrids in the case of slopes and walls or biaxial geogrids for embankments on soft ground, pavements on weak subgrade, and landfill liners. One alternative is to study the response of these structures by constructing full-scale structures in the field with extensive instrumentation. However, costs associated with prototype instrumentation may also be significant and the reliability of the collected data was often questionable (Zornberg and Arriaga, 2003). Among the different categories of reinforced soil structures, lack of field data is particularly striking for the case of geosynthetic reinforced slopes, which may result in significantly high factors of safety have been incorporated into current design methodologies. In such situations, application of the centrifuge modelling technique for studying the deformation behaviour of geosynthetic reinforced slopes is a viable option and moreover soil-geosynthetic interaction is highly influenced by the presence of prototype stress conditions.

At least ten to fifteen varieties of synthetic (many having different styles) geogrids are available. They differ considerably in geometry and mechanical properties. The manufacturers attempt to vary the typical geometrical characteristics and tensile load characteristics (as manufacturing process variable) in order to achieve a desired geogrid. As tensile load characteristic of synthetic material is mainly dependent on the composition and type of raw material, it has become one of the manufacturing process variables. The properties of these geogrids are specified based on: (i) rib cross-sectional area, (ii) grid opening size, (iii) tensile strength, and (iv) type of material composition. The criterion for scaling down the reinforcement function of the geocomposite in centrifuge based small scale physical modeling is based on: (i) tensile load-strain behavior and (ii) frictional bond behavior along soil-geogrid interface. Scaling considerations given by Viswanadham and König (2004); and Rajesh and Viswanadham (2012) were used to achieve the modeling geogrids. The frictional bond behavior along soil-geogrid interface was assumed to be achieved byensuring identical percentage open area f between model and prototype geogrids. The percentage open area f expressed in percentage is the ratio of area formed by grid opening sizes to area formed by grid opening sizes measured up to center of width of ribs. The expression for percentage open area can be written as $f = a_l a_t/[(a_l + b_l)(a_t + b_t)]$ (where, a_l and a_t are grid opening sizes in longitudinal and transverse directions and b_l and b_t the widths of the rib in longitudinal and transverse directions respectively), of the geogrid. Table 1 summarises general scaling relations relevant to centrifuge modelling.

3. MODEL TEST PACKAGE, TEST PROGRAMME AND TEST PROCEDURE FOR CLAY CONFINED COAL ASH EMBANKMENTS

The container used to build the models for this study was a strongbox having dimensions inside of 760 mm long, 200 mm wide, and 410 mm deep. All walls of the box except for the front are composed of steel plates. A cross-section of the model test package is presented in Figure 4. Considering symmetry, only right half section of bund was selected. The front wall is a viewing window, consisting of a thick Perspex glass sheet. To reduce any boundary effects caused by friction between the coal ash and container walls and to maintain plane-strain conditions, a thin layer of petroleum grease was applied to the container walls, followed by thin polythene sheet strips. This method was adopted by Viswanadham et al. (2009) to reduce friction effects and allow for plane-strain conditions to exist. A seepage tank was custom designed and calibrated to induce seepage in the model slopes. The seepage tank was made of 10mm thick aluminum plates. The internal dimensions of the seepage tank were 80 mm × 360 mm × 200 mm. One wall of the seepage tank was perforated to allow seepage of water. The perforated wall of the seepage tank was covered with a layer of non-woven geotextile

Table 1: Scaling Relationships for Modelling Geogrid

Parameter	*Units*	*Centrifuge Model/ Prototype*
a_l, a_t, b_l, b_t, t*	m	1/N
Cross-sectional area of rib/unit length A'	m	1/N
Perecentage Open area f	[%]	1
Tensile load T_g	[kN/m]	1/N
Strain ε_g	[%]	1
Secant modulus J_g	[kN/m]	1/N
Bond stress τ_b	[kN/m²]	1

*a_l, a_t = grid opening sizes in longitudinal and transverse directions; b_l, b_t = width of ribs in longitudinal and transverse directions; t = average thickness of the rib.

to prevent clogging of the perforations by the soil/coal ash particles. Table 2 shows the details of centrifuge model tests discussed in this paper. Two slope models REM11 and REM12 having 11 m height were presented. Dimensions of different components of rail embankment models constructed with coal ash core confined with a local soil and blanket layer are marked in Fig. 4 and given in model and prototype dimensions.

Slope models were H (m) in height which excludes a blanket layer thickness of t_b (m) and a base soil thickness of D (m). The slopes were constructed with a slope inclination of 26° (1V:2H). Drainage was provided at the downstream end by providing a drain at the right end of the container. After compacting base soil layer, a thin layer of sand of about 6 mm (~0.3 m) was placed in both the models. To prevent erosion of soil particles, a thin non-woven geotextile fabric was provided by wrapping back into the slope. In all models, thickness of top soil cover is provided as t_t mm and side cover thickness of t_c mm. In model REM11, the cover soil thickness above the second berm is 1 m and in model REM12 it is 1.5 m. In addition, model REM12 has also a provision of drain at mid-height (i.e. second berm from the toe). The models were instrumented using three Linearly Variable Differential Transformers (LVDTs) on the top surface of the slope to measure the displacements and out of these, one was placed on the loading plate. Five Pore water Pressure Transducers (PPTs) were used to measure the pore water pressure during the progress of the test.

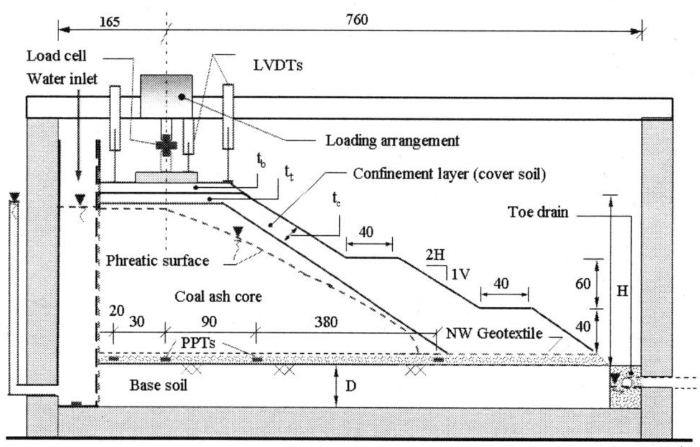

FIGURE 4: Cross-section of a model test package (all dimensions are in mm)

After the compaction of base layer was completed, temporary wooden supports were placed to achieve the desired slope inclination. This support has got a special provision to compact the side cover parallel to the slope surface with the local soil. The slope was constructed in layers. After completion of each layer, specially designed L-shaped plastic markers made of thin transparency sheets were placed to study the movement of coal ash and soil cover during the test. To allow free movement of the markers, white petroleum grease was applied on the sides of the markers in contact with the container and the soil. Colored food dye was placed at intervals on the top of each layer to visualize the movement of water during seepage. After construction of the slope, the entire slope along with top, side cover soil and blanket layer was allowed to inundate in water for about 24 hours at normal gravity in the laboratory. After saturation, a rigid stainless steel strip footing of 50 mm thick was placed with its center 165 mm from the inside left edge of seepage tank, and connected to a loading cell (Fig. 2). The crest of the slope was located 140mm from the left edge of seepage tank, so the load was applied at a setback distance d = 70 mm from the slope crest. After the ramping time, the centrifuge was allowed to rotate at a constant revolutions per minute (rpm) corresponding to a desired g-level, given in Table 2. After attaining a desired g-level, water was allowed to seep in through the coal ash core and its behavior was observed for global failure or localized failures, if any. After establishing steady state seepage conditions, with the help of pneumatic cylinder connected to the footing via a load cell, a static pressure was applied incrementally. Each increment of static pressure was maintained only for a short duration of 5 minutes in model dimensions and increased up to 700 kPa or failure, whichever occurred first. The proceedings of the test were grabbed through a Canon make digital photo camera mounted along with the model to view the front elevation of the model. The data from the LVDTs and PPTs was obtained using the on-board data acquisition system. Figure 5 shows the loading arrangement for inducing load at the top surface of the embankment along with the load cell.

3.1 Analysis and Interpretation of Centrifuge Test Results

Figure 6 gives variation of measured pore-water pressure with time in prototype dimensions for model REM11. Within 2 to 3 days of seepage time, steady state seepage conditions were found to be established. This was maintained to simulate raise of ground water table within the coal ash core due to rainfall water infiltration into rail embankment. During this stage, model was monitored for any global failure or localized failures within the soil cover of the soil confined coal ash embankment. Figure 7 shows measured variation of settlements with seepage time (from L1, L2, L3, as shown in Fig. 4) of the slope along with footing plate. In the case of model REM11, beyond 7.5 days of seepage time a steady increase in the footing plate was observed. Photographs captured during the various stages of seepage through model REM11 indicate side cover failure, as shown in Fig. 8a-8b. Results of these tests are discussed in brief by Viswanadham et al (2012).

As can be seen from Fig. 8, through the failed side cover, exposure of coal ash core can be noted. This necessitates a need for enhancing the

Table 2: Details of Centrifuge Tests

Test Legend	*N*	*H* (m)	*D* (m)	t_c (m)	t_t (m)	t_b (m)	*Mid-height Drainage*
REM11	50	0.22 (11)	0.074 (3.7)	0.02 (1)	0.012 (0.6)	0.006 (0.3)	No
REM12	54	0.204 (11)	0.090 (4.8)	0.027 (1.5)	0.012 (0.64)	0.006 (0.32)	Yes

t_c = thickness of side soil cover in the upper portion of the slope; Dimensions for H and t_c are given in corresponding prototype dimensions within the parentheses; Side berms were cut to shape from the initially compacted thick soil cover.

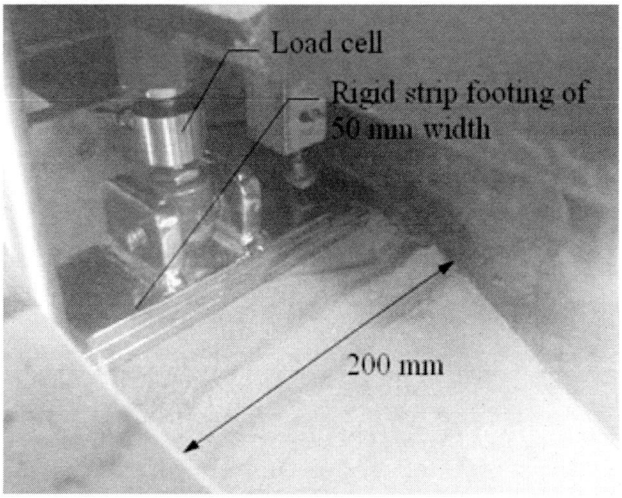

FIGURE 5: Loading arrangement for inducing load at the top surface of the embankment

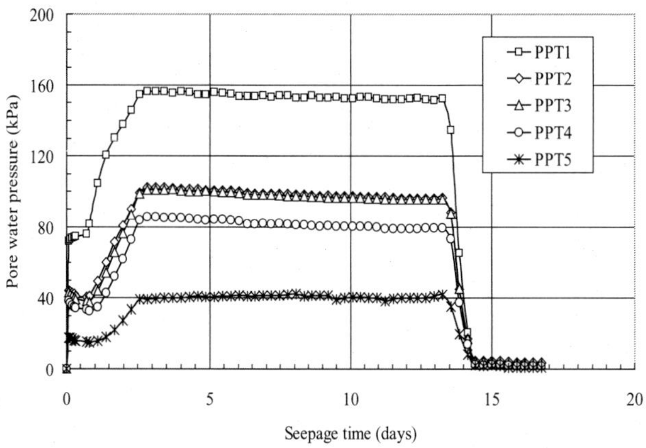

FIGURE 6: Variation of pore water pressure with seepage time for model REM11

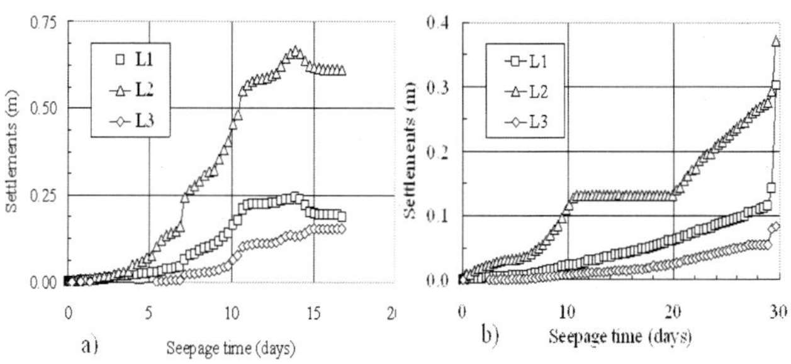

FIGURE 7: Variation of settlements with time for (a) Model REM11 and (b) REM12

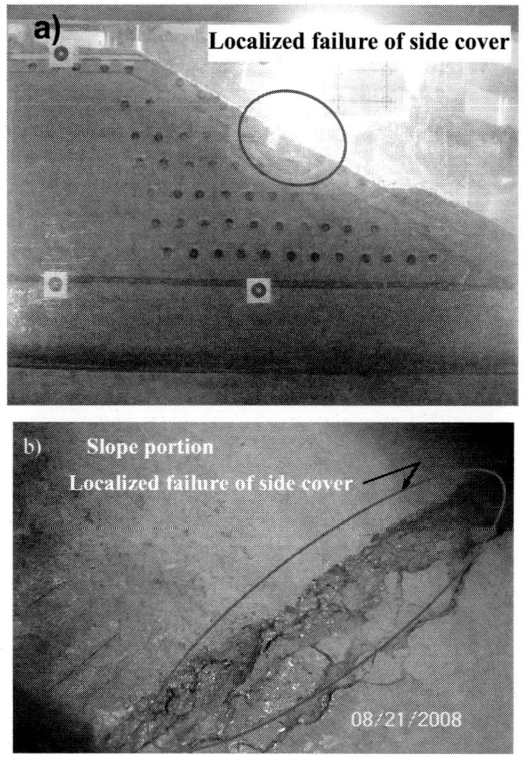

FIGURE 8: Localized failure of side soil cover for model REM11

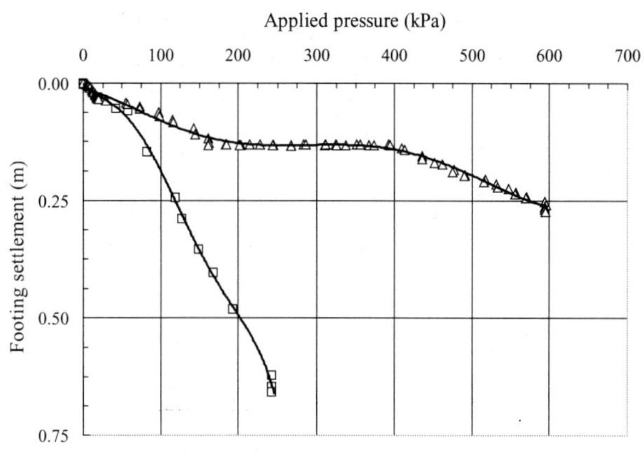

FIGURE 9: Variation of applied pressure with footing settlement

t_c from 1 m to 1.5 m and also to check the effect of provision of drain at mid-height. As mentioned in Table 2, the performance of a rail embankment constructed with coal ash as a structural fill material was addressed in model REM12.

3.1.1 Applied Pressure-Footing Settlement Variation and Phreatic Surfaces

Figure 9 presents variation of applied pressure with footing settlement for models REM11 and REM12. As can be noted, in the case of model REM11, which was provided with t_c = 1.0 m and without any provision for drainage at mid-height was found experience a footing settlement of 0.5

m at an applied pressure of 200 kPa. This was attributed to the observed behavior depicted in Fig. 8. In comparison, for model REM12, at an applied pressure of 200 kPa a footing settlement of only 0.125 m was observed (Fig. 9). Further increase in applied pressure beyond 400 kPa, the footing settlement was observed to increase to 0.25 m.

This applied pressure-footing settlement behavior implies that the model REM12 is superior in performance than the model REM11 and also bring-out the significance of adequate confinement of the side cover above berms and a need for drain provision at mid-height.

Figure 10 shows variation of head of water with the horizontal distance from the left edge of the seepage tank for models REM11 and REM12. These phreatic surfaces were plotted during penultimate stages of the centrifuge test and were obtained by converting the measured pore-water pressure at a specified stage of the seepage. As can be noted from Fig. 10, the presence of mid-height drain helps to drain the water rapidly and lead to a depletion of phreatic surface. However, a toe pressure of 40 kPa was observed to develop in both the models.

3.1.2 Stability Analysis of Model REM12

In the present study, slope stability analysis was performed using SLOPE/W software (Geostudio, 2007). Analysis was carried-out for a phreatic surface exhibiting highest PWP within the bund and with an increase in the applied footing pressure. Variation of factor of safety with applied pressure is plotted in Fig. 11 for model REM12. With an increase in the applied footing pressure, a decrease in the factor of safety was observed. At the onset of failure, at a particular magnitude of applied footing pressure, the factor of safety was observed to drop below 1. This implies that at that pressure the section is exhibiting limiting value of applied footing pressure on the bund section. As can be noted from Fig. 11, after application of a footing pressure of the magnitude of 400 kPa, the factor of safety of the embankment is 1. This is also found to be in agreement with the measured applied pressure-footing settlement variation (Fig. 9). This implies that the equivalent ultimate static pressure of 400 kPa can be applied on rail embankments constructed with coal ash core. Figure 12 show the status of model REM12 at the end of the centrifuge test.

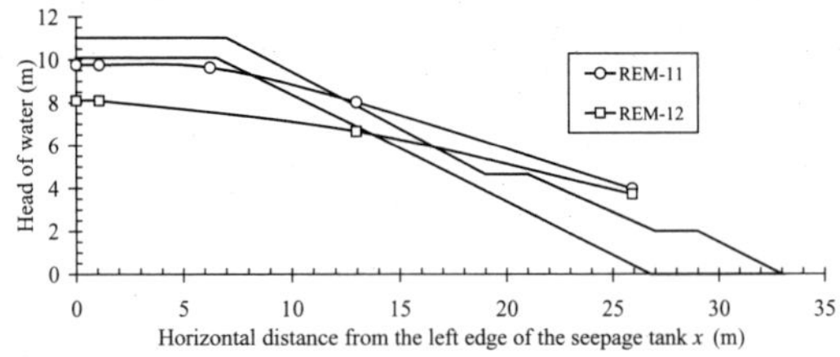

FIGURE 10: Plot showing efficacy of drain at the level of second berm from toe

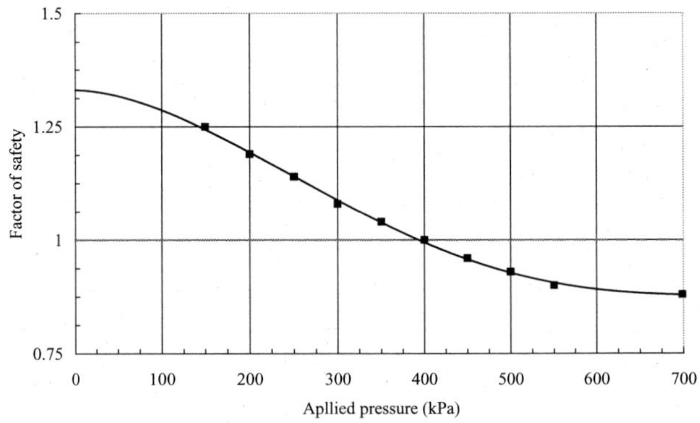

FIGURE 11: Variation of factor of safety with applied pressure for model REM12

FIGURE 12: View of model REM12 at the end of centrifuge test

4. MODEL TEST PACKAGE FOR GEOGRID REINFORCED COAL ASH EMBANKMENT

In the present study, a geogrid reinforced steep slope having 2V:1H with a flexible facing was modeled. Flexible facing was adopted using a wrap-around technique. Considering 4 m height and availability of model geogrids, it was decided to model reinforced coal ash embankments at a g-level of 20. Based on scaling considerations presented in Section 2.2.2, a model geogrid G1 was selected. Model geogrid G1 was tested for its wide width tensile strength, as per ASTM D: 4595 (1995). At 5 % strain, value of tensile strength in the longitudinal direction is 0.287 kN/m respectively. The secant stiffness (up to 5 % strain) is 5.74 kN/m in the longitudinal direction. In order to consider confining effect of geogrid, geogrid tensile load obtained from zero-grip tensile tests was adopted, and this corresponds to 0.56 kN/m in model dimensions for model geogrid G1.

The container used for constructing geogrid reinforced coal ash embankments is identical to the one used for clay confined coal ash embankments. A cross-section of the model test package is presented in Fig. 13. All slope models were H (mm) in height which excludes a blanket layer thickness of t_b mm and a base soil height of D mm. The slopes were constructed with a slope inclination of 63.4° (2V:1H). After compacting base soil height, first reinforcement layer was placed along with a form work for getting a desired slope inclination. In order to trace movements of geogrid layer, at the onset of seepage and loading, discrete markers were embedded onto geogrid at 20 mm c/c. Movements of markers during various stages of tests were traced to get displacement vectors and strain distribution in reinforcement layers.

The models were instrumented using three Linearly Variable Differential Transformers (LVDTs) on the top surface of the slope to measure the displacements and out of these, one was placed on the loading plate. Five Pore water Pressure Transducers (PPTs) were used to measure the pore water pressure during the progress of the test. One of them was placed at the bottom of the water tank and the other four were placed in the bottom coal ash layer (as shown in Fig. 13). The position of the LVDTs and pore water pressure transducers were also shown.

A rectangular grid of permanent markers was pasted firmly on the inner side of the Perspex sheet. The rectangular grid was 340 mm x 220 mm in size. The model was constructed in layers of 20 mm thickness using moist compacted coal-ash at its maximum dry unit weight and optimum moisture content. After completion of each layer, specially designed L-shaped plastic markers made of thin transparency sheets were placed to study the movement of geogrid layer during the test. To allow free movement of the markers, white petroleum grease was applied on the sides of the markers in contact with the container and soil. Coloured food dye was placed at intervals on the top of each layer to visualize the movement of water during seepage.

After saturation, a rigid strip footing of 50 mm thick was placed with its center 165 mm from the inside left edge of water tank, and connected to a loading cell for all the tests reported in this report. The crest of the slope was located 140mm from the left edge of water tank, so the load was applied at a setback distance x = 70 mm from the slope crest. For all the models, a crest width of 140 mm was maintained throughout the study. The loading arrangement for inducing load at the top surface of the embankment is shown in Fig. 5. With this arrangement, end effects due to loading plate can be minimized to a great extent (Sommers and Viswanadham, 2009).

The pore water pressures developed in the slope was measured using pore water pressure transducers placed in the water tank as well as in the base layer of the slopes. From these pore water pressures, the phreatic lines were obtained for all the slopes. Figure 14 gives variation of measure pore-water pressure with time in prototype dimensions and development of phreatic lines within the slope for model REM14. As can be noted from Fig. 15, highest value of measured PWP is that of a PPT placed within the water tank. Constant variation of PWP with time indicates the excellent establishment of steady state seepage conditions. In all the experiments, these types of conditions are established by facilitating a flow from left hand side to right hand side of the model. After establishing steady-state seepage conditions, load on the piston of the pneumatic cylinder increased in increments up to a maximum value or to a pressure, where considerable surface settlements are registered catastrophically.

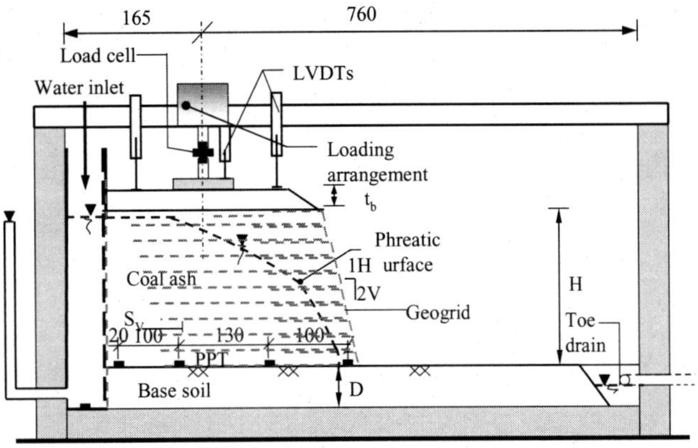

FIGURE 13: Cross-section of geogrid reinforced coal ash embankment for model REM14 (all dimensions are in mm)

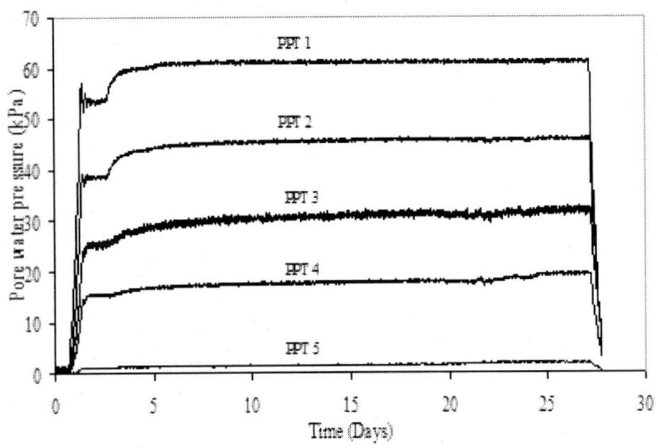

FIGURE 14: Variation of measured PWP with time for model REM14

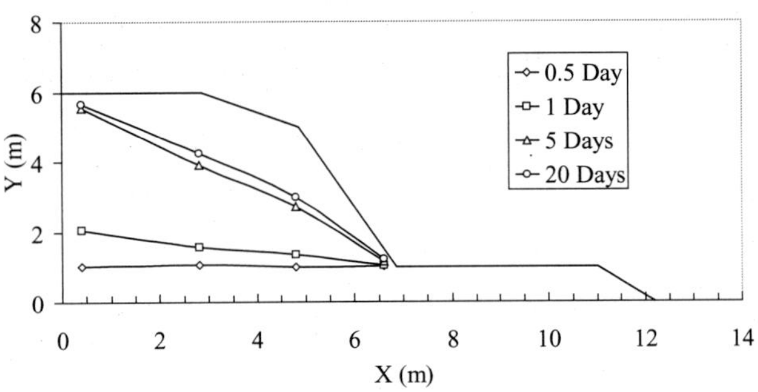

FIGURE 15: Measured phreatic surfaces for model REM14

Figures 16a-c show the front elevation of model geogrid reinforced coal ash slope at 20g for model REM14. As can be noted from Fig. 16c, at the onset of piston pressure $p = 400$ kPa, formation of crack at the left hand side of the strip footing can be noted. Figure 16c shows development of a failure surface extending up to toe of the slope. Plot showing displacement markers from beginning to end of the centrifuge test at 20g are shown in Fig. 17. Pattern of movement of markers indicate a clear and distinct formation of failure surface. Moreover, post-test investigations reveal, that 4 out five reinforcement layers in the top-half of the slope height were found to be ruptured completely. This implies that the mobilization of tensile strength is more than the available tensile strength of the geogrid reinforcement inclusions.

In the present study, slope stability analysis was performed using SLOPE/W software (Geostudio, 2007). Analysis was carried-out for a phreatic surface exhibiting highest PWP within the bund and with an increase in the applied footing pressure. Variation of FOS with applied footing pressure is plotted for number of cross-sections tested in this study. With an increase in the applied footing pressure, a decrease in the factor of safety was observed. At the onset of failure, at a particular magnitude of applied footing pressure, the factor of safety was observed to drop below 1. This implies that at that pressure the section is exhibiting limiting value of applied footing pressure on the bund section. For convenience, all intermediate layers were not considered in the analysis. Based on the observations made in model test REM14, innumerable numbers of failure surfaces within the reinforced zone were tried and the failure surface which gave a least value of factor of safety was considered. Herein, contribution of geogrid layers was considered and tensile strength values for model geogrid were given as discussed above in the slope stability analysis.

Figure 18 presents variation of factor of safety of a 4 m high geogrid reinforced coal ash embankment (with a blanket layer) with an applied footing pressure. As can be noted from Fig. 18, immediately after subjecting the slope to a load equivalent to 300 kPa, the factor of safety has attained a value equal to 1. Majority of reinforcement layers in the upper half zone were found to be strained and ruptured. This implies that there is a need for use of high strength geogrid reinforcement layers within in the geogrid reinforced slope is mandatory. This was found to be in agreement with the results published by

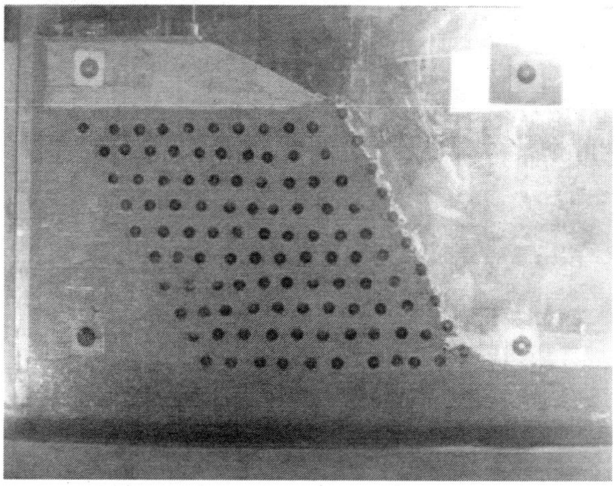

(a) Model REM14 20g (p = 10 kPa)

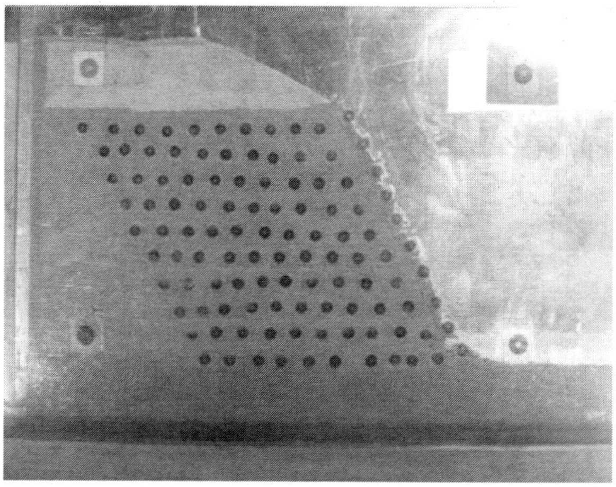

(b) Model REM14 20g (p = 350 kPa)

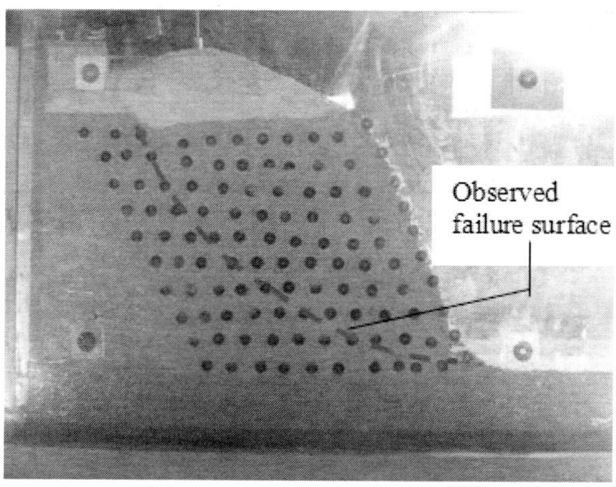

(c) Model REM14 20g (p = 450 kPa)

FIGURE 16: Front elevation of model REM14 during centrifuge test at 20g

Sommers and Viswanadham (2009). According to them, there is a need to strengthen the slope loaded externally loaded externally through a strip footing in the upper half to reduce straining and rupturing of reinforcement layers.

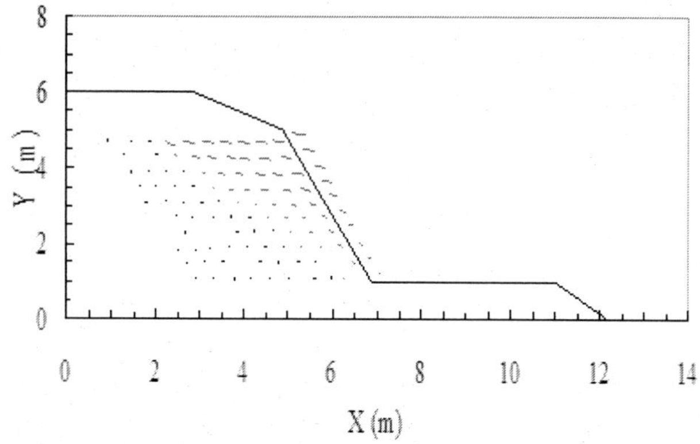

FIGURE 17: Plot showing displacement vectors from beginning to end of the test (for model REM14)

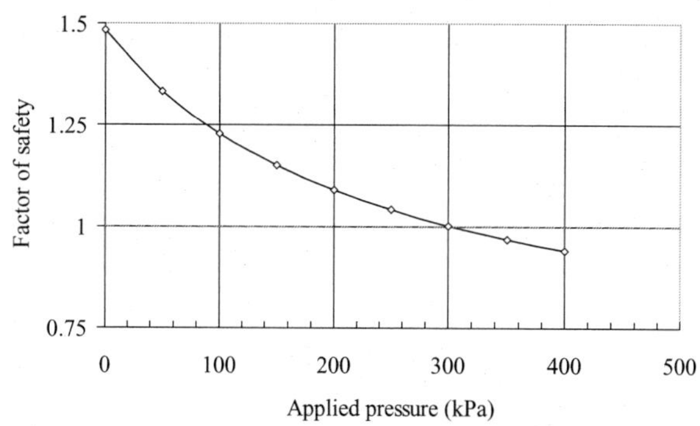

FIGURE 18: Variation of factor of safety with applied pressure for model REM14

5. CONCLUSIONS

Based on the analysis and interpretation of centrifuge test results, the following conclusions can be drawn:

The above information obtained from centrifuge tests indicate that coal ash can be used for constructing rail embankments. Adequate measures shall have to be taken to provide drainage, provision of a side soil cover of thickness not less than 1.5 m and suitable berms for embankments of heights of the order of 11 m. In the case of geogrid reinforced coal ash embankment, geogrid layers were noticed to subject to strains higher than their respective ultimate strain values within top-half zone. This implies that there is a need for use of high strength geogrid reinforcement layers within in the geogrid reinforced embankments is mandatory.

Geogrid reinforcement layers in a reinforced coal ash embankment with 2V:1H slope were observed to experience straining after applying a bearing pressure of 250 kPa in the top-half zone and indicating a need to provide high strength layers in top-half zone. Further, results of stability analysis of soil confined coal ash embankment models constructed with coal ash as a fill material were found to corroborate well with physically observed centrifuge test results.

However, it will be interesting to construct in the field on pilot scale basis for evaluating the performance of rail embankments subjected to real loading and climatic conditions.

ACKNOWLEDGEMENTS

The authors would like to thank the Centrifuge team at the National Geotechnical Centrifuge Facility of Indian Institute of Technology Bombay, Powai, Mumbai-400076, India for their untiring support throughout the present study. Thanks are also due to M/s NTPC Limited, Noida, India for sponsoring the study.

REFERENCES

Chandrasekaran, V.S. (2001) "Numerical and centrifuge modelling in soil structure interaction". Indian Geotechnical Journal, 31, Issue 1, pp 30–59.

Das, A., Jayashree, Ch.,Viswanadham, B.V.S. (2009) "Effect of randomly distributed geofibers on the piping behaviour of embankments constructed with fly ash as a fill material". Geotextiles and Geomembranes, 27, Issue 5, pp 341-349.

Faber, J. H., and DiGioia Jr., A. M. (1976) "Use of fly ash in embankment construction". Transportation Research Record, No. 593, pp 13-19.

Geostudio (2007) Geo-Slope, User's Manual for SEEP/W and SLOPE/W: Version 4, Geo-Studio International, Canada.

Gray, D. H., and Lin, Y. K. (1972) "Engineering properties of compacted fly ash," J. Soil Mech. and Found. Div., ASCE, 98, Issue SM4, pp 361-380.

Leonards, G.A., Balley, B. (1982) "Pulverized coal ash as structural fill." Journal of Geotechnical Engineering, ASCE, 108, Issue GT4, pp 517 - 531.

Lewis, T.S. (1976) "Construction of fly ash in roadway embankment in Illnois". Transportation Research Record ,No. 593, pp 20-23.

Martin J. P., Collins, A. C., Browning, J.S., and Biehl, F.J. (1990) "Properties and use of fly ash for embankments". Journal of Energy Engineering, ASCE, 116, Issue 2, pp 71-85.

NTPC report (2008) "Centrifuge model studies on the performance of rail embankments constructed with coal ash". Indian Institute of Technology Bombay, India.

Rajesh, S., and Viswanadham, B.V.S. (2012) "Modelling and instrumentation of geogrid reinforced soil barriers of landfill covers". Journal of Geotechnical and Geoenvironmental Engineering, ASCE, 138, Issue 1, pp 26-37.

Schofield, A. (1980) "Cambridge geotechnical centrifuge operations". Geotechnique, 30, Issue 3, pp 227-268.

Sommers, A.N., and Viswanadham, B.V.S. (2009) "Centrifuge model tests on the behaviour of strip footings on geotextiles reinforced slopes". Geotextiles and Geomembranes, 27, Issue 6, pp 497-505.

Sunaga, M., and Sekine, E. (1992) "Utilization of fly ash as materials for railway embankment". Japanese Railway engineering, 121, October, pp. 13.

Sridharan., A., Pandian, N.S., and Prasad.P.S. (2000) "Liquid limit determination of class F fly ash". Journal of Testing and Evaluation, ASTM, **28** , Issue 6, pp 455-461.

Toth, P.S., Chan, H. T., and Cragg, C.B. (1988) "Coal ash as structural fill with reference to Ontario experience". Canadian Geotechnical Journal, 25, Issue 4, pp 694 - 704.

Viswanadham, B. V. S., and König, D. (2004) "Studies on scaling and instrumentation of a geogrid". Geotextiles and Geomembranes, 22, Issue 5, pp 307-328.

Viswanadham, B.V.S., Jha, B.K., and Sengupta, S.S. (2009) "Centrifuge testing of fiber reinforced soil liners for waste containment systems". Journal of Practice Periodical of Hazardous, Toxic, and Radioactive Waste Management, ASCE, 13, Issue1, pp 45-58.

Viswanadham, B. V. S., Das, A., and Mathur, V. K. (2012) "Centrifuge Model Tests on Rail Embankments Constructed with Coal Ash as a Structural Fill Material". Geo-Congress 2012, ASCE, Special Publication No. 225, Oakland, CA, USA, pp 3786-3795.

Seismic Response of Composite Dams with Joints by Dynamic Centrifuge Tests

S.H. Kim, Y.W. Choo, D.S. Kim and J.G. Ha

Civil and Environmental Engineering, KAIST, Daejeon, Korea

E-mail: olive2020@kaist.ac.kr; ywchoo@kaist.ac.kr; dskim@kaist.ac.kr; jgha87@kaist.ac.kr

ABSTRACT

A composite dam which is a combination of a concrete dam and an earth or rockfill dam has been constructed over the decades and has been exposed to earthquakes. This paper aims to simulate and to evaluate the seismic behaviour of two composite dams with different shapes of the joint, interfaces between concrete dams and rockfill dams using dynamic centrifuge. Two composite dam models were identically constructed except the geometries of the joints. Tests were performed in series of stages as increasing shaking intensities with 40g centrifugal acceleration. The distributions of amplification ratio in the concrete and rockfill model of two composite dams were almost similar. However, the geometry of the composite dam influenced especially the amplification of the joint and the horizontal relative displacement on the top near the joint.

1. INTRODUCTION

1.1 Backgrounds

A composite dam is a combination of two types of dams: a concrete dam and earth or rockfill dams. The composite dams are generally built because of two reasons. The first reason is to buttress existing concrete dams by earth material. For the second reason, if one side of a rock foundation is deformable and not strong enough to support the concrete dam, the composite dam with combination of concrete and rockfill dams can be one of the solutions. In this case, a CFRD or an earth dam is possible to be constructed on the weak side of the dam foundation and a concrete dam will be on the sound foundation (Emiroglu, 2008). Among two types of the composite dams, this paper is focused on the second type of the composite dams.

Many engineers have tried to understand the seismic behavior of the joint in the composite dam using numerical analysis programs such as FLUSH and NIKE2D (Chang, 2000, Léger, 2007). However, programs' input parameters impose uncertainties and the analytic programs can not reflect complexities of real construction site (Sharp, 2006). Furthermore, previous research has been focused on the composite dam for the buttressing (Léger, 2007) and the composite dam with the concrete dam right next to the CFRD in three dimensions is rarely considered. Thus, a scaled physical modeling test is required to find the behavior of the joint in the composite dam and to compare with results of the numerical analysis.

To satisfy this need, a dynamic centrifuge test with in-flight shaking table has been an effective scaled physical modeling test in geotechnical area. Because the dynamic centrifuge reproduces in-situ stress in a scaled model under seismic loading, the centrifuge has been used to investigate dynamic behaviors of the dams. For the composite dam, it is important to reproduce ground contact pressure at the interface between the concrete material and rock-fill material. Therefore, the dynamic centrifuge test for the composite dam is necessary to evaluate the dynamic behavior of the composite dam.

1.2 Seismic Problem in Composite Dams

Over the past decades, various types of dams including the composite dams have been constructed. In Korea, the composite dams are sometimes constructed due to Korea's geological condition. In Korea, dams are constructed on rock foundation. In case of the foundation with discontinuities due to fault and joints, concrete dams cannot be constructed on the foundation because concrete dams are not bearable to deformation. If a left side of the dam foundation consists of sound rock whereas a right side of the foundation is deformable to support concrete dams, rockfill dams, which is less sensitive to settlement, should be on the right side and concrete dams are on the left. Some types of the joints have been considered for construct the composite dams and may catastrophically fail if they are not seismically stable. Furthermore, concrete dams and rockfill dams are moved separately while shaking under seismic loadings. This point makes a dynamic analysis of the composite dams more difficult and complicating. Therefore, to evaluate the seismic safety of the composite dam, amplification characteristics, phase difference, and displacements at joints between concrete parts and rockfill parts caused by dynamic loading should be investigated.

In the case of the composite dam combined by a concrete faced rockfill dam (CFRD) and a concrete dam, water tightness in an upstream of the dam particularly at the joint is a key issue to determine the stability of the composite dam during and after earthquakes. Water tightness of the composite dam is maintained by the concrete slab on the CFRD and it is connected to the concrete dam by the membrane in the composite dam. Because a joint of the composite dam, a transition part from the concrete part to the CFRD, consists of two different material, concrete and rockfill, relative displacements between the concrete face on rockfill and the concrete dam can cause separation with tearing of membrane and the composite dam cannot retain water tightness. Thus, it is important to analyze the seismic behavior of the joint and interface between the concrete and rockfill parts in the composite dam to verify the dynamic stability during earthquakes.

This paper aims to simulate the seismic behaviors of two composite dams with different joint geometries using dynamic centrifuge. The centrifuge tests were conducted with a scaling ratio 40. The composite dam models consisted of a concrete model and a rockfill model with a concrete faced slab on the upstream. Ofunato and Hachinohe earthquake record were applied to the dam models. Amplification characteristics of accelerations, phase difference, and horizontal displacements near the joint were investigated during earthquakes. Also, the seismic responses of the two composite dams were compared to find how the geometry of the joint affected the dynamic stability of the composite dam.

2. CENTRIFUGE TEST PROGRAM

2.1 KAIST Centrifuge and Earthquake Simulator

The maximum capacity of KAIST beam centrifuge shown in Figure 1 (a), with a 5m radius, is 2400kg for up to 100g of centrifugal acceleration (Kim et. al., KSCE, 2012). Earthquake loading can be simulated by an in-flight earthquake simulator equipped with an electro-hydraulic system shown in Figure 1 (b). It can generate random earthquake excitations lasting up to 2 sec with model frequency ranging from 40 to 300 Hz. The base shaking acceleration can be exerted to a maximum value of 20gh with a maximum payload of 700 kg, which corresponds to 0.5gh in prototype scale at 40gc centrifugal acceleration.

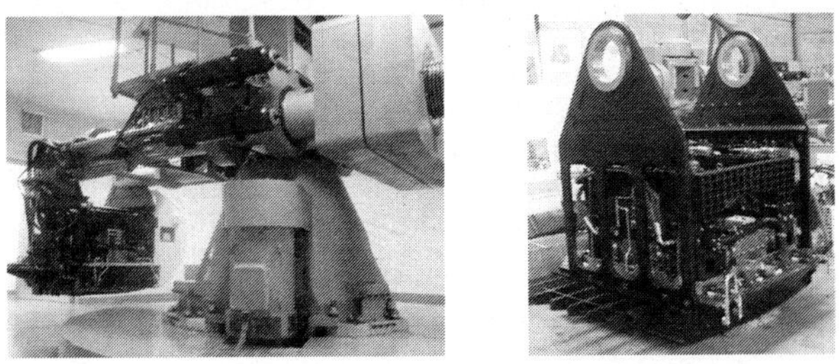

(a) KAIST centrifuge (b) Earthquake simulator

Figure 1: KAIST centrifuge and earthquake simulator

Using the KAIST earthquake simulator, real earthquake records can be simulated during the in-flight condition of the centrifuge. The scaling laws for centrifuge testing have been derived (Scofield, 1980). In addition, a dynamic balancing system of the earthquake simulator can remove undesired vibration from the main centrifuge equipment to decrease damages to the centrifuge while generating seismic loadings.

2.2 Model Preparation

In this study, two types of the composite dam were designed; Model A, and Model B. Cross-sections of the models were designed based on a database of dams in Korea. The cross-sections of the rockfill and the concrete model in the composite dams were exactly the same so that the only difference between Model A and Model B was the geometries of the joints in the dam models.

In Model A shown in Figure 2 (a), the concrete joint consisted of two retaining walls in upstream and downstream sides. The walls were inserted into rockfill part of the composite dam. However, Concrete dam of Model B was extended into the rockfill dam and the joint of Model B in downstream was shaped same as the concrete dam. The retaining wall was in upstream of the joint. The shape of the joint in Model B is provided in Figure 2 (b).

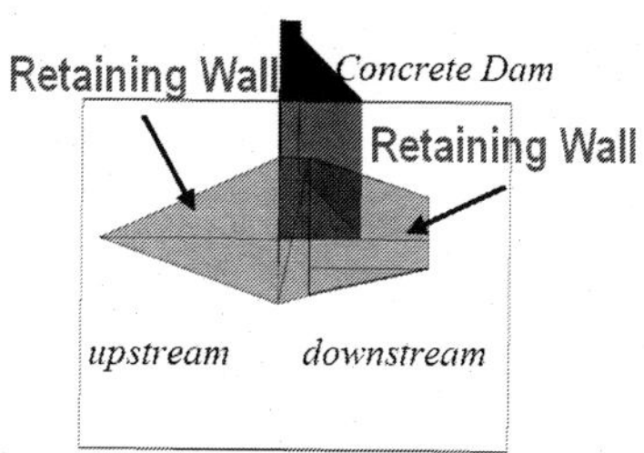

Model A

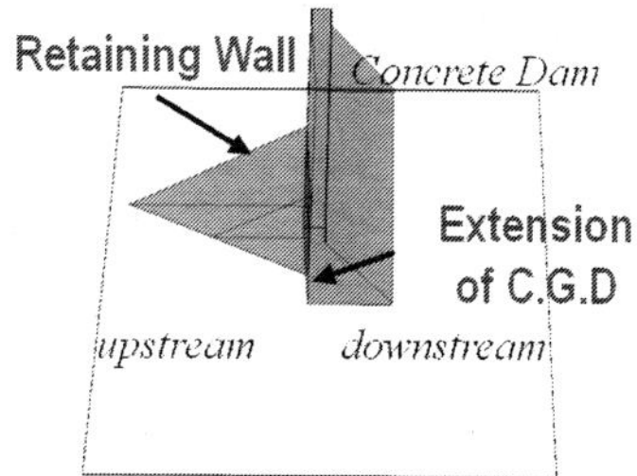

Model B

FIGURE 2: Geometries of the joints

The rockfill dam model had a slope of 1:2.0 and the concrete model had 1:0.7 as a downstream slope. In addition, to decide the dimensions of the composite dam models, two factors were considered; size of the test container, and centrifugal acceleration capacity. Due to these two factors, the dimensions of the prototype models were set as smaller than the dimensions of the actual dams and a scaling ratio N was decided as 40. The final height of the test models was 146mm and the prototypes were 5.84m high. The dimension difference was acceptable in this study because the study was planned to observe qualitative behaviors rather than quantitative behaviors of the composite dams under earthquakes.

2.2.1 *Preparation of the Concrete Model*

The concrete model was constructed to simulate a concrete gravity model in the composite dam. The mold for the concrete model was made of steel plates so that both the concrete dam model and the concrete joint were monolithically casted. A concrete paste was mixed with a ratio of water: cement: concrete = 1: 2: 6 and was poured into the mold to be casted. After a three week cure of the concrete, the steel mold was separated from the concrete model. The model was attached to a base-plate of the test container by adhesive so that the base-plate acted as the bedrock of a dam foundation. Furthermore, the cross-section of the concrete model was exactly met with one side of the container wall.

2.2.2 *Preparation of the CFRD Model*

In the case of the rock-fill model in the composite dam, an appropriate rock-fill material was obtained. To build the rock-fill model, a rock-fill material was collected from a dam construction site. The particle size of distribution curve showed that the rock-fill material was too large to be used in the dynamic centrifuge tests. To solve this problem, the size of the rock-fill material for the tests was reduced and grains passing through a No. 100 sieve were added to the test rock-fill material. Therefore, the particle size distribution curve of the test material was parallel to the particle size distribution curve of the material from the construction site as shown in Figure 3.

The rock-fill model was built in the test container with inner dimensions of 600mm height, 490 mm width, and 490mm length. Before the construction of the rock-fill model, sandpaper was attached to the base-plate where the rock-fill model was later placed because enough friction between the bedrock and the dam body must be simulated. Also, the side walls of the container were covered with grease to minimize side friction between the container and the dam body. The model was constructed with five layers to keep rock-fill densities uniform. Each layer was compacted with 95% relative compaction by a hand compactor with a compaction energy of 750kN·m/m^3. Those layers were constructed by uniform dry unit weight 2.05 g/cm^3 with 6% water content.

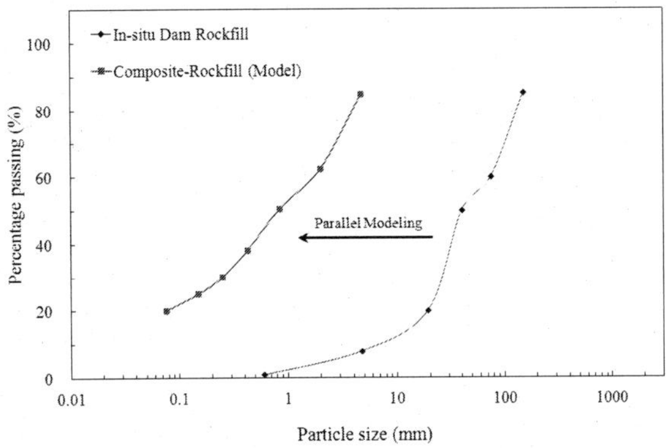

FIGURE 3: Particle size distribution curve

A face slab model was placed on the upstream slope of the rock-fill model to simulate a concrete face slab in a CFRD. A High-density polyethylene (HDPE) plate was used for the face slab model because concrete for the slab model was difficult to have same bending stiffness of the prototype slab. A thickness of the slab model was determined by Eq. (1) (Zhang L, 1994):

$$t_m = \sqrt[3]{\frac{E_p}{E_m}\frac{t_p}{N}} \quad (1)$$

Here, E_p, E_m, t_p and N are the elastic modulus of the prototype, the elastic modulus of the model, the thickness of the prototype slab and the scale ratio, respectively. In this study, E_p is 27GPa, E_m for the HDPE is 1.2GPa and N equals 40. Therefore, the thickness of the slab model was decided as 3mm based on the thickness of the prototype slab and equation (1).

An Ethylene Propylene Diene Monomer (EPDM) membrane was placed on the upstream slope of the rockfill model to prohibit water from penetrating into the dam body. The dam was pressured by a 135mm depth of water contained in the upstream of the dam model. Seepage flow inside the dam body was not considered in this study to estimate only the seismic response of the composite dams. Figure 4 shows the composite dam model for dynamic centrifuge tests.

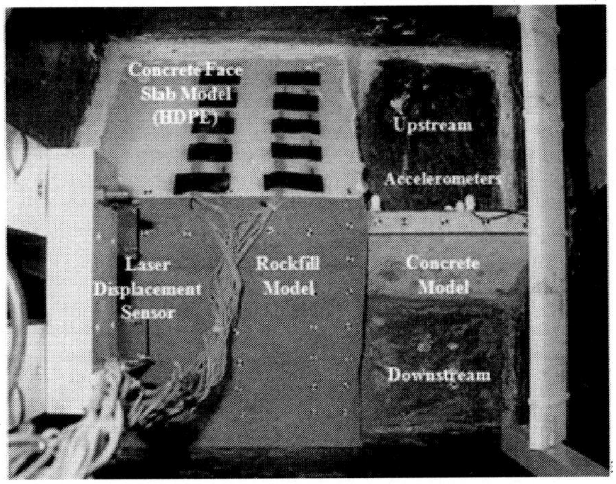

FIGURE 4: Composite dam model for the test

2.3 Testing and Monitoring Program

Two kinds of input motions were used to simulate earthquakes; Ofunato and Hachinohe. The Ofunato record is characterized by a short period dominant earthquake which is most probable in Korea. In contrast, the Hachinohe record is a long period dominant earthquake and used to compare the behaviour with the Ofunato record. Figure 5 contains earthquake records of Ofunato and Hachinohe which were used in the dynamic centrifuge tests. In all tests, shaking was imparted in the upstream-downstream direction. Stage tests were conducted for each dam model. The smallest peak acceleration of 0.04g was applied at the beginning and the peak acceleration was increased up to input intensity of 0.45g. Input accelerations used for the stage tests are provided in Table 1.

To monitor the acceleration amplification of the composite dams during earthquakes, single-axis accelerometers were embedded inside the dam models as shown in Figure 6. The accelerometers were placed at different heights and at different arrays. As shown in Figure 6 (b) and Figure 6 (d), three arrays of accelerometers were located in the dam models; the first array in the concrete model, the second array near the concrete joint in the rock-fill model, and the third array in the center of the rock-fill model. Three accelerometers were placed in each array at heights of 55mm, 94mm, and 146mm, respectively. For the array of the concrete model, it can be assumed that the seismic behavior at the center of the concrete model is the same as the behavior of the concrete joint because the concrete dam model and the joint were made monolithically. The arrays were used to estimate the comparative seismic responses at each location.

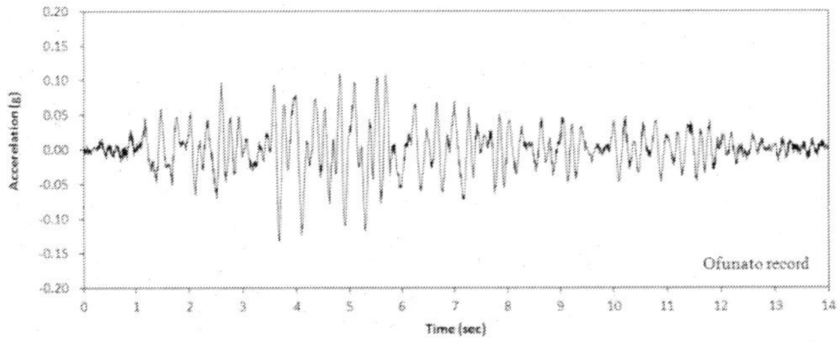

(a) Ofunato record

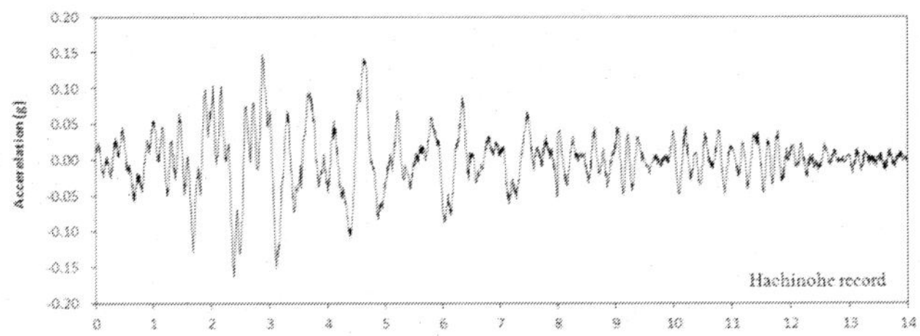

(b) Hachinohe record

FIGURE 5: Earthquake records

Table 1: Input Acceleration (Unit: g)

	Model A	Model B
Ofunato	0.047	0.044
	0.134	0.110
	0.212	0.223
Hachinohe	0.108	0.102
	0.164	0.147
	0.247	0.251
	0.451	0.439

Horizontal displacements of the dam crest were measured to compare the deformation of each dam models. To measure the horizontal displacements, a high-speed camera was placed at the frame of a centrifuge arm to capture the top view of the dam models during earthquakes. Targets for recording by high-speed camera were placed on the composite models and the location of targets was given in Figure 6 (b) and Figure 6 (d). The Horizontal displacement histories were recorded in photos taken by the high-speed camera and these photos were analyzed by image-processing software, TEMA 2.6. The software is capable to trace the targets installed on the models and compare location of the targets in target area in adjacent two photos. This procedure is automatically processed for all the pictures taken during each seismic loading. Through the procedure, displacement trace records are provided. The horizontal displacements were used to evaluate the relative displacement of the joint in the composite dam.

3. TEST RESULTS AND DISCUSSIONS

In this study, two composite dam models were simulated in the dynamic centrifuge to identify the seismic behaviours of the composite dams and to compare the dynamic characteristics of the composite dam with different joint geometries. To obtain amplification and phase characteristics, accelerometers were placed at the positions shown in Figure 6. The first array (C) was on the concrete model, the second array (J) was near the joint in the rock-fill model, and the third array (RF) was in the centre of the rock-fill model. Also, image processing of high-speed camera data was conducted to find horizontal displacement at the top of two composite dam models. These data indicate that there were different patterns in the seismic behaviours between Model A and Model B.

3.1 Phase Difference

In order to compare the behaviors among concrete model, joint, and rock-fill model effectively, records of the acceleration-time histories at the crest in Model A by Hachinohe earthquake are shown in Figure 7. There exists a phase difference in time signal, among the bedrock, the concrete model, the joint, and the rock-fill model. Same trend was also observed in Model B.

The magnitudes of the accelerations and the phase differences were increased in sequence as the bedrock, the concrete model, the joint and the rock-fill model. Especially the acceleration of the concrete model was almost synchronous with the acceleration of the bedrock. The acceleration record of the rock-fill model had the highest amplitude of the acceleration and the biggest phase difference. The phase difference and the

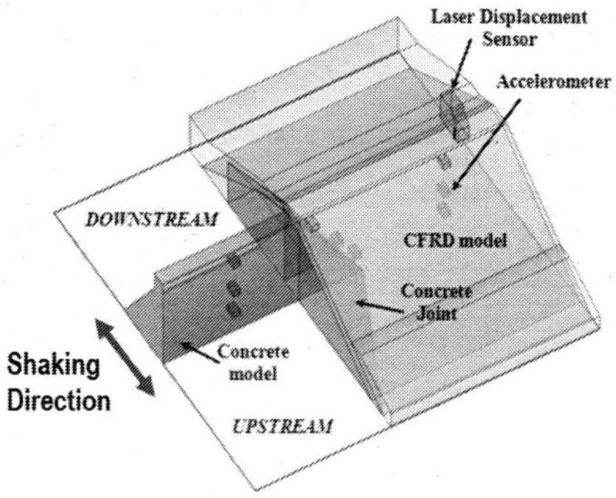

(a) Preparation of Model A

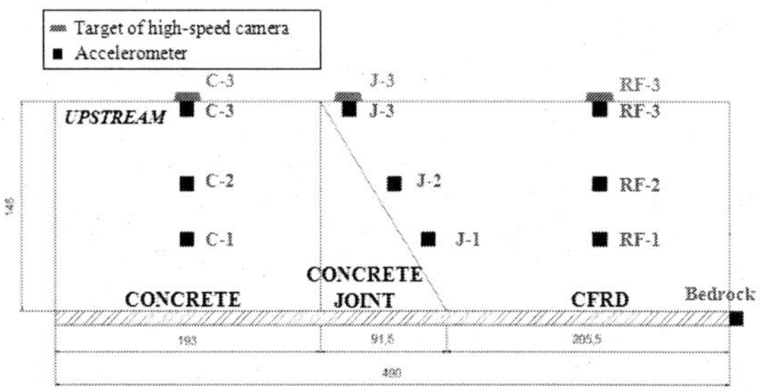

(b) Monitoring locations of Model A

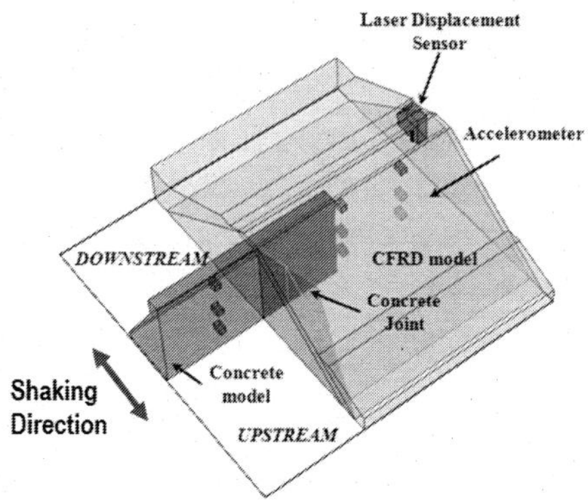

(c) Preparation of Model B

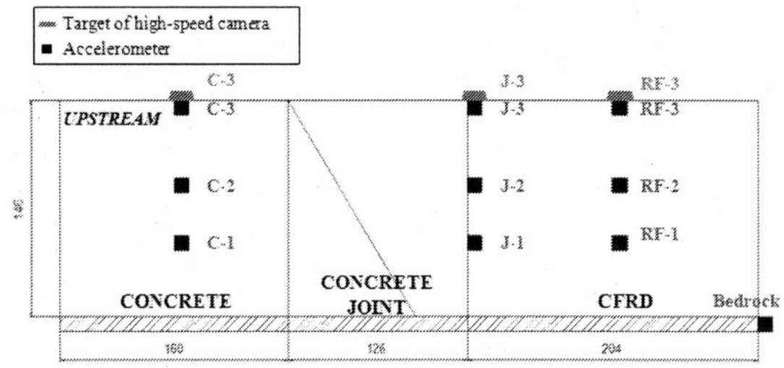

(d) Monitoring locations of Model B

FIGURE 6: Model preparation and monitoring locations

magnitude of the acceleration of the joint were intermediate between the joint and the rock-fill model. Therefore, it can be noticed that the joint can be considered to have a characteristics in

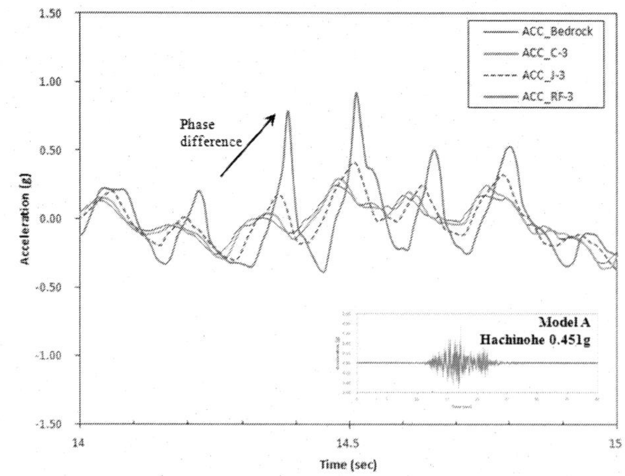

FIGURE 7: Phase difference at the crest (Model A, Hachinohe PGA .451g)

the seismic behaviour between the concrete dam model and the rockfill model.

3.2 Acceleration Measurements with Heights

Figure 8 (a) and (b) represent maximum accelerations at each array of Model A and Model B under Ofunato loadings with different magnitudes. As shown in Figure 8 (a) and (b), the peak acceleration of the joint and the rockfill model increased from the bedrock to the crest in both Model A and Model B. Moreover, the peak acceleration at the crest significantly increased as the bedrock acceleration increased.

The amplification trends were clearly found in normalized acceleration distributions of Model A and Model B, as can be seen in Figure 8 (c). In Model A and Model B, the concrete dam models were not amplified comparing the joint and rockfill dam model, and the amplification ratios of the concrete models were below 1.3. For the concrete model in Model B, the amplification ratios were almost 1.0 under three Ofunato loadings. On the other hand, the rock-fill dam model was the most amplified model compared to the concrete dam model and the joint. The average amplification ratio of the rock-fill model in Model A was 2.19 and the average amplification ratio of the Model B was 2.22. Comparing exiting normalized acceleration distribution of CFRD (Kim et. al., 2011), the amplification pattern of the rockfill

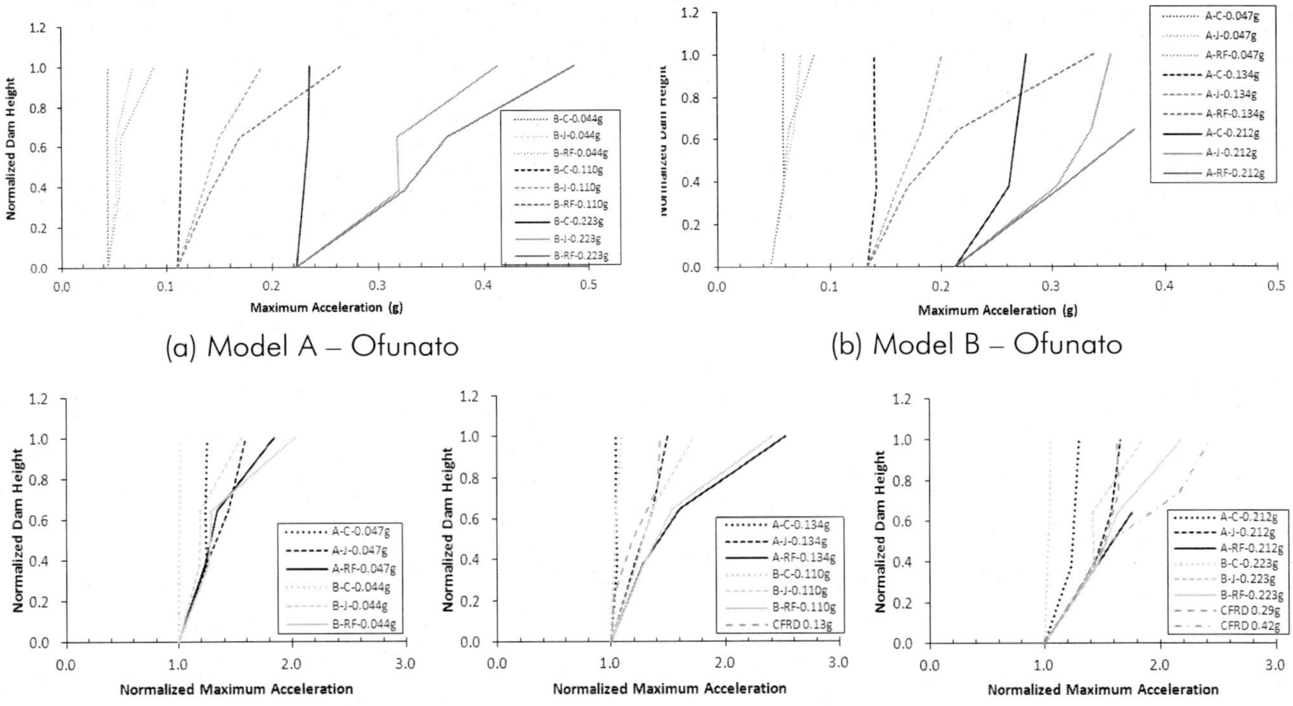

(a) Model A – Ofunato (b) Model B – Ofunato

(c) Normalized maximum acceleration with height

FIGURE 8: Acceleration characteristics under Ofunato earthquakes

model in the composite dam was consistent with the pattern of CFRD. Thus, this result affirms that the amplification characteristic at the centre of the rockfill dam in the composite dam is similar with the amplification characteristic of the CFRD.

In the case of the joint, the amplification patterns of the joint was in-between the patterns of the concrete model and the rockfill model. However, a difference in the seismic behaviours between A-model and B-model could be observed in the amplification of the joint given in Figure 8 (c). The amplification ratio of the top of the joint in the Model A was 1.6 when the bedrock PGA was 0.047g, 1.5 at 0.134g and 1.66 at 0.212g. The top of the joint in the B-model was a little increasingly amplified with the bedrock PGA; 1.0 at the bedrock PGA 0.044g, 1.74 at 0.11g, and 1.86 at 0.223g. This difference might come from the geometries of the two joints. The shape of the concrete joint in the Model A is a triangular dipyramid and two region exits together under the top of the rockfill model near the concrete joint in Model A. However, the region under the top of joint in Model B consists of only one rock-fill material. Because the concrete region is amplified less than the rockfill model under the same seismic loading, it is possible that the top of the joint in Model A has a lower amplification ratio than the top of the joint in Model B has.

Acceleration characteristics by Hachinohe earthquake records are given in Figure 9. Common characteristics of the amplification in both Model A and Model B were observed. The concrete dams were not amplified with the normalized dam height and with the magnitude of ground accelerations. In case of rockfill dams, the amplification was increased with height and the top of the rockfill dam was the most amplified. Moreover, both Model A and Model B were less amplified by 0.1g Hachinohe loadings than by other loadings.

Under Hachinohe loadings under 0.25g, the top of the joint in Model A was less amplified than the top of the join in Model B. when 0.15g Hachinohe record was applied the center of the rockfill dam in Model B was more amplified than the center of the rockfill dam in Model A.

As observed by Ofunato records, the joints were acted as in-between the rockfill dams and the concrete dams by Hachinohe shown in Figure 8 (c). Comparing Ofunato and Hachinohe records, the top of the joints (J-3) both in Model A and B were more amplified by Ofunato than by Hachinohe with PGA near 0.154g for seismic classes I structures (1000 year return period) at collapse level of earthquake based on the Korean seismic design code. In the case of the joint in Model A, the amplification ratio by Ofunato was 1.50 and 1.16 was by Hachinohe. Also, the joint of the model B was amplified with 2.41 times higher than the bedrock PGA by Ofunato whereas the amplification ratio by Hachinohe with PGA 0.149g was 1.78.

3.3 Horizontal Displacements

An essential factor to fail sustaining water tightness of the composite dam is a relative displacement caused by the horizontal displacement near the joint between the concrete model and the rockfill model. Figure 10 shows horizontal deformations near the joint at the crest in Model A and Model B. Horizontal deformations of top of the composite dams were measured by the high-speed camera and analyzed through image processing using TEMA 2.6. Two signals were analyzed under Hachinohe earthquakes along upstream-downstream direction including bedrock motion with the magnitude 0.451g for Model A and with the magnitude 0.439g for Model B. The names C-3, J-3, and RF-3 in Figure 10 are indicated depicted in Figure 6. J-C means relative horizontal displacement between the joint and the center of the concrete and RF-C means relative horizontal displacement between the center of the rockfill dam and the concrete dam. Furthermore, positive displacement means a movement toward downstream and negative displacement indicates the model moves toward upstream.

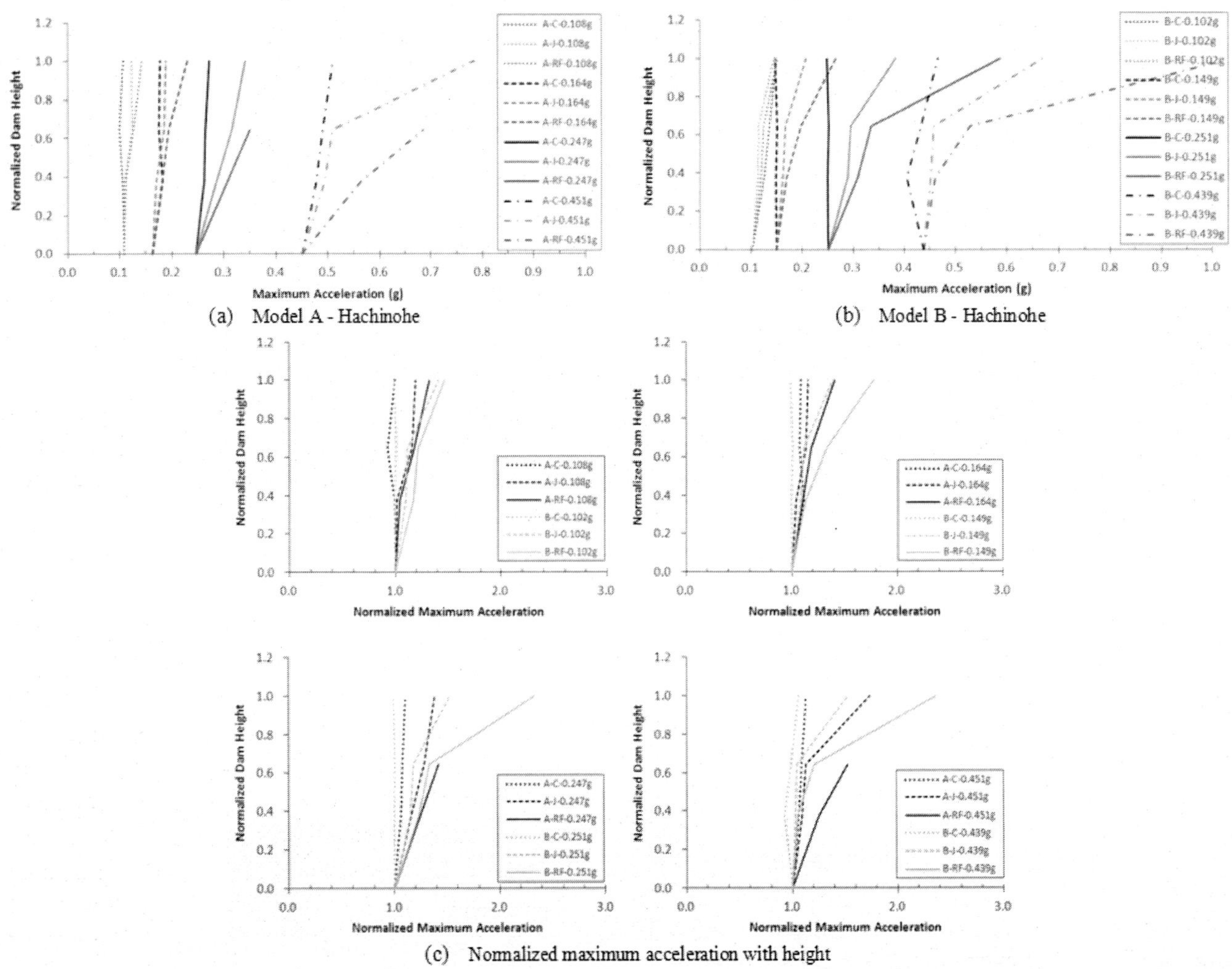

FIGURE 9: Acceleration charateristics under Hachinohe earthquakes

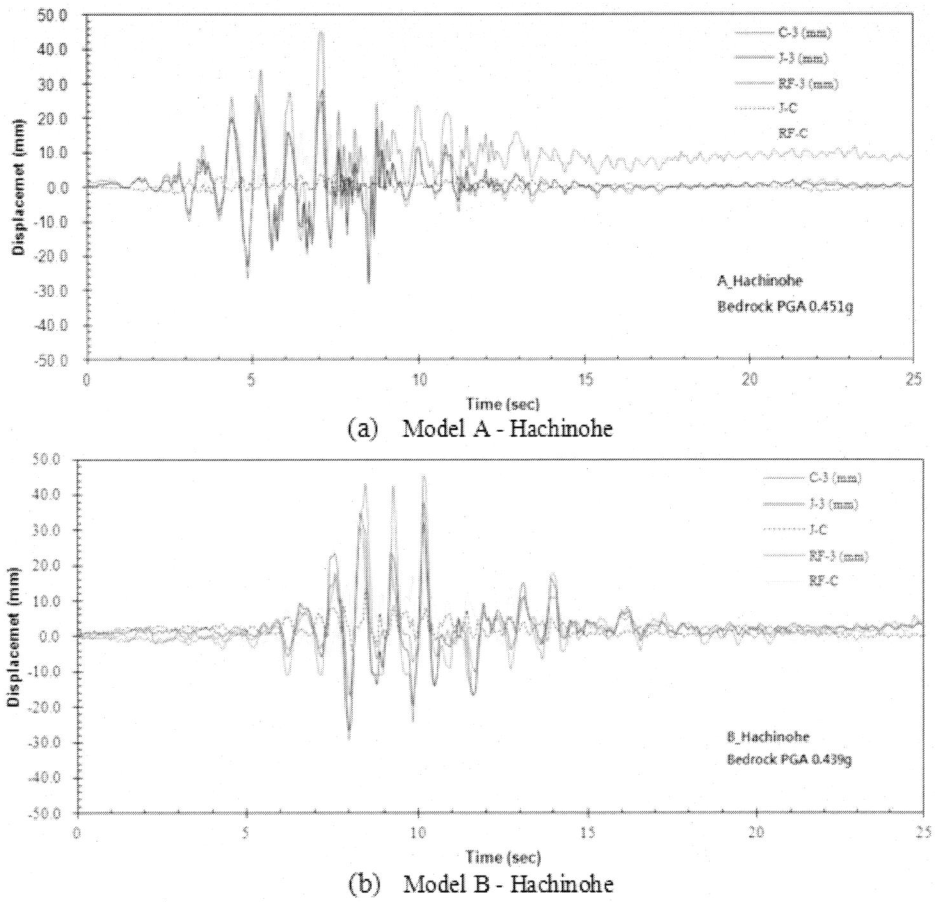

Figure 10: Acceleration characteristics under Hachinohe earthquakes

The common characteristic of the horizontal displacements was that the rockfill dam was the most horizontally deformed part in Both Model A and Model B. Also, displacements of the top in rockfill dam toward the downstream were bigger than the displacements toward the upstream direction under earthquakes with around 0.45g peak ground acceleration. Thus, rockfill material might be slid on the surface of the rockfill dam toward downstream by seismic loadings.

However, there were a few differences in records of the horizontal displacements in Model A and Model B. First, the relative horizontal displacement of Model A was smaller than the relative horizontal displacement of Model B: Maximum J-C of Model A was 3.84mm whereas Maximum J-C of Model B was 13.65. Second, a permanent deformation, 9mm, was occurred in the top of the rockfill dam in Model A although the rockfill dam of Model B was not deformed after the seismic loading.

4. CONCLUSIONS

This study aims to simulate two composite dams with different joints with different geometries to qualitatively present the seismic behaviors of the composite dams using the dynamic centrifuge under the scaling ratio 40. These two composite dam models had the same cross-sections, except the geometries of the concrete joints and the models were subjected to the same input motion; Ofunato and Hachinohe. The concrete models of both Model A and Model B were amplified until the amplification ratio 1.3. For the joint, the amplification ratio of Model A was smaller than the ratio of the Model B. Moreover, the relative horizontal displacement near the joint at the top of Model A was smaller than the displacement of the Model B. This study could offer basic information about the seismic behavior of composite dams. To verify quantitative behavior of the composite dams, numerical analysis is necessary because this research was planned to find qualitative trends and the dimensions of the models limited.

ACKNOWLEDGEMENTS

Authors would like to thank SAMSUNG C&T CORPORATION Engineering & Construction Group and Korea Water Resources Corporation for their funding and their efforts to collect valuable data.

Also, this research was supported by the Basic Science Research Program through the National Research Foundation of Korea (NRF) funded by the Ministry of Education, Science and Technology (Grant number: 2012-0000154).

The authors also acknowledge all of the students and colleagues who worked with the KOCED Geotechnical Centrifuge Centre at KAIST for their help in conducting the tests presented in this paper.

REFERENCES

Chang, N. Y., and Oncul, F. (2000) "A parametric study on seismic behavior of a composite dam". ASCE, Geo-Denver 2000, Soil Dynamics and Liquefaction 2000,GSP 107.

Emiroglu, M. E. (2008) "Influences on selection of the type of dam". International Journal of Science & Technology, 3, Issue 2, pp 173-189.

Kim, D.S., Lee, S. H., Choo, Y. W., and Rames, D. (2012) "Self-balanced earthquake simulator on centrifuge and dynamic performance verification". KSCE Journal of Civil Engineering. (Accepted)

Kim, M. K., Lee, S. H., Choo, Y. W., and Kim, D. S. (2011) "Seismic behaviors of earth-core and concrete-faced rock-fill dams by dynamic centrifuge tests". Soil Dynamics and Earthquake Engineering, 31, Issue 11, pp 1579-1593.

Léger P., and Javanmardi, F. (2007) "Seismic stability of concrete gravity dams strengthened by rockfill buttressing". Soil Dynamics and Earthquake Engineering, 27, Issue 3, pp 274-290.

Schofield, A. N. (1980) "Cambridge geotechnical centrifuge operations". Geotechnique, 20, Issue 1, pp 227-268.

Sharp, M. K., and Adalier, K. (2006). "Seismic response of earthdam with varying depth of liquefiable foundation layer". Soil Dynamics and Earthquake Engineering, 26, Issue 11, pp 1028-1037.

Zhang L, Hu T., and Zhang J. (1994) "Evaluation of the cut-off structures of a rockfill dam", Proceedings of the International Conference Centrifuge '94, Singapore, pub. Balkema, A.A., pp 593-598.

Infrastructures on Soft Sensitive Clays: Some Aspects Related to the Flow Behaviour of Remoulded Materials

V. Thakur

Faculty for Engineering and Natural Science, Aalesund University College, Norway
Geotechnical Division, Norwegian Directorate of Roads, Norway
E-mail: vikas.thakur@vegvesen.no

ABSTRACT

Soft sensitive clays are considered to constitute a major threat to nearby infrastructures due to their tendency to remould when subjected to excessive loading. According to the current practice in Norway, the soft sensitive clays that are prone to large landslides is characterised by a sensitivity (S_t) ≥ 15 and a remoulded shear strength (c_{ur}) ≤ 2.0 kPa. At present, there is a lack of substantial evidence supporting this practice Norway. The aim of this work is to initiate a discussion on a key question such as the following: Does clay that has c_{ur} > 0.5 kPa have the potential to retrogress to the same extent as quick clays (c_{ur} ≤ 0.5 kPa)? In this study, the author attempts to approach this question using data from documented landslides involving quick and sensitive clays and using the results from the small scale model tests carried-out on three different soft sensitive clays. The results from the small scale tests, so called quickness tests, shows that remoulded clays with c_{ur} ≥ 1.0 kPa are not as fluid as it originally thought that therefore are less prone to large landslides.

1. INTRODUCTION

It is a well-known fact that soft sensitive clays are associated with various hazards related to the loss of stability and bearing capacity, as well as substantial ground deformation, which can lead to structural damage and jeopardizes the overall stability of an area, see Figure 1. Over the decades, several large landslides in Norway have been affecting the infrastructure like, houses, roads and railways. Figure 2 and 3 shows example from the Lyngen slide and the Byneset slide which took place in the Northern Norway in 2010 and 2012, respectively. The Lyngen slide (2010) was initiated due to a fill work at the slope toe. The slide retrogressed to 230 m. The Byneset slide (2012) was probably initiated due to natural erosion at the toe of the slope. The slide retrogressed 400 m backward and the slide mass had run-out distance of 870 m from the slide location. Both the slides affected the infrastructures within its range.

The majority of Norwegian clays are notably soft and sensitive, and some are categorized as quick clays. Several classification systems for these clays have been proposed in the literature (Skempton and Northey, 1953; Rosenqvist, 1953; Norwegian Geotechnical Society (NGF), 1974). According to the current definition in the NGF (1974), a clay is said to be quick if it has a c_{ur} ≤ 0.5 kPa. However, the Swedish Geotechnical Institute (SGI) defines quick clay as clay with a sensitivity of 50 or greater and a c_{ur} ≤ 0.4 kPa (Rankka et al., 2004). The term sensitivity (S_t) is the ratio of the intact undrained shear strength (c_{ui}) to the corresponding remoulded undrained shear strength (c_{ur}) measured using the fall cone test at the same water content. In both cases, it has been stipulated that the remoulded material must behave as a viscous fluid rather than a plastic solid (Torrance, 1983).

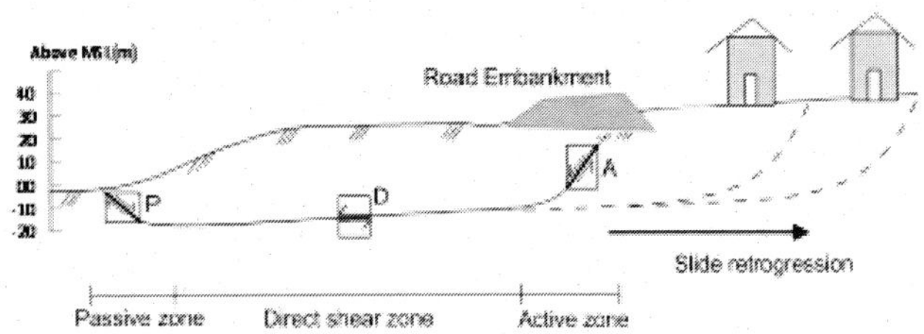

Figure 1 A schematic view of geotechnical structures like road embankment on a slope consist of soft sensitive clays. Here *A*, *D*, *P* corresponds to the stress directions under the active (compression), direct simple shear and passive (extension) condition. Slide retrogression (*L*) may be initiated due to the initial failure caused by the road embankment

FIGURE 2: The Lyngen slide (2010)

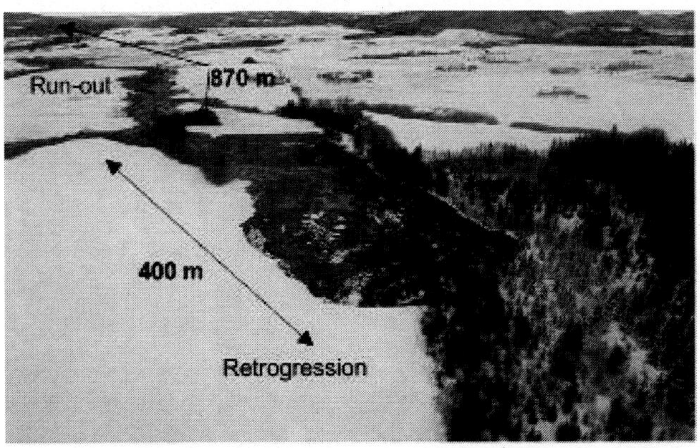

FIGURE 3: The Byneset slide (2012)

The extent of retrogression and the stability of the sliding mass depend on geometry, boundary and loading conditions. However, for landslides of large extent to occur, after an initial slide, it is important that at least the criteria related to the safety margin and the flow behaviour of materials are fulfilled (Mitchell and Markell, 1974; Lebuis and Rissmann, 1979; Tavenas et al., 1983; Karlsrud et al., 1985; Thakur et al., 2012; Thakur and Degago, 2012).

Tavenas et al. (1983) clearly indicates the following three conditions to get a large landslide in soft sensitive clays:

- to have a first failure;
- to have enough potential energy so that the soil can be remoulded; which is reflected by a stability number. It appears that the energy required for remoulding clays depends on the plasticity of the clay;
- to have a remoulded soil that flows and which is thus liquid-like.

Additional factors such as the topography can also influence the extent of retrogression. However, if the criteria mentioned above are not fulfilled, then landslides like the Byneset and the Lyngen slide, as shown in Figures 2 and 3, are unlikely to occur. Technological advances have expedited the construction process; accordingly soft sensitive clays are critical towards failure during construction phase i.e. under undrained loading. Therefore, two important aspects must be ensured; first, a sufficient safety margin must be taken in the design consideration; second, landslide susceptibility of this material must be well-understood.

The author believe that a better physical understanding of c_{ur} will help in explaining why flow type landslides is less likely to occurs even in soft sensitive clays having c_{ur} beyond a specific range. Thus, a simple test set-up, referred as quickness test, is developed to study this. In this paper, an attempt is made to demonstrate which sensitive clay can be susceptible to large slide using some simple examples and small scale model tests.

2. SOFT SENSITIVE CLAYS

The index properties of more than 500 samples of Norwegian soft sensitive clays taken from 130 different locations in Norway are presented in Figure 4. Majority of soft sensitive clays having, $S_t > 15$, are usually low plasticity materials, with plasticity index, PI, (I_p) values of less than or approximately 10, which transforms them into brittle materials when subjected to external loading. The clays follow the A-line with PI = 0.73 (LL - 20) on Casagrande's plasticity chart, as in Figure 4 (upper). These clays are uniquely related by their S_t and liquid limit, LL, (w_L) values, and the majority of highly sensitive clays have $w > w_L$, as in Figure 4 (middle). The friction angle φ varies between 25 - 28 degrees when these clays are normally consolidated, but φ decreases with increasing w/w_L, as in Figure 4 (lower). Norwegian soft sensitive clays with $w/w_L > 1.0$ indicate open void structures and can be susceptible to flow slides when their liquidity index (I_L) is greater than 1.2 (Burland, 1990; Lebuis et al., 1983, Lerouil et al., 1983).

It has been well documented by several researchers (e.g. Lacasse et al., 1985; Lunne et al. 1997) that routine sampling procedures (e.g. using 54 mm cylinders) lead to sample disturbance in soft sensitive clays. In turn, sample disturbance leads to, among other problems, the underestimation of preconsolidation pressure, c_{ui} and the rate of strain softening. Figure 5 illustrates the effect of sample disturbance in a soft sensitive clay from the Kløfta road project in Norway. The figure shows that 54-mm sampler induced a great amount of disturbance in the sample. Consequently, the stress-strain-deformation characteristics obtained from the laboratory are not representative to the true response of the material as seen in the field. Therefore, the larger diameter samplers like 76, 95 or 230 mm samplers are becoming increasingly popular in Norway. In addition to this the sample quality also influences S_t. Table 1 presents a comparison between the measured c_{ui} and S_t values from a block sample and from 54-mm diameter samplers from the Kløfta road project in Norway. The measured c_{ur} = 0.2 kPa is clearly unaffected by the type of sampling. The measured c_{ui} for a block sample is 27 kPa, whereas 54 mm diameter samples at the same depth have c_{ui} values of 8.6 kPa and 11.7 kPa. The range of error for the S_t values in Table 1 is in the order of 300%, which has a significant effect in the interpretation and the design procedures. Consequently, c_{ui} and S_t are influenced by sample disturbance, whereas c_{ur} is not dependent on the quality of the sample.

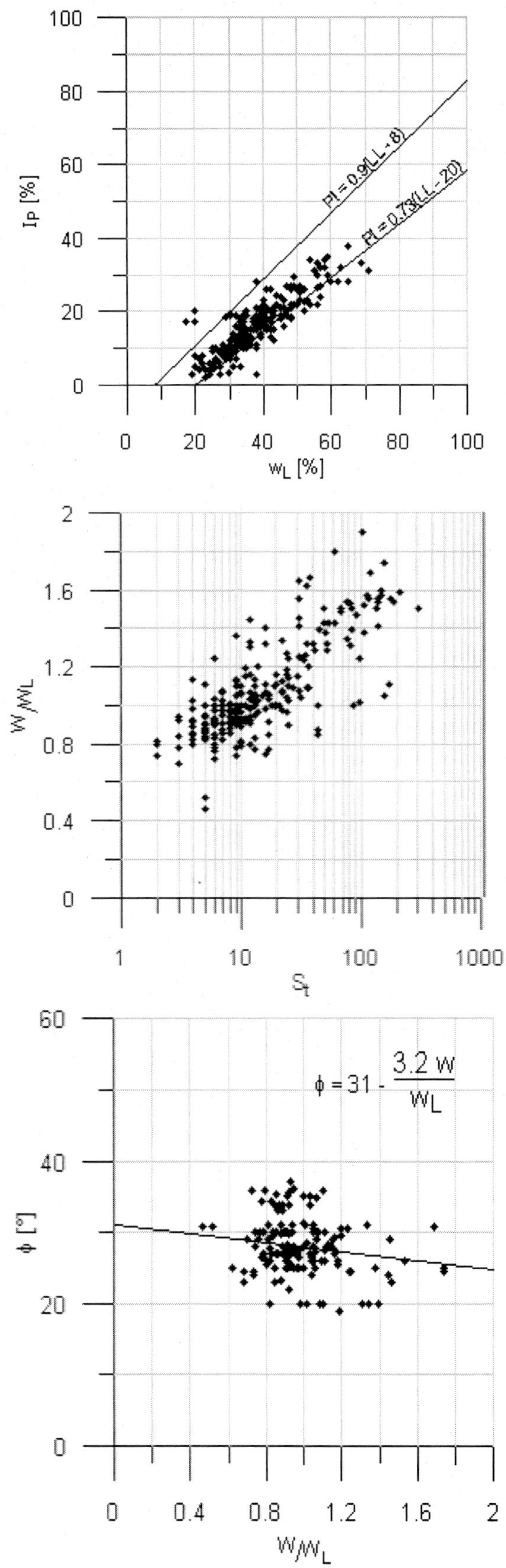

FIGURE 4: Norwegian soft sensitive clays

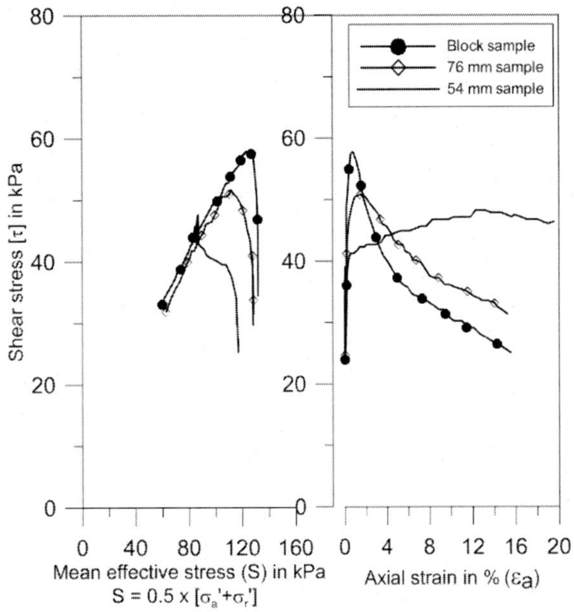

FIGURE 5: Effect of sample disturbance in quick clays. Here σ_a' and σ_r' are the effective stresses in the axial and the radial direction, respectively. The presented results are from a Kløfta road project in Norway. (Source; NPRA, 2009)

Table 1: The Sensitivities Measured for Block and 54-mm-Diameter Samples from the Kløfta Road Project

Depth [m]	c_{ui} *[kPa]*	c_{ur} *[kPa]*	S_t *[-]*
18.5[A]	27	0.2	135
18.5[B]	8.6	0.2	42
18.5[B]	11.7	0.2	58

[A] Block sample; [B] 54 mm diameter sample

This single example shows there is a danger that practitioners may be misguided by sample disturbance, especially when categorizing a material (for example, for S_t values between 10 and 14) as sensitive or non-sensitive. These results also raise concerns regarding the correct interpretation of S_t values.

Table 2: Selected Large Landslides in Norwegian Sensitive Clays (Thakur and Degago, 2012)

No.	*Year*	*Landslide*	L_T* *Total*	c_{ur}	S_t	I_L	I_P
			[m]	*[kPa]*	*[-]*	*[-]*	*[%]*
1	2012	Byneset	1170	0.12	120	3.8	4.8
2	2010	Lyngen	650	0.35	50	1.5	12
3	2009	Kattmarka	650	0.24	63	2.9	8
4	1978	Rissa	2000	0.24	100	2.2	6
5	1974	Baastad	930	0.53	35	1.8	8
6	1967	Hekseberg	1100	0.25	100	2.4	4
7	1965	Selnes	440	0.35	100	2.3	7
8	1963	Skjelstadmarka	2800	0.83	48	1.6	10
9	1959	Furre	332	0.2	85	1.3	6
10	1953	Bekkelaget	165	0.2	130	2.4	9
11	1953	Borgen	165	0.7	100	1.2	20
12	1953	Ullensaker	250	0.35	42	1.9	6.7
13	1893	Verdalen	7000	0.2	300	2.2	5

*Including the retrogression (L) and the run-out distance (L_R)

3. LANDSLIDE HAZARDS

According to the current geotechnical code, issued by the Norwegian Water Resources and Energy Directorate (NVE), Norwegian soft sensitive clays, that have an $S_t \geq 15$ and a $c_{ur} \leq$ 2.0 kPa are treated as quick clays. According to the code, the overall stability of areas that consist of these clays must be investigated. This overall stability assessment considers areas where there is potential for retrogression and/or propagation of landslides. It also considers areas that are outside the slide zone but may still be subjected to dangers, such as large movements and structural damages. Therefore, in comparison with local and global slope stability, the overall stability addresses a much wider perspective with respect to the stability of a vast area. It is important to note that slides, either initially or globally, may occur in any type of geomaterial, regardless of their S_t and c_{ur} values.

The definitions illustrated in Figure 6 are based on the assumption that sensitive clays are also prone to large landslides that may develop due to progressive failure, as is often seen in case of highly sensitive clays or quick clays (see Table 2).

Experience has shown that quick clays with low c_{ur} and a high S_t values may cause devastating landslides, such as the Verdalen (1893), the Rissa (1978) and the Byneset (2012) landslides. However, little documented information is available regarding large landslides in sensitive clays (0.5 kPa < c_{ur} < 2.0 kPa). One explanation for this phenomenon could be that sensitive clays with c_{ur} values that are relatively high but less than 2 kPa are not as susceptible to landslides, as has been assumed. Figure 7 shows a result from the run-out modelling carried-out for the Byneset landslide (2012) which occurred in quick clay. Consequently, this arises a key question i.e. does a clay with a c_{ur} > 0.5 kPa have the potential for a retrogressive and run-out landslide, as is often seen in the case of quick clays? These aspects are discussed herein.

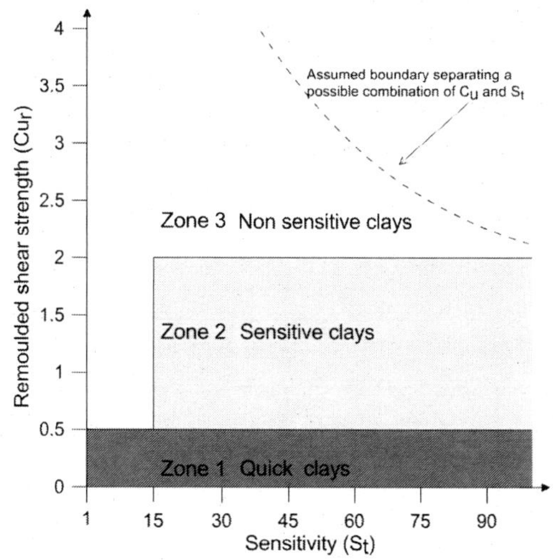

FIGURE 6: The classification of soft sensitive clays prone to landslides according to the NVE guidelines (2011) in Norway

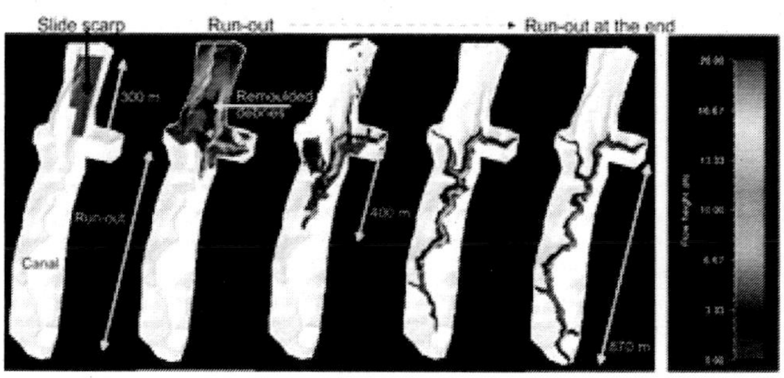

FIGURE 7: Run-out modelling of the Byneset landslide (2012)

4. SAFETY MARGIN

In order to trigger a landslide in soft sensitive clays, there must be an initial slide and there after a possibly for a successive development of the slide, Figure 8. Bjerrum (1955) suggests that the investigation of the danger of an initial slide at the toe of a natural slope is a typical long-term stability problem. Such analysis of the stability may be carried out in terms of effective stresses inserting pore-water pressures as measured in the field. However, he suggests if quick clay is encountered, an analysis in terms of total stresses may be preferred.

If the safety margin resulting from these analysis show that the danger of initial slides can be neglected, the whole slope will be stable, independent of the presence of quick clay. Bjerrum (1955) and Janbu (1979) suggest that the danger of initial slides includes an investigation of the overall stability of the slope. In short, it is important that a slope have a sufficient safety margin to avoid any landslide.

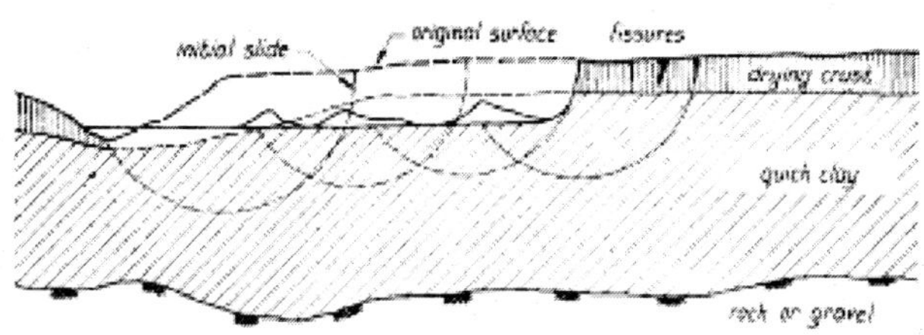

FIGURE 8: Illustration by Bjerrum (1955) of the course of a typical slide in sensitive clays

From 2010 Geotechnical design in Norway must follow the European standard, Eurocode 7 NS-EN 1997-1: 2004+NA: 2008 (hereafter, referred as EN 1997). According to the national annex (NA) to EN 1997 in Norway, the unified material factor or partial factor for soil parameters (γ_M), shown in Table 3, will replace the factor of safety.

Table 3: Material Factor (EN 1997)

Soil Parameters	*Symbol*	γ_M
Friction angle (tan φ')	γφ'	1.25
Effective cohesion (c')	γ_c'	1.25
Undrained shear strength (cu)	γ_{cu}	1.4

In practice this does not imply a significant change. For undrained conditions we define:

$$\gamma_M = \frac{c_u}{c_{u,d}} \tag{1}$$

Here c_u refers to undrained cohesion and subscript d refers to design value. For effective stress based conditions we define:

$$\gamma_M = \frac{c}{c_d} \qquad and\ \gamma_M = \frac{\tan\phi'}{\tan\phi'_d} \qquad ...(2)$$

Here c' and $\tan\varphi'$ respectively refers to effective cohesion and effective friction angle.

According to the EN1997, the minimum partial material factors to be accounted for the landslide hazard assessment are very important prior to constructing a structure in areas with soft sensitive clays. This is because a relatively small and local disturbance in a highly sensitive clay could create an extensive landslide often in areas with only modest inclination. Among other parameters, the critical slope angle (θ) is an important parameter used in hazard mapping. The θ primarily corresponds to γ_M and the soil condition, i.e. drained or undrained. θ can be obtained using the following equations for the drained and undrained conditions.

Drained condition:

$$\tan(\theta) = \frac{\gamma'}{\gamma} \times \tan(\phi) \times \frac{1}{\gamma_M} \qquad (3)$$

Undrained condition:

$$\sin(\theta) = \frac{\gamma'}{\gamma} \times \alpha \times \frac{1}{\gamma_M} \qquad (4)$$

Implications of Eqns (3) and (4) can be illustrated using a simple slope stability calculation of a long natural slope where the failure surface is parallel to the ground. The ground water level was assumed to be at the surface level. The input parameters are shown in Table 4, and the results are presented in Table 5.

Table 4: Input Parameters

γ *Bulk Unit Weight of Soil*	γ' *Effective Unit Weight of Soil*	φ *Friction Angle*	α *Strength Factor*	Cu_m *Average Undrained Shear Strength*
20 KN/m^3	10 kN/m^3	30	0.2	$\alpha.p_0'$

While a long natural slope may exhibit to be critical to slide when θ = 16.1° under drained condition, according to Table 4, any slope steeper than 5.7° could be subjected to failure under undrained loading if the stability of slope is not improved. Therefore, slopes steeper than 3.6° may require measures to improve γ_M prior to construction of any transport related structures.

Table 5: Critical Slope Angle Corresponding to Various Material Factors

γ_M	*Undrained* θ [°]	*Drained* θ [°]
1.0	5.7	16.1
1.4	4.1	11.7
1.6	3.6	10.2

In the Norwegian National Annex (NA) to EN 1997, it is stated that the material factors for the soft sensitive soils as, suggested by NVE, like quick clays should be increased relative to the tabulated values in Table 2. There is no specific suggestion on how much the partial factors should be increased. The Norwegian Public Roads Administration (NPRA) guideline (2010) suggests practicing an increased material factor to 1.6 when transport infrastructures like roads and bridges are constructed on soft sensitive clays. The suggested increase is about 14.2% increase on γ_{cu} and about 25.9% increase for the $\gamma_{\varphi'}$ and $\gamma_{c'}$ values, see Table 3. One can question if the increased value for γ_{cu} is sufficient or is too conservative. A preliminary study by Grimstad and Jostad (2011) and Jostad and Fornes (2012) shows that the classical theories based on limit equilibrium calculation, on average, over-predicts the factor of safety about 10% which is well within the suggested increased value by the NPRA guideline.

5. LANDSLIDE RETROGRESSIONS

In the quick clay deposits of Scandinavia and eastern Canada, landslides are particularly destructive, due to the possibility of small landslides initiating a fast and extensive retrogression process, which may involve massive soil movements in the order of millions of cubic meters (Figures 1). Figure 1 show how, an initial slide occurred in a local area, and then the slide gradually propagated retrogressively towards the existing road and houses in the vicinity of the slope.

An overview of Norwegian landslides as shown on Table 2 clearly indicates that soft sensitive clays must be handled carefully. For vast landslides, after an initial slide, to occur, it is important that at least the following two criteria are fulfilled:

- The slide debris should be completely remoulded.
- The slide debris should be able to flow out of the slide area if remoulded.

There may be additional factors, such as the topography and the stability of the area behind the initial slide zone. However, if the two criteria mentioned above are not fulfilled, then vast landslides, such as those listed in Table 2, are unlikely to occur. It is interesting to analyse sensitive clays from this perspective and observe whether sensitive clays with a $0.5 < c_{ur} < 2$ kPa have similarly to quick clays.

Landslides in soft sensitive clays have been studied worldwide, notably in Canada. Mitchell and Markell (1974) suggest a direct relationship between c_{ur}, S_t and the retrogression distance (L). Figure 9 presents a modified plot from Mitchell and Markell (1974). The plot clearly indicates that landslides with an $L > 100$ m are observed for $c_{ur} < 1$ kPa. Retrogression was zero for $c_{ur} > 1$ kPa. An L up to 100 m was observed for c_{ur} values between 1 and 1.5 kPa are reported by Mitchell and Markell (1974). But in such cases the retrogression length could be related to the length of the initial slide itself because the term L also includes the length of the initial slide as well. A relationship illustrated by Figure 9 has not yet been established for the Norwegian slides. However, the study by Mitchell and Markell (1974) has encouraged the author to focus on this issue. To facilitate their comparison, Figure 9 presents a plot of information from the selected Norwegian landslides listed in Table 2 along with the data reported by Mitchell and Markell (1974). This figure show that the most known Norwegian landslides with a $c_{ur} < 1$ kPa were subjected to large retrogression distances ($L > 100$ m). Notably, the finding by Mitchell and Markell (1974) appear to be in agreement with the Norwegian landslides. Based on this $c_{ur} > 1$ kPa seems to be the threshold limit above which the extent of the retrogression of a landslide can be limited to initial slide only. To explain this aspect further it is important to understand how quickly a slide mass will turn into a remoulded mass which again will depend on the available potential energy. The available potential energy is function of the slope geometry and the soil density. In landslides of sensitive clays, the change in potential energy before and after the slide is transformed to different form of energy that results in remoulding the slide debris to its remoulding state and slide movement (kinetic and frictional energy). For a given potential energy, sensitive clays with higher remoulding energy results in smaller slide movement (because of low kinetic energy available) than sensitive clays with lower remoulding (Thakur et al. 2012).

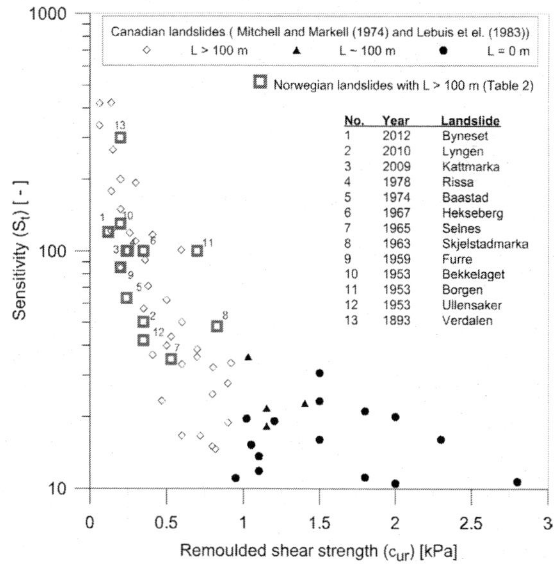

FIGURE 9: Comparison between the Norwegian landslides (with $L > 100$ m) and the findings from Michell and Markell (1983)

6. A PHYSICAL UNDERSTANDING OF THE REMOULDED SOFT SENSITIVE CLAYS USING SMALL SCALE TESTS

The author believes that one of the major challenges related to c_{ur} is the physical understanding of the parameter itself. Brittle clays, such as quick clays, are often presented as shown in Figure 10 in which quick clay becomes a semiliquid when subjected to remoulding. The question is; Do sensitive clays (0.5 kPa $< c_{ur} <$ 2 kPa) also become liquefied when they are subjected to remoulding? If not, then why should sensitive clays carry the same potential risk for retrogression as observed with quick clays?

FIGURE 10: Schematic presentation of quick clays at remoulded

In this section, we present simple test results to provide the basis for a physical understanding of c_{ur} values using a new type of geotechnical engineering test, known as the quickness test. The test methodology was firstly applied and presented by Thakur and Degago (2012). This small scale model test procedure was developed to understand the flow and spread of remoulded materials.

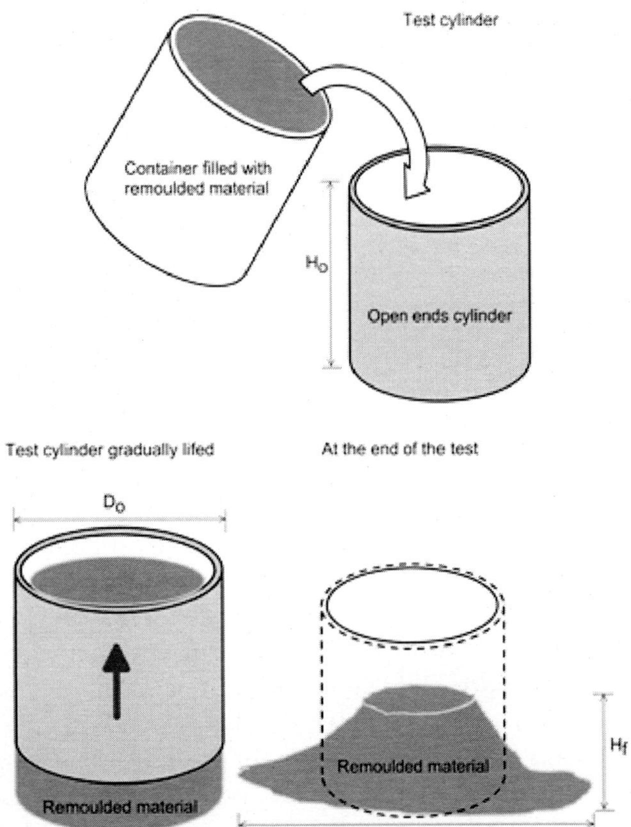

FIGURE 11: Quickness test (Thakur and Degago, 2012)

7. QUICKNESS TEST

Thakur and Degago (2012) developed a small scale test, the quickness test that is applicable in understanding the flow of remoulded materials. The quickness test is a simple procedure that is performed by filling an upright open cylinder with remoulded sensitive clay, slowly lifting the cylinder, and measuring the deformation (height and lateral spreading) as the material is subjected to flow due to its own weight. The test concept is similar to the slump test for concrete that has been used for years to measure the consistency of freshly mixed concrete. Two different cylinder sizes were used. The dimensions were the diameter (D_o) = 65 mm and height (H_o) = 45 mm for the small cylinder, and D_o = 100 mm and H_o = 120 mm for the large cylinder. The large cylinder has the same size as the cylinder used for the standard proctor tests. Figure 11 shows the concept of the proposed quickness test. The thoroughly remoulded material is placed into the cylinder, levelled off, and allowed to flow outward as the cylinder is slowly lifted upward with minimum disturbance to the sample. The difference in height between the cylinder and the

Table 6: Properties of the Tested Materials from Three Different Locations

Properties	*Lersbekken*	*Byneset*	*Olsøy*
Sampling depth (H) [m]	6 – 10	4 – 12	4 – 15
Clay fractions (< 2 μm) [%]	30	30 – 55	50 – 65
Water content (w) [%]	22 – 34	27 – 48	28 – 38
Plasticity index (I_P) [%]	5 – 7	3 – 15	3 – 10
Liquidity index (I_L) [-]	0.7 – 2.0	0.9 – 5.4	0.6 – 3
Undisturbed undrained shear strength (c_{ui}) [kPa]	12 – 58	5.2 – 72	60 – 100
Remoulded undrained shear strength (c_{ur}) [kPa]	0 – 2	0 – 3	0 – 2.1
Sensitivity (S_t) [-]	16 – 29	4 – 400	30 – 100
Over consolidation ratio (OCR) [-]	1.8 – 2.0	1.1 – 3.3	– 4

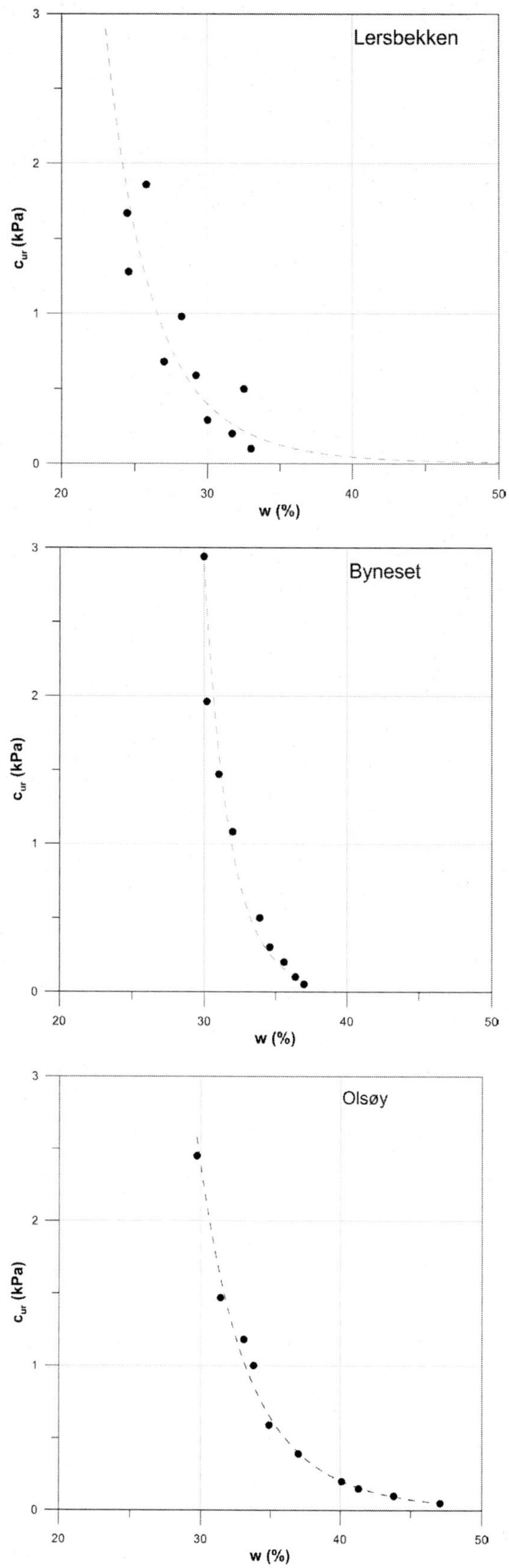

FIGURE 12: Remoulded shear strength (c_{ur}) as a function of water content (w)

slumped material is visually recorded throughout the test period, and the difference in height between the cylinder and the slumped material (H_o-H_f) is noted. The outward flow spread diameter (D_f) is also noted. The quickness (Q) in % is defined as $[1- H_f/H_o] \times 100$.

7.1 The Tested Material

Quickness tests were performed on several sensitive clay samples taken from three different locations in the central Norway. These sites have been studied extensively in connection to landslide hazards. Laboratory index properties of the sampled material are presented in Table 6. Liquid limit (w_L), c_{ui} and c_{ur} of the tested material were obtained using the fall-cone method as described by Hansbo (1957). Figure 12 shows c_{ur} of the tested materials as a function of w.

8. RESULTS AND DISCUSSIONS

Quickness tests were performed on 60 different samples. Sensitive clays from all three locations; Lersbekken, Byneset and Olsøy, with c_{ur} < 0.2 kPa seemed to be more like a soup as reported by Mitchell and Soga (2004). The slump and spread registered during the testing period are shown in Fig. 13 for two remoulded Byneset clay with c_{ur} = 0.1 kPa and c_{ur} = 1.0 kPa. The registered collapse of the remoulded material was negligible when the remoulded sensitive clay had c_{ur} > 1.0 kPa.

The slump and spread registered during various tests in Lersbekken clay presented in Fig. 14 shows that; the tested sensitive clay having a $c_{ur} \approx 0.5$ kPa was not as fluid as it was originally thought. The sensitive clays that had 0.5 kPa < c_{ur} < 1.0 kPa were semisolid in nature.

A typical flow and spread behaviours of two remoulded clays from the Olsøy site is presented in Figure 15. The tests on Olsøy clay shows that the remolded material with c_{ur} < 0.2 kPa seemed to be more like a soup, very thick viscous or semisolid behavior is observed when the remolded clays had 0.4 < c_{ur} < 1 kPa. Additionally, a remolded material that has a c_{ur} > 1.0 kPa has little or no flow. In addition, here, the observed slumps were not due to the collapse or flow of the remolded material but due to the removal of the lateral support artificially

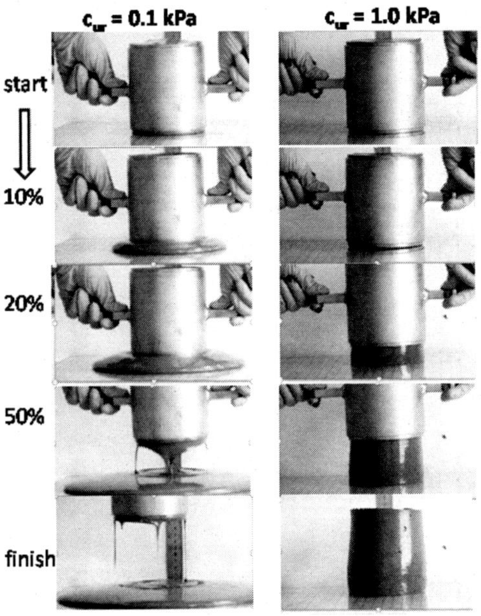

FIGURE 13: Slump and spread observed from the start to the end of the Quickness tests for remoulded materials from Byneset. The 10%, 20% and 50% indicates the test stage at which the photos were taken. The tests take between 3 to 5 seconds to finish

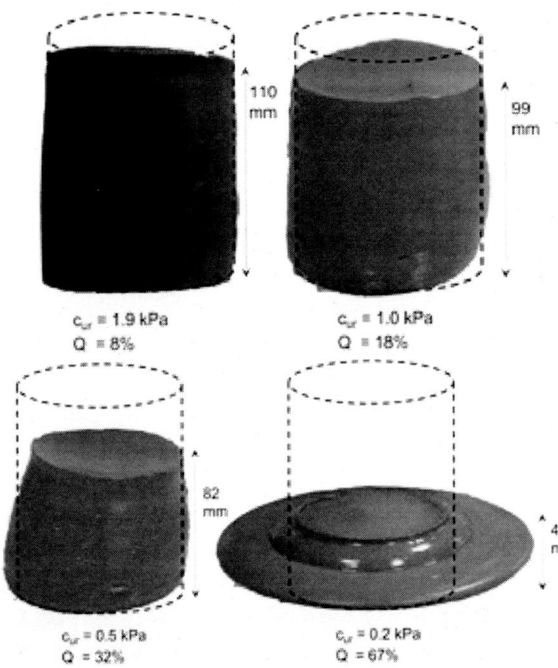

FIGURE 14: Slump and spread of the remoulded clays, from the Lersbekken site, observed from the quickness test using a 100 mm diameter and 120 mm high cylinder. c_{ur} for these samples are 1.9 kPa (upper left), 1.0 kPa (upper right), 0.5 kPa (lower left) and 0.2 kPa (lower right)

created by the test cylinders. This simple test indicates why soft sensitive clays with a $c_{ur} > 1$ kPa are not subject to large retrogression due to their inability to flow.

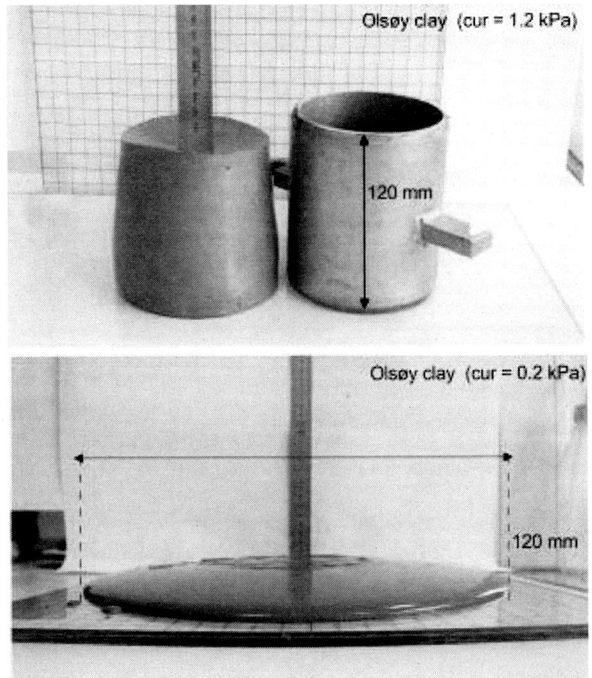

FIGURE 15: Flow and spread behaviour of the two remoulded clays from the Olsøy site. The observation is made at the end of the quickness test using a 100 mm diameter and 120 mm high cylinder. c_{ur} for these samples are 1.2 kPa (upper) and 0.2 kPa (lower)

Apparently, the flow behaviour of Lersbekken, Byneset and Olsøy clay is identical. Figure 16 presents Q versus c_{ur} for various sets of tests on the Lersbekken, Byneset and Olsøy clays performed with two different cylinder sizes, 100 mm x 120 mm, 65 mm x 45 mm. For the major area of interest, $c_{ur} > 1.0$ kPa, material flow was not registered irrespective of the test cylinder sizes. This study recommends using a cylinder size 100 mm x 120 mm and proposes some correlations based on this cylinder size because this cylinder size is vastly adopted and readily available for the geotechnical testings. The Figures presents the lower and the upper bound Q observed for various c_{ur} of the each tested material. A combined plot is shown in Figure 17. Thakur and Degago (2012) suggests considering the lower bound quickness in evaluating flow slides since it provides a conservative estimate. It can be noticed from the

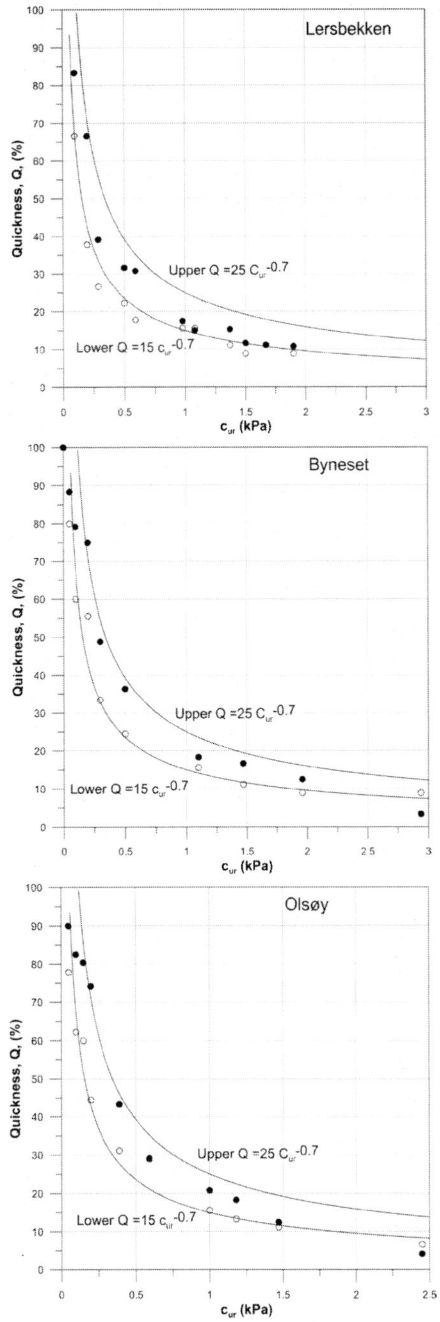

FIGURE 16: The quickness in % registered from the quickness test for various samples of different c_{ur} values. The upper, middle and the lower plots are representative to the Lersbekken clay, the Byneset clay and the Olsøy clay respectively. Here, the dark circles shows the results obtained using a 100 mm diameter × 120 mm high cylinder, and the empty circles belongs to the results obtained using a 65 mm diameter × 45 mm high cylinder. Here, the quickness in % is defined as $[1- H_f /H_o] \times 100$

quickness test results (Figure 16 and 17) that all Lersbekken, Byneset and Olsøy materials have nearly identical responses and the lower bound Q = 15 % corresponds to c_{ur} = 1kPa for all the three sensitive clays.

Fluidity of sensitive clays is difficult to interpret by small numerical values of c_{ur} because a seemingly small change implies significant alteration in flow behaviour of clay. For example, a significant change in the flow behaviour of the sensitive clay from the Lersbekken site is observed for a small variation of c_{ur}, i.e. from 0.5 to 0.2 kPa (Figure 14). In this case, the significant behavioural change is reflected by the quickness test visually as well as numerically, i.e. Q varies from 32% to 67%. The quickness test amplifies the small range of c_{ur}, i.e. from 0 to 2.0 kPa, to a scale that varies from 0 to 100%. Quickness test therefore gives a better visualisation of the flow behaviour of sensitive clays where small c_{ur} values have big implications in regards to understanding the potential for retrogressive landslides.

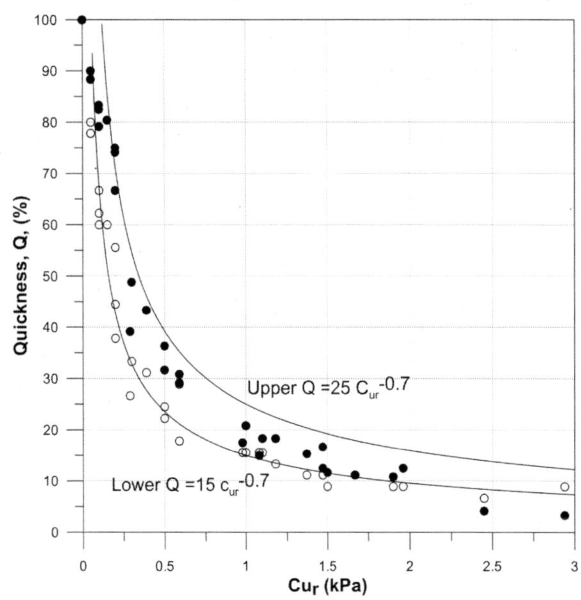

FIGURE 17: The quickness in % registered from the quickness test for various samples obtained from all three locations; Lersbekken, Byneset and Olsøy of different c_{ur} values. Here, the dark circles shows the results obtained using a 100 mm diameter × 120 mm high cylinder, and the empty circles belongs to the results obtained using a 65 mm diameter × 45 mm high cylinder. Here, the quickness in % is defined as $[1- H_f /H_o] \times 100$

Both c_{ur} and Q principally explains the same soil characteristic through different test approaches. The fall-cone test is a point specific method calibrated against the undrained shear strength of soil under undisturbed and remoulded state; whereas, the quickness test gives a value that is representative of the volume of the material tested. In contrast to the fall-cone test, the quickness test has an added advantage of qualitative description that can provide a better visualisation with respect to understanding flow slide. (Thakur and Degago, 2012)

There exist relationships between c_{ur} and S_t as well as between c_{ur} and I_L to evaluate the retrogression potential of soft sensitive clays. Significance of Q can also be incorporated in evaluating the retrogression potential where it has been suggested that an $I_L > 1.2$ to define occurrence of flow slides (Bjerrum and kjærnsli, 1957; Drury, 1968; Eide and Bjerrum, 1955; Huchinson, 1961; Kenney, 1967; Gregersen and Løken, 1979; Lebuis and Rissmann, 1979, Lebuis et al., 1983, Gregersen, 1981; Leroueil et al., 1983, Torrance, 1983; Karlsrud et al., 1985; Locat and Demers, 1988; Trak and Lacasse, 1996; Longva et al., 2003, Nordal et al., 2009, NPRA, 2010; Quinn, 2011). The conditions obtained in these studies are in good agreement with this study because $I_L >$ 1.2 is only possible when $c_{ur} < 1.0$ kPa and which puts a modified condition for flow slides using quickness, i.e. $Q > 15\%$.

Despite this agreement, quickness is believed to be a more convenient parameter as it is independent compared to that of I_L which demands determination of three parameters a priori, i.e. w_L, w_p and w. Notably the conventional thread-rolling method of determining w_p has a significant drawback as it can easily biased by subjective judgement (Whyte, 1982, Feng, 2000, Sivakumar et al., 2009, Thakur and Degago, 2012). In addition, the well-acknowledged relationship between I_L and c_{ur} is established by Leroueil (2001) over c_{ur} that ranges from 0.1 kPa to 100 kPa. However, from the authors point of view, there seems a relatively large scatter between I_L and c_{ur} for values of $c_{ur} < 1.0$ kPa. In this sense, Q is more reliable owing to its mechanical nature of determination as well as for the lower range of c_{ur}, for instance between 0 to 2 kPa.

Based on the quickness test results and the data from the Norwegian landslides it is possible to categorise the landslide based on the Q values. Figure 18 shows that large flow slides are less likely to occur when $Q < 15$ % (or $c_{ur} > 1$ kPa) and in this case the slide will be limited to an initial slide only. Flow slides having retrogression length (L) more than 250 m is possible when soil masses involved in the slide have $Q > 25$ % (or $c_{ur} < 0.5$ kPa). Finally, landslides up to 250 m long is possible for 15 % $< Q <$ 25 % (or 0.5 kPa $< c_{ur} <$ 1 kPa).

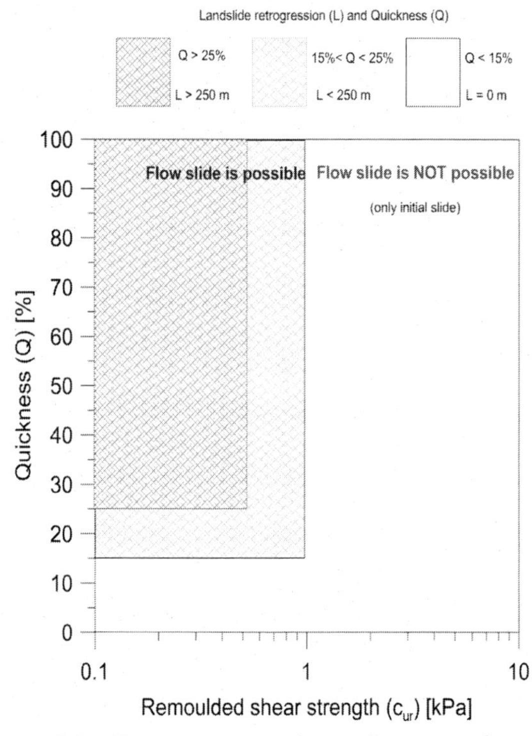

FIGURE 18: Quicness tests based criteria for the retrogression distance of landsdlises in soft sensitive clays

9. CONCLUSIONS

This study demonstrates a need for a better understanding of the mechanical properties of sensitive clays. Soft sensitive clays are critical to the undrained response and therefore even a moderate slope angle like 6° can be considered as critical slope angle. Evidences from the Norwegian landslides available in literature indicated that large landslides in term of flow slide do not happen for sensitive clay with $c_{ur} >$

1.0 kPa. This observation is well supported by the Canadian landslides data as well. This paper presents a laboratory procedure that focuses on remoulded behaviour of sensitive clays in terms of a numerical value referred as quickness (Q). The quickness test was used to demonstrate why sensitive clays with c_{ur} > 1.0 kPa are not susceptible to flow. This particular study shows that a Q < 15% or c_{ur} > 1.0 kPa seems to be the threshold limit where the extent of the retrogression of a landslide seemingly limited to the initial slide.

ACKNOWLEDGEMENTS

This study is a part of the National research program "Natural hazards: Infrastructure, Floods and Slides (NIFS)", initiated by the Norwegian Public Roads Authority, Norwegian Water Resources and Energy Directorate and Norwegian National Railways Administration. The author wishes to acknowledge Dr. Samson Degago and Mrs. Olga Lapkovski for their contributions in the testing and the discussions.

REFERENCES

Aas, G., (1979) "Quick clay slide". Lecture at NIF-course ,Skredfare og planlegging, Lofthus Hardanger , Norway.

Bjerrum, L. and Kjærnsli, B., (1957). "Analysis of the stability of some Norwegian natural clay slopes", Géotechnique, Issue 7, pp 1–16.

Bjerrum, L., (1955) "Stability of natural quick clay slopes". Géotechnique, 5, Issue 1, pp101–119.

Burland, J. B., (1990) "30th renkine lecture – on the compressibility and shear strength of natural clays". Géotechnique, 40, pp 329-378.

Drury, P., (1968) "The Hekseberg landslide, March 1967". NGI publication, 75, pp 27–31.

Eide, O. and Bjerrum, L., (1955) "The slide at Bekkelaget". Géotechnique, 5, Issue 1, pp 88–100.

EN. 1997. Euro code 7. Guideline.

Feng, T. W., (2000). "Fall cone penetration and water content relationship of clays". Géotechnique, 50, pp 181–187.

Gregersen, O. and Løken T., (1979) "The quick-clay slide at Båstad, Norway, 1974", Engineering Geology, 14, pp 183–196.

Gregersen, O., (1981) "The quick clay landslide in Rissa, Norway", Norwegian Geotechnical Institute (NGI) Publication, 135, pp1–6.

Grimstad, G. and Jostad, H. P., (2011) "Stability analyses of quick clay using FEM and an anisotropic strain softening model with internal length scale", Geotechnical day, Oslo Norway

Hansbo, S., (1957) "A new approach to the determination of the

shear strength of clay by the fall cone test", Stockholm: Royal

Swedish Geotechnical Institute, 14.

Hutchinson, D.J., (1961). "A landslide on a thin layer of quick clay at Furre, central Norway". Geotechnique, 11, pp69–94.

Janbu, N., (1979) "Failure mechanics in quick clays". Nordic Geotechnical Meetings. 50p.

Jostad, H. P. and Fornes, P., (2012) "Effekt av progressiv bruddutvikling for utbygging i områder med kvikkleire", The proceedings of the Quickclay conference, ISBN: 978-82-410-0821-4, Teknologidagene, Trondheim Norway, pp 111-123.

Karlsrud, K., (1983) "Analysis of a small slide in sensitive clay in Fredrikstad, Norway". In Proceedings of the Symposiumn Slopes on Soft Clays, Linköping, Sweden, 8-10 March 1982. Swedish Geotechnical Institute (SGI) Report, 17, pp175–184.

Karlsrud, K., Aas G. and Gregersen, O., (1985) "Can we predict landslide hazards in soft sensitive clays? Summary of Norwegian practice and experience" Norwegian Geotechnical Institute, Publication No. 158.

Kenney, T. C., (1967) "Slide behaviour and shear resistance of a quick clay determined from a study of the landslide at Selnes", Norway. NGI publication, 76, pp 57–64.

Laccase, S., Berre, T. and Lefebvre, T., (1985) "Block sampling of sensitive clays", International Conference of Soil Mechanics and Foundation Engineering, 2, pp 887-892.

Lebuis, J. and Rissmann, P, (1979) "Les coule´es argileuses dans le re´gion de Que´bec et de Shawinigan. In: Argiles sensibles, pentes instables, mesures correctives et coule´es des re´gion de Que´bec et Shawinigan". Geological

Association of Canada Guidebook. St Johns, Newfoundland: Geological Association of Canada, pp 19–40.

Lebuis, J., Robert, J.-M. and Rissmann, P., (1983) "Regional mapping of landslide hazard in Quebec", Proceedings of the Symposiumn Slopes on Soft Clays, Linköping, Sweden, 8-10 March 1982. Swedish Geotechnical Institute (SGI) Report, 17, pp 205–262.

Leroueil, S., (2001) "Natural slopes and cuts: movement and failure mechanisms". Géotechnique, 51, Issue 3, pp 197-243.

Leroueil, S., Locat, J., Vaunat, J. Picarelli, L., Lee, H., and Faure, R., (1996) "Geotechnical characterisation of slope movements", Seventh International Symposium on Landslides, Trondheim, pp 53–74.

Leroueil, S., Tavenas, F. and Le Bihan, JP., (1983) "Propriétés caracteristiqués des argiles de I'est du Canada".Canadian Geotechnical Journal, 20, pp 681–705.

Locat, J. and Demers, D., (1988) "Viscosity, yield stress, remolded strength, and liquidity index relationships for sensitive clays".Canadian Geotechnical Journal, 25, pp 799–806.

Locat, P. Leroueil, S. and Locat J., (2008) "Remaniement et mobilitet des bedris de glissements de terrain in dans les argiles nesibles de l'est du Canada", In : J. Locat, D. Perret, D. Turmel, D. Demers et S. Leroueil, (2008). Comptes rendus de la 4e Conférence canadienne sur les géorisques: des causes à la gestion. Proceedings of the 4th Canadian Conference on Geohazards: From Causes to Management. Presse de l'Université Laval, Québec, 594 p.

Longva, O., Janbu, N., Blikra, L. H. and Bøe, R., (2003) "The 1996 Finneidfjord Slide: seafloor failure and slide dynamics. Submarine Mass Movements and their Consequences", Proceedings First International Submarine mass movements and their consequences: proceedings of the 1st international Symposium (Locat & Mienert (eds)). Kluwer Academic Publishers, Dordrecht, The Netherlands, pp 531–538.

Lunne, T., Berre, T. and Strandvik, S., (1997) "Sample disturbance effects in soft low plastic Norwegian clays". Conference on recent development in soil and pavement mechanics, pp 81-102.

Mitchell, J. K. and Soga, K. (2004) Fundamentals of soil behaviour. 3rd eds. John Wiley & Sons.

Mitchell, R. J. and Markell, A. R., (1974) "Flow slides in sensitive soils". Canadian Geotechnical Journal, 11, Issue 1, pp11–31.

NGI (1988). Quick clay mapping. NGI report 81074.

Nordal, S., Alen, C., Emdal, A., Jendeby, L., Lyche, E. and Madshus, C., (2009) "Landslide in Kattamrkavegen in Namsos 13. March 2009", A report to the transportation Ministry of Norway, ISBN 978 82-92506-71-4 (paper version) and ISBN 978-82-92506-72-1 (electronic version).

Norsk Geoteknisk Forening, (1974) "Retningslinjer for presentasjon av geotekniske undersøkelser" Oslo, Norway, 16.

NPRA guidelines, (2010) "geoteknikk I vegbygging", The Norwegian Public Roads Administration, Hb016.

NPRA report, (2009) "Geotechnical Investigation report- Lyngen", The Norwegian Public Roads Administration (Statens vegvesen) internal NPRA report.

NVE (2011) "Vurdering av områdestabilitet ved utbygging på kvikkleire og andre jordarter med sprøbruddegenskaper", Vedlegg 1 til NVEs retningslinjer: Flom- og skredfare i arealplaner.

Quinn, P. E., Diederichs, M. S., Rowe, R. K. and Hutchison, D. J., (2011) "A new model for large landslides in sensitive clay using a fracture mechanics approach". Canadian Geotech-nical Journal, 48, pp 1151-1162.

Rankka, K., Ansresson, S., Y., Hulten, C., Larsson, R., Leroux, V., and Dahlin T., (2004) "Quick clay in Sweden", Report 65, Swedish Geotechnical Institute.

Rosenqvist, I. T., (1953) "Considerations on the sensitivity of Norwegian clays", Geotechnique, 3, pp 195-200.

Sivakumar, V., Glynn, D., Crains, P and Black, J. A., (2009) "A new method of measuring plastic limit of fine materiales". Géotechnique, 59, pp 813–823.

Skempton, A. W. and Northey, R. D., (1953) "The sensitivity of clays". Geotechnique, 3, pp 30–53

Tavenas, F., Flon, P., Lerouil, S. and Lebuis, J., (1983). "Remolding energy and risk of slide retrogression in sensitive clays", Proc. Symp. Slopes on Soft Clays, Linkoping, Swedish Geotechnical Institute (SGI) Report, 17, pp 423 – 454.

Thakur V and Degago S. A., (2012). "Quickness of

sensitive clays". Géotechnique Letters. 2, Issue 3, pp87-95.

Thakur, V, Oset, F., Degago S., Aaboe R. and Watn A., (2012) "Infrastructures on soft sensitive clays: some essential aspects and examples", Second International Conference on Transportation Infrastructure, Sapporo, Japan. pp 836-842.

Thakur, V. and Degago S., (2012) "Utbredelse av skredet i sensitive leirer". The proceedings of the Quickclay conference, ISBN: 978-82-410-0821-4, Teknologidagene, Trondheim Norway, pp 139-153.

Thakur, V., Oset, F., Aabøe, R., Berg, P. O., Degago, S. A., Wiig, T., Lyche, E., Haugen, E. E. D., Saeter, M. B. and Robsrud, A., (2012) "A critical appraisal of the definition of Brittle clays (Sprøbruddmateriale)", Proceedings of the 16th Nordic Geotechnical Meeting Copenhagen, pp 451-462.

Torrance, J. K., (1983) "Towards a general model of quick clay development". International Journal of Sedimentology, 30, Issue 4, pp 547–555

Trak, B. and Lacasse, S., (1996) "Soils susceptible to flow slides and associated mechanisms", Proceedings of the Seventh International Symposium on Landslides, Balkema, Rotterdam, pp 497–506.

Whyte, I. L., (1982) "Soil plasticity and strength: a new approach using extrusion". Ground Engineering, 15, Issue 1, pp 16–24.

Swelling Pressure and Uplift of Piles in Expansive Soils

B. Soundara and R.G. Robinson[1]

SKR Engineering College, Chennai
[1]*Department of Civil Engineering, Indian Institute of Technology Madras, Chennai*
E-mail: robinson@iitm.ac.in

ABSTRACT

Pile foundations are widely used in swelling soils in order to anchor the structure down at a deeper depth (Stable zone) where the uplift pressure on the pile is negligible. Estimates of depth of stable zone, also called as active zone, is very important for the design of piles in expansive soils. The depth of active zones is usually estimated from a measure of water content or suction profiles with depth for different seasons. This paper attempts to estimate the depth of active zone from the swelling pressure of soils. The stress level that is adequate to prevent swell is nothing but the swelling pressure of the soil. Hence the uplift pressure on the pile will be negligible at the depth corresponding to swelling pressure of the soil. This in turn is confirmed via laboratory model pile uplift tests and oedometer swelling pressure tests for a natural soil. From the laboratory results it is observed that the surcharge pressure, at which the uplift pressure on the pile is zero, is same as that of the swelling pressure of soil.

1. INTRODUCTION

Soils that have potential for shrinkage or swelling under changing moisture conditions are called expansive soils. Due to climatic changes, these soils undergo significant volume changes. In India, the area covered by expansive soils is nearly 20 percent of the total area. The problem is generally dealt with soil treatment or foundation alternatives. The lightly loaded structures are more susceptible to damage by the differential movements caused by the swelling and shrinkage of expansive soils. Piles have been used extensively for foundations in swelling soils in order to anchor the structure down at a depth where changes in moisture content are negligible. The anchorage may be effectively done by providing enlarged base or bulbs (Under reamed piles).

Chen (1988) and O'Neill (1988) presented analysis of rigid piers in expansive soil. They considered that the total uplift force is the integration of the uplift skin friction over the area of the pier within the active zone. This uplift force is resisted by the skin friction in the anchorage zone beneath the active zone, uplift resistance of the bell (in the case of belled pier) and the applied load. This is depicted in Fig. 1.

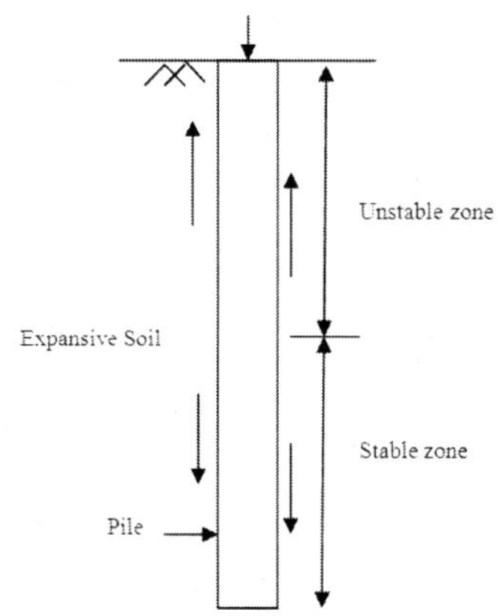

FIGURE 1: Forces acting on a pile in expansive soil

The depth of active zone is defined as the thickness of the soil layer in which moisture deficiency exists. Thus the soil within the active zone will be subjected to seasonal moisture changes and cause differential movements. The depth of active zone is influenced by the soil type, soil structure, topography and climate. The actual measurement of moisture or the suction values in the site may be used to determine the active zone depth. Fig. 2 shows the actual variation of moisture measured at a site (Komornik and Zeitlen, 1961).

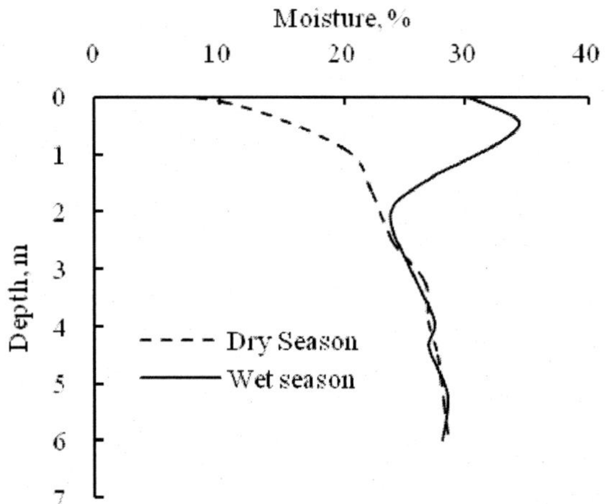

FIGURE 2: Moisture content profiles for different seasons

The depth of active zone can be also calculated by knowing the heave profile of the soil with depth. The depth at which the heave is zero is taken as the depth of active zone. Gupta *et al.*, (1983) revealed that the heave reduced to a negligible value at about 5 m depth. The rate of reduction could be defined by the exponential curve. At a depth of 3.75 m (3/4 of that to zero heave), heave was only about 10% of the surface value, which seems to be acceptable. Hence Gupta *et al.*, recommended that the under reamed piles should be taken at least to a depth at which 10% of maximum heave occurred.

The recommendation by Indian standards (IS: 2911- Part-III) regarding the minimum founding depth for under reamed pile in deep deposits of expansive soil is about 3.5 m based on the field measurements of ground movements, carried out in the fifties. Prakash *et al.*, (1988) carried out field study on full scale instrumented piles of 3.5 m deep in expansive soil deposit having active zone thickness of more than 3.5 m. They suggested that in most cases 3.5 m deep underreamed piles may provide a safe foundation system.

Bhandari *et al.*, (1987) measured the pile head displacements and uplift loads on the instrumented 3.5 m deep uniform diameter, single and double under reamed piles in the expansive clay layer extending to a depth of 4.5 m. Based on the field monitoring, variation in moisture content between dry and wet seasons was inappreciable below 4.5 m, indicating depth of active zone of about 4.5 m. The results from the instrumented piles clearly show that all piles uplifted by almost similar amounts of about 20 mm. This is contrary to the belief that 3.5 m deep underreamed piles are not susceptible to the influence of ground movement irrespective of the depth of active zone.

From the literature review, it is observed that in most cases, the depth of active zone is evaluated based on moisture content variation with the seasons. The monitoring of moisture variation and creating the profiles for different seasons is time consuming task. Therefore, alternate methods of evaluation of depth of active zones are necessary. In the present investigation, the active depth evaluation based on swelling pressure of the soil is discussed.

Swelling pressure is the pressure required to hold the soil at constant volume when water is added. Knowledge of swelling pressure is essential for the design of a variety of geotechnical structures on expansive soils. The swelling pressure is evaluated in the laboratory by a number of testing methods which include oedometer method, suction measurements, triaxial methods, *etc.* (Brackley,1973; Abduljauwad and Al-Sulaimani, 1993; Al-Shamrani and Al-Mhaidib, 2000; Thompson *et al.*, 2006).

Out of all the methods, laboratory oedometer testing method is extensively used to determine the swelling pressure due to its simplicity and operational ease. Brackley (1973) lists three different oedometer methods for the determination of swelling pressure as explained below (Soundara and Robinson, 2009).

- **Method A** – The sample is inundated and allowed to swell vertically at a small seating pressure until primary swell is completed. The sample is then loaded in intervals similar to the procedure of conventional consolidation testing until the specimen reaches its initial thickness. The pressure required to bring back the sample to its initial thickness is regarded as the swelling pressure. This method is also often termed as Swell-consolidation method.

- **Method B** – Three identical samples are loaded with different pressures near the expected swelling pressure and submerged in water. The vertical movements were plotted against the applied pressure and the pressure corresponding to zero volume change is taken as swelling pressure. While only one sample is enough to determine the swelling pressure in method A, at least three identical samples are needed in method B. This method is also often called as Different pressure method.
- **Method C** – In this method, also called as Constant volume method, a specimen is maintained at constant height by adjusting the vertical pressure after the specimen is inundated in free water. The pressure required to maintain constant volume is the swelling pressure. The details of the test methods are given in Fig. 3.

Each of the methods is equally sensible, but gives entirely different swelling pressure values for the same placement conditions of the soil. A number of investigators have attempted to study the cause for the variation of the swelling pressure values by these methods.

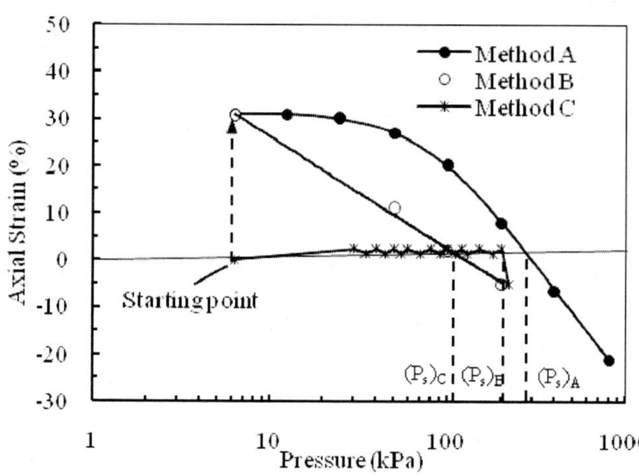

FIGURE 3: Determination of swelling pressure (Brackley 1973)

Johnson and Snethen (1978) compared the swelling pressure values by different oedometer methods and found that the magnitude of swelling pressure depends on the degree of confinement. Ali and Elturabi (1984) conducted methods A and C for the measurement of swelling pressure of expansive soils. Results obtained show that method A gives higher swelling pressure values than method C.

Sridharan et al. (1986) compared the results from the three oedometer methods (Methods A, B and C) to determine the swelling pressure and concluded that method A gives an upper bound value, method B gives the least value and method C gives intermediate values. As shown in Table 1, they also found no definite relation between the three methods.

Table 1: Typical Swelling Pressure Values (Sridharan *et al.* 1986)

Soil Sample	*Initial Conditions*		*Swelling Pressure (kPa) by Method*		
	e_i	*w (%)*	*A*	*B*	*C*
BC1	0.893	0	300	160	190
BC4	1.002	0	390	160	220
BC5	0.742	0	510	140	310
BC8	0.656	20.8	150	–	70

Soundara and Robinson (2009) discussed the influence of test method on swelling pressure measurements in the laboratory. An attempt was made to find out the reasons for different swelling pressure values for the same soil, with identical test specimens, when different test methods are adopted. The possible reason for difference in the swelling pressure values was attributed to the fabric change that occurs in the sample, which depends on the test method. The fabric change was evaluated through permeability measurements and Scanning Electron Microscopy (SEM). Typical SEM pictures of samples after testing by Method A and Method C are shown in Fig. 4 (a) and (b), respectively.

The SEM pictures clearly show that the structure of the clay gets changed from a flocculated fabric to a dispersed fabric when method A is adopted for measuring the swelling pressure. This change is expected to occur when the soil undergoes large swelling. But, in the case of method B and C, volume change is not permitted during the saturation process and the fabric change is minimum. Therefore, the difference in swelling pressure value between methods is a direct consequence of fabric change that occurs during the test.

(a)

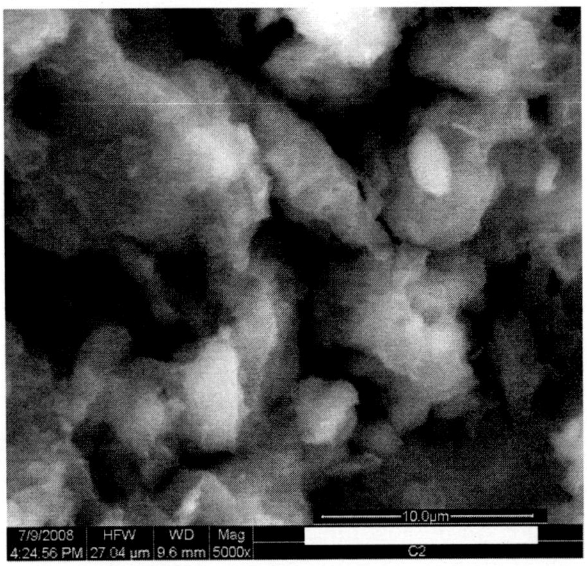

(b)

FIGURE 4: SEM pictures of samples in (a) Method A and (b) Method C (Soundara and Robinson, 2009)

As the values of swelling pressure strongly depend on the method of testing, it is essential to adopt appropriate values depending on the field situations. In the present study, the swelling pressure for the soil is determined by all the three methods. The method that can be used to evaluate the depth of active zone correctly is identified through an apparatus developed for the study.

2. EXPERIMENTAL PROGRAMME

2.1 Properties of Soil Used

A natural expansive soil from Siruseri (Tamil Nadu, India) was collected from a depth of 1.5 m from ground level and air dried. The air dried natural soil was pulverized and sieved through 1mm sieve. The index properties and compaction test results of the soil were arrived based on IS code of practices. The compaction curve obtained is shown in Fig. 5. The other properties are listed in Table 2. The soil is classified as Clay of High compressibility (CH). From shrinkage limit and differential free swell values, the soil is further classified as soil of high swelling potential.

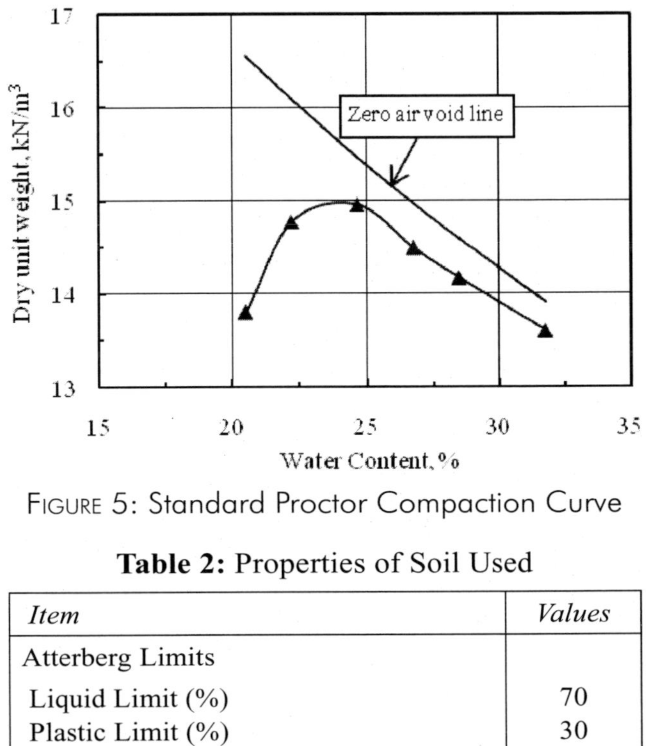

FIGURE 5: Standard Proctor Compaction Curve

Table 2: Properties of Soil Used

Item	*Values*
Atterberg Limits	
Liquid Limit (%)	70
Plastic Limit (%)	30
Shrinkage Limit (%)	8
Specific gravity	2.60
Grain size distribution	
Clay sixe (%)	69.5
Silt size (%)	24.0
Sand (%)	6.5
Compaction test characteristics	
(a) Maximum dry unit weight (kN/m^3)	14.96
(b) Optimum moisture content (%)	24
Differential free swell (%)	80

2.2 Swelling Pressure Test

In order to achieve larger swelling it was decided to prepare the samples in the dry side of optimum moisture content. The water content and dry unit weight adopted are 20% and 14.2 kN/m^3, respectively.

The soil was mixed with the required quantity of water and kept for 12 hours in an airtight bag for moisture equilibrium. Required quantity of wet sample was then pressed, in a static compaction device to a thickness of 14 mm by placing a spacer directly into the oedometer ring of 60 mm diameter and 20 mm height so as to achieve the required unit weight. The inner sides of the rings were smeared with silicon grease to reduce side friction before preparing the soil sample by static compaction. The swelling pressure of the soil was determined by all the three methods discussed earlier using the conventional one-dimensional consolidation apparatus.

- **Method A** – The sample was inundated and allowed to swell vertically at a seating pressure of 6.25 kPa until primary swell is completed. The sample was then loaded in intervals similar to the procedure of conventional one-dimensional consolidation testing until the specimen reaches its initial thickness. Sufficient time was allowed under each pressure increment till the consolidation was over. The pressure required to bring back the sample to its initial thickness is regarded as the swelling pressure.
- **Method B** – Four identical samples were loaded under different pressures of 6.25, 100, 200 and 400 kPa and inundated with water. If the pressure is less than the swelling pressure, the sample exhibit swelling where as the sample consolidates if the pressure is more than the swelling pressure. Under each pressure, the volume change of the sample is recorded upon inundation. The pressure corresponding to zero volume change in a plot of vertical movement versus applied pressure is regarded as the swelling pressure.
- **Method C** – In this method, also called as Constant volume method, the specimen was maintained at constant height by adjusting the vertical pressure after inundating with water. The pressure required to maintain constant volume is the swelling pressure.

The results of the swelling pressure test for the soil sample are shown in Fig. 6. Clearly, the results are different as expected. The swelling pressure values obtained by methods A, B and C are 195 kPa, 140 kPa and 115 kPa, respectively. The swelling pressure by method A is the highest and that obtained by method C is the lowest, similar to that reported in the literature.

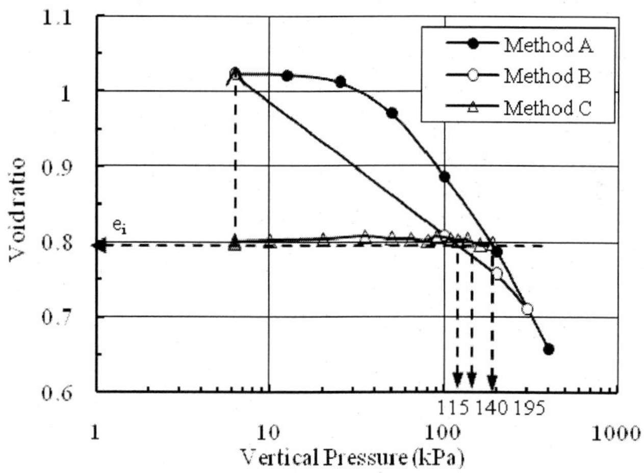

FIGURE 6: Results of the swelling pressure test by the three methods

2.3 Test Apparatus for Pile Tests

Gadre and Chandrasekaran (1994) carried out centrifuge model tests on an expansive soil to simulate a prototype soil deposit of 1.875 m thickness. Tests were conducted on a model clay bed of 12.5 mm thickness at 150 g. The magnitude of swelling observed in the centrifuge model was compared by integrating the 1g swelling experiments in the oedometer at different surcharge pressures as shown in Fig. 5.1. The results show that it is possible to compute the magnitude of swell of soil deposits by integrating the resuts of 1g swelling experiments on models subjected to different surcharge pressures. Guided by this, the pile uplift tests were conducted in the laboratory using a CBR mould of 150 mm internal diameter and 180 mm height and cement mortar pile of 30 mm diameter. The tests were carried out for different surcharge pressures to simulate different depths in the field.

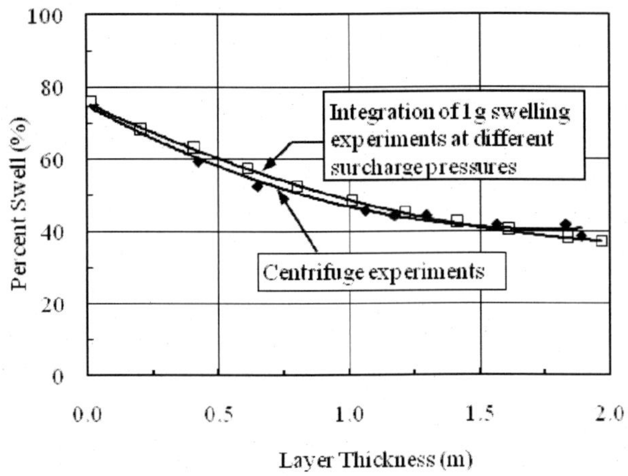

FIGURE 7: Variation of percent swell with layer thickness (Gadre and Chandrsekaran, 1994)

The uplift force acting on the shaft of the pile is due to the relative movement of pile and soil. If the pile is restrained at the top, then the force on the pile will act on the restrained end. Hence in the model tests, a load cell was fixed at the restrained end and the uplift force on the pile is monitored using this load cell. The materials and methodology for the tests are explained in the following sections.

2.3.1 Model Pile

The model pile was made of cement mortar with a mix proportion of one part of cement and two parts of sand with water cement ratio of 0.45. The diameter of the pile is 30 mm. The cement mortar pile was reinforced with a central rod of 5 mm diameter, that in turn was welded with metal strips so as to obtain proper bonding between the mortar and the pile. The central rod was welded to a mild steel rod of 30 mm diameter and 100 mm long as shown in Fig. 8a. The rod was placed inside a PVC pipe and the mortar is poured around the reinforcement (Fig. 8b) and tamped appropriately so as to remove the air voids. After 24 hours of curing, the PVC pipe was cut from the cement mortar pile and the pile was soaked in water for further curing. The curing was done for 28 days, before the tests. The casted cement mortar pile, which was used for the pile uplift tests is shown in Fig. 8c. The mild steel rod has appropriate threads for connecting to the load cell.

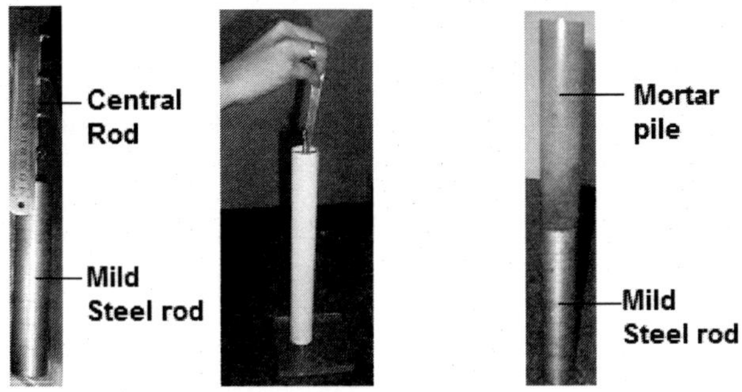

FIGURE 8: (a) Reinforcement for pile welded on MS rod (b) Casting in PVC casing and (c) Casted cement mortar pile

2.3.2 Sample Preparation

A cylindrical mould (CBR mould) of 150 mm internal diameter and 180 mm height was used for the study. The soil was mixed with the required quantity of water and kept in a desiccator for about 12 hours for moisture equilibrium. The cement mortar pile of 30 mm diameter was placed at the centre of the CBR mould. The bottom of the pile rests on an acrylic plate with a hole, so that the uplift of the pile at the tip is eliminated. Therefore, the uplift is only due to the shaft. The soil was statically compacted around the pile in

FIGURE 9: Photographic view of the apparatus

layers at moisture content of 20% and a dry unit weight of 14.2 kN/m^3. A sample thickness of 100 mm was achieved using suitable spacers. The photographic view of the test set-up is shown in Fig. 9.

The required surcharge was applied over the soil sample through the lever arm arrangement similar to that in the conventional consolidation set up (as shown in Fig. 10). After reaching equilibrium under the applied surcharge, the sample was inundated with water. The laboratory pile uplift experiments were conducted for different surcharge pressures of 1, 6.25, 20, 35, 70, 105 and 140 kPa for Soil B and 6.25, 20, 50, 70 and 105 kPa for bentonite, to simulate different depths in the field. For all tests, the soil movement and total uplift force on the pile were monitored at different time intervals. The vertical movement of soil was monitored using dial gauges and the uplift force on the pile by a load cell which is connected to the pile and restrained by a fixed frame. The pile derives the uplift force only from the shaft since the pile base is not in contact with the soil. Sufficient time was allowed for the swelling to take place. A swelling pressure of about 150 kPa was obtained from the tests.

3. RESULTS AND DISCUSSIONS

3.1 Effect of Surcharge on Interface Shear on Pile

When the expansive soil swells on addition of water, shear stress develops at the interface between the soil and the pile material. If the pile is restrained, the shear stress generates uplift force on the piles. The interface shear stress developed along the shaft of the pile due to the relative movement of pile and soil was monitored with time using the load cell fixed at the top of the pile.

The variation of interface shear stress, defined as the ratio of the uplift force on the pile to the circumference area of pile, with surcharge pressure is shown in Fig. 11. The results show that the interface shear stress initially increases to reach a maximum value and then decreases with the increase in the surcharge pressure on the sample for the same placement condition of the soil sample. This indicates that the variation of interface shear stress on the pile with depth is not constant but follows a trend as shown in Fig. 11.

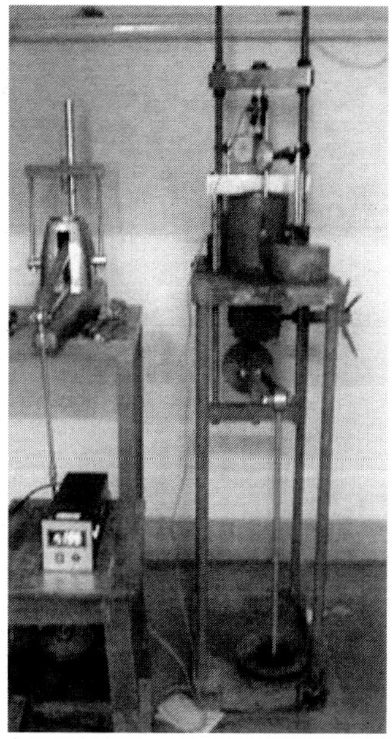

FIGURE 10: Photographic view of the complete set-up

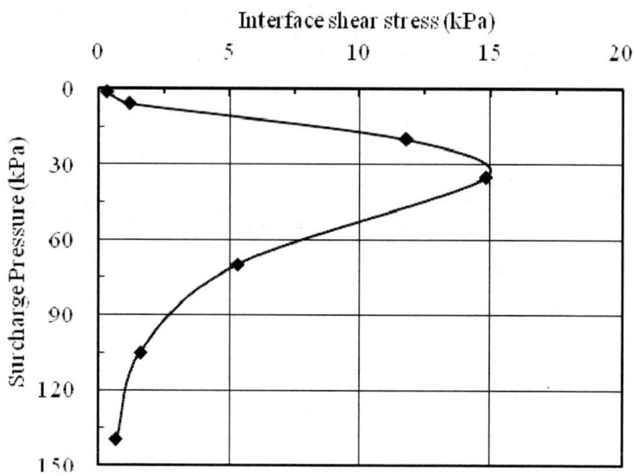

FIGURE 11: Effect of surcharge on interface shear stress on pile

3.2 Estimation of Depth of Active Zone

When the surcharge pressure on the soil is less, soil heave is large leading to a higher relative movement between pile and soil. This results in full mobilisation of the interface shear stress between the soil and the pile. However, the mobilised interface shear stress also depends on the normal stress on the pile. At smaller surcharge pressures (representing shallow depth), the normal stress on the pile is also less. Therefore, even though the shaft friction is fully mobilised, the magnitude is relatively less. As the surcharge pressure on the soil increases, the normal stress on the pile also increases resulting in higher interface shear stress on the pile. However, the soil movement gets reduced with increase in surcharge pressure. Though the available shaft friction is much higher at higher surcharge pressures, the soil movement is not sufficient for the full mobilisation of the shaft friction. This results in lesser interface shear stress on the pile. Therefore, near the soil surface the soil movement is large but the available shaft friction is less leading to a smaller interface shear stress. At deeper depths, the soil movement is not sufficient for mobilising the shaft friction. This results in an uplift pressure distribution shown in Fig. 11

In the field, the surcharge pressure is related to the depth of the soil bed. The surcharge pressure (σ_v) is given by Eq. (1) as:

$$\sigma_v = \gamma \times d \quad (1)$$

γ – bulk unit weight of soil;
d – depth of soil

The bulk unit weight for the soil is 17 kN/m^3. Typically, the overburden pressure is converted into depth of soil and plotted the variation of uplift pressure distribution with depth in Fig. 12.

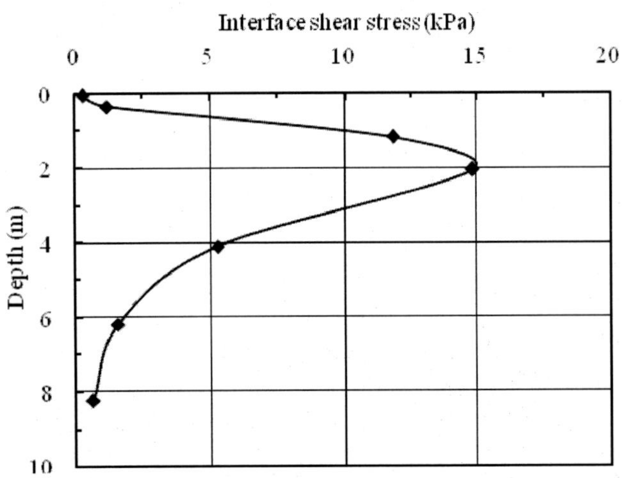

FIGURE 12: Variation of interface shear stress with depth

It may be noted from Fig. 12 that the depth at which the uplift pressure is close to zero is about 8.5 m. The swelling pressures determined by methods A, B and C are 195 kPa, 140 kPa and 115 kPa. The depths required to induce these pressures are 11.5 m, 8.2 m and 6.8 m. The swelling pressure value by the different pressure method (method B) matches well with the surcharge pressure required to reduce the uplift on the pile close to zero.

4. CONCLUSIONS

Based on the experimental investigation, it is observed that the variation of interface shear stress on the piles in expansive soil during swelling with depth is not constant. The interface shear stress initially increases to reach a maximum value and then decreases with the increase in normal stress on the soil and become zero when the overburden pressure on the soil is equal to the swelling pressure of the soil, determined by Method B. The value of swelling pressure determined by method B may be used to estimate the depth of active zone in the swelling soil. The depth of active zone is the depth at which the overburden pressure equals the swelling pressure of the soil.

REFERENCES

Abduljauwad, S. N., and Al-Sulaimani, G. J. (1993) "Determination of swell potential of Al-Qatif clay". Geotechnical Testing Journal, ASTM, 16, Issue 4, pp 469-484.

Ali, E. F. M., and Elturabi, M. A. D. (1984) "Comparison of two methods for the measurement of swelling pressure." Proceedings of the 5th Int. Conerence on Expansive soils, Adelaide, Australia, 21-23 May, pp 72-74.

Al-Shamrani, M. A., and Al-Mhaidib, A. I. (2000) "Swelling behavior under oedometric and triaxial loading conditions". Proceedings of Sessions of Geo-Denver 2000 - Advances in Unsaturated Geotechnics, Geotechnical Special publication 99, Volume 287, pp 344-360.

American Standards for Testing Materials (2003) Standard Test methods for One-dimensional swell or settlement potential of cohesive soils, ASTM D4546.

Bhandari, R. K., Prakash, C., Balodhi, G. R. and

Chandra, R. (1987) "Behaviour of short bored piles in expansive soils during the stage of swelling", Proceedings of the 6th International Conference on Expansive soils, New Delhi, India, pp 295-300.

Brackley, J. J. A. (1973) "Swell pressure and free swell in compacted clay." Proceedings of the 3rd International Conference on Expansive soils, Haifa, 1, pp 169-176.

Chen, F.H (1975) Foundations on expansive soils, Amsterdam, New York, Elsevier Scientific Publishing Company.

Gadre, A.D. and Chandrasekaran, V.S. (1994) "Swelling of black cotton soil using Centrifuge modeling". Journal of Geotechnical Engineering, ASCE, 121, Issue 5, pp 914-919.

IS: 2911(PART-III)-1980: Indian Standard code of practice for design and construction of under-reamed piles, BIS.

Komornik, A. and Zeitlen, J. G. (1961) "Deformation and moisture movements in expansive clays", Proceedings of the 5th International Conference on Soil Mechanics and Foundation Engineering, Paris, 1, pp 873-879.

O'Neill, M.W. and Poormoayed, N. (1980) "Methodology for foundations on expansive clay". Journal of Geotechnical and Geoenvironmental Engineering, ASCE, 106, Issue GT 12, pp 1345-1367.

Prakash, C., Wardle, I.F, Chandra, R., Balodhi, G.R., and Price, G (1988) "Performance of instrumented under-reamed pile foundation supporting a single storey structure in expansive soil". Indian Geotechnical Journal, 18, Issue 4, pp 340-355.

Soundara, B. and Robinson, R.G. (2009) "Testing method and swelling pressure of clays". International Journal of Geotechnical Engineering, USA, 3, Issue 3, pp 439-444.

Sridharan, A., Rao, S.A. and Sivapullaiah, V. (1986) "Swelling pressure of clays". Geotechnical Testing Journal, ASTM, 9, Issue 1, pp 24-33.

Thompson, R. W., Perko, H. A. and Rethamel, W. D. (2006) "Comparison of constant volume swell pressure and oedometer load-back pressure". Proceedings of the 4th International Conference on Unsaturated Soils, Carefree, AZ, United States, pp 1787-1798.

Centrifuge Modelling on Construction and Long-term Settlement of Reclaimed Island- Practical Approaches

X.F. Ma[1,2], L.Q. He[2], C.L. Zhang[2], X.Y. Wu[3] and X. Zhang[3]

[1]*Key Laboratory of Geotechnical and Underground Engineering of Ministry of Education, Tongji University, Shanghai, China*
[2]*Department of Geotechnical Engineering, Tongji University, Shanghai, China*
[3]*Shanghai Harbour Engineering Research Institute, Shanghai, China*
E-mail: xf.ma@tongji.edu.cn

ABSTRACT

Reclaimed islands are often constructed as the transition part between bridges and tunnels in cross-sea projects. One of the major concerns is the stability of the reclaimed island in both construction and post-construction period, during which complicated construction processes are usually practiced. In order to obtain reasonable prediction and further understanding on the behavior of reclamation in near sea, centrifuge model test was conducted based on conditions of a real project going on in the near sea south of China. Main aspects of the project, including large diameter steel pipes, sand compaction piles, plastic drain board and underground water pumping etc. were simulated with practical approaches. Settlement of the reclaimed island, dissipation of pore water pressure, lateral displacements of the steel pipe were observed and discussed, providing valuable references to similar projects.

1. INTRODUCTION

Rapid development of infrastructure has been seen increasingly in recent years, among which cross-water roads have been particularly developed in cities near sea or rivers. For cross sea projects, reclaimed islands are often constructed as the transition part to connect bridges and tunnels. Besides the function of composing a key part in the project and guaranteeing the safety both during the construction and future operation period, the reclamation also provides an important working platform for the whole construction which includes a series of complicated procedures. Obviously, it is extremely important for the reclaimed islands function normally as designed and the safety would be ensured. To achieve this, the behaviour of the reclamation, i.e. the stability of retaining structure and settlement of the whole reclamation etc., should be properly predicated prior to the construction works on site. It is understandable that as the in-situ conditions are usually complicated, it is usually difficult to get precise calculation of the settlement of the reclaimed island just through numerical modeling. However, conducting a centrifuge model test on a typical section of the reclaimed island would be a reasonable alternative to obtain the settlement of the reclaimed island both during the construction and afterwards. The following part of this paper will show an example of the simulation, particularly the methodology of modeling complicated works by practical approaches.

The work in this paper is based on an ongoing project in the south of China, the Hong Kong-Zhuhai-Macau cross sea project which connects Zhuhai city, Macau and Hong Kong (Figure 1). It is a 35km long project, in which bridge part is about 22km and tunnel occupies 6km in length. According to the design, two reclaimed islands will be constructed somewhere east to the mid point of the project, spacing about 5.6km between each other and accommodating an immersed tunnel just under the water in the location between the two islands. The tunnel rises up to the surface of the island, transforming to the ground road and then connecting to the bridge part transitionally along the long axis of the island. Clearly, the reclaimed islands are the key part in the whole project, connecting the bridge structure and the tunnel structure into one whole passage and forming the foundation to support neighboring bridge structure and enclose the tunnel structure, There are two problems particularly crucial to the success of the project: the stability of the sea wall enclosing the whole reclamation and the settlement of the islands both during the construction and afterwards. The test in this paper thus make main efforts towards observing the two issues bearing in mind the essential aspects of the situation should be simulated reasonably since the whole work is too complicated to be followed in every detail.

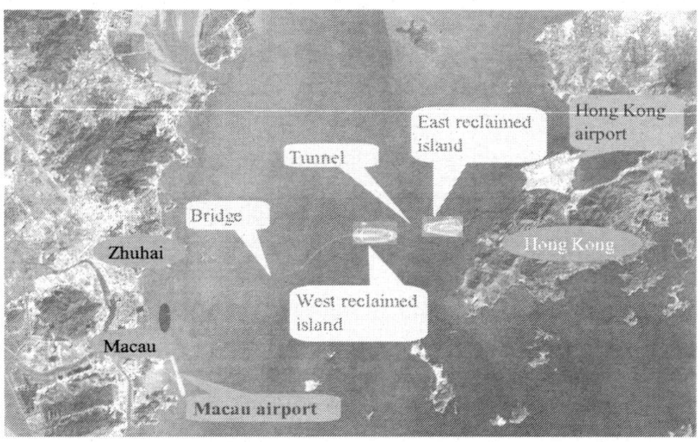

FIGURE 1: Plan view of Hong Kong-Zhuhai-Maucau cross sea project

2. RECLAIMED ISLANDS AND THE CONSTRUCTION PROCEDURE

The reclaimed islands are similar in the shape and design, featuring a narrow ellipse in the plan. The right one in Figure 1 (the east island) is taken as the studied object in the test, which is 625m in length and 215m in width at the most(Figure 2). The depth of the sea bed at the location of the reclamation is about 10m below the average sea level and the designed formation of the reclaimed ground surface is about 4m above the average sea level.

The geological profile on the site features a very soft ground formation (Figure 3), which includes soft silt layer extending from the sea bed down to 10m or 20m below the sea bed. The second group of layers underlying the silt layer is basically silty clay whose thickness varies from 5m to 10m. Beneath the silty clay, stiff clay or sandy clay dominates until the depth of -40m and the ground transforms to sand layers beyond the depth of -40m. The base rock is about 70m below the sea level. The main properties of the ground is shown in Table 1.

In order to retain the reclaimed ground, steel pipes with diameter of 22m and the thickness of the pipe skin of 16mm are employed as sea walls enclosing the island. In the east island, altogether there are 63 pipes installed to function as retaining structures, with connecting joints between two neighboring pipes to seal water from flowing into the island side. Prior to the main construction, ground replacement is carried out by excavating the original silt layer until -18m both inside and outside the island.

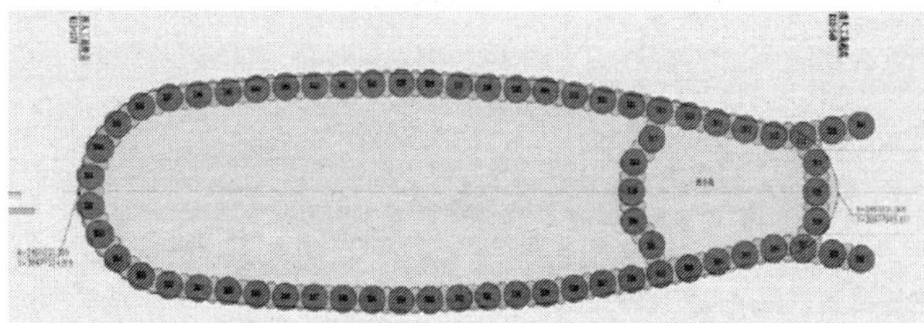

FIGURE 2: Plan view of the east reclaimed island

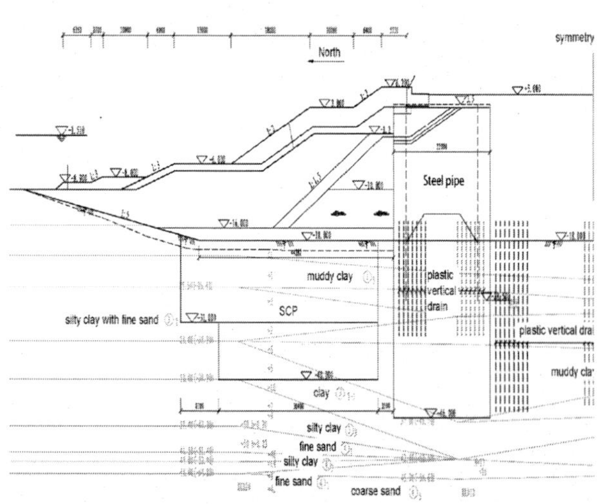

FIGURE 3: Ground profile and section of the embankment

Table 1: Physico-Mechanical Parameters of In-site Soil

Soil layer		*Silt*	*Muddy Clay*	*Silty Clay*	*Fine sand*
Water content (%)		61.3	34.4	35	19
Wet density (kN/m³)		16.7	18.5	18.1	20
Direction CU	Cohesive strength (kPa)	12	17	16	6
	Angle of friction (°)	14.4	15.5	17.3	34.2

Steel pipes are then installed until the designed depth of about -35m surrounding the boundary of the reclaimed island. Sand layers are backfilled inside the island till the depth of -6m, above which ground treatment is carried out. Outside the island, sand compaction piles (SCP) are installed down to a depth of about -31m, and gravel backfill was carried out to form the embankment outside the steel pipe in the designed shape. The inner part of the island was also treated but plastic vertical drain board was adopted which is different from the sand compaction piles. Above the top of the drain board treated ground, 5 sand layers were banked in sequence to form the ground and function as preloading layer at the same time. To accelerate the settlement of the reclaimed ground, groundwater pumping is also designed during the embankment of the layers, and continues to the point when the consolidation ratio reaches 80%.

To observe the behaviour of the reclaimed island, a typical section of the reclamation is choosen perpendicular to the long axis of the island. As the long axis is well beyond the short axis, the problem can be treated as a plane strain type and the width of about 110m is decided.

3. TEST SCHEME AND MODEL PREPARATION

3.1 Strong Box and Model Simplification

The strong box used in the test has an inner size of 900mm(L)*700mm(W)*700mm(H), and the centrifugal acceleration of 120g is thought to be

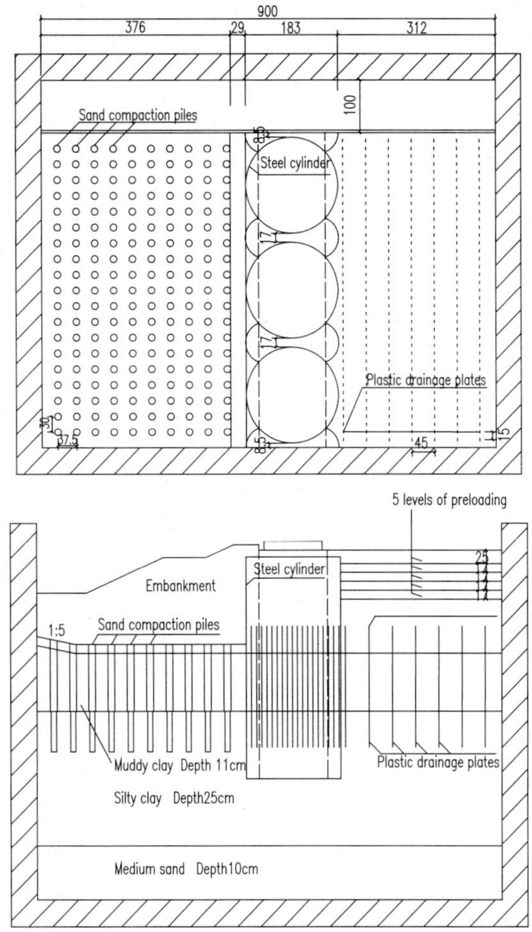

FIGURE 4: Layout of the reclaimed island typical section model

suitable based on the previous discussion. Figure 4 shows the top view and the front section view of the model, which is a simplified version to the prototype one. Due to the restriction of the box, an area of about 110m long and 72m wide is chosen to be simulated, which contains three model pipes in the area. This is acceptable as the prototype structure can be regarded to be a plane strain problem in the direction perpendicular to the long axis of the island, especially when the stability of the sea wall structure is concerned. The length of the model is about 110m, which covers the whole sea wall structure, the whole area of improved ground using SCP outside the island and leaves 40m for the ground inside the sea wall. As the island would settle evenly on a large area, the settlement of the ground inside the sea wall in the model would not differ much from that on site when the friction between the box wall and the soil is deal with by pasting the lubricant given other main factors are suitably simulated.

3.2 Sand Compaction Piles Preparation

Sand compaction piles (SCP) are widely used in practice in order to improve soft ground, such as soft clay or peat. In particular, when less settlement-sensitive structures are designed, the more cost effective method of sand piles is often preferred.

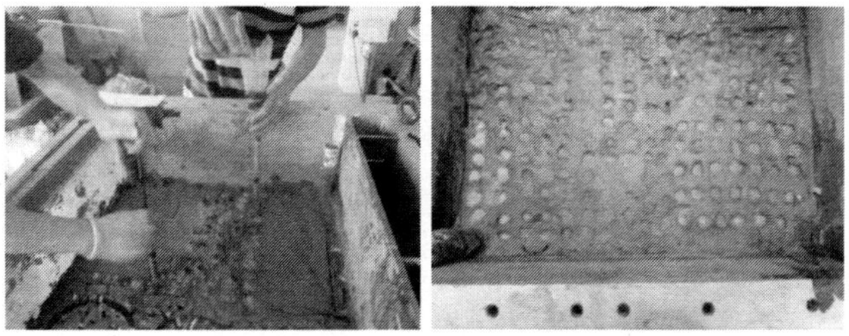

FIGURE 5: Installing of Sand compaction piles

Lee (2001, 2004) and Weber(2004, 2005) has investigated varied ways to simulate SCP in centrifuge modeling, and inflight installation methods have been developed. However, in this test, due to the limited equipment and conditions, the installation of the SCP was conducted manually under 1g condition. The installation was fulfilled using a simple set of tools which include a aluminum tube and a piston-like rod able to penetrate into the tube and compact the sand filled into the tube. The inner diameter of the tube is designed considering the required diameter of the SCP and the expansion effect during the installation. The rod penetrating into the tube has a piston in the end that suit the inner diameter of the tube so that it can be used to compact the sand inside the tube smoothly. Figure 5 shows the installation of the piles into the model ground. To achieve better simulation of SCP, preliminary trial installation has been carried out to decide suitable diameter of the tube and compaction ratio of the filled sand column. The trials shows that a SCP with diameter of 14mm can be reasonably achieved by a tube of 12mm in the diameter when compacting the sand filled into the tube at a ratio of 15/10, which means compacting the sand column of 15mm high inside the tube to be a sand column 10mm high.

3.3 Steel Pipe Modelling

The steel piles enclosing the reclaimed island has a large diameter up to 22m while a very thin skin whose thickness is only 16mm, making it very difficult to model the pipe just by scaling down the size in the same similarity law. For a quick estimation by imagining the condition under 100g, the diameter can be scaled down to 22cm in the model, however, the thickness become 0.16mm using the same material, which is extremely difficult to machine and install. However, if the material can be changed to a soft one, the thickness of the pipe can be reasonably large. In this paper, polymethyl methacrylate plastics (organic glass) is chosen to be used to simulate the steel pipe. According to stiffness equivalence, the size of the organic glass can be obtained by the formula as follows.

$$d_p/d_m = N \quad (1)$$

$$\frac{E_p I_p}{E_m I_m} = N^4 \quad (2)$$

As the width limitation of the test box, only three steel cylinders were placed in the box. The parameters are shown in Table 1 and Figure 6 shows the photo of the model.

Table 1: Parameters of the Prototype and the Model

Parameters	*Prototype*	*Model*
Elasticity modulus (GPa)	210	3
Outer diameter (mm)	22000	183.3
Wall thickness (mm)	16	10
Length (mm)	49500	412.5

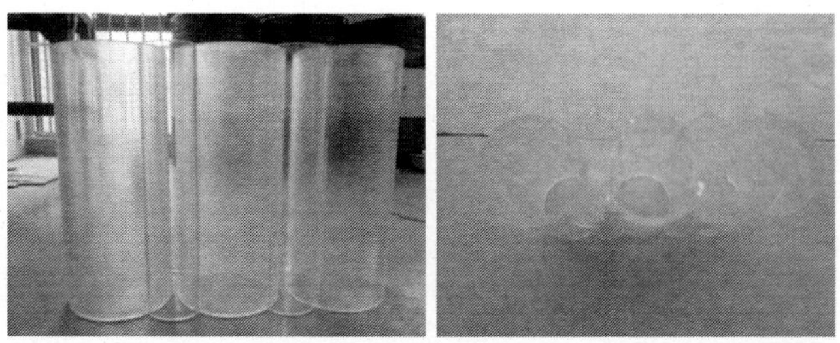

FIGURE 6: Photos of the steel pipes

3.4 Plastic Drain Board Preparation

Non-woven fabric was used to simulate the plastic drainage board because of its similarity in permeability. The non-woven fabric was cut into strips of 2mm in the width and 233mm in the length each. These non-woven fabrics were used to accelerate the consolidation of the artificial island. Figure 7 shows the example of cut strips and Figure 8 shows the scene of installing them into the model.

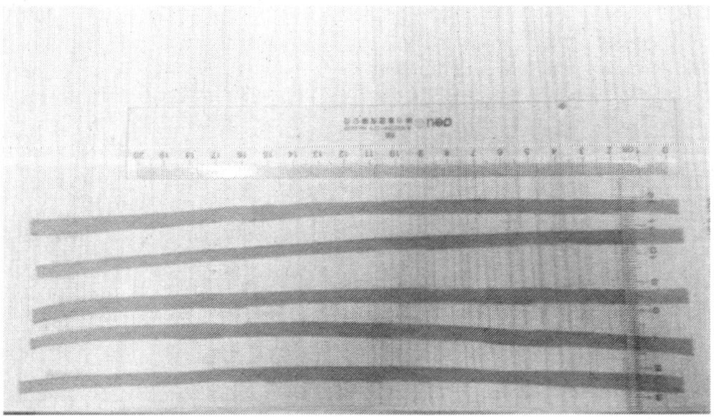

FIGURE 7: Non-woven fabric strips

FIGURE 8: Inserting non-woven fabric strips

3.5 Pumping of Underground Water

Pumping out underground water and lowering the underground water level is an important procedure in the construction on site for this project. During the banking of 5 layers of backfill inside the reclaimed island, underground water level is designed to be lowered to -18m to accelerate the consolidation. It would be easily understood that it is almost impossible to put a small pump in the model and operate it under centrifugal acceleration as high as 120g. In this paper, a simple method of draining water through porous stone placed at the designated level is workable to achieve the dewatering results. A sand layer in the model is placed in the required depth and porous stone is inserted in the box wall connected by a plastic tube controlled by solenoid valve draining the water out. Porewater pressure reading during the test proved that the method was workable.

4. CENTRIFUGE MODEL TESTS

The centrifuge model test was performed at an acceleration 120-times of the gravity. The model has been described in the previous parts in which the ground has been simplified into three layers: silt layer, silty clay layer and sand layer, the property parameters of the former two are shown in Table 2.

Selected soil parameters are stated in Table 2.

Table 2: Physico-Mechanical Parameters of In-site Soil

Soil Layer	*Silt*	*Silty Clay*
Depth (cm)	11	25
Void ratio	1.4	1.2
Water content (%)	50	44
Density (kN/m$^{3)}$	17.0	17.8
Cohesive strength (kPa)	12	14
Angle of friction (°)	12.5	19.2
Compression modulus	3.02	3.04
Permeability coefficient	1.6×10^{-7}	9×10^{-8}

4.1 Test Procedure

The procedure of the model test started with an initial consolidation phase at planned g level in the centrifuge. The following steps of the tests basically follow the real procedure on site, in which the three steel pipes were firstly installed into the clay ground in the middle of the test box, splitting the box into 2 sections by the three model pipes. Then the sand compaction piles were installed along with the plastic board drainage in these 2 sections respectively. A gravel embankment was located above the sand compaction piles reinforced area next to the steel pipes. Then 5 layers of sand were loaded in sequence, followed by the dewatering procedure . When the consolidation degree of the reclaimed island meets the requirement, raise the underground water table up to the initial level as unloading effects. The construction period ended at this point and then the centrifuge continued running for long-term observation.

4.2 Measurement

Three pore water pressure transducers were placed in the test box, which can be named as PPT1, PPT2 and PPT3. PPT1 was placed in the silty clay layer and PPT2 was put into the area where the plastic drain board was improved, and PPT3 stands for the transducer in the sand compaction pile reinforced area.

Two displacement transducers were used to measure the settlement of the reclaimed island, named as AI-DT1 and AI-DT2 respectively. Another two displacement transducers were located above the steel pipes to measure the settlements of the steel cylinders, and were named SC-DT1 and SC-DT2. Also, another three displacement transducers were used to measure the lateral displacements of the steel pipes and named as SC-HDT1, SC-HDT2 and SC-HDT3 (see Figure 9).

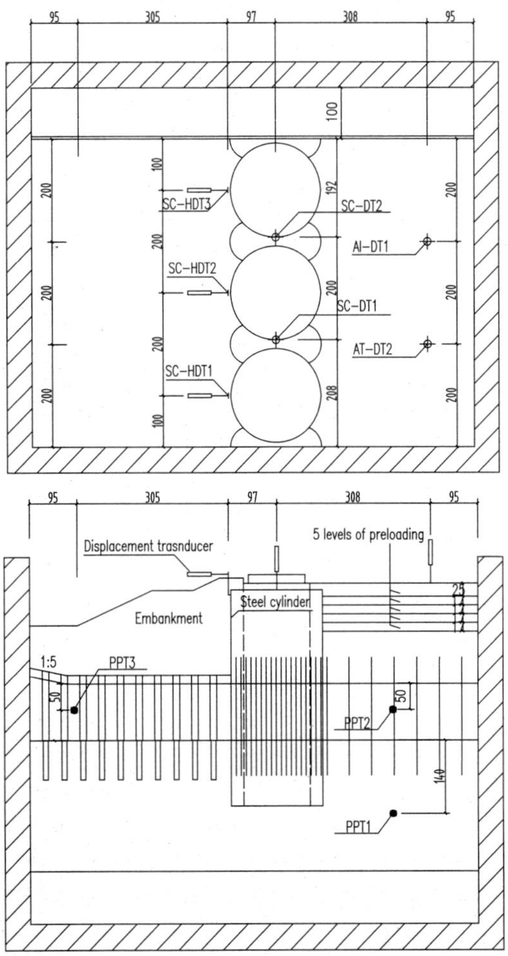

FIGURE 9: Layout of the measuring points in the model

5. OBSERVATION ON THE RESULTS

Some results are picked out to show the observation. All values are given in the following part in prototype scale except where stated otherwise.

5.1 Preloading Period

Figure 10 shows the progressive changes of the lateral displacements of the steel pipes during preloading period. The numbers above the graph indicate the stage number of the preloading, or banking. The trend is clear that the continuous loading caused development of later displacement on the top of the pipe. However, the sequence of magnitude of the displacement is meaningless as it would be influenced by many local conditions. It should be noted that the magnitude of the displacement in all of the three curves is small, indicating the sea wall is very stable and safe. Figure 11 and Figure 12 show the settlements of the steel pipes and the reclaimed island during the preloading stage respectively. They both indicate that under loading conditions in each step, the settlements of steel pipes and the reclaimed island developed substantially in a sudden drop and a smooth curve afterwards. After unloading which

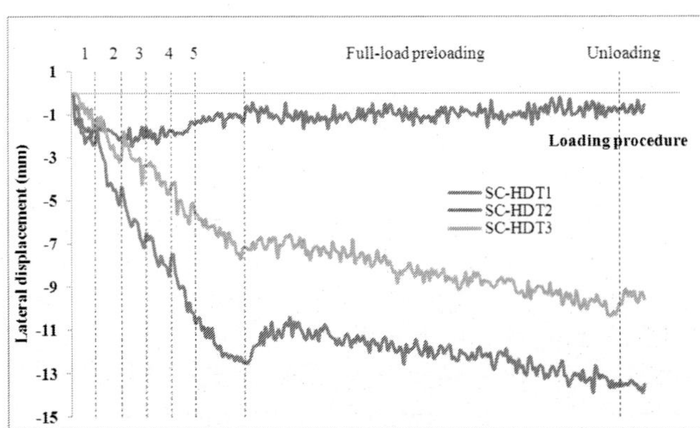

FIGURE 10: Lateral displacements of steel cylinders during preloading period

is actually restoring underground water level, the settlements of the two figures rebounded instantly for about 60mm both in average.

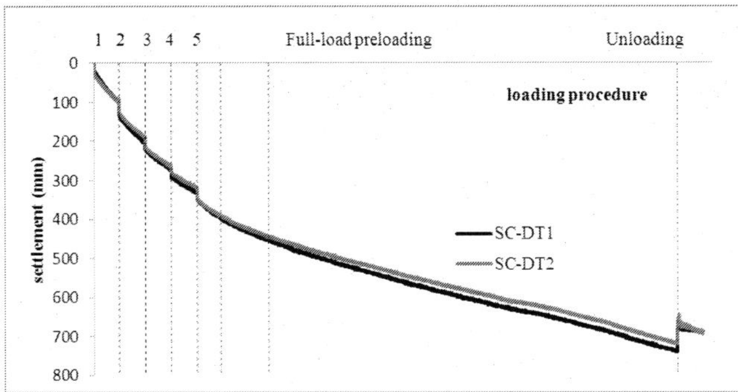

FIGURE 11: Settlements of steel cylinders during preloading period

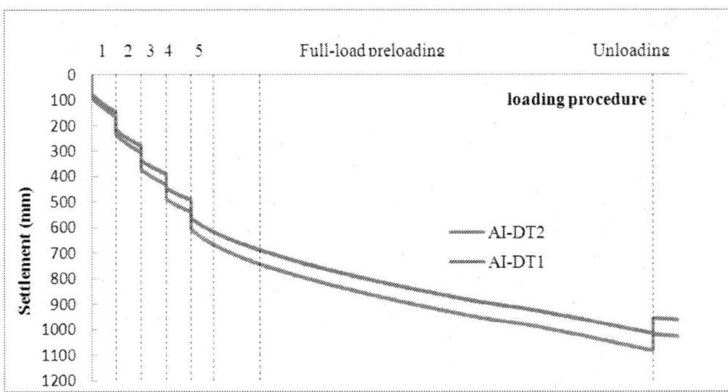

FIGURE 12: Settlements of the reclaimed island during preloading period

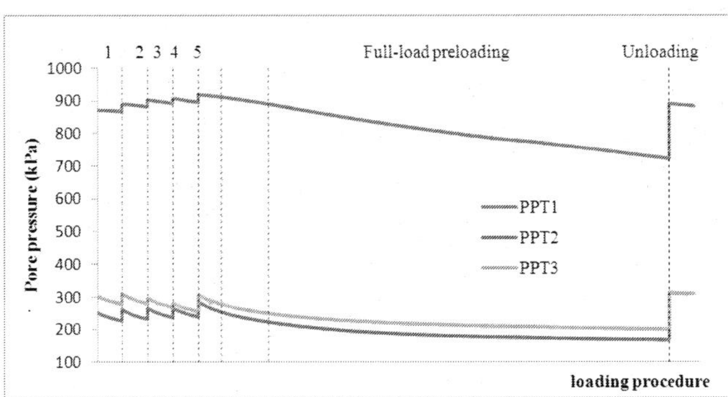

FIGURE 13: Pore pressure changes during preloading period

Figure 13 shows that the pore water pressure also had a sudden increase instantly when loading is applied, which fits perfectly the effective stress theory. When the time elapses, excess porewater pressure dissipated gradually, and the consolidation of the ground progressed. While the underground water table recovered, the pore pressure had an immediate increase. All of this indicate that the centrifuge test has got a reasonable results and that the simulation expressed the real construction process well.

5.2 Post-Construction Period

The test lasted 16 hours altogether at the acceleration level of about 120-times of the gravity after the preloading stage to observe the long-term stability of the reclaimed island.

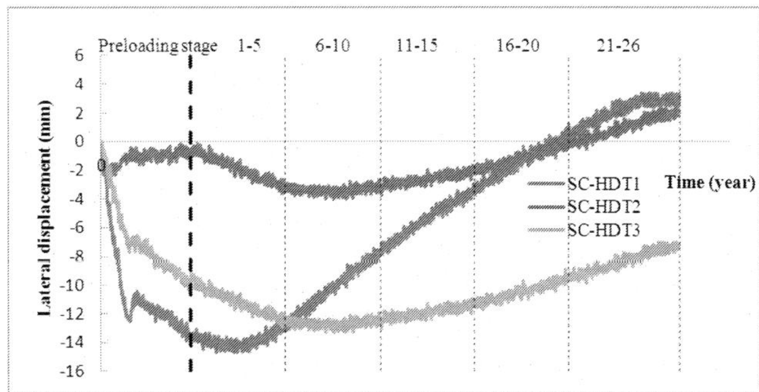

FIGURE 14: Lateral displacements of steel pipes after construction

Figure 14 shows that the total lateral displacements of steel cylinders are rather small. The steel cylinders tend to incline against the reclaimed island for about 2mm in 26 years' time in average. Thus, the steel cylinders are considered to be stable after the construction of the artificial island.

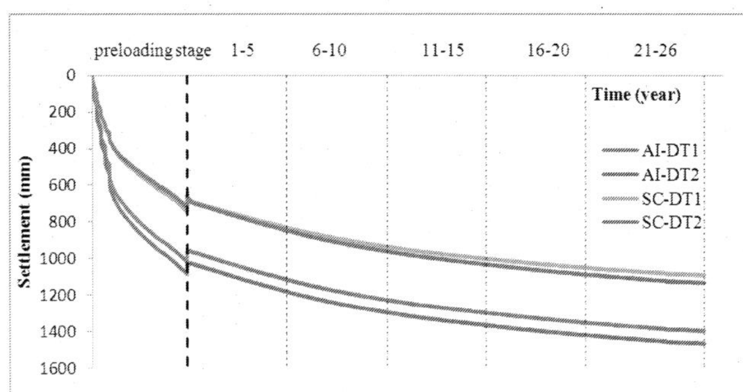

FIGURE 15: Settlement of the reclaimed island and steel cylinders after the construction

Figure 15 shows that the settlements of both the reclaimed island and the steel pipes tend to be stable in 26 years, although the settlements of them are both still slowly increased. In average, the total settlement of the artificial island and the steel cylinders are 1433mm and 1100mm respectively.

Figure 16 indicates that the pore pressure changes after the construction. As the pore pressure transducer located in the sand compaction piles reinforcement area broke down, only two curves were obtained during the test. In 26 years, the pore pressure in the area containing plastic drain board was almost stable. But the pore pressure in the silty clay layer continued to decrease even after such a long time, which means the settlement of the silty clay layer still account for a part of the total settlement. Probably that is the reason why the settlements of both the reclaimed island and the steel pipes were still developed after 26 years.

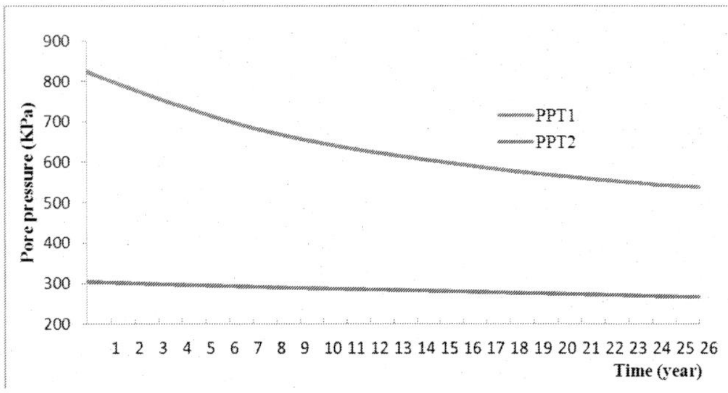

FIGURE 16: Pore pressure changes after the construction

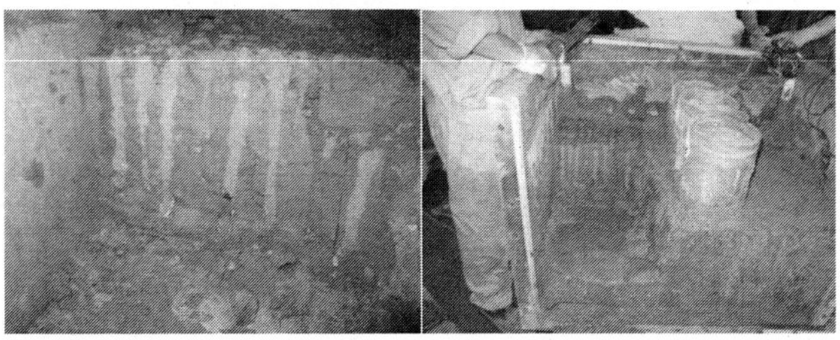

FIGURE 17: Installed SCP section checked after the tests

6. DISCUSSION AND CONCLUSIONS

In this paper, the behaviour of a reclaimed island was studied by centrifuge modeling. The nature that the construction of the reclaimed island consists of complicated procedures makes the test challenging in suitably simulating the major aspects deciding the safety and serviceability of the structure. To carry out the test, the studied area was chosen and simplified, large-scale complicated structures were modeled by simple ways while scaling laws was strictly followed. Installation of SCPs has been fulfilled by manually operated tools and trials has shown the reasonable results. Figure 17 shows the section of installed SCPs after cut out during the check following the test and the SCP looks well except that the verticality was not perfect. Plastic drain board was simulated by non-woven fabric strips and dewatering was fulfilled by draining underground water from seepage routes consists of pipe and solenoid valve. However, banking of sand layers has been conducted manually after stopping the centrifuge and then spin it again..

Test results showed that the lateral displacements of the steel pipes are very small during the preloading and post-construction period, the steel pipe's stability meets the demands. The settlements of the steel pipes and the reclaimed island were obtained and discussed was made. The results indicate the safety of the structure can be ensured and the long-term settlement is still within the predicated range.

Further researches would be encouraged concerning how to simulate the real construction situation more precisely. To achieve this, test scheme should be studied carefully to address the essential influencing factors and trials should be encouraged to develop practical approaches that include reasonable simplification. More cases and further improvement are expected.

REFERENCES

Lee, F. H., Ng, Y. W. and Young, K. Y. (2001) "Effects of Installation Method on Sand Compaction Piles in Clay in the Centrifuge". Geotechnical Testing Journal, ASTM. 24, Issue 3, pp 314-323.

Lee, F.H., Juneja, A. and Tan, T.S. (2004) "Stress and pore pressure changes due to sand compaction pile installation in soft clay". Geotechnique. 54, Issue 1, pp 1-16.

Weber, T. M., Laue, J. & Springman, S.M. (2005) "Modelling the inflight construction of sand compaction piles in the centri-fuge", XVI International Conference on Soil Mechanics and Geotechnical Engineering, Osaka, Japan, pp 1291-1294., Millpress: Rotterdam.

Weber, T. M. (2004) "Development of a sand compaction pile in-stallation tool for the geotechnical drum centrifuge.". In Brandl, H. & Kopf, F. (Eds), XVI European Young Geo-technical Engineer's Conference, Vienna, Austria, pp 391-400.

Tunnelling Induced Ground Deformation in Reinforced Soil

A. Juneja

Department of Civil Engineering, Indian Institute of Technology Bombay, Mumbai
E-mail: ajuneja@iitb.ac.in

ABSTRACT

Results of centrifuge tests and finite element program using Plaxis 2D of tunnel excavation adjacent to vertical reinforcement in sand beds are discussed in this paper. The model tunnel walls were made of pressurized rubber liner. Depth of the tunnel and the spacing between the reinforcement was investigated. To simulate the tunnel excavation, the pressure on the walls of the liner was decreased in steps whilst the centrifuge was in-flight. Magnitude and extent of settlement trough with increase in ground loss was captured in the vicinity of the tunnel axis. From the limited test observations, reinforcement using vertical rods have shown to affect the extent of settlement trough especially when the volume loss is less than 5%. Finite element results seem to reasonably agree to the centrifuge results although there is significant differences possibility because of the use of simplified assumptions in numerical modelling.

1. INTRODUCTION

Ground deformations due to shallow tunneling in urban areas have to be minimized to avoid damage to surface structures. Ground movement due to tunnelling is normally evaluated using empirical procedures or by physical and numerical modelling. Physical modeling can be used to continuously monitor the soil deformation during tunnel collapse. Instability due to urban tunnelling can occur when the soil is over-excavated during boring or when the tunnel support is inadequate during construction. Pre-reinforcement techniques such as jet grouting, forepoling and face bolting have been used to reduce inward soil movement during the tunnelling. Relatively little attention has been paid to the use of vertical reinforcement embedded from the surface down to the tunnel level. Vertical reinforcements can improve face stability when installed ahead of the tunnel in rows perpendicular to the tunnel axis (Kamata and Mashimo 2003). This type of reinforcement technique can be fast and economical especially when tunnelling is done at a shallow depth. Numerical tools are now common methods used in investigating geotechnical engineering problems owing to user friendly interface. These have been used to calculate and predict soil movement due to tunnel construction.

The purpose of this paper is to examine the influence of vertical reinforcements on settlement trough due to tunneling in model sand beds. However, creating a model that would predict the soil movement accurately is difficult. The emphasis is on physical modeling which has been backed up by numerical analysis using Plaxis 2D and empirical equations for comparison of the results. Numerical simulation can be used to predict the settlements without having to do physical modeling. Because of the simplicity in tunnel geometry in sand bed, 2D simulation is thought to be sufficient for the present study. The model tests were carried out at the Indian Institute of Technology Bombay (IITB) Geotechnical Centrifuge. Details of the IITB centrifuge and its operation are available in Chandrasekaran (2001). All dimensions in the figures are in mm unless and otherwise stated.

2. EXPERIMENTAL SETUP

Nine model tests (C1 to C9) were conducted at 50g environment to investigate the effect of reinforcement on ground deformation during tunnel collapse in sand beds. The soil cover and diameter of the reinforcement rods were the variables in the tests. 3mm to 10mm diameter rods were inserted up to nearly the entire thickness of the soil cover. The model tunnels were 40mm and 60mm in diameter which correspond to 2m and 3m in prototype scale. The purpose of choosing this diameter was to not model any specific tunnel but to observe the effect of tunnel depth on the extent of ground deformation. Model sand beds were prepared in centrifuge strongbox of internal dimensions 720mm x 450mm x 410mm which had its front wall made of transparent perspex sheet. This large sized box was taken primarily because of its availability in the laboratory. Inside walls of the strongbox were first smeared with water-resistant grease and then coated with thin polyethylene sheets to reduce friction between the soil and the sidewalls

Dry beach sand was then pluviated at a constant flow rate from 400- to 600mm height using a hopper. Physical properties of the sand are shown in Table 1. A regular path was followed during pluviation. In each soil model, the pouring height was maintained to achieve relative density between 0.4 and 0.7. Pluviation was intermittently interrupted to inspect the density of the sand bed. In this procedure, the density of each layer was directly measured using a 15mm high cup of known volume placed on top of the preceding layer and away from the tunnel location. Sand was then rained and stopped only to weight the cup when it was full. The flow was tuned at the end of each step.

Table 1: Physical Properties of Sand Used in the Tests

d_{30} [mm]	d_{50} [mm]	d_{60} [mm]	e_{max}	e_{min}	G_s	E_u/s_u	φ' [°]
0.19	0.21	0.28	0.83	0.53	2.63	130	31

Tunnelling was simulated using pressurized rubber tube which was hand-placed after laying the bottom 5 to 7 layers. Low stiffness rubber tubes were made using dipping process described by Venis (1963). The rubber tube was filled with water and placed centrally along the width of the container with its closed-end facing the front window. Rear end of the tube was connected to a standpipe placed outside the soil container. Constant water level was maintained in the standpipe which could balance the increasing overburden during the centrifuge spin-up. Outlet from the standpipe was connected to a normally-open solenoid valve. The water in the tunnel could be deflated to simulate the tunnel excavation when the power to the valve is shut.

After preparing the dry sand bed, 3 to 10mm diameter rods of flexural rigidity (EI) between 4.4 $\times 10^{-1}$ and 3.43 $\times 10^{1}$ Nm^2 were pushed from the surface in a line perpendicular to the tunnel axis. A guide block was used to maintain the position of the rods during jack-in. Bottom ends of the rods terminated close to the elevation of the tunnel crown. The top ends of the rods were flushed up to the ground surface. The rods were free-floating columns because emphasis of this study was to examine restriction to the flow of sand till the instant when the soil failure had reached the ground surface. Thereafter, the rods were likely to follow deformations similar to that of the soil (Chen et al. 1999). Table 2 shows the soil cover (C) and clear spacing between the rods (S) at model scale in all the tests. As the table shows, the rods were installed up to 90 to 140mm depth and at 6 to 15mm clear spacing in different tests. The

Table 2: Details of Various Tests Under Study

Test	*D [mm]*	*C [mm]*	*Diameter of Rod, d [mm]*	*S [mm]*	*Length of Penetration [mm]*
C1	60	90	10	10	90
C2	40	140	6	12	128
C3	60	140	10	25	130
C4	60	140	10	25	140
C5	40	160	6	12	80
C6	60	120	12	25	114
C7	40	80	3	12	80
C8	40	110	3	18.5	110
C9	40	114	3	25	114

potentiometers were installed in-line with the rods and dictated the clear spacing between them.

Water was then slowly seeped from the bottom to submerge the sand with about 20mm of water cover at the top. Cabrera et al. (2006) suggested high-g saturation of sand-beds prepared by hand. It is uncertain if Cabrera et al.'s (2006) findings are also applicable to beds prepared by dry pluviation. Nonetheless, 1g saturation has practical advantage in that the flow can be viewed and if necessary altered since the container is kept on the laboratory floor. Fig. 1 shows the typical layout in each test.

After completion of model preparation, the strongbox with standpipe assembly was mounted on the centrifuge platform and accelerated to 50g. About 10 minutes after reaching equilibrium, the water pressure in the tunnel was gradually reduced to simulate loss in internal support leading to tunnel collapse. Potentiometer readings were concurrently taken to measure the surface settlement. Figure 2 shows the test setup on the centrifuge bucket.

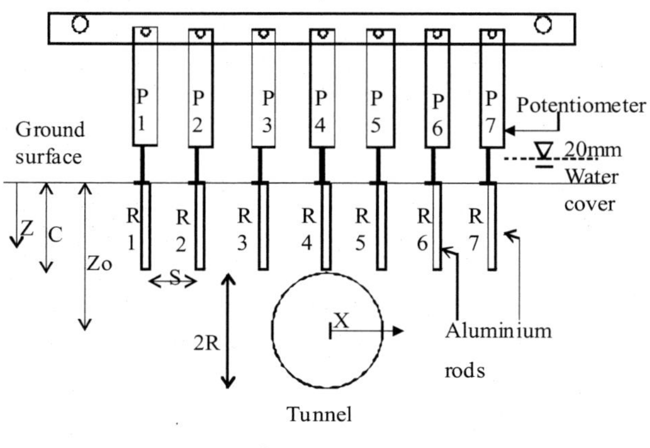

FIGURE 1: Layout of tunnel and reinforcement in high-g tests in centrifuge.

3. NUMERICAL ANALYSIS

Plaxis 2D, a finite element program, was used to model the behavior of ground during tunnelling. The 3D interaction was idealized into a 2D plane strain problem. The use of plain strain condition results in a finite element model with only two degrees of freedom per node. Well known Mohr-Coulomb's model with associated flow rule was used to model the soil. This elastic and perfectly plastic model is considered suitable to model medium to dense cohesionless soils which have tendency to dilate prior to failure state. Five input parameters were required in the program, namely, Young's modulus (E), Poisson's ratio (η), cohesion (c), friction angle (ϕ), and dilatancy angle (ψ). These parameters were obtained in another investigation and are summarized in Table 3. In order to compare results of the physical and numerical modeling, the boundary conditions and soil properties in computation were taken the same. All the tests were modeled by means of a plane strain model, which composed of fifteen nodded triangular elements. The subsoil was entirely of submerged sand. For mesh refinement and sufficient accuracy a separate cluster was created around the tunnel and the reinforcements. Standard fixities were used as boundary conditions. Pore pressures were generated on the basis of phreatic line at the ground surface and the initial stresses then generated by gravity procedure. Figure 2 shows the mesh before collapse in test C2 (15-noded elements). The vertical reinforcements were modeled as one-dimensional elastic beam elements.

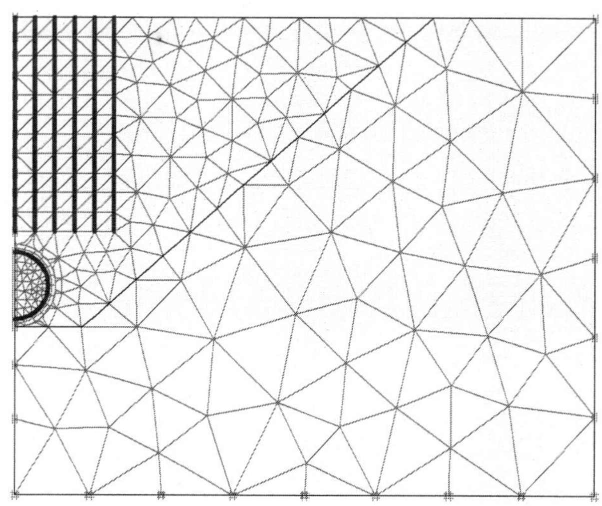

FIGURE 2: Mesh in Test C2 (C/D=3.5, S/d=2)

Table 3: Input Parameters for Sand

Material	*Model*	γ_{sat} *(kN/m³)*	E_{ref} *(kPa)*	η	k_o
Sand	Mohr-Coulomb	20	13500	0.270	0.485

4. RESULTS AND DISCUSSION

For numerical simulation, the collapse was simulated in three phases. In the first phase, the tunnel lining and vertical reinforcements were activated. In the second phase, the soil within the tunnel was deactivated and internal support pressure equal to the overburden pressure at the centre of the tunnel was applied within the tunnel cluster. In order to minimize the effect of application of hydrostatic pressure within the tunnel, all subsequent deformations were measured using local coordinates at the end of this stage. This was necessary in order to keep the tunnel stable prior to the excavation. In the third phase, the tunnel support pressure was reduced to zero in 250 steps and collapse initiated. Fig. 3 shows the deformed mesh for test C2 in 150 steps.

Figures 4a-d show observed vertical settlement, z plotted against distance away from the tunnel centre, x measured at the start of the tunnel collapse. The dark circles in the figures represent the settlements as obtained from the test. Continuous line in these figures shows the settlement profile as obtained from the Plaxis 2D simulation. The broken lines are the normal distribution curve obtained from the equation

$$\frac{z}{z_{max}} = \exp\left[\frac{-1}{2}\left(\frac{x}{i}\right)^2\right] \quad \text{...(1)}$$

where z_{max} is the maximum vertical settlement and i is the point of inflection.

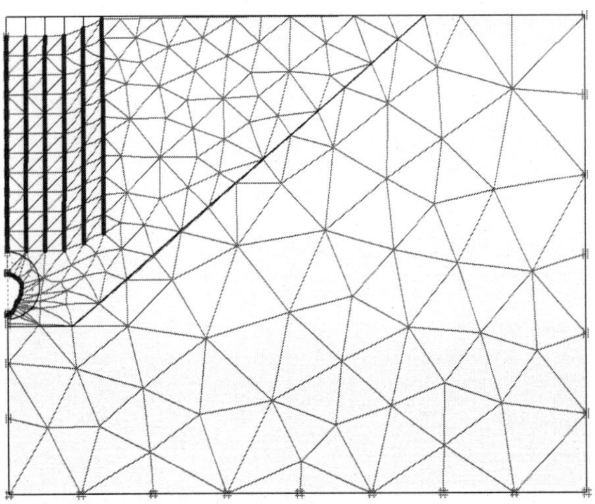

Figure 3: Deformed mesh in Test C2

It is very difficult to model the tunnel collapse using numerical simulation. Although the trend of the observed settlements and the results of FE analysis appear to be similar, there is significant difference between the observed settlement and that obtained using Plaxis. It is evident that the curve obtained through 2D numerical simulation is much wider than the actual profile.

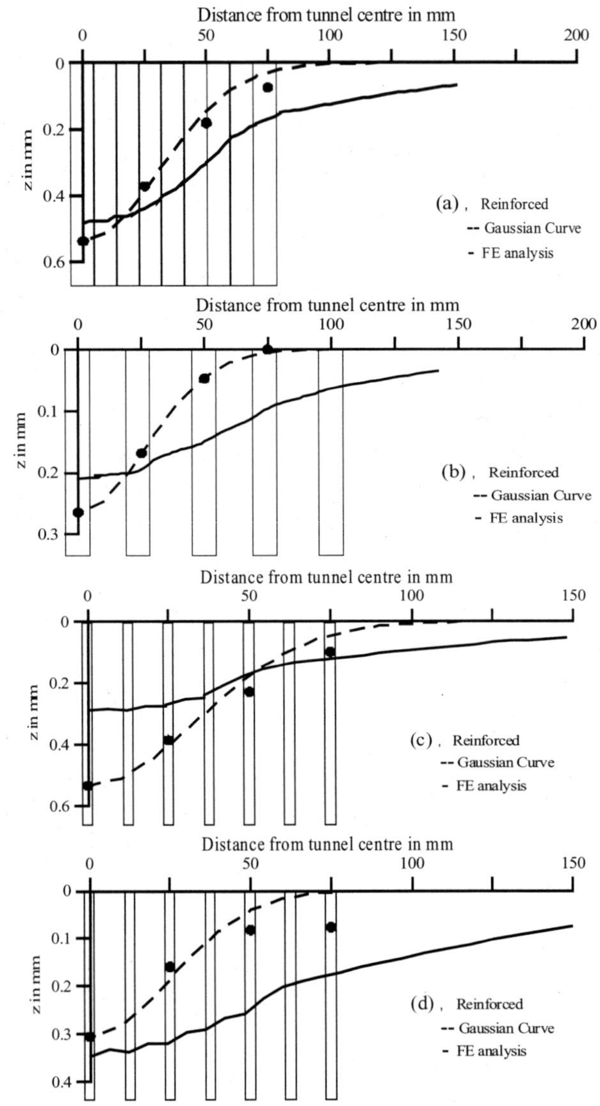

FIGURE 4: Observation of vertical settlement at the start of tunnel collapse. (a) Test: C1 Rods at: 1d c/c (b) Test: C5 Rods at: 2.5d c/c (c) Test C2 Rods at: 2d c/c (d) Test C4 Rods at: 2d c/c

The positioning of reinforcements is also shown in the figures. Furthermore, Fig. 4a indicates that the settlement profile was narrower when compared to other results. The use of 10mm diameter rods in test C1 was able to reduce the width of settlement trough. At 1-d spacing, the behavior of rods is likely to be similar to that of a diaphragm wall. The results of FE modeling as shown in the figures was observed to be much wider than the observed measurements. The Normal distribution curve was able to predict closely the greenfield settlement trough for the tests. The figures show that the results of two-dimensional finite element model are not able to estimate the settlement near and above tunnel crown accurately and tend to overestimate them further away. This is because the squeezing of soil into tunnel void cannot totally be simulated directly in 2D finite element (FE) mesh even with Lagrangian mesh update (Karakus 2007). Notwithstanding the above, the figure seems to indicate that the settlement trough obtained from Eq. 1 seems to predict the experimental results better than numerical analysis for volume loss corresponding to the start of the tunnel collapse. The presence of reinforcements reduced the maximum settlement of the trough but did not seem to reduce the total volume loss. Results of FE analysis support this. The FE analysis suggested that not only the width of the settlement profile but also the deformations at various depths below the surface were more than that at the surface. Table 4 shows the comparison of tunnel pressure at the start of collapse.

Figure 5 shows the effect of tunnel depth, z_0 on i calculated using Eq. 1 for surface settlement in tests using rods. The figure also shows the data o Peck (1969) and Lake et al. (1992). Relative density of the sand varied between 0.4 and 0.7. There are two points worth noting in Fig. 5.

First, both reinforced and greenfield results significantly underestimate Peck's (1969) trough width parameter in submerged sands. Although much of Lake et al.'s (1992) data are of deep tunnels, yet the results of greenfield tests at IITB all lie within the bounds suggested by Mair and Taylor (1997) and has an average $i = 0.28z_0$. Likewise, i for reinforced tests is further reduced. Although the findings are not fully conclusive because of insufficient data points, yet the results seem to suggest that the bars were able to partially restrict the flow along the circumference in the direction of tunnel collapse. Average i in reinforced test is equal to $0.25z_0$.

Figure 6 shows the variation of volume loss, V with decrease in tunnel support pressure, P_0 obtained from test results. In the figure, the support pressure is normalized by γz_0, where γ is the total

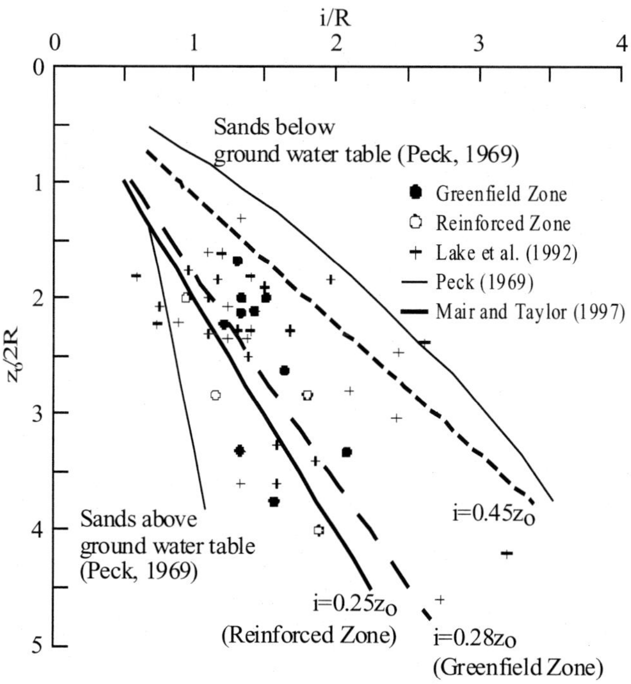

FIGURE 5: Effect of tunnel depth on I for surface settlement for greenfield zone

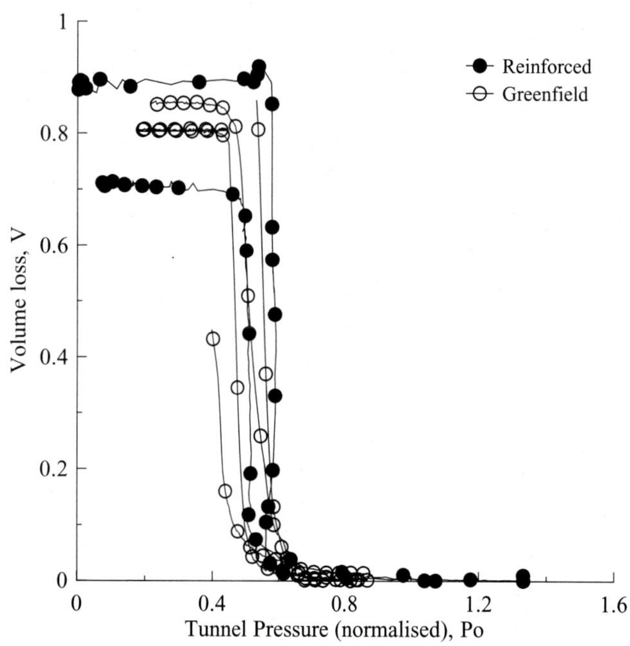

FIGURE 6: Variation of ground volume loss with loss in tunnel support pressure in the reinforced zone

unit weight of the soil and z_0 is the tunnel depth [= C + 0.5D]. The figure shows that P_0 of 0.4 to 0.6 times the overburden pressure was required to maintain tunnel stability. The figure also compares the reinforced to the greenfield results of an earlier investigation at IITB. As can be seen, the reinforcement bars seem to delay the volume loss at the early stage of tunnel movement. However, there is no difference between the Greenfield and reinforced test results at volume loss more than about 5% and till the ultimate state that is, when complete collapse of the tunnel occurs. The results of FE analysis can be summarized in Table 4.

Table 4: Comparison of Collapse Pressures

Test No.	*Experimental Results (P_l) [kN/m²]*	*Numerical Results (P_{le}) [kN/m²]*
C1	64.5	130
C2	108	145
C3	119	125
C4	117	185
C5	98	108
C6	120	130

Table 4 shows the comparison of the collapse pressures obtained from the tests (P_l) with the results of the numerical analysis (P_{le}). The number of iterations and tolerance values were kept close to 140. Knowing the value of P_l it was possible to calculate the number of iterations required. It was assumed that in numerical modeling, at the start of the tunnel collapse, total plastic failure of soil occurs, the mesh fails and the program stops to run.

5. CONCLUSIONS

The discussion above presented data relating to settlement observed due to tunnel deformation in shallow sand beds when tested at 50g. Vertical rods were pushed from the surface and near to the tunnel crown before the start of the tests. The data seem to suggest that rod spacing-to-diameter ratio of up to 2 helps reduce the volume loss measured at the surface. These observations were corroborated using data on support pressure measured at the instant when large deformations were observed. The data however was inconclusive for high volume loss measured close to the total tunnel collapse.

Numerical simulations were conducted with FE method to analyze the set of tests done. The deformations were predicted using phased excavation scheme. It was observed that the intermediate phases have a great influence on tunnel collapse pattern. The deformations created by volume of soil squeezing into the tunnel cavity cannot be accurately modeled using plane strain two dimensional finite element analysis. The settlements profile obtained are observed to be shallower and wider than the actual data. This has been suggested by authors like Karakus and Fowell (2003). Hence, use of more advance soil models and also 3 dimensional modelling is suggested.

ACKNOWLEDGMENTS

The data presented in this paper has been collected by Roshan N.S, Amarnath Hegde and Subhra Dutta during their graduate research programme at IIT Bombay.

REFERENCES

Kamata, H. and Mashimo. (2003) "Centrifuge model Test for face reinforcement by bolting" Tunnelling and Underground Space Technology, 18, pp 205-212.

Chandrashekharan, V.S. (2001) "Numerical and centrifuge modeling in soil structure interaction". Indian Geotechnical Journal, 31, Issue 1, pp 1-59.

Venis, C. G. (1963) "A simple laboratory technique for the making of rubber objects". J. Sci. Instrum., p. 331.

Chen et. al. (1999) "Pile responses caused by tunnelling". Journal of Geotechnical and Geoenvironmental Engineering, March 1999, pp 207-215.

Cabrera et. al. (2006) "Inclined load capacity of buried mass concrete thrust blocks", Proc. Int. Conf. On Physical modeling in Geotechnics, Taylor and Francis, London, pp. 777-782.

Karakus, M. (2007) "Appraising the methods accounting for 3D tunnelling effects in 2D plane strain FE analysis". Tunnelling and Underground Space Technology, 22, pp 47-56.

Peck, R.B. (1969) "Deep excavations and tunneling in soft ground", Proceedings of the 7th International Conference on Soil Mechanics and Foundation Engineering State of the art volume, Mexico City, pp 225-290.

Lake et. al. (1992) "Prediction and effects of ground movements caused by tunnelling in soft ground beneath urban areas", Construction Industry Research and Information Association, CIRIA Project Report 30, London.

Mair, R.J. & Taylor, R.N. (1997) "Theme lecture: Bored tunnelling in the urban environment", Proceedings of the 14th International Conference on Soil Mechanics & Foundation Engineering (Publication committee of XIV ICSMFE eds.), Hamburg, Taylor and Francis, pp 2353-2385.

Karakus, M. & Fowell, R. J. (2003) "Effects of different tunnel face advance excavation on the settlement by FEM". Tunnelling and Underground Space Technology, 18, pp 513-523.

Celestino et al. (2000) "Errors in ground distortions due to settlement tough adjustment". Tunnelling and underground Space Technology, 15, Issue 1, pp 97-100.

Generalized Scaling Law for Flat Layered Sand Deposit

T. Tobita, S. Escoffier[1], J.L. Chazelas[1] and S. Iai

Disaster Prevention Research Institute, Kyoto University, Japan
[1] *IFSTTAR, Nantes, France*
E-mail: tobita.tetsuo.8e@kyoto-u.ac.jp

ABSTRACT

To verify the generalized scaling law, dynamic centrifuge tests under two different centrifugal accelerations of 25 g and 50 g are conducted. The model ground constitutes of a flat dry sand layer. With the scaling law, a prototype ground is scaled down to 1/100. A sinusoidal input acceleration of frequency 1.0 Hz, maximum amplitude 0.5 g, and duration 14 sec in prototype scale is applied to the model ground. Each model is exposed to the identical input motion sequentially 10 times. In total nine accelerometers are installed in the model. Surface settlements are measured by laser displacement transducers. Settlements at three different depths (300, 200 and 50 mm – model scale - from the surface) are measured by settlement gauges. Measured settlements after the initial shake in prototype scale show significant agreements between the two models when the intensity of shaking is nearly identical.

1. INTRODUCTION

In the centrifuge model testing, although structural models have to be small and simplified, the prototype behaviour is approximated in accordance with scaling laws (e.g., Garnier et al., 2007), and it qualitatively represents prototype behaviour. One of the major obstacles for application of physical modelling results to performance-based design practice is that a specific prototype cannot be tested due to restrictions associated with experimental conditions, such as the size of the model container and scaling effects on materials. For 1-g model testing, to overcome these restrictions, the size of the experimental facility has become larger and larger so that real-scale models can be tested [e.g., E-defense (Tokimatsu et al., 2007), NEES@UC San Diego Large High Performance Outdoor Shake Table (Einde et al., 2004)]. However, for geotechnical structures, development of larger research facility may still have limitations because, even with such a large facility, physical modelling with foundations and surrounding ground has to be reduced due to factors inherent in a large facility, such as the capacity of the shake table and budget.

Demands for the testing of large prototypes are increasing under the restrictions mentioned above. To resolve such demands and restrictions, Iai et al. (2005) generalized the scaling law by combining the scaling law for centrifuge testing with the one for 1-g dynamic-model testing. They call it the "generalized scaling law" in dynamic centrifuge modelling. Tobita et al. (2011) investigated its applicability with a flat saturated sand bed. They conducted a series of centrifuge model tests to verify and find issues on the generalized scaling law under the scheme of the modelling of models technique. In a series of dynamic tests, four different centrifugal accelerations from 5 g to 70 g are applied to the scaled models for which the prototype is uniquely given. With the scaling law, the prototype is scaled down to 1/100. The models are exposed to sinusoidal input accelerations with 0.65 Hz and amplitudes of 2.1 m/s^2 and 3.1 m/s^2

in prototype scale. For response during shaking, nearly identical accelerations and excess pore-water pressure buildups are recorded for all the cases in the prototype scale. Discrepancies are found on surface settlements and duration time for dissipation of excess pore-water pressure. The major cause of the discrepancy of the latter "duration time" may be associated with low confining stress in model ground under low centrifugal acceleration. The cause of discrepancy of surface settlement has been yet to be identified.

In this study, the applicability of the generalized scaling law, in particular, the scaling law for displacement is investigated.

1.1 Brief Review of the Generalized Scaling Law

Scaling factors for physical model tests can be introduced in general forms by choosing a set of basic physical properties to be independent and deriving the scaling factors for other properties via governing equations of the analysed system. In the concept of the generalized scaling law, a model on a shaking table in a geotechnical centrifuge is considered to be a small scale representation of a 1-g shaking-table test. Figure 1 visualizes this concept by introducing a virtual 1-g model to which the prototype is scaled down via a similitude for 1-g shaking-table tests (Iai, 1989). The virtual 1-g model is subsequently scaled down by applying a similitude for centrifuge tests to the actual physical model. In this way, the geometric scaling factors applied in 1-g tests (μ) [row (1) of Table 1] can be multiplied with those for centrifuge tests (η) [row (2) of Table 1], resulting in much larger overall scaling factors $\lambda=\mu\eta$ [row (3) of Table 1].

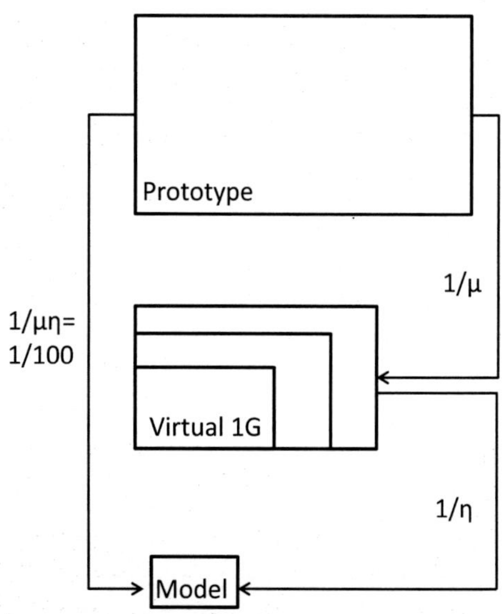

FIGURE 1: Relationship among prototype, virtual 1G model and centrifuge model for the case of scaling factor of 1/100 ($\lambda=\mu\eta=100$)

1.2. Issues Pointed Out by the Past Studies (Tobita et al., 2011)

Tobita et al. (2011) applied the generalized scaling law to the cases of a flat saturated sand layer. In a rigid sand box filled with saturated sand (Dr=30 to 40%), accelerometers, pore water and earth pressure transducers, thermometers, and laser displacement transducers were implemented as shown in Fig. 2. To satisfy the scaling law for the duration time of the pore water pressure dissipation, the methylcellulose (Metolose) solution was used for saturation of the model ground by adjusting its viscosity so that required viscosity was attained at a certain centrifugal accelerations (Fig. 3). In what follows, some of

Table 1: Scaling factors in physical model testing (Iai, 1989, and Iai, et al. 2005)

	(1) Scaling factors for 1g test	(2) Scaling factors for centrifuge test	(3) Generalized scaling factors
Length	μ	η	$\mu\eta$
Density	1	1	1
Time	$\mu^{0.75}$	η	$\mu^{0.75}\eta$
Frequency	$\mu^{-0.75}$	$1/\eta$	$\mu^{-0.75}/\eta$
Acceleration	1	$1/\eta$	$1/\eta$
Velocity	$\mu^{0.75}$	1	$\mu^{0.75}$
Displacement	$\mu^{1.5}$	η	$\mu^{1.5}\eta$
Stress	μ	1	μ
Strain	$\mu^{0.5}$	1	$\mu^{0.5}$
Stiffness	$\mu^{0.5}$	1	$\mu^{0.5}$
Permeability	$\mu^{0.75}$	η	$\mu^{0.75}\eta$
Pore pressure	μ	1	μ
Fluid Pressure	μ	1	μ

the significant results and issues pointed out by Tobita et al. (2011) are presented.

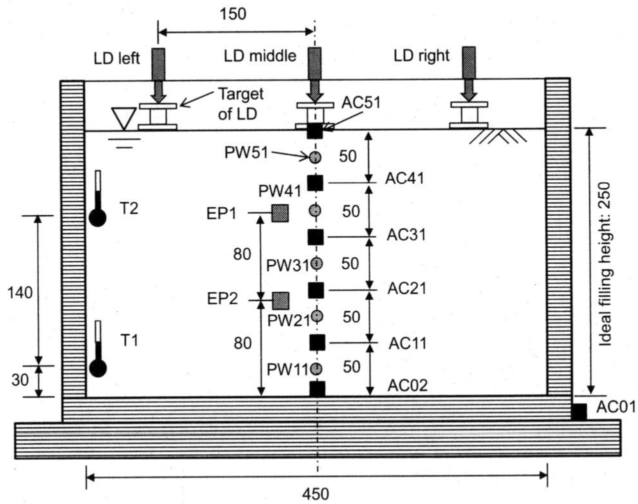

FIGURE 2: Test setup. LD: Laser displacement transducers. AC: Accelerometers. PW: Pore-water pressure transducers. EP: Earth pressure transducers. T: Thermometers [Tobita et al (2011)]

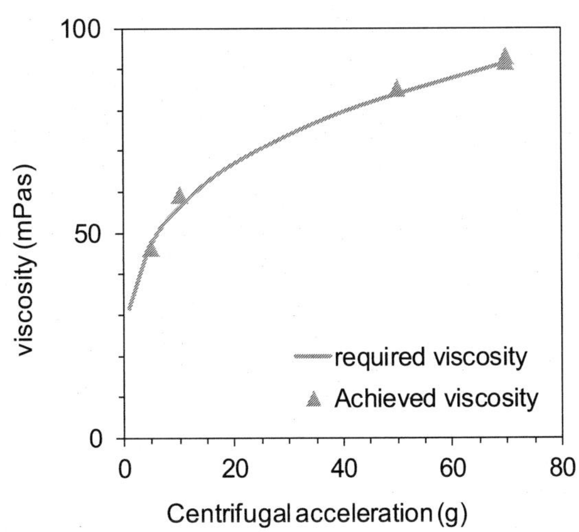

FIGURE 3: Achieved and required fluid viscosity [Tobita et al (2011)]

Figure 4 compares acceleration response of the model ground in prototype scale. Applied centrifugal acceleration is 5 g and 70 g. Nearly identical input acceleration (AC02) was given at the base of the sand box. As shaking continues, the amplitude of acceleration in ground and at the surface are reducing due to liquefaction. In both cases, the shape and peak values of the acceleration time histories are similar, validating the applicability of the generalized scaling law for acceleration.

Time histories of excess pore water pressure build-up for the applied centrifugal acceleration of 5, 10, 50, and 70 g, are plotted in prototype scale in Fig. 5. These figures also validate the applicability of the scaling for pressures.

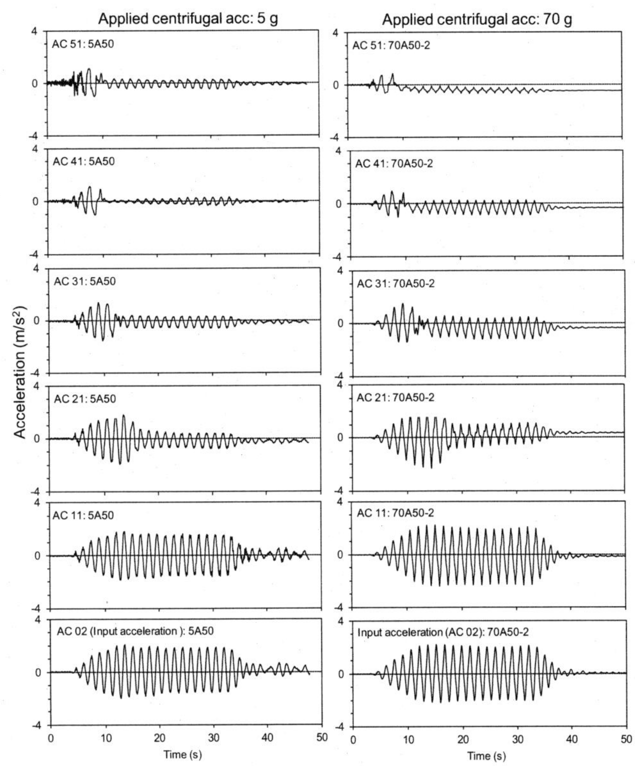

FIGURE 4: Measured time histories of acceleration in prototype scale: Cases 5g and 70 g [Tobita et al (2011)]

What have not confirmed in this series of tests are the scaling low for (1) the duration time of excess pore water pressure dissipation (time for consolidation) and (2) the ground surface settlements (displacement). For (1), the duration time of dissipation took long in the case of the low centrifugal accelerations (less than 10 g). This fact may suggest the limitation of the applicability of the scaling law. It is known from the theory of 1D consolidation that small elastic stiffness leads to a long duration time for consolidation. Thus in the paper, influence of the scaling of stiffness on the time of the pore water dissipation was investigated. Another cause of this is thought to be associated with the chemical properties of viscous water. After conducting falling head permeability tests with the viscous solution, we found (Fig. 6) its permeability decreases with duration time of flow, i.e., volume of fluid passing through the specimen. We suspect that long chains of polymer constituting methylcellulose ether are caught in the pores or absorbed on sand particles.

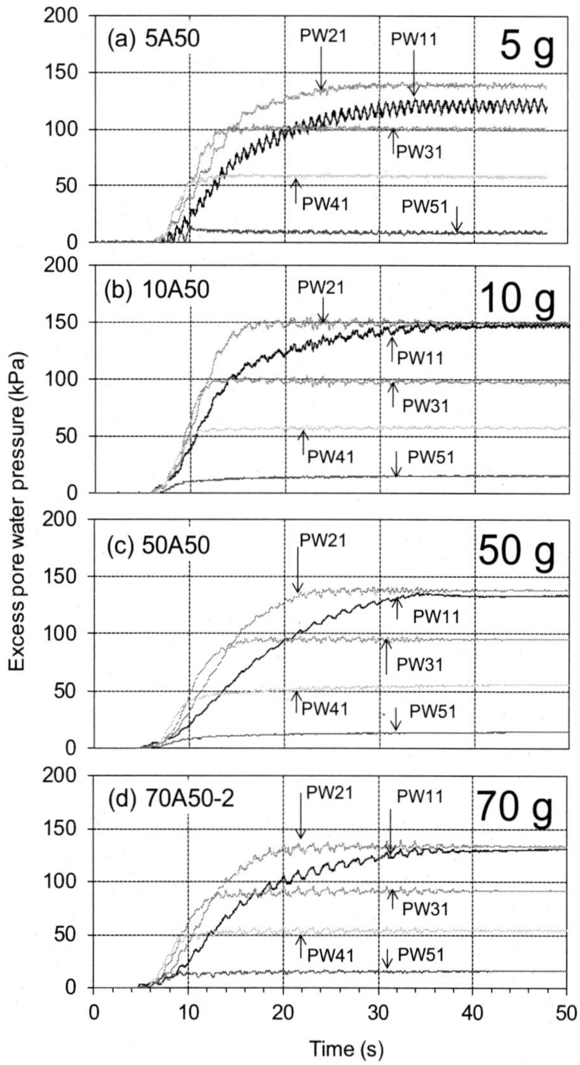

FIGURE 5: Measured time histories of excess pore-water pressure in the phase of pressure buildup in prototype scale (0–50 s) [Tobita et al (2011)]

For (2), issues are found that the generalized scaling law requires precise measurement of displacement because as shown in Table 1, the scaling factor of "displacement" is expressed as $\mu^{1.5}\eta$, which becomes 447 and 120 for given centrifugal accelerations of 5 g and 70 g, respectively. Compared with other physical parameters, these scaling factors of displacement are quite large. In the test shown in Fig. 2, surface displacements were measured by the laser displacement transducers with targets placed on the ground surface. With this test setup, surface settlements (displacements) could not properly be measured, because the dissipating water due to liquefaction made the target position unstable. Therefore, the validity of the generalized scaling law for displacement was left out of the conclusions.

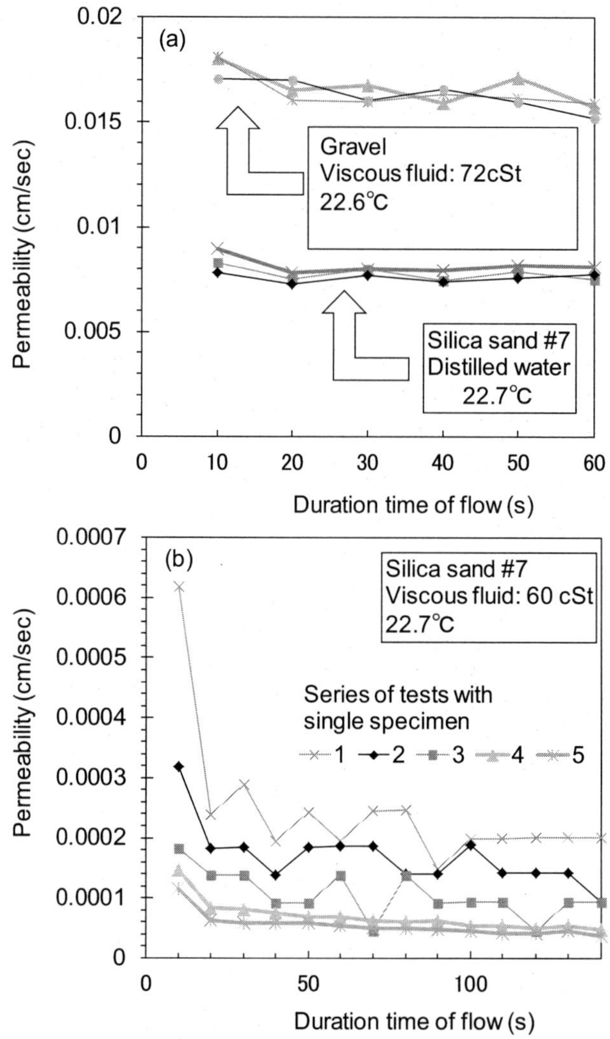

FIGURE 6: Variation of the permeability with duration time of flow: (a) Silica sand (Dr = 40%) and distilled water, and Gravel (Dr = 40%) and viscous water, (b) Silica sand (Dr = 40%) and visocus water

2. DYNAMIC CENTRIFUGE TESTS ON FLAT, LOOSE, DRY SAND DEPOSIT

To investigate the applicability of the generalized scaling law described above, a series of dynamic tests was conducted following the principle of "modelling of models." This technique was introduced by Schofield (1980) to assess the behaviour of a prototype through repetition of the test at different scales and comparison of the results in prototype scale. In the present study, without changing the actual size of the physical model but varying the virtual 1-g dimension, the overall scaling factor ($\lambda=\mu\eta=100$) is kept constant (Fig. 1). Here, it is set to a fixed value comprising different combinations of the scaling factors for 1-g model testing, μ, and centrifuge testing, η. Table 2 lists the applied geometric scaling factors as well as frequencies and amplitudes of the input motions employed in the study. As shown in Table 2, the scaling factors of displacement are relatively larger than the other physical quantities (200 in 25 g and 141.42 in 50g). As mentioned earlier, this fact demands precise measurements in displacement. In total 5 tests [3 tests in 25 g (25 g_1, 25g_2, and 25g_3) and 2 tests in 50 g (50g_1 and 50g_2)] are conducted. In what follows, units are in prototype unless otherwise specified.

Table 2: Test Cases and Scaling Factors Used in the Present Study

	Case 1	25 G		Case 2	50 G	
Quantity	scaling factor 1g test	scaling factor centrifuge test	generalized scaling factors	scaling factor 1g test	scaling factor centrifuge test	generalized scaling factors
Length	4.00	25.00	100.00	2.00	50.00	100.00
Density	1.00	1.00	1.00	1.00	1.00	1.00
Time	2.83	25.00	70.71	1.68	50.00	84.09
Frequency	0.35	0.04	0.01	0.59	0.02	0.01
Acceleration	1.00	0.04	0.04	1.00	0.02	0.02
Velocity	2.83	1.00	2.83	1.68	1.00	1.68
Displacement	8.00	25.00	200.00	2.83	50.00	141.42
Stress	4.00	1.00	4.00	2.00	1.00	2.00
Strain	2.00	1.00	2.00	1.41	1.00	1.41
Stiffness	2.00	1.00	2.00	1.41	1.00	1.41
Permeability	2.83	25.00	70.71	1.68	50.00	84.09
Pore pressure	4.00	1.00	4.00	2.00	1.00	2.00
Fluid Pressure	4.00	1.00	4.00	2.00	1.00	2.00

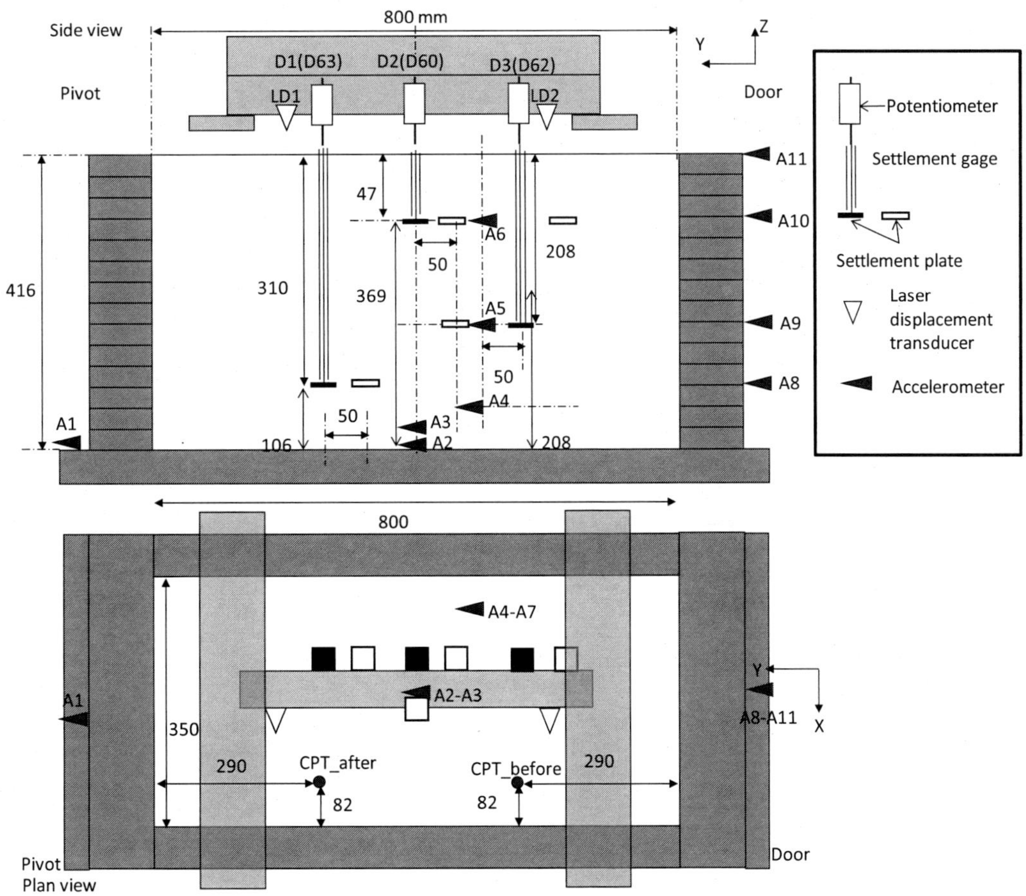

FIGURE 7: Schematic view and sensor location of the model

2.1 Test Setup

A series of dynamic tests under two different centrifugal accelerations of 25 g and 50 g are conducted with the geotechnical centrifuge (arm length=5.0 m) at the IFSTTAR (Institut français des sciences et technologies des transports, de l'aménagement et des réseaux), Nantes, France. The model ground constitutes of a flat dry sand layer, which is constructed with air-pluviation method (pluviation height=0.6 m, slot width=4 mm) to form the relative density of 50% of the Fontainebleau sand NE34 (e_{min}=0.545, e_{max}=0.866). With the scaling law, a prototype ground is scaled down to 1/100. Thus, the depth of the ground in prototype scale is 41.6 m. The flexible ESB (equivalent shear beam) box whose inside dimension is 800 (W) x 416 (H) x 340 (D) (mm) in model scale is employed (Fig. 7). A sinusoidal input acceleration of frequency 1.0 Hz, maximum amplitude 0.5 g, and duration 14 sec is applied to the model ground. Each model is exposed to the identical input motion sequentially 10 times in order to increase the number of measurements.

In total nine accelerometers are installed in the model (Fig. 7). Surface settlements are measured by laser displacement transducers. Settlements at three different depths (300, 200 and 50 mm – model scale - from the surface) are measured by settlement gauges, which are made of a plate, and a rod connected to potentiometers (Fig. 8, 9).

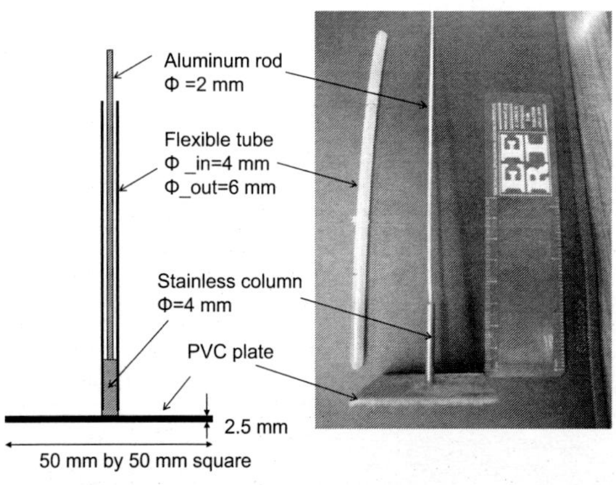

FIGURE 8: Detail of a settlement gauge

Settlement gauges are carefully placed at the specified depth (Fig. 9) with fishing strings. The PVC plates without attaching the potentiometers are installed for comparison purposes. Three potentiometers are mounted after completing model ground.

(a) Installation of settlement gauge

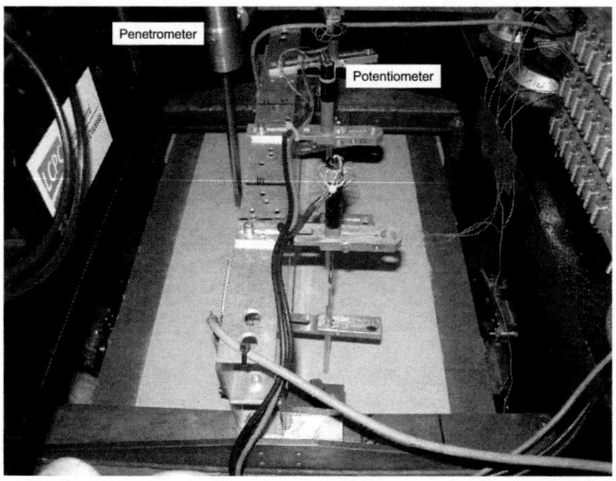

(b) Setup of penetrometer and potentiometers

(c) Settlement gauges and PVC plates after the test

FIGURE 9: Model Setup

3. RESPONSE OF THE MODEL GROUND

3.1 Input and Ground Acceleration

As shown in Fig. 10, nearly identical input accelerations are given to the model ground. Figure 11 summarizes intensity of input motion in the form of Arias intensity (Eq. 1)(Arias 1970) for all the cases employed in the present study.

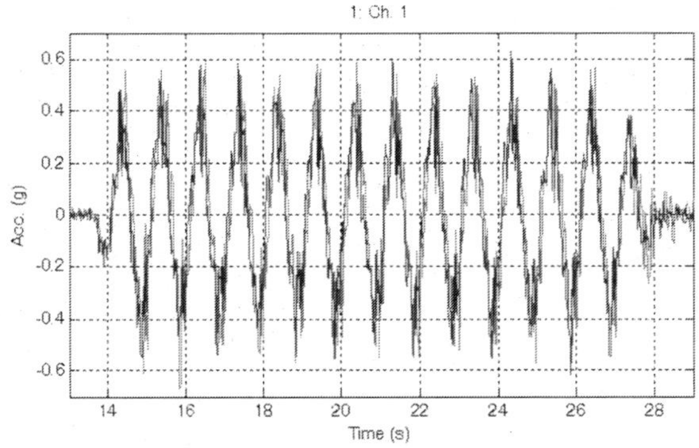

FIGURE 10: Time histories of input acceleration in prototype scale (red 25g_1, blue 50g_1)

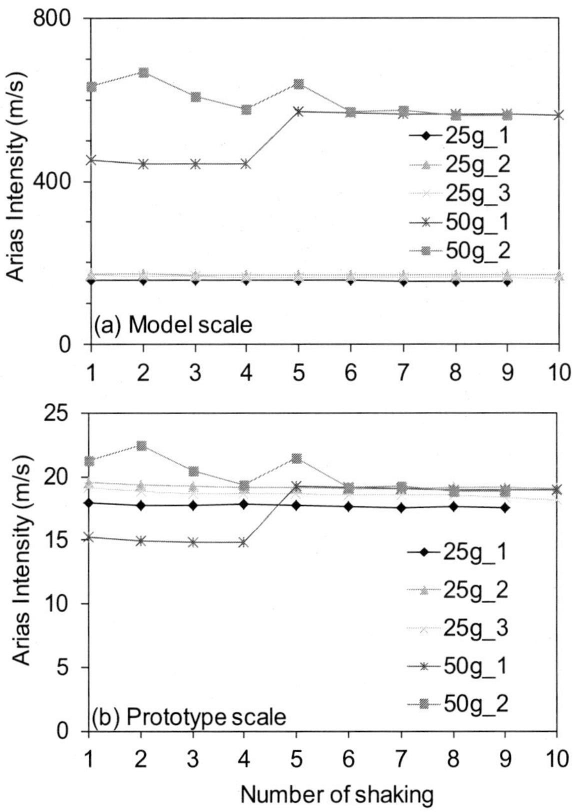

FIGURE 11: Arias intensity of the recorded input acceleration in prototype scale

$$I_A = \frac{\pi}{2g}\int_0^T a(t)^2\,dt \quad \text{(m/s)} \qquad (1)$$

where I_A: Arias intensity (m/s), $a(t)$: acceleration, and T: duration of shaking (16 sec in this study). As shown in Fig. 11, at all the shaking, the intensity is almost identical, except for the first 4 cases in 50 g tests. This variation may be due to the instability of shake table controller. As explained later, this variation might cause significant difference on ground settlements in the case of 50g_1.

3.2 Penetration Resistance

Before the initial shaking and after the 10th shaking, the soil penetration resistance of the model ground was measured by the miniature penetrometer. As shown in Fig. 12(a), the penetration resistance in depth under 50 g in model scale is, as it is expected, larger than that of 25 g. While the profiles of the resistance in prototype scale [Fig. 12(b)] show systematic variation between 25 g and 50 g, i.e., the resistance measured under 50 g is systematically smaller than the ones obtained under 25 g. Cause of this is unknown yet, and further investigation is necessary to know the limit of application of the generalized scaling law.

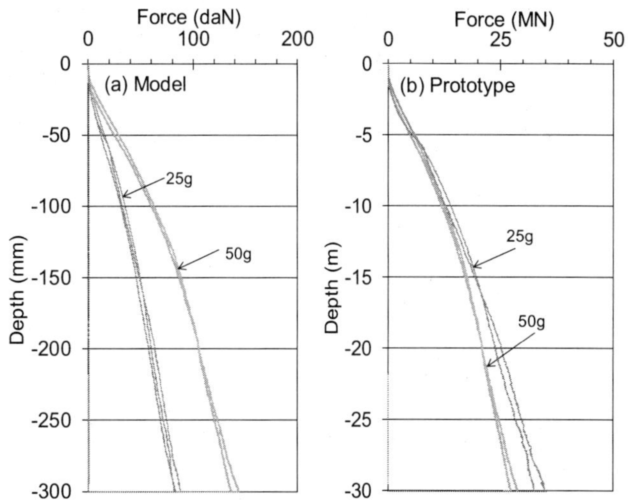

FIGURE 12: Penetrometer resistance after the 10th shaking in model scale (a) and prototype scale (b)

The same trend can be observed in the profile after the 10th shaking (Fig. 13). Degree of coincidence of the profile becomes lower especially in deeper depth. Small difference on shaking intensity in each shot can be accumulated to make such a large difference on the penetration resistance.

Although there are some variations, profiles of the resistance in prototype scale show more or less coincidence in prototype scale.

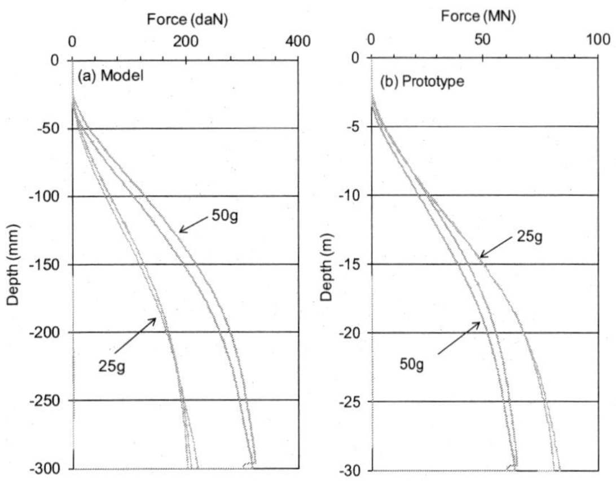

FIGURE 13: Penetrometer resistance before the initial shaking in model scale (a) and prototype scale (b).

3.3 Ground Settlements

Settlements at the ground surface are measured by laser displacement transducers, and those in the ground are by settlement gauges. Before shaking, the ground is consolidated by applying the specified centrifugal accelerations consecutively 3 times to stabilize the ground.

This process is manifested in Fig. 14 (model scale). For example, as shown in Fig. 14(a), as centrifugal acceleration increases, the ground surface settles about 1.1 mm (D122) to 1.3 mm (D116). After the centrifugal acceleration of 25 g being kept about 5 min, centrifuge rotation is stopped (1 g). At this moment, the ground settlements are reduced to about 1 mm. Then again the centrifugal acceleration is applied, and as shown in Fig. 14(a), the surface settlements further increases up to about 1.5 to 1.7 mm. This process is repeated 3 times. Common trend shown in Fig. 14 is that the ground settles as centrifugal acceleration increases and in every application of centrifugal acceleration, the amount of ground settlements gradually increases. However, the rate of settlement is decreasing, i.e., the amount of settlement asymptotically approaches to some maximum values.

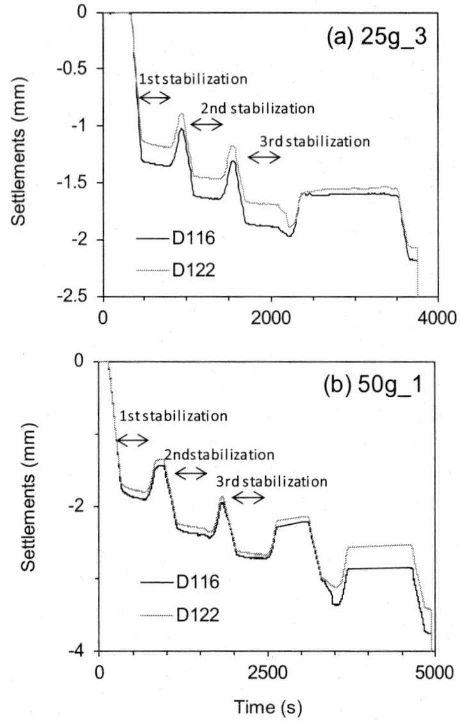

FIGURE 14: Time histories of ground settlements due to consolidation before shaking (model scale): (a) 25g_1 and (b) 50g_1

Figure 15 shows time histories of the ground settlements at all shaking stapes. The curves indicate stepwise increase of settlements at each shaking. In Fig. 15(b), the curve of D63 is increasing, i.e. settlement gauge is uplifted. This may be because of malfunctioning of the sensor or sands near the PVC plate were flowing under the plate.

Amounts of settlements after each shaking are summarized in Fig. 16. The amounts of settlement shown in this figure simply correspond to the height of steps found in Fig. 15. If the generalized scaling law works correctly, settlements of the same sensors should be identical. Results show [see Fig. 16(a) for surface settlements], for example, that settlement after the 1st shaking is about 1,000 mm (2.4% of the total depth of 41.6 m) and values recorded by each sensor (D122 and D116) match well. However, as number of shaking increases, the difference of settlements measured under 25 g and 50 g seems to be increasing.

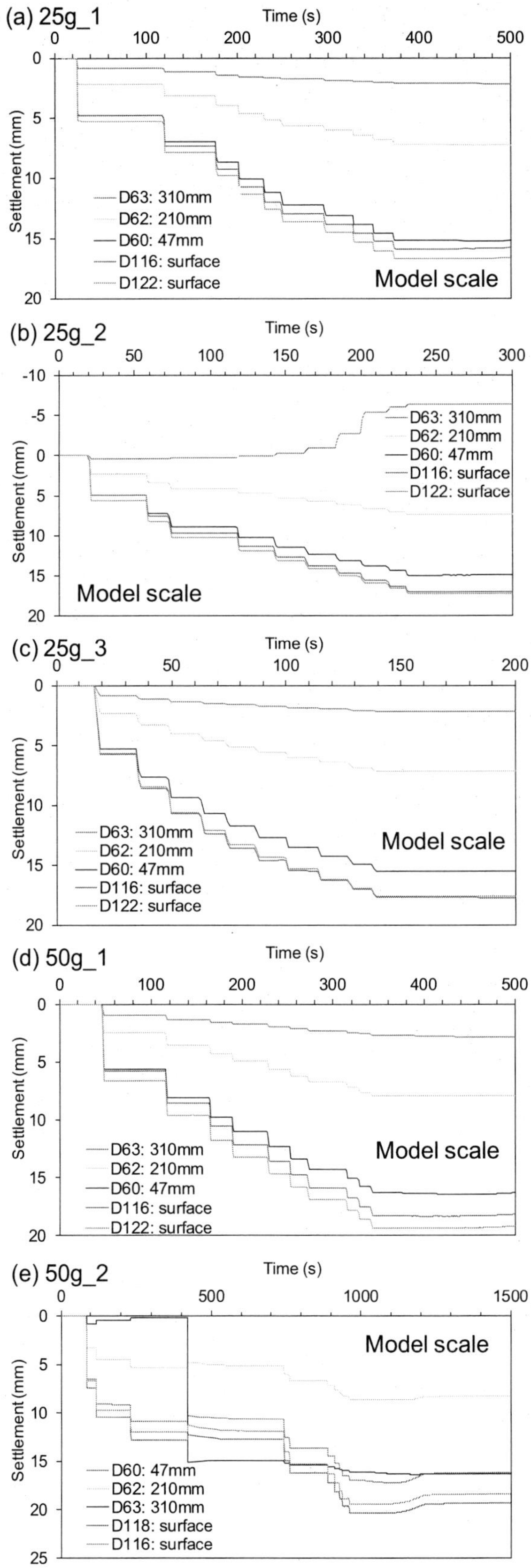

FIGURE 15: Time histories of ground settlements due to shaking (model scale)

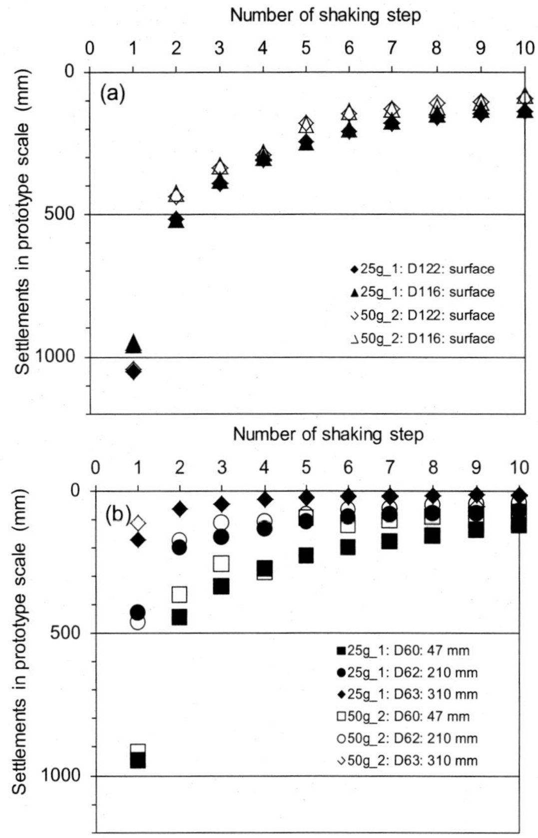

FIGURE 16: Settlements after each shaking in prototype scale: (a) surface, (b) in depth

Since the amount of settlements after the initial shaking matched well compared with those in subsequent shaking, in what follows, settlements after the 1st shaking are investigated in terms of validation of the scaling law. Figure 17 compares settlements of all the sensors recorded after the 1st shaking for all cases. As mentioned earlier that due to the variation of input intensity, case 50g_1 had lower and case 50g_2 had slightly larger intensity of shaking. This trend is found in the amount settlement shown in Fig. 17. Considering that the intensity of case 50g_2 is close to the ones in 25 g [Fig. 11(b)], the amount of settlement measured in the case of 50g_2 seems to be matching quite well with the cases of 25 g. The scaling factor of settlement (displacement) is as large as 200 for 25 g and 141 for 50 g, and considering possibility to have minor variation in constructing the model ground, in sensor setups, and in the input accelerations, these differences may be regarded as minor. Thus, the scaling law for settlement (displacement) may well be validated.

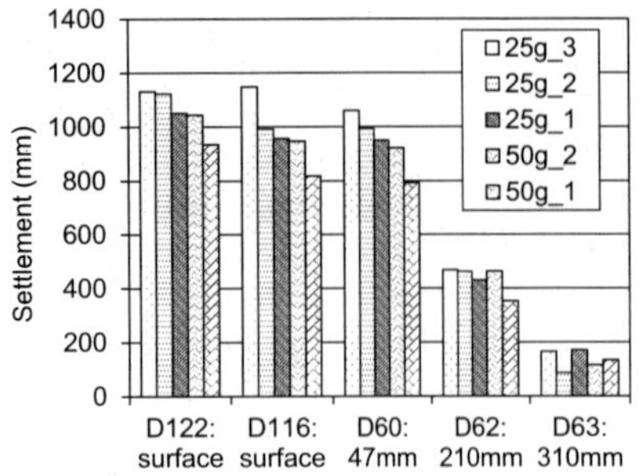

FIGURE 17: Summary of settlements measured after the 1st shaking in all the test cases in prototype scale

4 CONCLUSIONS

To examine the applicability of the generalized scaling law, a series of dynamic tests under two different centrifugal accelerations of 25 g and 50 g were conducted under the scheme of "modelling of models." In the ESB box, the model ground consisted of a flat dry sand layer with the relative density of 50% of the Fontainebleau sand NE34. A prototype ground and input accelerations were scaled down to 1/100 according with the scaling law. A sinusoidal input acceleration of frequency 1.0 Hz, maximum amplitude 0.5 g, and duration 14 sec in prototype scale was applied to the model ground. Each model was exposed to the identical input motion sequentially 10 times.

In terms of the Arias Intensity, the input accelerations given at the beginning of the cases of 50 g were slightly larger than that of the cases of 25 g. Although effect of this is manifested in the amount of settlements, measured settlements about 1 m (2.4% of the total depth of the model ground) after the initial shake in prototype scale showed significant agreements between the two models when the intensity of shaking was nearly identical. Thus, the applicability of the generalized scaling law for displacement is validated under 25 g and 50 g.

The soil resistance measured by the miniature penetrometer in prototype scale showed systematic variation between 25 g and 50 g, i.e., the soil resistance measured under 50 g is systematically smaller than the one obtained under 25 g. Cause of this is unknown yet, and further investigation is necessary to know the application limit of the generalized scaling law.

ACKNOWLEDGEMENTS

The authors would like to thank laboratory technicians at IFSTTAR, Nantes, France: Mr. Philippe Audrain, Mr. Claude Favraud, Mr. Patric Gaudicheau, Mr. Damien Macé, Mr. Alain Neel, and Dr. Gerard Rault, for their sincere assistance for conducting the centrifuge experiments reported in the present paper.

REFERENCES

Arias, A. (1970) A Measure of Earthquake Intensity. R.J. Hansen, ed. "Seismic Design for Nuclear Power Plants". MIT Press, Cambridge, Massachusetts, pp 438-483.

Einde, V.D., Restrepo, L., Conte, J.P., Luco, E., Seible, F., Filiatrault, A., Clark, A., Johnson, A., Gram, M., Kusner, D., and Thoen, B. (2004) "Development of the George E. Brown Jr. network for earthquake engineering simulation (NEES) large high performance outdoor shake table at the University of California, San Diego". Proceedigns of the 13th World Conference on Earthquake Engineering, Vancouver, BC, Canada, 1-6 August, Paper No. 3281.

Garnier, J., Gaudin, C., Springman, S. M., Culligan, P. J., Goodings, D., Konig, D., Kutter, B., L., Phillips, R., Randolph, M. F., and Thorel, L. (2007) "Catalogue of scaling laws and similitude questions in geotechnical centrifuge modelling". International Journal of Physical Modelling in Geotechnics, 7, Issue 3, pp 1–23.

Iai, S. (1989) "Similitude for shaking table tests on soil-structure-fluid model in 1g gravitational field". Soils and Foundations, 29, Issue 1, pp 105–118.

Iai, S., Tobita, T., and Nakahara, T. (2005) "Generalized scaling relations for dynamic centrifuge tests". Géotechnique, 55, Issue 5, pp 355–362.

Schofield, A.N. (1980) "Cambridge geotechnical centrifuge operations". Géotechnique, 30, Issue 3, pp 227–268.

Tobita, T., Iai, S., von der Tann, L., and Yaoi, Y. (2011) "Application of the generalised scaling law to saturated ground". International Journal of Physical Modelling in Geotechnics, 11, Issue 4, pp 138-155.

Tokimatsu, K., Suzuki, H., Tabata, K., and Sato, M. (2007) "Three dimensional shaking table tests on soil-pile-structure models using E-Defense facility". 4th International Conference on Earthquake Engineering, June 25–28, Thessaloniki, Greece, pp 1-11.

Shaking Table Tests on Reinforced Soil Retaining Structures

A. Murali Krishna

Department of Civil Engineering, Indian Institute of Technology Guwahati, Guwahati
E-mail: amurali@iitg.ernet.in

ABSTRACT

Shaking tables are used for physical modeling in earthquake geotechnical engineering. Shaking table research provides valuable insight into liquefaction, post earthquake settlement, foundation response and soil-structure interaction and lateral earth pressure problems. This paper describes about the shaking table testing facility available at the Indian Institute of Science, Bangalore and its associated instrumentations. Details of the dynamic shaking tests conducted on wrap faced reinforced and rigid faced reinforced/unreinforced soil retaining wall models are presented. Typical results obtained from the shaking tests indicated that wall face deformations vary non-linearly with number of cycles of dynamic excitation and higher deformations occur at higher elevations. Relative density of the backfill material play important role in the seismic behavior. It is also presented that even a low strength reinforcement material is effective in receding the face deformations significantly under seismic conditions.

1. INTRODUCTION

Studies in earthquake geotechnical engineering can be carried out either through element tests or model tests in the laboratory. Model tests are very much essential when the prototype behaviour is very complicated to understand. The use of scaled models in geotechnical engineering offers the advantage of simulating complex systems under controlled conditions, and the opportunity to gain insight into the fundamental mechanisms operating in these systems. In model testing, the boundary conditions of a particular problem are reproduced in a small scale model of the prototype. Studies on physical model help in providing better understanding of the behaviour of the prototype and the process leading to failure of the model at a real time can be directly observed. Different model studies available in geotechnical engineering are: Tilt table test, Centrifuge test (high-g tests), Hydraulic gradient tests (Yan and Byrne 1989) and Shaking table tests (1-g tests).

In the early years of geotechnical earthquake engineering, virtually all physical model testing was performed on shaking tables. Shaking table tests involve testing physical models, under normal (1-g) gravitational field, subjected to seismic shaking using earthquake simulators. With this approach, data from laboratory scale models can be obtained and analysed to predict the behavior of geotechnical structures. Shaking table research has provided valuable insight into liquefaction, post earthquake settlement, foundation response and soil-structure interaction and lateral earth pressure problems (Kramer 1996). Wood et al. (2002) discussed a number of issues associated with the use of shaking tables for geotechnical earthquake engineering, concentrating in particular on: the application of scaling and modelling laws; the development of containers for testing geotechnical systems; and the control of seismic motions. Iai (1989) derived a similitude for the shaking table tests on saturated soil-structure-fluid model in 1g gravitational field. Severn et al. (2012) discussed about the contribution of shaking tables to earthquake engineering since 1900, referring to the shaking tables at major research institutions. Wood et al. (2002) highlighted the following three attractions associated with 1g model tests:

1. As with other laboratory modelling, the boundary conditions in shaking table tests are, in principle, well defined and well controlled. The physical modelling provides a source of reliable data for supporting numerical modelling and back analysis.
2. The size of the models can be quite large – limited only by available space and loading devices - so that the linear model scale from typical prototypes may be low. The degree of extrapolation required for the numerical modelling may be low - and other undesirable effects that may be associated with small models (for example particle size problems) may be somewhat avoided.
3. As the models are large, the space available for instrumentation and actuators will be greater and more subtlety in loading, control and observation will be feasible. The disturbance to the soil arising from the finite size of instruments will be correspondingly lower.

As the most dominating feature of earthquakes is the horizontal shaking, most of the shaking tables around world are of uni-axial horizontal shakers. However, shaking tables of multi degree freedom are also under function in some places. Shaking tables are usually driven by servo hydraulic actuators; their dynamic loading capacities are controlled by the capacity of the hydraulic pumps that serve the actuators. However, small capacity manual shaking tables are also reported (Prasad et al. 2004). For the models used in shaking tables, soil can be placed by different methods (wet or dry pluviation) to achieve uniform density or compacted and instrumented relatively easily. After preparing the complete instrumented model, the shake table is to be operated for required dynamic motion (pure sinusoidal, variable amplitude harmonic or random real earthquake motion) and behavior of the model can be observed and analysed through physical observations and instrumental recordings. Developed technologies in photography and videography could also be used for better visual observations and analysis.

Several seismic studies using shaking table are reported on various geotechnical structures and soil structure related problems namely: Kovacs et al. (1971) on clay banks; Koga and Matsuo (1990) on embankments resting on liquefiable sandy ground; Ramakrishnan et al. (1998) on geotextile wrap-faced segmental retaining walls; Orense et al. (2003) on wall type gravel drains as liquefaction countermeasure for underground structures; El Emam and Bathurst (2004) on reinforced soil walls; Toyota et al (2004) on flow dynamics in liquefied slope and Cubrinovski et al. (2006) on piles undergoing lateral spreading in liquefiable soils etc.

This paper describes about the shaking table testing facility available at the Indian Institute of Science, Bangalore and its associated instrumentations. Details of the dynamic shaking tests conducted on wrap faced reinforced and rigid faced reinforced/unreinforced soil retaining wall models are presented.

2. DESCRIPTION OF SHAKING TABLE FACILITY AT IISC BANGALORE

A computer controlled servo hydraulic shaking table facility of single degree freedom (horizontal) is available at Geotechnical Engineering laboratory, Indian Institute of Science (IISc) Bangalore. The facility is intended for simulating horizontal shaking action, associated with seismic and other vibration conditions on test specimens. The shaking table is of a loading platform of 1 m × 1 m size and payload capacity of 1000 kg. Shaking action is provided by a digitally-controlled servo-hydraulic actuator with 450 mm stroke and of 30 kN force rating. The shake table itself is mounted on a pair of linear-bearings to friction free movement at any given payload. A dedicated control room housing the control system includes a host computer to facilitate testing under both constant amplitude as well as pseudo-random conditions. The shaking table can be operated over an acceleration range of 0.05g to 2g and frequency range of 0.05 Hz to 50 Hz with a maximum amplitude of ± 200 mm.

Various instrumentations are procured and facilitated for the dynamic testing of model structures on shaking table. They include:

accelerometers, pressure sensors, ultrasonic non-contact displacement transducers and strain gauges. All the instrumentations are of analog voltage output type and will generate continuous data to the application software as the test progresses. Accelerometers are of analog voltage output type with a full-scale acceleration range of ±2g along both the x and y axes, with sensitivity of 0.001g. The sensing range of the ultrasonic displacement transducers is 30 mm to 300 mm and output response time of 30 ms.

FIGURE 1: Laminar box mounted on the shaking table at IISc

Two model containers are associated with the shaking table facility: a rigid container and a flexible laminar container. The rigid container was made of perspex material (acrylic plastic) of 8 mm thick sheeting and was rectangular in shape with inside dimensions of 500 mm × 1100 mm and 800 mm deep. The laminar container is a flexible boundary container that minimises the boundary effects on model, which are of major concern in laboratory model studies related to earthquake engineering. The Laminar box is a large sized shear box consisting of several horizontal layers, built such that the friction between the layers is minimum. The layers move relative to one another in accordance with the deformation of the soil inside. The laminar box is rectangular in cross section with inside dimensions of 500×1000 mm and 800 mm deep consisting fifteen rectangular hollow layers machined from solid aluminum compose. The gap between the successive layers is 2 mm and the bottommost layer is rigidly connected to the solid aluminum base of 800 mm × 1200 mm in section with 15 mm thickness. The layers are separated by linear roller bearings arranged to permit relative movement between the layers with minimum friction. Laminar box mounted on the shaking table is shown in the Figure 1.

3. SHAKING TABLE TESTS ON RETAINING STRUCTURES

A series of shaking table tests were conducted on wrap faced reinforced retaining structures and rigid faced unreinfroced and reinforced retaining structures models. Effects of the frequency, acceleration of the base sinusoidal motion, surcharge pressure and number of layers on the accelerations, horizontal face displacements and soil pressures in retaining walls models were investigated.

The details of the experiments: materials used, model preparation, instrumentation and testing procedure are discussed by Krishna and Latha (2007, 2009) Latha and Krishna (2006, 2008).

3.1 Backfill Material

Locally available dry sand was used as the backfill material. Figure 2 shows the grain size distribution of the sand. The sand is classified as poorly graded sand (SP) using the Unified Soil Classification System. The sand has a maximum dry density of 18 kN/m^3 in a vibration test and a minimum dry density of 14.3 kN/m^3. The specific gravity of the sand is 2.64 and the friction angle was measured as 43.50 in direct shear tests conducted on the sand at 63% relative density (RD).

3.2 Reinforcement

Two types of reinforcement materials were used in the present study: geotextile and geogrid. The geotextile was a polypropylene multifilament woven fabric. The ultimate tensile strength of the geotextile was determined as 55.16 kN/m (as per ASTM D 4595-05). The tensile strength of the geotextile material corresponding to 2% strain is

3.04 kN/m and the corresponding stiffness is 152 kN/m.

The geogrid was a biaxial geogrid (BX) made of polypropylene. Ultimate tensile strength of the geogrid was obtained from a standard multi-rib tension test (as per ASTM D 6637-01) as 26.3 kN/m. Tensile strength of the geogrid material corresponding to 2% strain is 4.43 kN/m and the corresponding stiffness is 221 kN/m.

3.3 Model Construction and Testing Procedure

Model retaining walls were constructed in each model container to a size of 750 mm × 500 mm in plan and 600 mm height. The models were constructed in lifts of equal height while reinforcing each lift with a layer of reinforcement material. For the wrap-faced walls (Fig. 2), each geotextile layer was wrapped at the facing for a length of 150 mm. For rigid-faced walls (Fig. 3), the facing was constructed using 12 hollow rectangular steel box sections of 50 mm height by 25 mm width each, bolted together to form a 600 mm high rigid panel of 25 mm thickness. The rigid-faced reinforced model walls were reinforced with four layers of reinforcement that ran through the bolts, which form a part of the facing system. The unreinforced soil walls were built similar to the rigid-faced walls but without reinforcement.

To achieve uniform density, sand was placed in the laminar box using the pluviation (raining) technique. The fall height to achieve the desired relative density was determined by performing a series of trials with different fall heights. However, the actual relative densities achieved in each test were measured by collecting samples in small cups of known volume placed at different locations and levels during the model retaining wall preparation. The model walls were built to the same 'medium dense' relative density. The average unit weight and relative density achieved were within the range of 16.35 – 16.50 kN/m3 and 62–65% respectively for the same height of fall.

The retaining wall models were constructed using wooden plank formwork for each lift. After the completion of all lifts up to full height of the wall (H, 600 mm), a nominal surcharge of 0.5 kPa (in the form of concrete slabs) was applied at the top. Then the supporting formwork was carefully removed lift-by-lift sequentially from bottom to top. Figures 2 and 3 show photos of finished model walls constructed in the perspex container with wrap and rigid facings respectively. The length (L) of geotextile reinforcement at the interface of sand layers was kept the same in all tests at 420 mm. This length (L) corresponds to a L/H ratio of 0.7 which is the minimum required for reinforced earth structures.

FIGURE 2: Wrap faced wall in the prespex container

FIGURE 3: Rigid faced wall in the prespex container

Typical model configurations along with the layout of the instrumentation are shown schematically in Figs. 4 and 5 for wrap-faced and rigid-faced model walls, respectively. In the schematic diagram, Accelerometers (A0, A1, A2 and A3), Pressure sensors (P1, P2, P3 and P4), Ultrasonic non-contact displacement transducers (U1, U2 and U3) installed at different elevations are also shown.

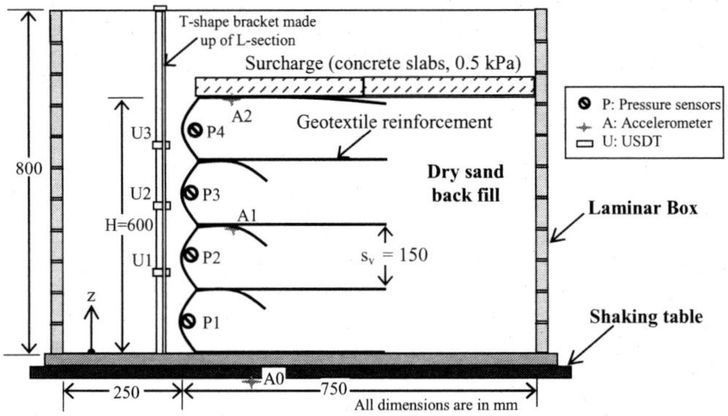

FIGURE 4: Schematic diagram of typical wrap-faced wall (4 layers)

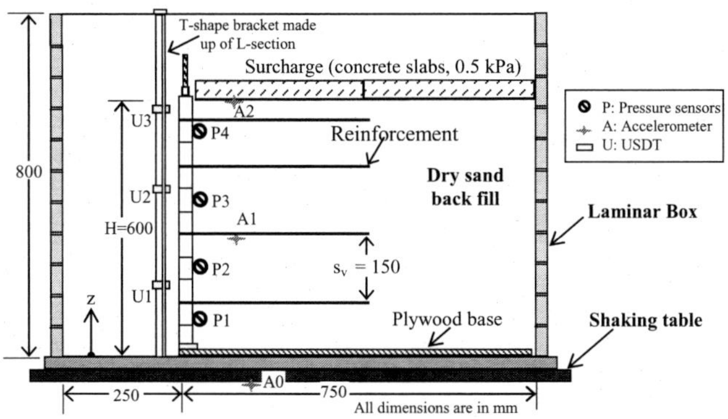

FIGURE 5: Schematic diagram of typical rigid-faced wall (4 layers)

4. RESULTS AND DISCUSSIONS

Typical results obtained from the dynamic tests (20 cycles at 0.2g acceleration at 3 Hz frequency) on rigid faced soil retaining walls are shown in Figures 6 and 7. Figure 6 shows horizontal displacement variation with number of cycles for the test on rigid faced unreinforced soil wall. The figure depicts a nonlinear increasing trend of displacements with increasing number of cycles and greater displacements at higher elevations. The maximum displacement recorded at the end of 20 cycles of shaking was about 11, 7.2 and 2.8 mm at elevations of 550, 350 and 150 mm, respectively.

Figure 7 shows the acceleration – time histories at various elevations of the rigid faced reinforced wall model with four layers of reinforcement. It can be seen that the accelerations were amplified at higher elevations.

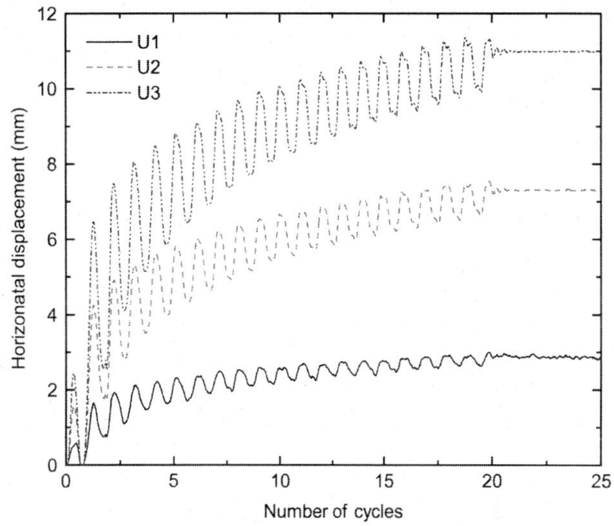

FIGURE 6: Typical displacement histories at different elevations for the test on unreinforced soil wall (0.2g acc. at 3Hz freq.)

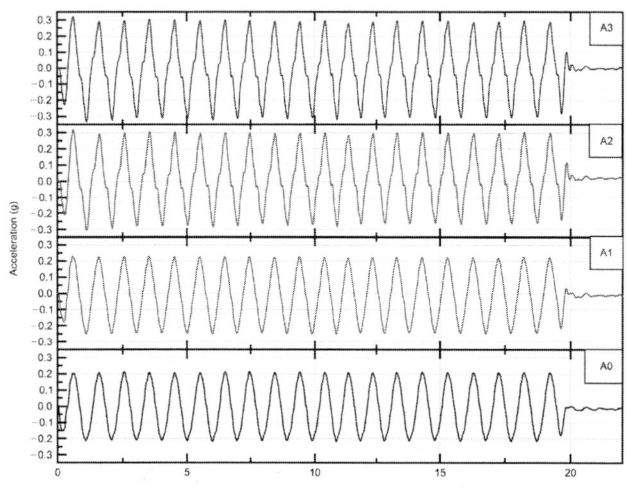

FIGURE 7: Typical acceleration histories at different elevations for the test on rigid faced reinforced (4 layered) soil wall (0.2g at 3Hz freq.)

4.1 Influence of Backfill Relative Density

Figure. 8 (a), (b) and (c) presents the displaced face profiles of the model walls for various tests that are subjected to 20 cycles of sinusoidal motion at 0.1 g acceleration at 1 Hz frequency (Latha and Krishna 2008). Each figure provides

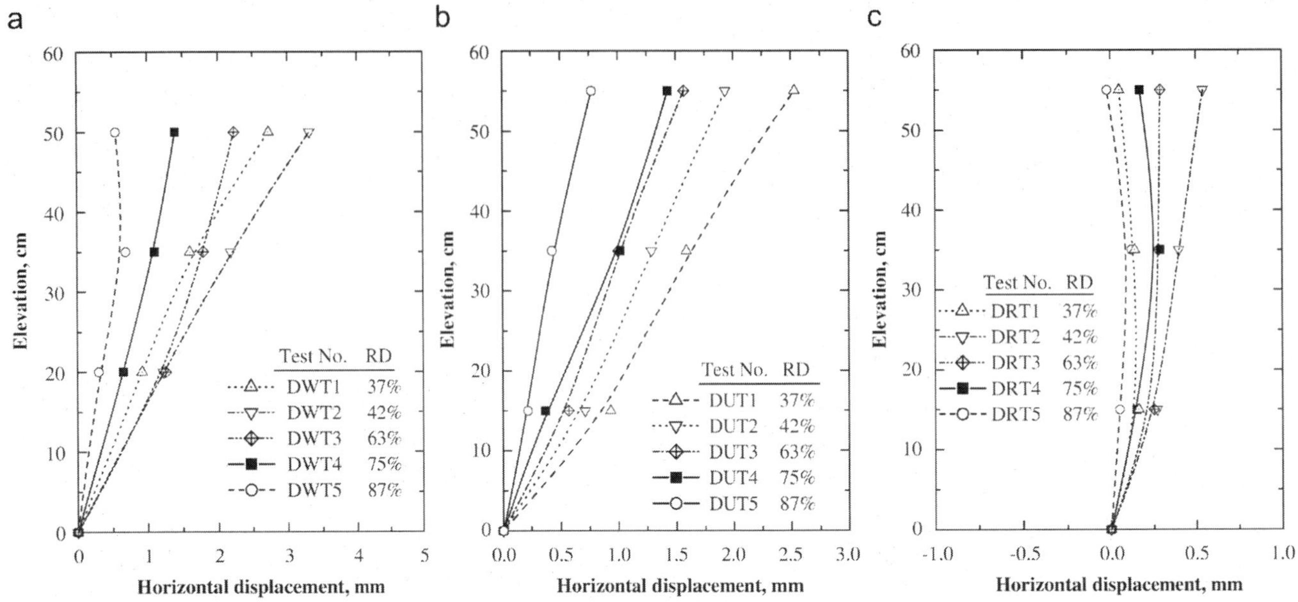

FIGURE 8: Influence of Relative density of backfill soil: Displacement profiles at the end of 20 cycles of dynamic motion (0.1g at 1 Hz): (a) wrap-faced walls, (b) rigid-faced unreinforced walls and (c) rigid-faced reinforced walls. (after Latha and Krishna 2008)

the comparison among displacement profiles for the model walls with same facing and reinforcement configuration but constructed to different backfill relative densities in the range of 37–87%. Fig. 8(a) shows the displacement profiles for the wrap-faced model walls and Fig. 8(b) shows the displacement profiles of the rigid-faced unreinforced model walls. Fig. 8(c) shows the displacement profiles for the rigid faced reinforced retaining walls. The displacements observed are very low in magnitude; in the range of 0–3.3 mm. As seen from these three figures, in general, the displacements reduced with increase in backfill relative density irrespective of the facing type and reinforcement. For example, the displacements measured in case of wrap-faced retaining wall at a relative density of 37% is 2.73mm, where as it is reduced to 0.55mm at a relative density of 87%. Similar trend is observed for other cases also.

4.2 Influence of Type of Reinforcement

Figure 9 shows the response of rigid faced unreinforced and reinforced model walls, with different reinforcement materials, after the dynamic testing for 20 cycles of 0.2g acceleration at 3 Hz frequency (Krishna and Latha 2009). Figure 9a presents the displacement profiles of these model walls after testing. The effect of reinforcement tensile strength on the seismic behaviour of rigid-faced walls can be understood from this figure. The maximum displacement of about 11 mm for the case of unreinforced model wall (UT2) was reduced to 3.5 mm for the model wall reinforced with WGT and it was further reduced to 1.63 mm for the model wall reinforced with geonet. The displacements measured in the case of geogrid reinforcement were as low as 0.5 mm. However, the effect of reinforcement ultimate tensile strength was insignificant for the tests RT4 and RT5 with BX1 and UA reinforcement materials, giving almost similar displacements in comparison with the unreinforced model wall. These materials have similar stiffness values corresponding to strain levels of 1 to 2%, which was the typical range of strains that were observed in these model tests. Hence the ultimate tensile strength of the reinforcing material has no bearing on the performance of the walls, but the secant modulus value at the low strains that will be developed in the walls is more important. Figure 9 b shows the

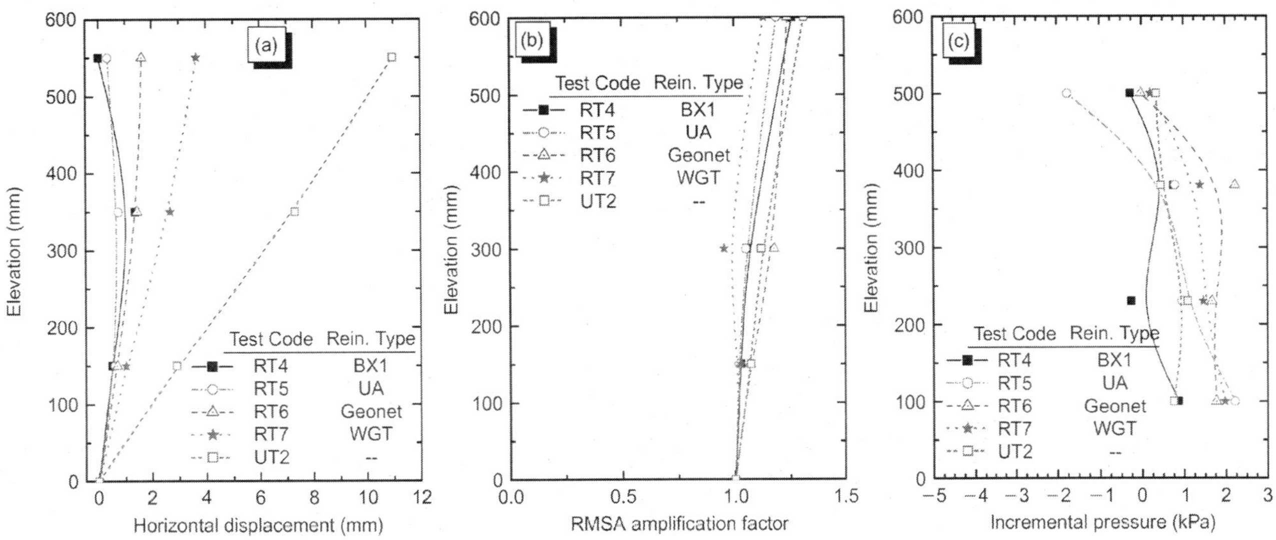

FIGURE 9: Response of rigid-faced model walls with different types of reinforcement after 20 cycles of 0.2g at 3 Hz dynamic motion: (a) displacement profiles; (b) RMSA amplification factors; (c) incremental pressures (after Krishna and Latha 2009)

RMSA amplification factors along the height of the model walls after 20 cycles of shaking. From the figure, it can be observed that the accelerations were amplified more on top of the wall in all the tests. However, there was no significant change in amplification values in the case of reinforced soil walls in comparison with the unreinforced soil wall. Maximum acceleration amplification values at the top were in the range of 1.15 to 1.30 for unreinforced and reinforced model walls. The incremental residual pressures observed at the end of dynamic excitation along the height of the wall for different tests, are presented in Figure 9c. Although, the trend in the pressure distribution along the height of the wall was not consistent among the various model walls, these values yielded the range of incremental pressures that can be expected for the tests conducted.

4.3 Response of Wrap Faced Wall in Random Dynamic Event

Wrap faced model wall with four layers of geotextile reinforcement is tested for random dynamic event. The random event selected as the input motion for this test is Bhuj earthquake that occurred in the Ahmedabad region, India (Latitude & Longitude 23 02 N, 72 38 E Comp: N 78 E) on 26 January 2001. Displacement and acceleration time histories of the event are shown in Fig. 10.

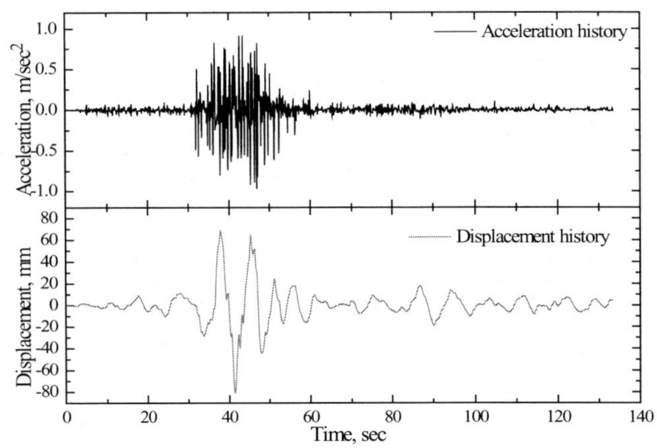

Figure 10 Displacement and acceleration time histories of Bhuj earthquake event

Figure 11 shows the accelerations at different accelerometer locations A0, A1, A2 and A3 at 0, 150, 300 and 600 mm elevations respectively. Acceleration at A0 location corresponds to the actual acceleration recorded during the Bhuj earthquake. The peak acceleration values are observed to be about 0.1g. Acceleration records at A1, A2 and A3 locations are amplified slightly by 1.01, 1.09 and 1.026 respectively.

Figure. 12 shows the displacement variations with time at different USDT locations U1, U2 and

U3 at 200, 350 and 500 mm elevations respectively. From the figure it is observed that the maximum displacement is measured at U3 location and the corresponding value is about 2.5mm. The trend of displacements, accelerations and pressures observed in the random dynamic event is similar to the one observed in tests with sinusoidal base motion, indicating that the observations made from the model tests are valid for equivalent dynamic event in any case.

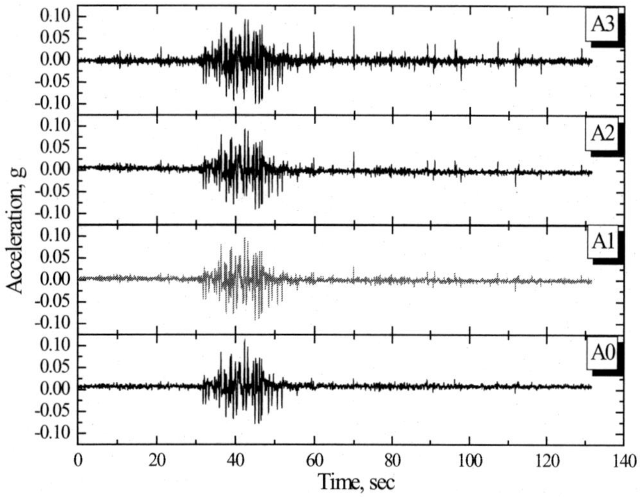

FIGURE 11: Acceleration histories at different elevations during the Bhuj event simulation

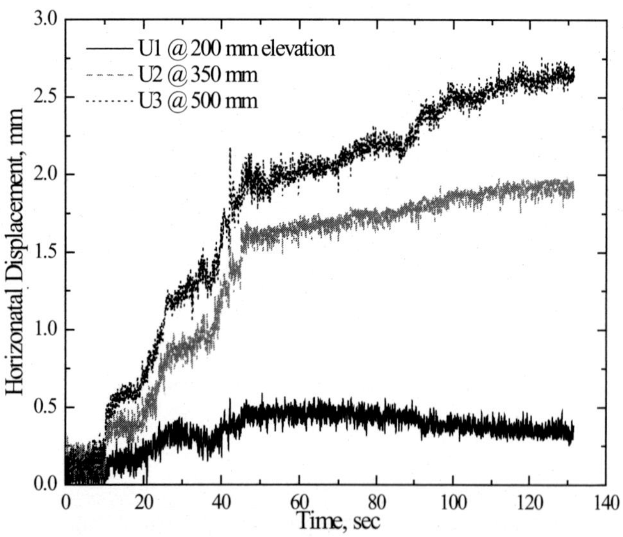

FIGURE 12: Displacement histories at different elevations during the Bhuj event simulation

5. CONCLUSIONS

The shaking table facility available at the Indian Institute of Science is described along with its various components and instrumentation. The whole facility can be used to test various model structures to verify and analyze the behavior of the same under a variety of dynamic conditions. Typical results obtained from the shaking table tests conducted on wrap-faced and rigid faced model walls are presented. These results showed a nonlinear increase in face displacements with time. Maximum displacement was measured at the top of the wall and accelerations are observed to be amplified at higher elevations. It is presented that the displacements of the facing decrease with the increase in backfill relative density. The effect of relative density is more significant at higher seismic excitation. Results from model tests on wrap faced reinforced soil retaining walls with Bhuj Earthquake motion are also presented.

REFERENCES

ASTM D 4595-05. (2005) "Standard Test Method for Tensile Properties of Geotextiles by the Wide Width Strip Method". ASTM standards, American Society for Testing and Materials (ASTM), West Conshohocken, Pennsylvania, USA.

ASTM D 6637–01. (2001) "Standard test method for determining tensile properties of geogrids by the single or multi-rib tensile method". ASTM standards, American Society for Testing and Materials (ASTM), West Conshohocken, Pennsylvania, USA.

Cubrinovski, M., Kokusho, T., and Ishihara, K. (2006) "Interpretation from large-scale shake table tests on piles undergoing lateral spreading in liquefied soils". Soil Dynamics and Earthquake Engineering, 26, Issues 2-4, pp 275-286.

El-Emam, M., and Bathurst, R. J. (2004) "Experimental design, instrumentation and interpretation of reinforced soil wall response using a shaking table". International Journal of Physical Modelling in Geotechnics, 4, Issue 4, pp. 13–32.

Iai, S. (1989) "Similitude for shaking table tests on soil-structure-fluid models in 1g gravitational field". Soils and Foundations, 29, Issue 1, pp 105–118.

Koga, Y., and Matsuo, O. (1990) "Shaking table tests of embankments resting on liquefiable sandy ground". Soils and Foundations, 30, Issue 4 , pp 162–174.

Kovacs W. D, Seed H. B., and Idriss, I. M. (1971) "Studies of seismic response of clay banks". J. Soil Mech. Foundation Div., ASCE, 97, Issue SM2, pp. 441-55.

Kramer, S. L. (1996) " Geotechnical Earthquake Engineering". Prentice Hall, Upper Saddle River, NJ.

Krishna, A. M., and Latha, G. M. (2007) "Seismic response of wrap-faced reinforced-soil retaining wall models using Shaking table tests". Geosynthetics International, 14, Issue 6, pp 355-364.

Krishna, A. M., and Latha, G. M. (2009) "Container boundary effects in shaking table tests on reinforced soil wall models", International Journal of Physical Modeling in Geomechanics, 9, Issue 4, pp 1-14.

Latha, G. M., and Krishna A. M. (2006) "Shaking table studies on reinforced soil retaining walls". Indian Geotechnical Journal, 36, Issue 4, pp 321-333.

Latha, G. M., and Krishna, A. M. (2008) "Seismic response of reinforced soil retaining wall models: Influence of backfill relative density". Geotextiles and Geomembranes, 26, Issue 4, pp 335-349.

Orense, R. P., Morimoto, I., Yamamoto, Y., Yumiyama, T., Yamamoto, H., and Sugawara, K. (2003) "Study on wall-type gravel drains as liquefaction countermeasure for underground structures". Soil Dynamics and Earthquake Engineering, 23, Issue 1 , pp 19–39.

Prasad, S. K., Towhata, I., Chandradhara, G. P., and Nanjundaswamy, P. (2004) "Shaking table tests in earthquake geotechnical engineering". Current Science, 87, Issue 10, pp 1398-1404.

Ramakrishnan, K., Budhu, M., and Britto, A. (1998) "Laboratory Seismic Tests on Geotextile Wrap-Faced and Geotextile-Reinforced Segmental Retaining Walls". Geosynthetics International, 5, Issues 1-2, pp 55-71.

Severn, R.T., Stoten, D. P., and Tagawa, Y. (2012) "The Contribution Of Shaking Tables To Earthquake Engineering". Proceedings of the 15th World Conference on Earthquake Engineering, Lisbon, Portugal, 24-28 September 2012, Paper No: 0211.

Toyota, H., Towhata, I., Imamura, S., and Ken-Ichi, K. (2004) "Shaking table tests on flow dynamics in liquefied slope". Soils and Foundations, 44, Issue 5, pp 67-84.

Wood, D. M., Crewe, A., and Taylor, C. (2002) "Shaking table testing of geotechnical models". International Journal of Physical Modelling in Geotechnics, 2, Issue 1, pp 1-13.

Yan, Li., and Byrne, P. M. (1989) "Application of hydraulic gradient similitude method to small-scale footing tests on sand". Canadian Geotechnical Journal, 26, Issue 2, pp 246-259.

Resilient Behavior of Planar Geo-jute Reinforced Sand Subgrades

Sireesh Saride, Vijay Rayabharapu and Suraj Vedpathak

Department of Civil Engineering, Indian Institute of Technology Hyderabad, Hyderabad

E-mail: sireesh@iith.ac.in, vkraya@gmail.com and vedpathaksd@gmail.com

ABSTRACT

This study illustrates different tests required to perform for the material characterisation including the interface shear tests followed by the series of large scale dynamic model tests which are conducted in a controlled laboratory environment. A series of experiments were conducted on waste geo-jute, extracted from package bags, reinforced and unreinforced sand beds under repeated loading. The effect of normal stress and displacement rates of geo-jute were investigated. All tests are conducted on homogeneous sand sub grades with and without geo-jute reinforcement configurations. Large scale direct shear box tests were also conducted to obtain interfacial shear properties of the geojute sand mixture. The loading was applied through a circular steel plate through a sophisticated double acting linear dynamic actuator which is attached to a reaction frame. The number of loading cycles and the number of reinforcement layers on the elasto-plastic behavior of geo-jute reinforcement was investigated.

1. INTRODUCTION

Ground improvement techniques such as soil reinforcement are certainly a great remedy to unstable soil problems. Thus, soil reinforcement such as application of geotextiles can be fully benefited and blend in for modern construction. To fully explore the benefits and functional mechanism of geotextiles, their interfacial behavior with infill soil is to be understood thoroughly. The behavior is usually being investigated in terms of shear properties between the geotextiles and sand.

About 95% of road network is comprised of rural roads stated by a survey whose performance is always questionable which forces engineers to seek alternative designs using different materials, commercial construction aids, and innovative design practices. Among all of commercial construction aids is utilization of geosynthetics, whose increased popularity is due to their time tested versatile characteristics. Geosynthetics include a large variety of products manufactured of different polymers which are adopted in numerous geotechnical and transportation applications. Often, it is important to estimate the efficacy of such inclusions in natural ground to improve the design methodologies and construction practices prior their utilization. In this study, in particular, the natural, waste and green material obtained from used jute bags are adopted in reinforcement applications under repeated loading and for all practical purposes, the jute material used in this study is called as geojute material which resembles coir geotextile in nature.

Studies on the geojute/coir geotextile reinforcement are being carried out for about half a century. Geojute is a natural and biodegradable material like coir geotextile. Increased bearing capacity, stiffness, and tensioned membrane effects were identified as the major reinforcement mechanisms for geotextile reinforcement (Giroud et al. 1981). Following sections discuss the interface behaviour of geosynthetic reinforcement and factors affecting the performance of the geosynthetics under repeated loading.

2. BACKGROUND

The use of geo-materials for reinforcement as a reinforced soil structures is attractive from an environment as well as economic viewpoint. Since the reinforcement forms ever used, many different kinds of geosynthetics have been used and the foremost kind of reinforcement is geotextile (fiber) reinforcement. These continuous fiber reinforcements with soil may be of synthetic or natural materials. The natural fibers like jute, coir etc. being cost effective and of environmental friendly, in contrast to that of synthetic fibers, can be effectively used for low traffic volume unpaved roads like rural roads. It has the properties of biodegradability, and is the strongest and most durable material among other natural materials (Girish et al. 2000). In the case of geotextiles, the interface shear resistance against soil results solely from the shear resistance between the geosynthetic surfaces against soil particles. In contrary, soil particles are not interlocked with aperture openings as in case of geogrids. Cancelli et al. (1992) investigated the shear strength of soil-geogrid interfaces using direct shear tests. Similar observations were also made by Chia-Nan et al. (2009). Although coir is biodegradable, due to its high lignin content, its degradation takes place much more slowly compared to the other natural materials. Geotextiles are particularly effective in roads constructed over weak subgrade soils. To quantify the benefits, from the geosynthetics especially in pavement applications, a non-dimensional parameter called traffic benefit ratio (TBR), which is defined as the ratio number of load cycles obtained in reinforced bed to unreinforced bed to obtain same amount of permanent deformation, has been introduced and is expressed in terms of extension of life or by saving in base course thickness. Several researchers worked on the cyclic behavior of geosynthetic reinforcements using geogrids observed a TBR of about 3.3 in a large test tank (Haas et al. 1988). Similarly geogrid reinforcement under a moving single wheel system observed a TBR of 1.2 (Barker et al. 1987). Similar observations were made by many other researchers where the TBR varied from 1 to 4 under single axle wheel loads (Cancelli et al. 1992). Sreerama Rao A. (2003) studied the effective application of jute geotextiles over weaker pavement subgrade having lesser CBR value. Recently Senthil et al. (2011) have conducted few preliminary California Bearing Ratio (CBR) tests on coir and jute geotextiles and found that the CBR values were higher for jute geotextile than non-woven coir textile. Extensive applications of jute geotextiles including rural roads were summarized by Abdullah (2008) in a technical document on submitted to Jute Diversification Promotion Centre (JDPC), Dhaka, Bangladesh. It is understand from the literature study that there is a research knowledge gap in understanding the natural geojute as reinforcement under repeated traffic loading.

In this study, an attempt has been made to understand the resilient behavior of geo-jute reinforced sand subgrades under repeated traffic loading conditions which can be preferably used in increasing the life cycle of the unpaved roads.

3. MATERIALS AND METHODS

3.1 Index Properties

The sand used in this investigation was dry sand. The particle size distribution of the sand was determined by dry sieve analysis as per IS 2720 (part-IV)-1985. The particle size distribution of the sand is shown in Figure 1. The sand is classified as poorly graded sand with letter

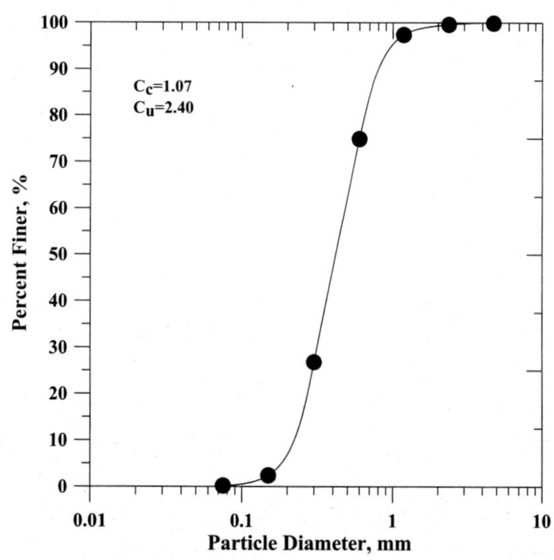

FIGURE 1: Particle size distribution curve

symbol SP according to the Indian Standard Soil Classification System (ISSCS). The physical properties such as specific gravity, maximum and minimum void ratios of sand were determined according to IS: 2720 (Part-III) – 1980 and IS: 2720 (Part-XIV) – 1983 respectively. The maximum and minimum void ratios of the sand used are determined from a vibrating table tests and the results are presented in Table 1. The specific gravity of sand is found to be 2.63.

Table 1: Properties of Sand Used in Investigation

Properties	*Values*
D_{10}, mm	0.20
D_{30}, mm	0.32
D_{60}, mm	0.48
c_u	2.40
c_c	1.07
Specific gravity	2.63
e_{max}	0.74
e_{min}	0.51
Φ	37.0

3.2 Direct Shear Tests

Direct shear tests conducted on unreinforced sand are performed in a small shear box of dimensions 100mm x 100mm x 30mm. The sand is compacted at its 70% relative density. These tests were conducted at 40kPa, 80kPa and 120kPa of normal stresses to obtain the angle of shearing resistance of sand. Direct shear tests and Interface shear tests are performed according to the American Society for Testing and Materials (ASTM) for determining the angle of shearing resistance and interfacial friction of only sand and sand-jute materials respectively.

4. LARGE SCALE TESTS

4.1 Interfacial Shear Tests

In this study, large scale direct shear tests were carried out to determine the angle of shearing resistance (Φ) by shearing geojute with sand material and only with sand material. The tests were conducted under normal pressures at 40kPa, 80kPa and 120kPa respectively. The tests were ended at the horizontal displacement of 10mm.

The interface shear strength parameters are calculated using fully automated direct and residual shear tests on Shear Trac-II Systems under laboratory controlled conditions. The Shear Trac-II system consists of a Shear load frame of capacity 44 kN with a computer attached network card for test control and data acquisition. Microsoft Windows application software, called SHEAR is used for running the test. The large Shear box is of dimension 300mm× 300mm×200mm as shown in the Figure 2. Figure 3 shows the sand specimen prepared in large scale shear box with interfaced geojute material. The geojute material

FIGURE 2: Direct shear test apparatus

FIGURE 3: Geojute specimen under testing for interfacial shear

used in the test is of dimension 320mm×500mm. whose weight is 80 gms and it was placed between compacted sand specimen in upper and lower direct shear box. During tests, the sand is forced to slide along geojute under constant rate of displacement of 1mm/min, while a constant load is applied normal to the plane of relative movement. The normal and shear forces applied to determine the point of failure.

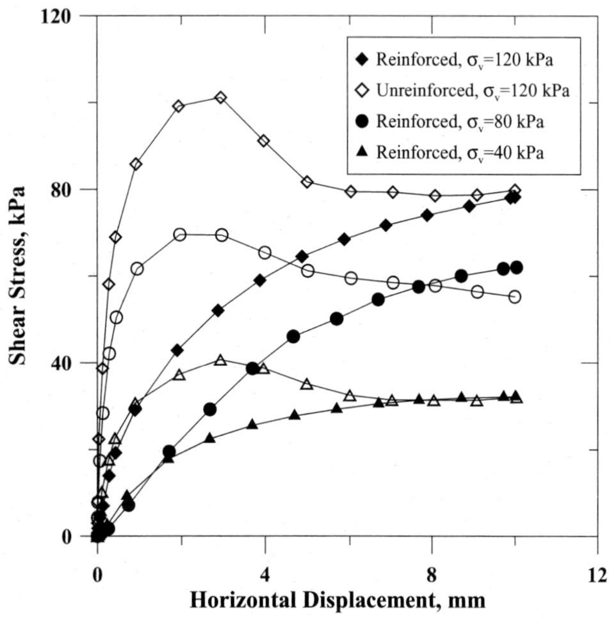

FIGURE 4: Variation of shear stress with horizontal displacement

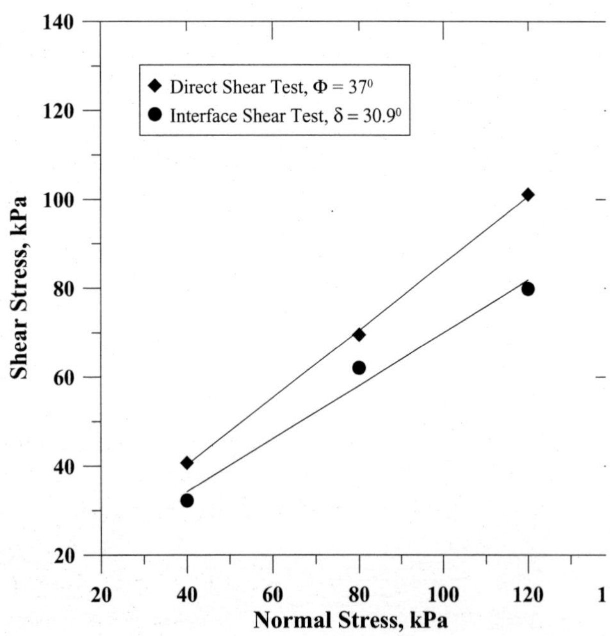

FIGURE 5: Variation of shear stress with normal stress

It can be seen from figure 4 that the shear stress values have reached to its peak and then reduced to constant value for unreinforced case whereas for geojute reinforced case, the shear stress was showing increasing trend until the end of the test. From the tests it is observed that the angle of shearing resistance of the unreinforced sand is 37.0° and for the geojute reinforced sand specimen the interfacial shear angle, δ is observed to be 30.9°. The angle of shearing resistance and interfacial shear angle are presented in Figure 5 The interfacial shear angle is found to be well within the range of values commonly observed for geotextile materials.

4.2 Cyclic Loading Tests

Cyclic load tests are very important to understand the resilient behaviour of a pavement layer under traffic loading. The increased resilient behaviour of the pavement will provide longer life to the pavement structure as well as reduced rut depth (permanent deformation) on the pavement surface. In this study, cyclic load tests were conducted on sand subgrade with and without geojute reinforcement. The sand beds were prepared at its 70% relative density in a test bed-

FIGURE 6: Test-cum-loading system used in the present study

cum-loading tank with internal dimensions of 1m × 1 m x1 m (length x width x height). A rigid steel plate of 150 mm diameter (D) and 30 mm thickness was used to apply the cyclic loading. Loading to the plate was given by graphical user interfaced MTS MPT software with the help of hydraulic power unit (HPU), hydraulic service manifold (HSM) and sophisticated double acting linear dynamic 100 kN actuator which is attached to a 3.5 m high reaction frame. The test setup is as shown in the Figure 6.

Figure 7 illustrates the schematic of the geojute reinforced test bed. In this study, a circular loading plate was used to apply wheel load whose diameter, D is 150 mm. All the test variables are presented in terms of non-dimensional form by dividing the quantity with the plate diameter viz. width ratio of the geojute as b/D and placement depth of first jute layer as u/D. In Figure 7, N is number of geojute layers.

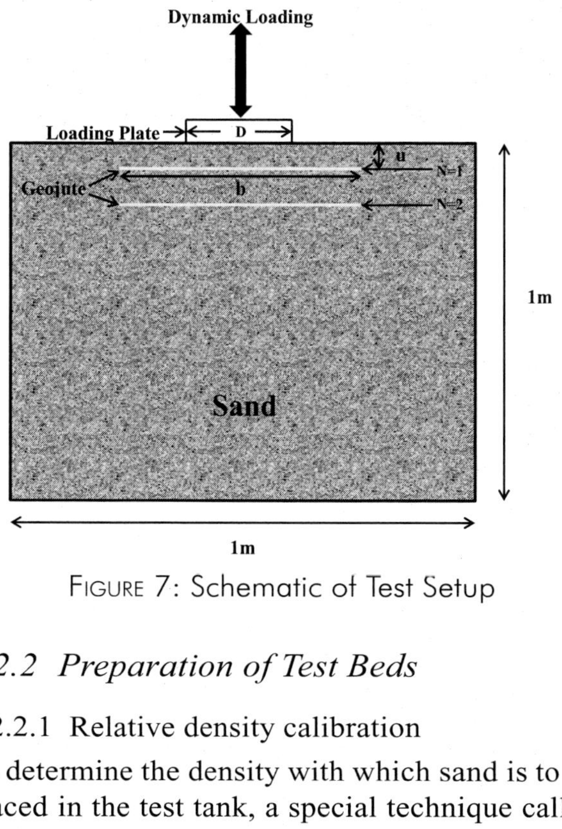

FIGURE 7: Schematic of Test Setup

4.2.2 Preparation of Test Beds

4.2.2.1 Relative density calibration

To determine the density with which sand is to be placed in the test tank, a special technique called sand raining or pluviation technique is used. To achieve this, a special device is designed. It consists of a long steel pipe of diameter 30mm with a cone fixed at the bottom. Apex of the cone is pointed up with cone apex angle of 60°. This pipe is fitted with a movable scale to arrange different heights.

Calibration tests were conducted to obtain a relation between relative density and height of fall as a calibration chart. Placement densities were measured physically by collecting samples in small containers whose weights and volumes were known. With the known values of the minimum and maximum void ratios of sand taken in the investigation, a calibration chart was prepared for the height of fall against the corresponding relative density. For any required relative density corresponding height of fall can be read from calibration chart shown in Figure 8.

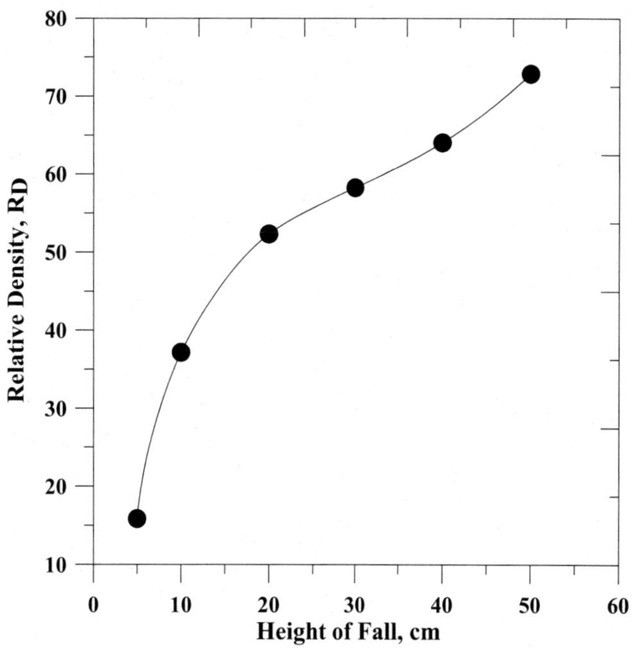

FIGURE 8: Calibration curve for the sand used in the study

4.2.2.2 Sand bed preparation

The sand was placed in the test tank using a raining technique. This device has a hopper with a pipe welded to its bottom and a 400 mm long pipe with an inverted cone welded at its bottom. The sand passes through the 30 mm internal diameter pipe and disperses at bottom by a 60° inverted cone. The height of fall to achieve the desired relative density was determined by performing a series of trials with different heights of fall earlier. In each trial, the densities were monitored by collecting samples in small cups of known volume placed at different locations in the test tank. With the known values of the minimum and maximum void ratios of sand in the study, a calibration chart was prepared for the height of fall against the corresponding relative density as shown in Fig. 8. The height of fall can directly read from the graph corresponding to the required relative density. In all tests, the relative density of sand was kept constant at 70 %. The corresponding height of fall is 50cm.

4.2.2.3 Cyclic load tests

The procedure to test the beds under cyclic loads was in accordance with the Indian Standard code IS 1880: 1982 (reaffirmed 1998). Upon filling the test tank up to the desired height, the fill surface was levelled and the footing was placed on a predetermined alignment such that the loads from the actuator applied would be transferred concentrically to the footing. A recess was made into the footing plate at its centre to accommodate a ball bearing through which vertical loads were applied to the loading plate.

The plate was located carefully at the centre of the loading actuator against the reaction frame to avoid eccentric loading. The cyclic load was then applied on loading plate using a computer-controlled servo hydraulic actuator, with a maximum load of 3 kN and a minimum on 0.3 kN, as a seating load which is 10% of the actual load. The load was estimated based on the field data using a strain type total pressure cells buried under subbase layer just above the subgrade. Since the intermediate layers have not been simulated in this model tests, the pressure exerted on to the subgrade was directly applied through a circular plate. A 10% of load (0.3 kN) was constantly applied on the plate to make the cycle a closed loop. The load form was applied at a frequency of 0.77 Hz. Multi-Purpose Test Ware (MPT) software was set up to control and acquire the applied load data as well as the deformation data. A limited number of repeated load tests were conducted on geojute reinforced sand beds to verify the efficiency of the geojute layers in the subgrade. These tests include single and double geojute layers at different sizes (b) with respect to the plate diameter (D). The width of the geojute was varied as b/D = 2 and 3. The number of layers (N=1 and 2) were also varied for b/D = 3 case. The spacing between the two geojute layers was maintained at 0.33D (i.e. h/D = 0.3). The spacing between consecutive jute layers were obtained from previous literature (Dash et al. 2001, Sitharam and Sireesh, 2004, 2005). The relative density of the sand was maintained at 70% in all tests. In all the tests, one particular parameter was varied, while the other parameters are kept constant, to understand influence of that particular parameter on the overall cyclic behaviour of the sand bed.

5. RESULTS AND DISCUSSION

As discussed, series of cyclic load tests were conducted on unreinforced and geojute reinforced sand beds. Figure 9 shows a typical bearing pressure – plate settlement (presented as a ratio of s/D) response for an unreinforced sand bed. All the tests were conducted until about 20% of the plate diameter (i.e. s/D = 20%). The plot shows the cyclic behaviour under repeated loads as specified earlier.

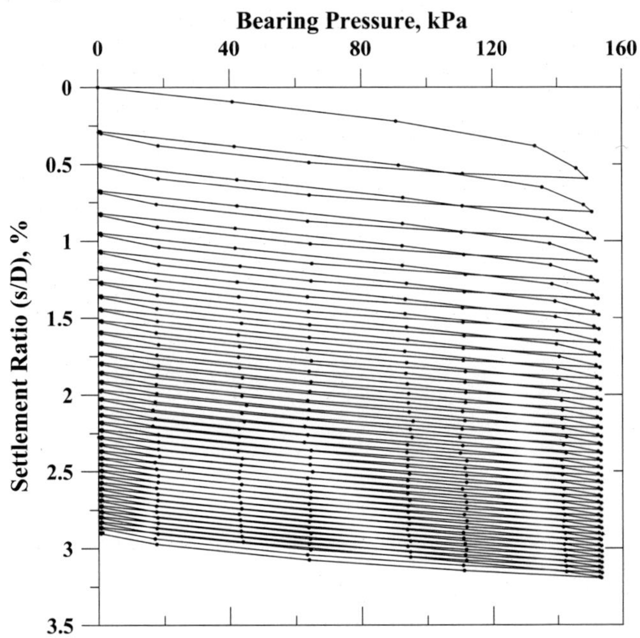

FIGURE 9a: Typical pressure-settlement patterns from cyclic load test

Settlement ratio is defined as the ratio of settlement to the plate width expressed in percentage. It can be observed that initially, at low number of loading cycles (n < 10), the total settlement ratios are higher for a given cycled and the settlements get attenuated with increase in the number of loading cycles. This is more predominant in case of unreinforced case where one can expect more plastic deformations than elastic rebound. To further visualize the data, the results were plotted in terms of elastic deformations and plastic deformations against the number of loading cycles as shown in Figure 10. The elastic and plastic deformations were extracted from the observed cyclic load v/s settlement data (as shown in Figure 9b).

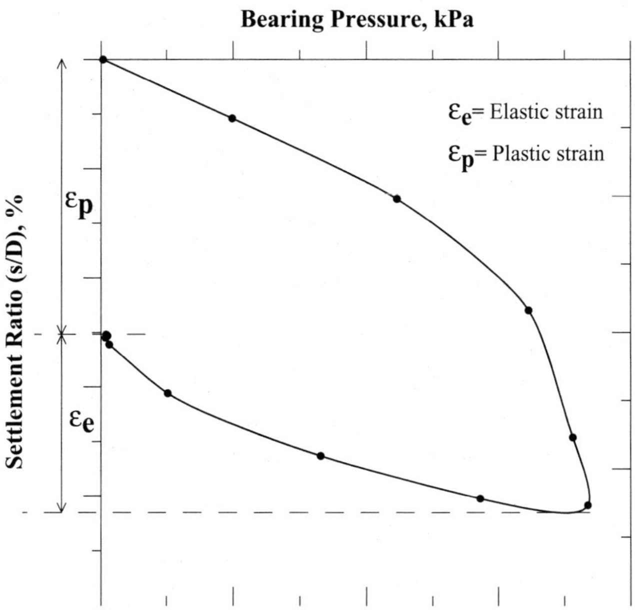

FIGURE 9b: Schematic showing the elastic and plastic component of strains for a typical cycle

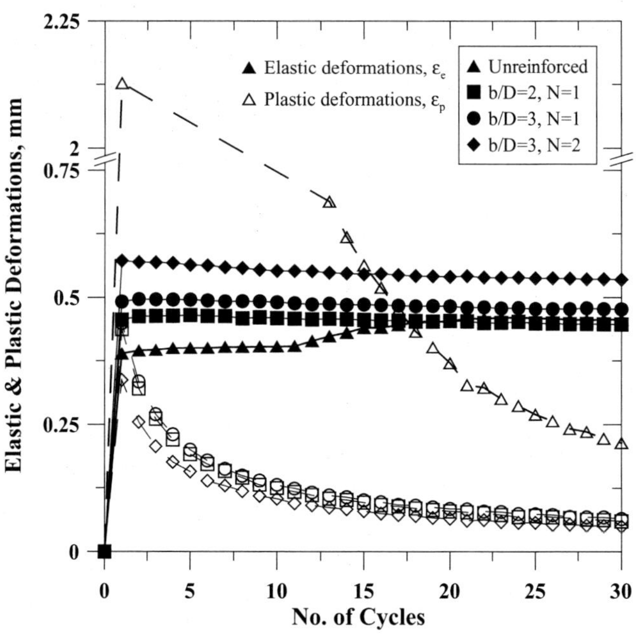

FIGURE 10: Variation of elastic and plastic deformation with number of loading cycles

From Figure 10, it is clear that the plastic deformation of unreinforced sand bed is very high (2.2 mm) at the first loading cycle. The high permanent deformations observed in most of the tests can be attributed to the immediate settlement that occurs in sand beds due to sudden dynamic loading. The order of permanent deformations observed in this study is common for sandy subgrades. In the case of reinforced subgrades, the permanent deformation in the first load cycle is also high, however, these values are very small when compared to the unreinforced case. Thereafter, with increase in the number of loading cycles, the plastic deformations have continuously decreased maintaining the elastic deformations constant. It is to be noted that a slight increase in elastic settlement in unreinforced bed is observed beyond 15 cycles which can be attributed to the densification of the top layers due to cyclic loading.

It can also inferred that as the amount of reinforcement increases from b/D = 2 to 3 and N = 1 to 2, the elastic settlements in reinforced sand beds have increased. This implies that the waste geojute is effectivce in terms of load carrying mechanism. The increased performance of the geojute can also be attributed to the high interfacial shearing angle between the jute and the sand material.

To further quantify the efficacy of the geojute material, the dynamic elastic modulus, M_r which is a ratio of deviator stress to the recovered strain or elastic strain, of the highly reinforced bed with highly recovered strain can be calculated. For the case of b/D = 3, N = 2, the recovered elastic strain is about 0.6 mm at a maximum load of 3 kN. In this case, the maximum vertical cyclic load acting on the plate is taken as deviatoric stress as the horizontal stress or confining stress at the surface level is negligible. However, this value may increase with depth. Therefore, the dynamic elastic modulus calculated from the cyclic load test would be equal to 45MPa against 54MPa for unreinforced bed. The elastic modulus of unreinforced bed is found to be marginally higher than the reinforced bed as the recovered strains in geojute reinforced beds are higher as shown in Figure 10. Mengelt et al. (2006) reported similar trend by conducting controlled resilient modulus tests on Rodefeld sand confined and unconfined in cellular reinforcements. The present calculated values are lower for a given subgrade as the effect of confining pressure is not considered in this study. With the increase in confining pressure, the dynamic elastic modulus would increase. Further elastic modulus improvement can be achieved by increasing the amount of geojute material under loading area.

Further study is required to perform to understand the optimum benefits from the geojute material. It is also important to determine the depth of this kind of reinforcement and number of layers of reinforcement for optimum performance. The longevity of the jute material is also an issue.

6. CONCLUSION

From a limited number of repeated load model tests on geojute material, following conclusions can be drawn:

1. Geojute packing bags abundantly available in the market can be used as a reinforcement which increases the life cycle of the unpaved roads.
2. The interfacial shearing resistance between geojute and sand was found to be as high as 30.9°.
3. Geojute reinforcement reduces the plastic deformations with number of loading cycles which reduces the rutting depths on the pavement surface. Further reducing in plastic deformation can be achieved by increasing the amount of reinforcement in the form of number of reinforcement layers and width of the layer.
4. The elastic deformations or recovered deformations in geojute reinforcement beds increased with increase in the amount of reinforcement. The elastic deformations become constant as number of loading cycles increases.
5. The resilient modulus of unreinforced and geojute reinforced beds are found to be 54MPa and 45MPa respectively. These values are found to be lower for a given subgrade to support unpaved road. Hence, further subgrade improvement is required, which can be obtained from increasing the amount of reinforcement in terms of increase in number of reinforcement layers or width of the geojute layer.
6. Further systematic study is required to completely understand the longevity of the jute material in pavement sub layers.

REFERENCES

Abdullah, A. B. M (2008) "Jute Geotextiles and their Applications", Technical Report submitted to Jute Diversification and Promotion Centre (JDPC), Dhaka, Bangladesh. P. 29.

ASTM, (2002), Standard test method for determining the coefficient of soil and geosynthetic or geosynthetic and geosynthetic friction by the direct shear method, D5321-02, ASTM, West Conshohocken, PA.

Barker, W.R., (1987) "Open-Graded Bases for Airfield Pavements", Technical Report GL-87-16, USAE Waterways Experiment Station, Vicksburg, MS, USA.

Cancelli, A., Rimoldi, P., and Togni, S. (1992) "Frictional characteristics of geogrid by means of direct shear and pullout tests, Proc., Int. Symp. On Earth Reinforcement Practice", Kyushu University, Fukuoka, Japan, pp 51-56.

Chia-Nan Liu, Jorge g. Zornberg, Tsong-Chia Chen, Yu-Hsien Ho and Bo-hung Lin., (2009) "Behavior of Geogrid-Sand Interface in Direct Shear Mode", Journal of Geotechnical and Geoenvironmental Engineering, 135, Issue12, pp 1863–1871

Dash S.K., Krishnaswamy, N.R. and Rajagopal K. (2001) "Bearing capacity of strip footings supported on geocell-reinforced sand", Geotextiles and Geomembranes, 19, Issue 4, pp 235-256.

Girish, M.S., and Ramanatha Ayyar, T.S. (2000) "Improvement of durability of coir geotextiles, Proc.", Indian Geotextiles Conf., Vol. 1, Bombay, India, pp 309–310.

Giroud, J.P. and Noiray, L. (1981) "Geotextile-reinforced unpaved road design", Journal of the Geotechnical Engineering Division, 107 , Issue 9, pp 1233-1254.

Haas R., Wall, J., and Carroll, R.G. (1988) "Geogrid

Reinforcement of Granular Bases in Flexible Pavements", Transportation Research Record", TRB, National Research Council, Washington, DC, USA, pp 19-27.

Bureau of Indian Standards (BIS) (1970) "Classification and identification of soils for general engineering purposes", IS-1498

Bureau of Indian Standards (BIS). (1980) "Methods of test for soils: Determination of specific gravity for fine, medium and coarse grained soils", IS 2720 part3

Bureau of Indian Standards (BIS). (1985) "Methods of test for soils: Grain size analysis", IS: 2720 part4.

Bureau of Indian Standards (BIS). (1988) "Method of load test on soils", IS 1880.

Mengelt, M., Edil, T. B. & Benson, C. H. (2006) "Resilient modulus and plastic deformation of soil confined in a geocell", Geosynthetics International, 13, Issue 5, pp 195–205.

Senthil Kumar, P., and Pandiammal Devi, S. (2011) "Effect of Needle Punched Nonwoven Coir and Jute Geotextiles on CBR Strength of Soft Subgrade", ARPN Jl. of Engineering and Applied Sciences, 6, Issue 11, pp 114-116.

Sitharam, T. G., and Sireesh Saride (2004) "Model Studies of Embedded Circular footing on Geogrid Reinforced Sand Beds", Ground Improvement, 8, Issue 2, pp 69-75.

Sitharam, T. G. and Sireesh, S. (2005) "Behaviour of Embedded Footings Supported on Geocell Reinforced Foundation Beds", Geotechnical testing Journal, ASTM, 28, Issue 5, pp 452-463.

Sreerama Rao. A., (2003) "Jute Geotextile application in Kakinada Port Area", Proc. National Seminar on Jute Geotextile & Innovative jute products, New Delhi.

Assessment of the Deformation Behaviour of Geogrid Reinforced Soil Barriers using Instrumented Model Geogrids

S. Rajesh and B.V.S. Viswanadham[1]

Department of Civil Engineering, Indian Institute of Technology Kanpur, Kanpur
[1] Department of Civil Engineering, Indian Institute of Technology Bombay, Mumbai
E-mail: hsrajesh@iitk.ac.in; viswam@iitb.ac.in

ABSTRACT

The objective of this paper is to study the deformation behaviour of a geogrid reinforced soil barriers (GRSB) of landfill covers subjected to differential settlement through centrifuge model studies. Differential settlements were induced using a motor-based differential settlement simulator at 40 gravities. The model soil barrier and model scale – down model geogrids were selected such that they represents real-field conditions when projected to prototype scale. Digital image analysis was performed to measure the strain induced in soil and geogrid layers. The mobilized tensile load of geogrids along the length of the soil barrier has been determined using strain gauge based instrumentation on model geogrids. Centrifuge test results reveal that performance of a GRSB was found to vary with the magnitude of geogrid stiffness for identical test conditions. When the soil barrier was reinforced with low stiffness geogrid, soil barrier was found to crack thereby losing its integrity. However, crack free soil barrier was noticed even after subjecting to a distortion level of 0.125 when the soil barrier was reinforced with high stiffness geogrid. The maximum mobilized tensile load of low stiffness and high stiffness model geogrids were found to be 37 kN/m and 120 kN/m respectively. The cracking of GRSB reinforced with low stiffness geogrid could be due to the insufficient mobilisation of tensile load of geogrid.

1. INTRODUCTION

Waste disposal through engineered landfills is one of the most popular waste disposal systems that are currently in practice in most of the countries. Engineered landfill comprises of various layers of soil and geosynthetics, based on the type of waste, thickness of waste to be contained, desirable period of post-closure monitoring, geological and geotechnical constraints, financial constraints etc. However, every landfill should have an impervious layer in all lining systems (top, bottom and side) intended to control infiltration of water into or from the waste. Clay rich soils can be considered as an effective and economical barrier material provided it has a coefficient of permeability of 1 x 10^{-9} m/s or less. However, a soil barrier is susceptible to cracking due to moisture fluctuations (desiccation cracks) and differential settlements (mainly due to readjustments and decomposition of the contained wastes). Excessive differential settlements can lead to cracking of the clay barrier and impair the main function of a soil barrier. The problem due to differential settlements is more pronounced in case of a cover systems than the bottom lining system because the former is subjected to low overburden pressure as well as settlement due to readjustments and ongoing bio-degradation of the wastes that lies underneath. A soil barrier tends to crack, if the induced tensile strain due to differential settlement exceeds the permissible tensile strain of the barrier material. Several researchers such as Jessberger and Stone (1991), Viswanadham and Mahesh (2002), Viswanadham and Rajesh (2009) carried out elaborate centrifuge studies on un-reinforced soil barriers and found that commonly adopted thicknesses of the soil barrier (0.6 m to 1.2 m) tend to experience severe cracking and lose their integrity at low distortion levels. Cracking of the soil barrier may be primarily due to the low tensile strength of the soil barrier material. This led several researchers to work towards the development of a crack free barrier at the onset of differential settlements by varying the thickness of soil barriers, the consistency of barrier materials, the compaction energies, the overburden pressures, the type of soil barrier materials and the strengthening of soil barriers in the form of fiber reinforcement and geogrid reinforcement (Jessberger and Stone 1991; Viswanadham and Mahesh 2002; Viswanadham and Jessberger 2005; Gourc et al. 2010; Viswanadham et al. 2011; Rajesh and Viswanadham 2012). The deformation behaviour of a geogrid reinforced soil barrier (GRSB) was found to be many times superior to the respective unreinforced soil barrier (URSB). However, performance of the GRSB was highly influenced by factors such as geogrid strength, overburden pressure and soil-geogrid compatibility. As the knowledge pertaining to the mobilization of the tensile load of the biaxial geogrid on these influencing factors were limited, in the present study, strain gauge based instrumented scale-down model geogrids of varying stiffness were introduced within the soil barrier and the mobilised tensile load of geogrids were quantified for various ranges of distortion level at 40 gravities.

2. CENTRIFUGE MODELLING TECHNIQUE

Centrifuge model testing involves the study of geotechnical events using small–scale models subjected to acceleration fields that provides a magnitude of many times the earth's gravity (g). The self-weight stresses and gravity dependent processes are correctly reproduced and observations from small–scale models can be related to the full-scale prototype situation using well-established scaling laws. Application of centrifuge modelling technique to the present study is relevant because the loss of integrity of soil barrier is highly influenced by the presence of prototype stress conditions. The basic principles of centrifuge modelling for geotechnical purposes has been described in detail by Schofield (1980) and Taylor (1995). In order to replicate gravity-induced stresses of a prototype in a 1/Nth model, it is necessary to test the model in a gravitational field N times larger than that of prototype. This can be achieved by rotating a model in a horizontal plane about a vertical axis at a desired angular velocity. When the mass density of the material involved for constructing model soil barriers is identical to that in the prototype, then to ascertain identical stress condition in model and prototype at homologous points, thickness of the model soil barrier d_m should be equal to d_p/N, where, d_p = thickness of the soil barrier in prototype; N = scale factor or g-level, which is equivalent to $R_e\omega^2/g$ (where R_e = effective radius, ω = angular rotational speed of the centrifuge and g = acceleration due to gravity). Centrifuge model tests reported in this paper were performed at 40g on soil barriers subjected to continuous differential settlements using a 4.5 m radius beam centrifuge having a capacity of 2500 g-kN available at IIT Bombay.

2.1 Scaling Considerations for Modelling Geogrids

In recent years, centrifuge modelling technique has gained popularity and recognised as one of the reliable tools for investigating geotechnical problems due to its ability to reproduce the same stress levels in a small-scale high gravity model as those present in a full-scale prototype. In order to properly replicate a prototype response in a small-scale model, it is necessary to develop scaling relationships, which link the model behaviour to that of the prototype (Schofield, 1980). Scaling laws can be derived using dimensional analysis or from governing differential equations (Langhaar, 1951). Several investigators like Springman et al. (1992), Viswanadham and König (2004) attempted to model the geogrid reinforcement for understanding the behaviour of reinforced soil structures at normal gravity conditions and / or at accelerated gravity conditions. Major difficulty encountered in model studies involving geogrid is the selection of scale down model geogrid.

Contrary to soils, the similitude condition does not allow the use of identical geogrid materials in model and prototype studies. Viswanadham and König (2004) derived and summarized various scaling relationships for modelling geosynthetic materials in normal gravity and high gravity model studies. The scaling of frictional bond behaviour and tensile-strength strain behaviour of the geogrid are considered to be the two basic requirements for modelling geogrid. To avoid any danger of losing contact between soil-geogrid-soil, the percentage open area of the model geogrid should be identical to that of prototype geogrid. The tensile strength of model geogrid should be 1 / N times as that of prototype geogrid (i.e., $(T_g)_m$ = $(T_g)_p$ / N where N – acceleration level); if not up to ultimate strength, at least up to the serviceable strain level (Viswanadham and Jessberger, 2005). Similar to tensile strength, the secant modulus (J_g) of model geogrid should also be 1/N times as that of prototype geogrid. In order to have identical bond stress in model and prototype, the soil-geogrid interface stiffness in a centrifuge shall have to be high when compared with the soil-geogrid interface stiffness in the prototype (Zornberg et al., 1997 and Viswanadham and Mahajan, 2007). Moreover with the occurrence of very small shear stress displacements itself, mobilisation of bond stress equivalent of the prototype can get mobilised. Summary of scaling relationships relevant for modelling geogrid in a geotechnical centrifuge is listed in Table 1.

Table 1: Summary of Scaling Relationships for Modelling Geogrid in a Centrifuge

Parameter	*Units*	*Scale Factor* [a]
Percentage opening area f	[%]	1
Tensile strength T_g	[kN/m]	1/N
Geogrid strain ε_g	[%]	1
Secant modulus J_g	[kN/m]	1/N
Bond stress τ_b	[kN/m²]	1
Soil-geogrid interface friction angle	[°]	1

[a] Scale factor is defined as a ratio of model dimension to prototype dimension.

2.2 Selection of Model Geogrids

The prototype geogrids are available in the wide range with differences in: i) type of raw material used and its composition, ii) grid opening area, iii) rib cross-section area, iv) tensile load and strain behaviour, hence, modelling of geogrid, which suits prototype geogrids is very difficult. In the present study, nine scale-down geogrids were procured from various agencies (Legends: GR1 to GR9). Basic properties of scale-down geogrids like rib dimensions of geogrids in machine and cross-machine direction, thickness of the geogrid, size and shape of the apertures, percentage opening area and wide-width tensile load were determined. A series of wide-width tensile tests were performed close to the procedure outlined in ASTM D 4595 (2005) to assess tensile load-strain characteristics of scaled-down geogrids. For each type of scaled-down geogrid, five samples were tested in both machine and cross-machine directions. The tensile load of geogrid is designated as GR1-MD and GR1-XMD (GR denotes Geogrid; 1 denotes scaled-down geogrid number) for machine (MD) and cross-machine (XMD) directions respectively. Samples of size 200 mm x 200 mm were carefully cut and the clamped with the hydraulic grip of 50 mm each subjected to a pressure of nearly 200 kPa and it is subjected to tensile force in a universal testing frame at a constant strain rate of 10 mm/min. In order to protect scale-down geogrids within the gripping zone, cushion was provided on both sides and at both ends. The variation of measured tensile load (expressed as force per unit meter width of model scale-down geogrid) with strain for all scale-down geogrids (GR1 to GR9) is shown in Fig. 1. In the present study, it has been decided to work at 40 gravities (i.e. N = 40), hence the tensile load of scale-down geogrids are to be multiplied by N (refer Table 1). Properties of prototype geogrids, which are commonly used in the field, are collected from various agencies. Nearly 120 prototype geogrids data was collected, analysed and compared with nine scale-down geogrids in prototype dimensions.

Scale-down model geogrids namely, GR5, GR6 and GR7 were found to have high extensible material property in machine directions; hence,

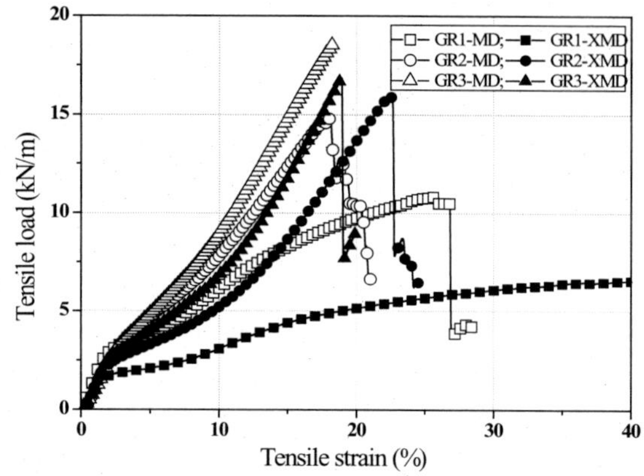

(a) Scale-down geogrids: GR1, GR2 and GR3 (MD and XMD)

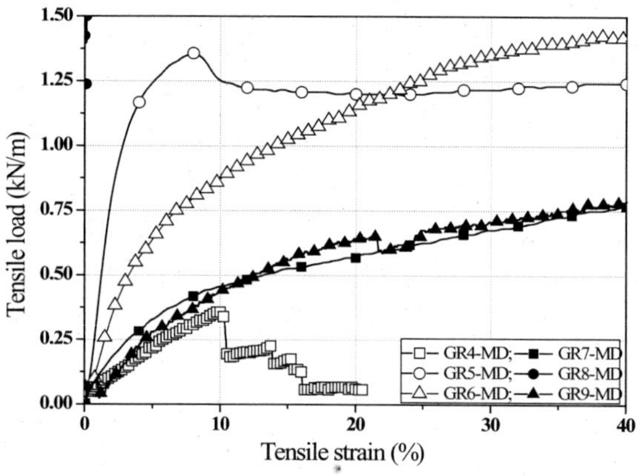

(b) Scale-down geogrids: GR4 – GR9 (MD)

FIGURE 1: Tensile load – strain behaviour of scale-down geogrids (in model dimensions)

may not represent prototype geogrids. Scaled-down geogrids namely, GR1, GR3 and GR8 were found to have very high tensile load and after projecting it at 40g, it was found that these geogrids could not represent prototype geogrids, hence, these geogrids are not selected as model geogrids. Scaled-down geogrids namely, GR2, GR9 and GR4 represent the bandwidth of properties of the commercially available geogrids having high, medium and low tensile load-strain characteristics. However, in this study, GR9 and GR2 are considered as model gogrids representing low and high stiffness geogrids. Table 2 summarises properties of the selected model geogrids along with the prototype geogrids. The properties of the model geogrid GR2 (in MD) was similar to a medium strength geogrid used by Viswanadham and Jessberger (2005) which has performed well in suppressing cracks in soil barriers, while being subjected to differential settlements in a high-gravity environment. In the present study, both the selected model geogrids were placed in machine direction while preparing geogrid reinforced soil barriers in centrifuge tests, such that maximum mobilised tensile load are carried by the members in machine direction. Average wide-width tensile load of model geogrids GR2, and GR9 corresponding to 5% strain in machine direction was found to be 4.51 kN/m and 0.287 kN/m respectively (in model dimensions).

Table 2: Properties of Selected Model Geogrids

Geogrid Properties	*Model GR2*	*Model GR9*	*Prototype Geogrid*
Rib dimensions (mm × mm)	1×0.5	0.2×0.2	-[e]
Aperture size (mm x mm)	3.5×3.5	3.5×3.5	-[e]
Percentage opening area (%)	68	93.4	60-90
Average tensile load (kN/m) at 5% strain – MD			
(i) In model dimensions	4.51	0.287	75-250*
(ii) At 40g	180	11.2	
Average tensile load (kN/m) at 5% strain – XMD			
(i) In model dimensions	3.45	0.31	75-250*
(ii) At 40g	138	12.4	

* Ultimate strength; -[e] Not reported.

2.3 Strain Gauge Based Instrumentation for Model Geogrids

In the present study, an attempt has been made to measure the mobilized tensile load of the geogrid layer at various ranges of distortion levels using calibrated strain gauge based instrumented model geogrids. Strain gauges of 0.6 mm in length, 0.8 mm in width with a base of 5.3 mm x 1 mm having a nominal resistance of 120Ω, gauge length of 0.6 mm, gauge factor of 2.24 and strain limit of 3% were used. As the average width of ribs of model geogrids (ranging from 0.5 mm to 1.5 mm) are very small, pasting of strain gauges on to ribs will not be possible and even if it is done, the response could be highly localized. Hence, few researchers preferred to fill geogrid opening sizes using epoxy type backing material with strain gauges pasted on to a hardened backing material (Springman et al., 1992; Bolton and Sharma, 1994; Viswanadham and König, 2004). In this study, a 25 mm x 25 mm square portion in the centre of the selected geogrid sample was filled with 2 mm thick rubber-based backing material as shown in Fig. 2a. This backing material was chosen due to its tough, flexible, film-forming characteristics with good resistance to heat and ageing. In order to ascertain the influence of thickness of backing material on the tensile load - strain characteristics of the geogrid with backing material, four different thicknesses, namely 0.5, 1, 1.5 and 2 mm were considered. For each sample, backing material was allowed to cure for about 48 hours in room temperature. Figure 2b presents variation of tensile load with tensile strain for model geogrid with a backing material thickness ranging from 0.5mm to 2mm. As can be observed from Fig. 2b, the presence of backing material was not found to have any significant effect on tensile load-strain behaviour. In addition, increase in the thickness of the rubber-based adhesive has not contributed to an increase in stiffness of the model geogrid GR2. Strain gauges were oriented in such a way that they can only measure the tensile strain. In order to avoid bending effect, two active strain gauges were pasted on both sides of the backing material. Two dummy strain gauges for each channel were pasted on to a 6 mm thick Perspex sheet coated with a rubber-based backing material, identical to the one used on model geogrids, to facilitate temperature compensation. A water-proofing sealant was applied on the strain gauges and the soldered ends of lead wires to avoid any contact of water with the electrical connections. A custom designed and developed load-based calibration test setup was used to calibrate instrumented model geogrids for measuring the change in resistance, which in-turn can be used to determine mobilized tensile load of geogrid under various magnitude of applied loads. The obtained calibration factors were used for determining the mobilized tensile load of the geogrid embedded in the soil, while inducing differential settlements in a centrifuge. The detailed description on strain gauge based instrumentation and the methodology for obtaining mobilised tensile load of model geogrids can be obtained from Rajesh and Viswanadham (2012).

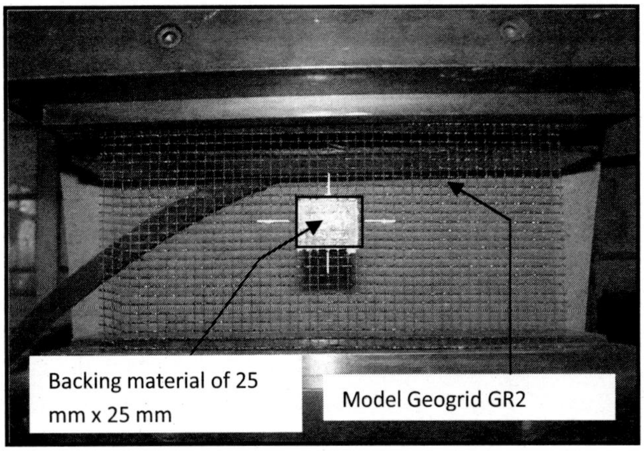

(a) Close view of the model geogrid GR2 with a backing material during wide-width tensile test

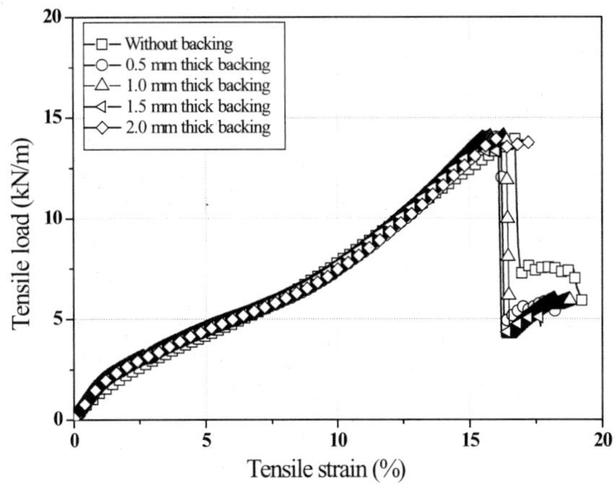

(b) Tensile load-strain behavior of model geogrid GR2 with and without a backing material

Figure 2 Influence of thickness of a backing material on the tensile load - strain behaviour of model geogrid (GR2)

2.4 Selection of Model Soil Barrier

The soil used for constructing the model soil barrier in the present study has properties similar to typical fine-grained soils used in landfill covers (Benson et al. 1999). It is composed of a mixture of kaolin and sand in the ratio of 4:1 by dry weight. The model soil barrier material was found to have a liquid limit of 38%, plastic limit of 16%, coefficient of permeability of 0.4×10^{-9} m/s, maximum dry unit weight of 15.9 kN/m^3 and optimum moisture content of 22%. The selected soil is classified as CL type according to Unified soil Classification system.

3. MODEL PREPARATION AND TESTING PROCEDURE

A custom designed motor-based differential settlement simulator (MDSS) was used for inducing differential settlements at 40 gravities. The MDSS system works on a simple principle: the rotational movement of the motor shaft is converted to translational movement of the central platform through a screw jack and series of gears. The description of MDSS setup, model preparation and testing procedure has been explained extensively by Rajesh and Viswanadham (2010). Figure 3a shows the status of the GRSB and its dimensions before inducing differential settlements. By using this arrangement it is possible to model a landfill cover area as large as 415 m^2 in the prototype scale (at 40g). A 30 mm thick model soil barrier was prepared at its wet side of optimum (OMC+5%) and corresponding unit weight (14.2 kN/m^3) above pre-saturated and drained coarse and fine layers of sand having 30 mm thickness. The calibrated instrumented scale-down model geogrid with discrete markers glued on it was placed at d_g distance and then the remaining thickness of the soil barrier was constructed. In addition, discrete markers are placed on the top surface of the soil barrier to capture the deformation pattern of the soil barrier as shown in Fig. 3a.

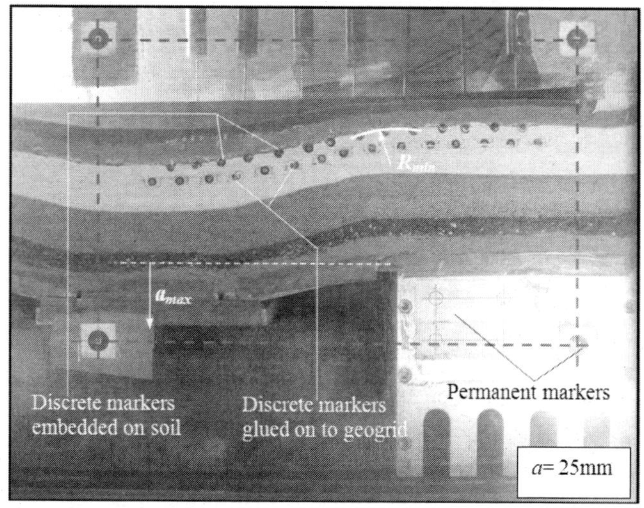

(a) Front elevation of the soil barrier before inducing differential settlements at 40g

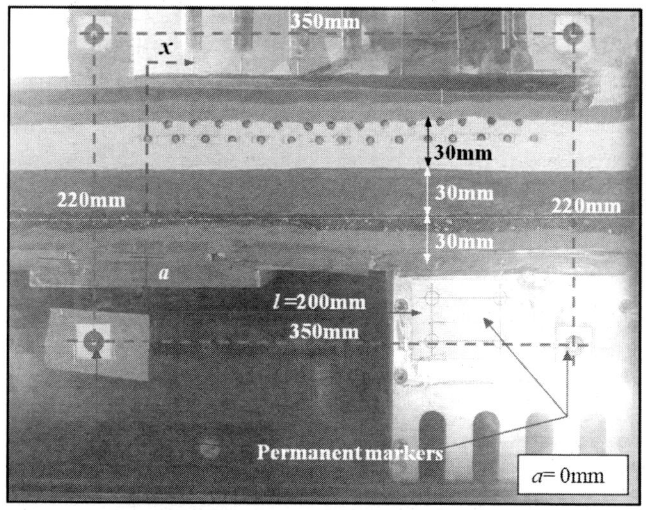

(b) Front elevation of the soil barrier subjected to a maximum central settlement at 40g

FIGURE 3: Snap shot of the soil barrier before and after being subjected to differential settlement

Various sensors like miniature pore pressure transducers (PPTs), linearly variable differential transformers (LVDT) and strain gauges were used to measure water breakthrough, displacement profiles of the soil barrier and mobilized tensile load of model geogrids respectively. Motor of the MDSS system was operated at a settlement rate of 1 mm/min using thyristor controller at 40g with a maximum central settlement of 25 mm. The nature of induced deformation in the soil barrier can be explained using the parameters like settlement ratio and distortion level. When the horizontal distance from centre of the soil barrier x is zero, the value of settlement is termed as a central settlement a (refer Fig. 3a); maximum central settlement induced in the present study is 25 mm (1 m at 40g). Settlement ratio, a/a_{max} is defined as the ratio of central settlement at any stage of deformation a to the maximum central settlement a_{max}. Distortion level, a/l is defined as the ratio of central settlement a to the influence length l (which is defined as a distance over which induced settlements cease to zero) within which differential settlements are induced (refer Fig. 3a). At various stages of central settlements and distortion levels, photographs were captured using digital photo camera placed on the front side of the model to view right half of the front elevation of the soil barrier and were later used for marker-based digital image analysis (DIA) to compute displacement profiles and strain distribution along the top surface of the soil barrier.

4. RESULTS AND DISCUSSION

In this study, two centrifuge test on 1.2 m thick soil barrier reinforced with instrumental model geogrids GR2 and GR9 subjected to an overburden pressure of 25 kPa (equivalent to that of the cover soil) were analysed. Prime focus of analysis is given to the strain distribution and the mobilised tensile load distribution of the model geogrids. From the marker based digital image analysis, coordinates of discrete markers are determined. The measured co-ordinates of a row of markers placed in the soil barrier as well as markers glued on to geogrid layer during various stages of central settlement were approximated with an exponential equation of the normal distribution to obtain deformation profiles of soil barrier and geogrid layer. Since the deformation profiles were almost identical to normal distribution curve, it was decided to adopt a normal distribution curve for fitting the points rather than adopting higher order polynomial. The difference between the initial coordinates of the discrete markers and the displaced coordinates at various stages of central settlements yield respective settlements of the soil barrier and geogrid layer. Deformation behaviour of the soil barrier reinforced with a model geogrid GR2 with an overburden pressure of 25 kPa is shown in Fig. 4. It can be observed that both the geogrid layer and the soil barrier experienced almost identical deformation profiles at various stages of central settlements. However, it can be observed that geogrid layer tends to deform more when compared to the soil layer. This slight extension of the geogrid can help in mobilization of the soil-geogrid interaction in restraining cracking of the soil barrier at the onset of differential settlements. Moreover, from the post-test examinations, it was observed that both GRSB models exhibited good compatibility between the soil and geogrid, without being detached.

Strains along the top surface of the soil barrier and geogrid layer were analyzed by measuring the displacement of discrete markers both in horizontal and vertical directions. The computation of the strain along the top fiber of the soil barrier can be arrived using combined bending and elongation method. The elongation strain is related to the in-plane deformation of the soil barrier. It is assumed to be constant throughout the thickness of the soil barrier. Bending strain or curvature strain is not constant throughout the thickness but varies along the horizontal distance and depth. In case of bending strain, neutral layer has zero strain, while the top and bottom fibers have tensile and compressive strains respectively. These strains can be calculated from the deformed profile of the soil barrier (Lee and Shen, 1969; Tognon et al., 2000). Let $w_s(x)$ be the function of deformation profile of the top surface of the soil barrier (equation of deformation at various stages of central settlement). Consider an element of length dx in the original undeformed state of the soil barrier. Let the element is subjected to a differential settlement of du in the vertical direction. The change in length due to differential settlements can be written as:

$$\delta x = \sqrt{dx^2 + ((du/dx)dx)^2} - dx \qquad (1)$$

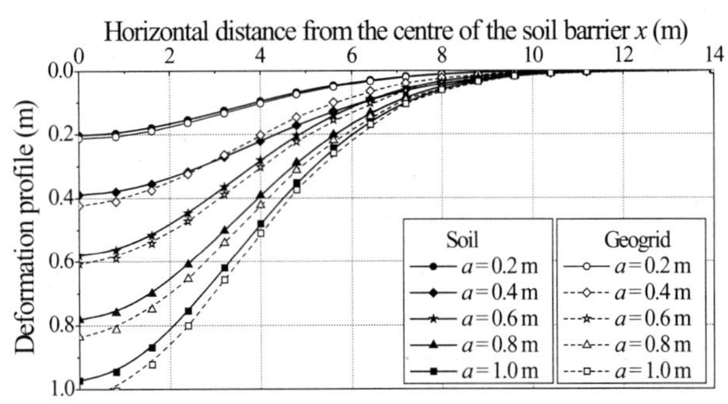

FIGURE 4: Variation of measured deformation profiles for geogrid and soil surface for soil barrier reinforced with model geogrid GR2

Outer fiber strain along the top surface of the soil barrier ε_{of} is the algebraic summation of elongation strain and bending strain of the soil barrier. Elongation strain of the deformed soil barrier is due to the change in length and it can be computed from $\delta x/dx$. By simplifying the Equation 1, the elongation strain of the soil barrier or the strain due to change in length ε_{ls} can be computed as follows:

$$\varepsilon_{ls} = \sqrt{1+[w_s'(x)]^2} - 1 \tag{2}$$

Bending strain or curvature strain along the top surface of the barrier can be determined using theory of bending as given in Equation 3:

$$\varepsilon_{bs} = R_{of}\kappa_s d \tag{3}$$

where, d = thickness of the soil barrier; R_{of} = neutral axis factor; κ_s = Curvature of the soil barrier at a horizontal distance . It can be obtained by differentiating the function of displacement profile of the soil barrier. It can be simplified as given in Equation 4.

$$\kappa_s = \left(\frac{w_s''(x)}{(1+w_s'(x)^2)^{3/2}} \right) = w_s''(x) \tag{4}$$

Radius of the soil barrier R is the reciprocal of curvature = $1/w_s''(x)$.

In case of pure bending, the value of R_{of} is 0.5. In the present study, neutral layer is assumed to be at two-thirds of the thickness of the soil barrier from the top surface of the soil barrier. This is based on the assumption that the mobilisation of shear stresses at the interface of the top surface of the sand layer (placed below the soil barrier) and the base of the model barrier lead to a downward shift of neutral layer from the mid-depth of the soil barrier (Lee and Shen, 1969). However, this assumption needs to be validated by performing Particle Image Velocimetry (PIV).

Geogrid strain is computed in the similar way as that of the soil barrier with only difference is in computing the bending strain. Since the geogrid was placed at a distance d_g from the top surface of the soil barrier, bending strain was computed as the numerical difference between neutral axis factor and d_g times the curvature and thickness of the soil barrier. Geogrid strain can be calculated from Equation 5:

$$\varepsilon_g = \varepsilon_{lg} \pm (R_{of} - d_g)\kappa_g d \tag{5}$$

ε_{lg} and κ_g is computed in a similar way as that of ε_{ls} and κ_s, in which deformation function of geogrid $w_g(x)$ was used instead of $w_s(x)$. The soil strain along the soil-geogrid interface can be determined by assuming that strain in the soil barrier varies linearly from the top surface of the soil barrier, i.e.

$$\varepsilon_{sg} = \left(\frac{R_{of} - d_g}{R_{of}} \right) \varepsilon_{of} \tag{6}$$

Figure 5 shows the variation of strain distribution along the length of the soil barrier, for a 1.2 m thick soil barrier reinforced with a model geogrid GR2 with an overburden pressure equivalent to that of cover systems. In the present study, sign convention adopted for tensile strain is positive and negative value for compressive strain. The outer fiber soil strain along the length of the

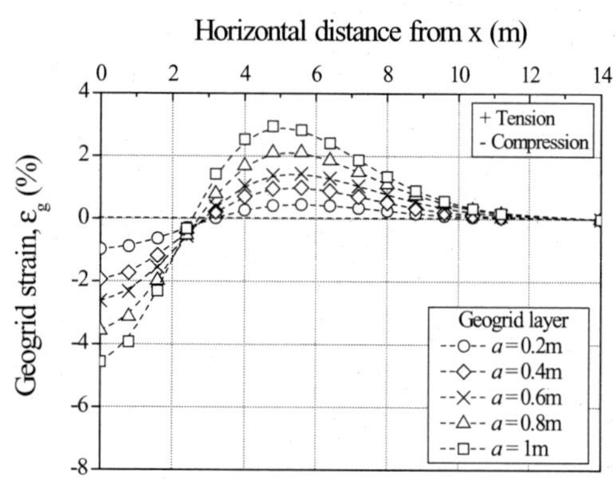

(a) Variation of outer fiber strain along the length of the GRSB reinforced with model geogrid GR2

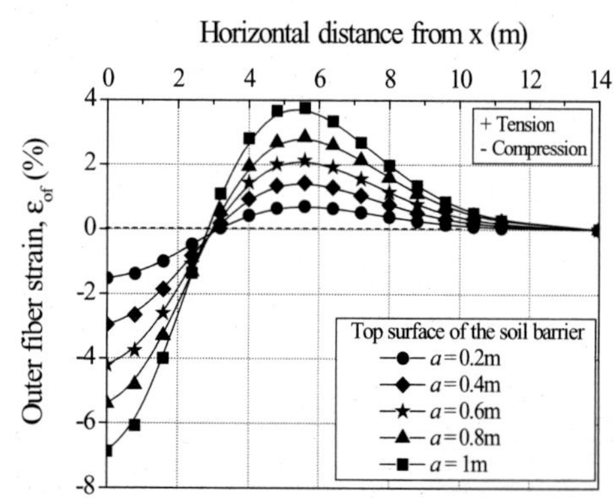

(b) Variation of geogrid strain along the length of the GRSB reinforced with model geogrid GR2

FIGURE 5: Variation of strain distribution along the length of the GRSB reinforced with model geogrid GR2

soil barrier was found to increase with an increase in the central settlement. It can be noticed that the maximum value of outer fiber strain and geogrid strain was found to be at the zone of maximum curvature. At lower central settlements, the value of soil strains and geogrid strains are almost identical; however, at higher central settlements, geogrid strains are found to experience higher strains when compared to soil strains.

Figure 6 shows the variation in maximum ε_g and maximum ε_{sg} at the zone of maximum curvature for various ranges of settlement ratio and distortion level for both the GRSB models. With an increase in the deformation of the GRSB, in other words, with the increase in the settlement ratio (from 0 to 1) and distortion level (from 0 to 0.125), maximum geogrid strain was also found to increase irrespective of the geogrid stiffness. A slight difference between values of ε_g and ε_{sg} can be noticed for both the GRSB models. This difference was observed to increase with an increase in a/a_{max} and a/l. Geogrid strain was found to be higher compared with the soil strain at the interface irrespective of distortion level and geogrid type. The higher geogrid strain value implies the participation of geogrid in the load-transfer mechanism, which confirms the existence of soil-geogrid compatibility.

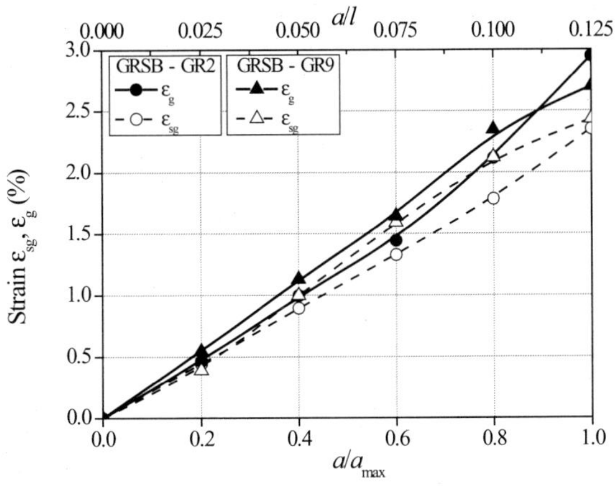

FIGURE 6: Variation in maximum geogrid strain and maximum soil strain at the geogrid interface with a/a_{max} and a/l

Figure 7 shows the status of the GRSB reinforced with GR9 and GR2 at the end of centrifuge test (a/l = 0.125 and a/a_{max} = 1). It can be noticed that soil barrier reinforced with low stiffness geogrid (GR9) was found to experience partial penetration of cracks at the zone of maximum curvature. However, with an increase in the geogrid stiffness (GR2), crack free soil barrier can be noticed, which implies that the mobilised tensile load of geogrid is sufficient enough to resist the bending stresses induced due to a distortion level of 0.125.

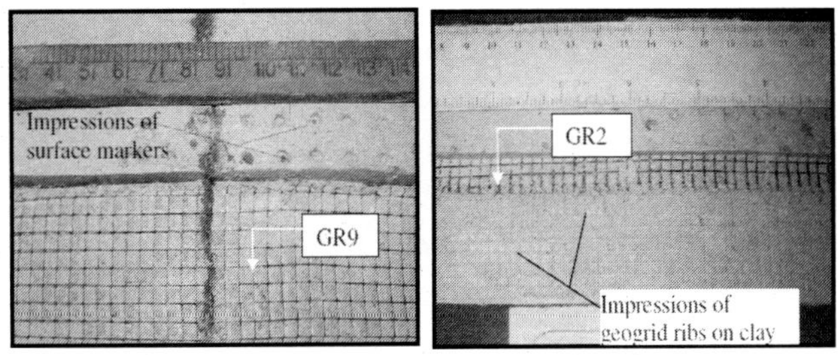

(a) GRSB with GR9 (b) GRSB with GR2

FIGURE 7: Status of GRSB reinforced with model geogrids (GR9 and GR2) at the end of centrifuge test

The mobilisation of the tensile load with time in model dimensions for a centrifuge test on a 1.2 m thick soil barrier reinforced with an instrumented low stiffness geogrid (GR9) is shown in Fig. 8. The location of various strain gauge pairs is also depicted in Fig. 8. It can be observed that as soon as the centrifuge has started, strain gauges commenced to respond and once the differential settlements were initiated by starting the motor of MDSS system, the mobilisation of tensile load in the model geogrid (model dimension) can be noticed by all strain gauges. As the central settlement increases, the response of strain gauges was observed to depend on their location along the geogrid layer.

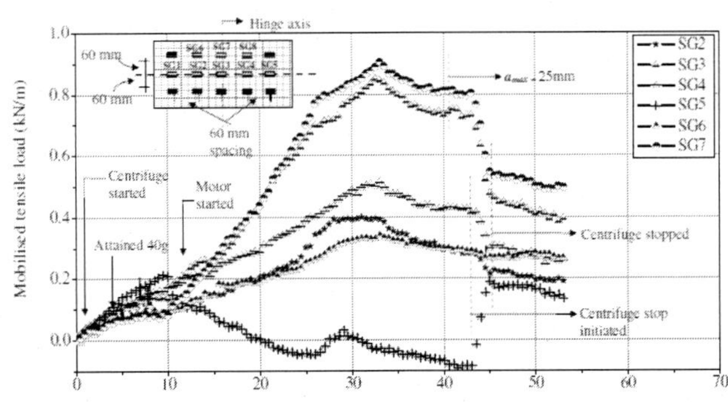

FIGURE 8: Variation in mobilization of tensile load of geogrid with time during centrifuge test for a soil barrier reinforced with GR9geogrid

Figure 9 shows the variation of mobilised tensile load distribution of geogrid for various stages of central settlements for model geogrid GR9. The strain gauges were located on the left hand side of the soil barrier, hence, measured mobilised tensile load distribution was plotted against the horizontal distance from the centre of the soil barrier (x) along the negative axis. It can be observed that with an increase in the central settlement, the mobilisation of tensile load was found to increase irrespective of its location. However, the mobilisation of tensile load at the zone of maximum curvature was found to be maximum, when compared to the other locations. The maximum mobilised tensile load at the zone of maximum curvature at $a/a_{max} = 1$ and $a/l = 0.125$ was found to be 37 kN/m for model geogrid GR9 in prototype dimensions. However, from Fig. 7a, it can be noticed that the GRSB reinforced with GR9 was found to experience partial penetration of crack which indicates clearly that the mobilised tensile load by the geogrid GR9 is not adequate to ensure zero infiltration through the soil barrier.

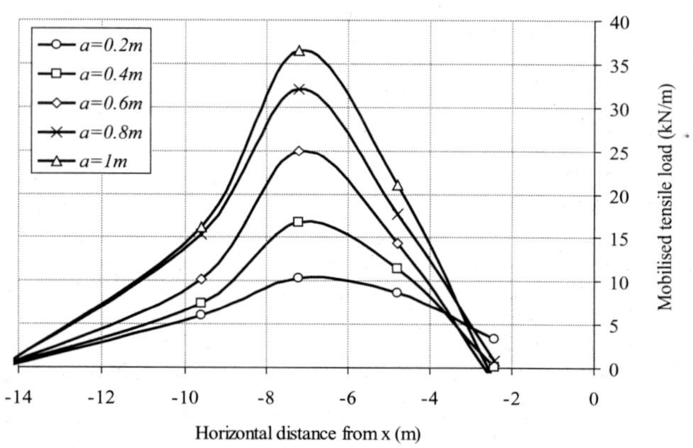

FIGURE 9: Variation of mobilised tensile load distribution of geogrid GR9 for various stages of central settlements (in prototype dimensions)

The tensile strain distribution in microstrain along the geogrid layer can be computed using

$$\varepsilon_g = \frac{4(e_0)}{(G)(n)(GF)(E_x)} \tag{7}$$

where, ε_g - measured geogrid tensile strain of the geogrid layer (%); e_0 - output voltage (V); GF- gauge factor; E_x – excitation voltage (V); G – gain set during calibration and centrifuge tests; n - number of active gauges (in the present study, two active strain gauges were glued on to the upper and lower sides of the geogrid). In the present study, FBY-06-11 TML make strain gauges with a GF of 2.24 were used. Throughout the study, gain of 300 was used. The excitation voltage of $2V$ was used. At a particular location, there are two active strain gauges (i.e. one on the top and lower sides of the geogrid). The variation of tensile strain distribution (in microstrains) from centre of the soil barrier for model geogrid GR9 reported in terms of micro strains with respect to the horizontal distance for various stages of central settlement is shown in Fig. 10. The response of tensile load and strain distribution from the centre of the soil barrier was observed to have an identical trend.

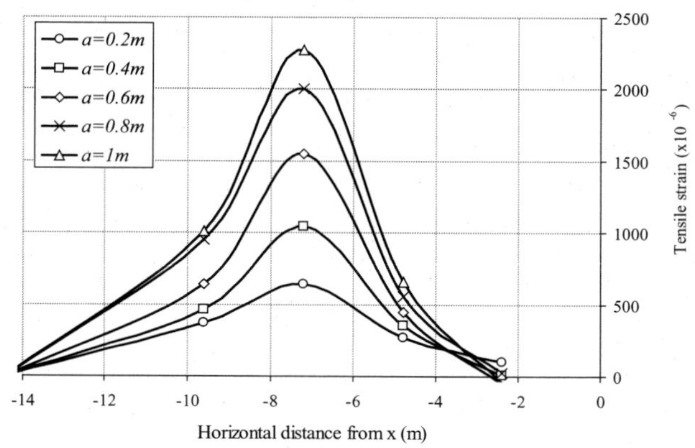

FIGURE 10: Variation of mobilised tensile strain distribution of geogrid GR9 for various stages of central settlements

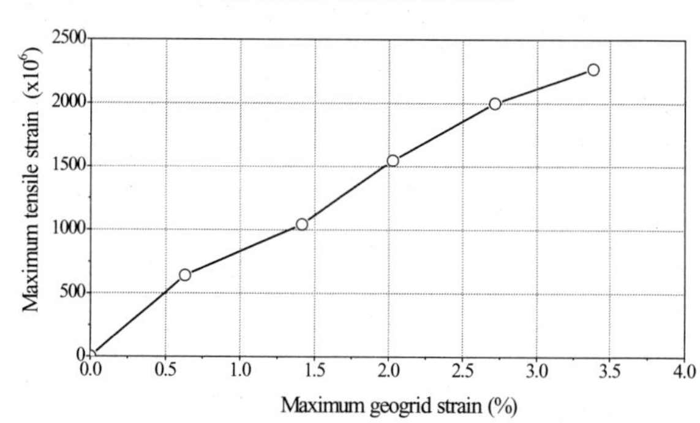

FIGURE 11: Variation of maximum tensile strain measured from DIA and strain gauge based study at the zone of maximum curvature for model geogrid GR9

Figure 11 presents variation of the maximum geogrid strain with tensile strain in microstrains measured at the zone of maximum curvature for model geogrid GR9. As can be noted from

Fig. 11, the observed variation was found to be almost linear and consistent with the computed geogrid strains using marker based digital image analysis.

Figure 12 shows the variation of measured mobilised tensile load distribution for central settlements ranging from a = 0.2 m to 1 m, for a model geogrid GR2. A distinct increase in the maximum mobilised tensile load at the zone of maximum curvature can be noted for the high stiffness geogrid (GR2). At all locations, mobilisation of tensile strength was observed to increase with an increase in central settlements and distortion levels. Figure 13 shows the mobilization of maximum tensile load for an identical soil barrier reinforced with low and high stiffness geogrids at various ranges of settlement ratio and distortion level. It can be noticed the maximum mobilised tensile load for a central settlement of 1 m (a/l = 0.125) for a high stiffness geogrid was found to be 114 kN/m. It can also be noticed that variation in the maximum mobilised geogrid

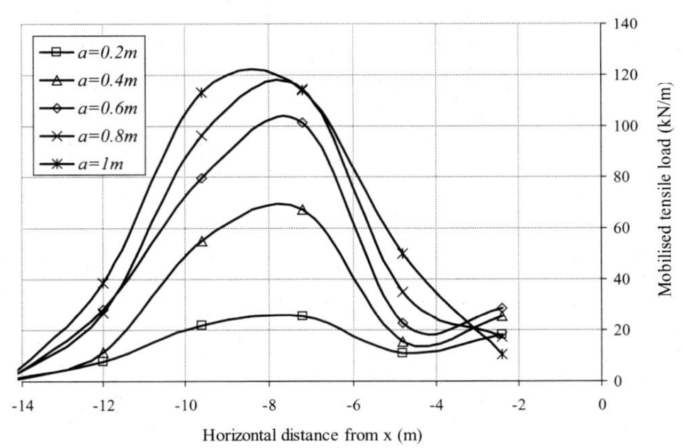

FIGURE 12: Variation of mobilised tensile load distribution of geogrid GR2 for various stages of central settlements (in prototype dimensions)

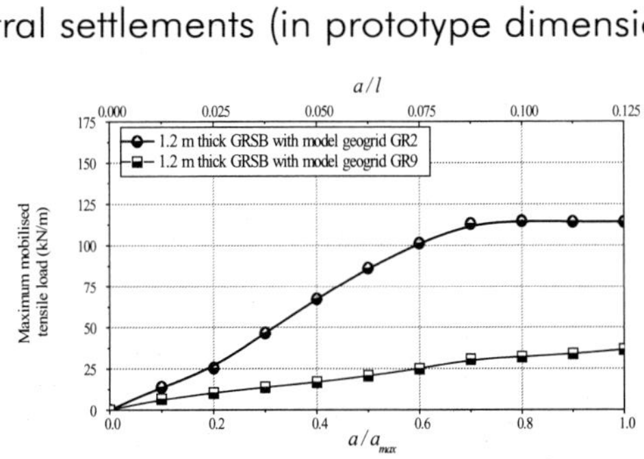

FIGURE 13: Variation in maximum mobilized tensile load at the zone of maximum curvature with a/a_{max} and a/l

tensile load for a = 0.8 m and 1 m was found to be insignificant, which signifies that geogrid layer has generated the maximum possible mobilisation of the tensile load. At a central settlement of 0.4 m, the low stiffness geogrid GR9 could generate a maximum tensile load of 16 kN/m in prototype dimensions. In comparison, the high stiffness geogrid GR2 embedded within the soil barrier could generate a maximum tensile load of 72 kN/m at a = 0.4 m. Despite of maintaining identical boundary conditions, the significance of influence on the geogrid stiffness in mobilising tensile load can be noted.

This study reveals that the beneficial nature of the inclusion of a geogrid layer as reinforcement within the soil barrier can only be seen when the geogrid has adequate stiffness. The presence of the geogrid in restraining cracking of the soil barrier can be idealized as a presence of a fictious overburden on the surface of the soil barrier.

5. CONCLUSIONS

From the results and discussion on two centrifuge test, it can be inferred that the performance of a GRSB was vary with geogrid stiffness. Crack free soil barrier has been achieved even after inducing a distortion level of 0.125 when an geogrid has adequate stiffness. From the digital image analysis, it can be noticed that geogrid strain was found to be higher value compared with the soil strain at the interface, irrespective of the distortion level and geogrid stiffness. The higher value of the geogrid strain implies the participation of the geogrid in the load transfer mechanism. From the strain gauge based instrumentation on model geogrids, it can be noticed that with an increase in the distortion level, a substantial increase in the mobilized tensile load of the model geogrid was observed, irrespective of geogrid stiffness. In both the cases, higher mobilization of tensile load was noticed from strain gauges placed at the zone of maximum curvature. The maximum mobilized tensile load of low stiffness geogrid for a 1.2 m thick GRSB at a distortion level of 0.125 was found to be 37 kN/m. However, for identical conditions with high stiffness geogrid showed an increased maximum mobilized tensile load of 120 kN/m which proves the influence of geogrid stiffness on the integrity of the geogrid reinforced soil barriers.

ACKNOWLEDGEMENTS

The authors would like to thank Centrifuge staff at the National Geotechnical Centrifuge Facility of IIT Bombay for their untiring support throughout the present study.

REFERENCES

American Society for Testing and Material (ASTM). (2005) Standard test method for tensile properties of geotextiles by the wide-width strip method Annual Book of ASTM Standards, Section 4, Vol. 04.13, Geosynthetics. ASTM, Philadelphia. D4595.

Benson, C. H., Daniel, D. E., and Boutwell, G. P. (1999) "Field performance of compacted clay liners", J. Geotech. and Geoenvir. Engrg., ASCE, 126, Issue 5, pp 390–403.

Bolton, M. D., and Sharma, J. S. (1994) "Embankments with base reinforcement on soft clay". Proc. Centrifuge '94, Singapore, pp 587–592.

Gourc, J. P., Camp, S., Viswanadham, B. V .S., and Rajesh, S. (2010) "Deformation behaviour of clay cap barriers of hazardous waste containment systems: full-scale and centrifuge tests". Geotextiles and Geomembranes, 28, Issue 3, pp 281-291.

Jessberger, H. L., and Stone, K. J. L. (1991) "Subsidence effect on clay barriers". Geotechnique, 41, Issue 2, pp 185-194.

Langhaar, H. L. (1951). "Dimensional analysis and theory of models." John Wiley, New York.

Lee, K. L., and Shen, C. K. (1969) "Horizontal movements related to subsidence." Journal of Soil Mech. and Found. Div., ASCE, 94, Issue 6, pp 139-166.

Rajesh, S., and Viswanadham, B.V.S. (2010) "Development of a Motor-based differential settlement simulator setup for a Geotechnical Centrifuge." Geotechnical Testing Journal, ASTM, 33, Issue 6, pp 507-513.

Rajesh, S., and Viswanadham, B.V.S. (2012) "Modelling and instrumentation of geogrid reinforced soil barriers of landfill covers". J. Geotech. and Geoenvir. Engrg., ASCE, 138, Issue 1, pp 26-37.

Rajesh, S., and Viswanadham, B.V.S. (2012) "Performance of geogrid reinforced soil barriers of landfill covers." Verlag: LAP LAMBERT Academic Publishing GmbH & Co. KG, Germany. (ISBN: 978-3-659-14981-8).

Schofield, A. N. (1980). "Cambridge geotechnical operations." Geotechnique, 30, Issue 3, pp 227-268.

Springman, S., Bolton, M., Sharma, J., and Balachandran, S. (1992) "Modelling and instrumentation of a geotextile in the geotechnical centrifuge." In: H., Ochiai, S., Hayashi and J., Otani (eds.), Proc. Earth reinforcement practice, A.A. Balkema, Rotterdam, pp 167-172.

Taylor, R. N. (1995) "Centrifuges in modelling: principles and scale effects." Geotechnical Centrifuge Technology, R. N. Taylor, ed., Blackie Academic and Professional, Glasgow, U. K., pp 19-33.

Tognon, A. R., Rowe, R. K., and Moore, I. D. (2000) "Geomembrane strain observed in large-scale testing of protection layers." J. Geotech. and Geoenvir. Engrg., ASCE, 126, Issue 12, pp 1194-1208.

Viswanadham, B. V. S. and Mahesh, K. V. (2002) "Modeling deformation behaviour of clay liners in a small centrifuge." Can. Geotech. J., 39, Issue 6, pp 1406-1418.

Viswanadham, B. V. S., and König, D. (2004) "Studies on scaling and instrumentation of a geogrid." Geotextiles and Geomembranes, 22, Issue 5, pp 307-328.

Viswanadham, B. V. S. and Jessberger, H. L. (2005) "Centrifuge modeling of geosynthetic reinforced clay liners of landfills". J. Geotech. and Geoenvir. Engrg., ASCE, 131, Issue 5, pp 564-574.

Viswanadham, B. V. S., and Mahajan, R. R. (2007) "Centrifuge model tests on geotextile reinforced slopes." Geosynthetics International, 14, Issue 6, pp 365–379.

Viswanadham, B. V. S. and Rajesh, S. (2009) "Centrifuge model tests on clay based engineered barriers subjected to differential settlement." Applied Clay Science, 42, Issues 3-4, pp 460-472.

Viswanadham, B. V. S., Rajesh, S., Divya, P. V. and Gourc, J. P. (2011) "Influence of randomly distributed geofibers on the integrity of clay-based landfill covers: a centrifuge study." Geosynthetics International, 18, Issue 5, pp 255–271.

Zornberg, J. G., Mitchell, J. K., and Sitar, N. (1997) "Testing of reinforced slopes in a geotechnical centrifuge." Geotechnical Testing Journal, ASTM 20, Issue 4, pp 470–480.

Field Scale Model Tests for Determination of Pullout Capacity of Suction Pile Anchors under Varying Loading Conditions

Vijaya Ravichandran, R. Ramesh, S. Muthukrishna Babu, G.A. Ramadass, M.V. Ramanamoorthy and M.A. Atmanand

National Institute of Ocean Technology, Chennai
E-mail: vijaya@niot.res.in

ABSTRACT

Full scale model tests have been conduction on suction pile anchors for testing the pullout behaviour under various loading conditions. Models with varying diameters and constant length to diameter ratios have been used in the tests. The tests have been carried out in the field within breakwaters where however, natural environmental factors like winds, lower waves and tidal influence exists. Comparison has been made between static pullout and dynamic loading. It is found that the pullout capacity of the model suction piles varied with respect to pile geometry, angle of pullout and nature of pull (static/dynamic loading).

1. INTRODUCTION

Suction anchors are widely used in mooring applications for floating production units and find widespread applications in the offshore oil industries. Suction caissons or anchors are large cylindrical (inverted bucket type structure) open at the bottom and closed at the top. Presently these caissons are being increasingly used for offshore foundations in deep waters.

NIOT proposes to install a desalination plant offshore in 1000m water depths and therefore there is a requirement to test the possibility of providing anchoring through suction pile foundation. Objective of the field scale model tests is to assess the response of suction pile models to various loading conditions so that design criteria for mooring rope and prototype suction pile is established.

The mooring / pullout capacity of the suction pile anchors are influenced by geometry, aspect ratio and installation methods. Suction pile anchor pullout capacities are generally determined from scaled model tests in the laboratory as field tests using larger model piles are prohibitively expensive and highly involved. The instrumentation and equipment requirements for field tests do not significantly vary between model and prototype. The variations in environmental parameters like waves, currents, winds and tides can severely impact the testing conditions.

Several scaled down model tests are conducted in clay as there is no requirement for modeling the overburden pressure in clays Review of literature reveals that several laboratory tests have been on pullout behaviour primarily with focus on soil resistance offered to the suction pile during pullout. Generally the impact of environmental conditions has not been tested in the laboratory. Goldberg (1994) has conducted laboratory scale model tests on small instrumented tubes (4.29 and 5 inch diameter) for assessing pullout capacity of pile and concluded that suction forces contributes over 75% of the pullout resistance. Tjelta (2001) suggested some design methods to predict capacities for suction piles, which are not standardized, to the same extent as for ordinary piles. While it is possible to design suction piles in a variety of soil conditions using the present knowledge, combined (horizontal & vertical) capacity prediction is a coupled problem. According to Tjelta (2001) low aspect ratio is required in sands and hard clays in order to limit extremely high suction values during installation. Aspect ratio greater than 1.5 is recommended in dense sands

Higher aspect ratios may be adopted in soft clays as the anchor will provide large penetration for the same weight thus reaching stronger clay strata.

Higher aspect ratio anchors may be easy to handle and more robust but often limited by resistance to penetration due to high external friction and upward movement of soil plug. Optimal aspect ratio for a suction pile with inclined load may theoretically be when the failure mode is combined vertical and horizontal translation with no rotation, which requires an aspect ratio in the range 2-3.

Luke et al. (2005) have conducted laboratory scale model tests for measurement of uplift capacity of suction piles in normally consolidated clay. Pull out tests were carried out with and without internal soil plug and the axial uplift capacity was evaluated. Axial capacities were considerably larger when the caisson was sealed during extraction, because of the large end bearing capacity and the weight of the interior soil plug.

Bang et al. (2005) have investigated the relationship between applied positive pressure inside the pile and the resulting pile movement. This was used to calculate the mobilized effective and friction angle ratios, which describes the reduction of the interface friction between pile and soil during retrieval.

Bang et al. (2006) have developed analytical solution to estimate the vertical pullout capacity of embedded suction anchors. They found that the primary factors influencing the vertical pullout capacity include loading point, load inclination angle, the soil type, embedment depth and addition of flanges. Narasimha Rao (2006) studied the effect of inclination of the mooring chain. Inclination is found to significantly influence the uplift capacity of the suction anchor. It is found that as inclination changes from 0° to 90°, with respect to horizontal, the resistance changes from passive earth pressure to skin friction.

2. DESCRIPTION OF FIELD TESTS

2.1 Site Description

Testing of piles was carried out at the Royapuram fishing harbour located in Chennai. The test site was located within breakwaters where water depths of 4 to 5m were available at all times. The variation in water depths was due to tidal variations off Chennai coast. Since the site was located within breakwaters, tests could be carried out without the impact of environmental conditions like wave action, rolling of deployment boat, currents and other uncertainties normally encountered in offshore locations. Waves of upto 0.3m could be experienced within the breakwater and winds of upto 2m/sec velocities were experienced during the trials. Tidal difference of up 0.6m to 1.0m was observed within the breakwater at the test location.

2.2 Soil Data

Soil at the site was predominantly soft clay in the top 0 to 1m layers followed by clayey silt. Undisturbed soil samples were collected at the testing site and tested in the laboratory. Hydrometer analysis and unconsolidated undrained triaxial tests were conducted on the soil. The undrained shear strength was 7.5 kPa and Es was 1500 kPa

2.3 Model Suction Pile Data

Diameters of suction piles used were 300mm, 500mm, 750mm and 1000mm. Studies have been carried out on model piles of four different diameters by keeping the length to diameter (L/D) ratio constant L/D ratio of 1.5 was used for all the piles. A padeye is placed on both the side walls of the pile at a distance of 1/3 of the pile from bottom, for enabling attachment of mooring rope using which pullout load is applied.

2.4 Deployment Methodology

The pump skid is a frame structure for housing the pump and related electronics used for deployment. The pump has been developed for provided suction pressure of upto 300 kPa when working underwater. A motor is also provided for delivering into pile at the rate of 720 litres/ minute. Pump is pressure compensated and is therefore capable of operating underwater in water depths of upto 300m.

The electronics arrangement for operating the pump and motor are provided within enclosures which are also pressure compensated and capable of working underwater

The suction pile model assembly was deployed at the test location using a winch operated barge. Schematic diagram of the deployment procedure is shown in Figure 1 and Figures 2 and 3 shows the assembly of model suction pile and deployment arrangements from the shore at the Royapuram test site.

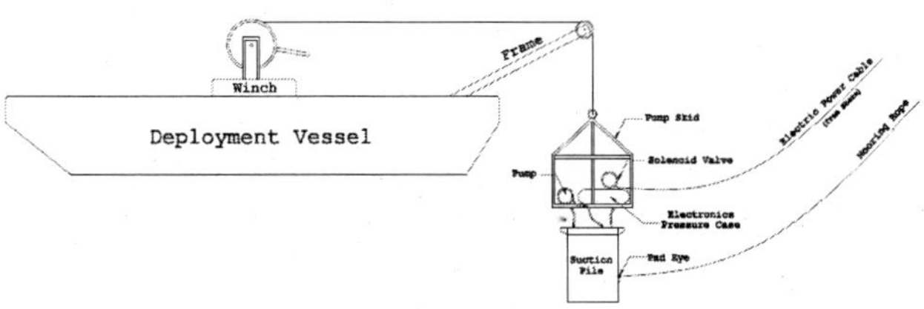

FIGURE 1: Schematic arrangement for deployment

The assembly of electronics and pump/motor was carried out onshore and completely housed on the pump skid. Power supply is provided from the shore using diesel generator using an aramid umbilical cable. The entire assembly was lowered

FIGURE 2: Model suction pile geometry and accessories

into the deployment barge using a shore crane and attached to the suction pile model using hooks. Pump/motor were connected through hoses and then the entire assembly was lowered into water.

Installation is carried out by pumping out water from the inside of the pile using the suction pump.

Once lowered into water the excess air was removed using relief valve and then entire assembly was lowered into water and the suction pile model was placed on the seabed.

FIGURE 3: Arrangements for transfer at test site

Divers were sent in to assess the verticality of the pile when seated on the seabed. After initial penetration was achieved under self weight, the pumping out was started and monitored onshore using computer software exclusively developed for this purpose. The entire data acquisition system is shown in Figure 4. The pumping was continued till the models completely penetrated the seabed. At the end of each test, divers where sent in to ensure complete penetration. During all tests, the mooring ropes were held away on a smaller boat.

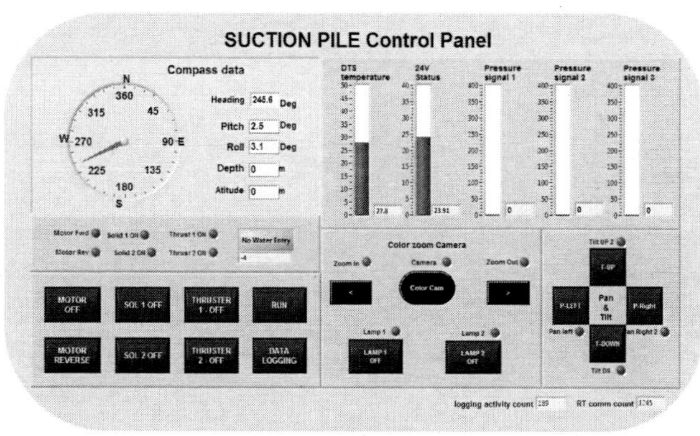

FIGURE 4: Data Acquisition System developed

2.5 Pullout Testing Methodology

Pullout loading is applied using the mooring rope which is attached to the sides at the padeye location at 1/3rd distance from the bottom of the model suction pile. In order to test the response of the model piles to various loading conditions, following types of loads were applied.

- Pumping in water into the suction pile using the motor in the pump skid arrangement
- Static pullout from the shore using a shore crane (static loading)
- Dynamic pullout by including the environmental conditions at the site by applying pullout load from the barge at two different angles. (a) Loading applied at an angle of 20° to horizontal and (b) Load application at an angle of 75° to horizontal

A calibrated load cell of 50 MT capacities with a least count of one kg was used for measuring the pullout load. The load cell was placed along the mooring rope between the pullout load application point and the rope.

After the pullout load is applied using the delivery pump, the connections between the pump skid and pile are completely removed before applying mooring load from the shore crane and barge.

Figure 5 shows schematic arrangement of the various load application points for the pullout tests. For pullout using the barge, the barge position was calculated and placed exactly to achieve the required degree of pullout.

Table 1: Summary of Pullout Capacity Tests

Diameter (m)	*Penetration (m)*	*Uplift Capacity (kN) during Pump Reversal*	*Lateral Capacity using Shore Crane (kN) (static loading)*	*Max. Lateral Capacity from Barge (kN) (dynamic loading)*	
				20	75
0.3	0.45	2.83	4.7	9.3	4.7
0.5	0.75	7.85	22	18	8.6
0.75	1.125	20	58.3	36.6	22.5
1.0	1.5	31.42	63	40	33

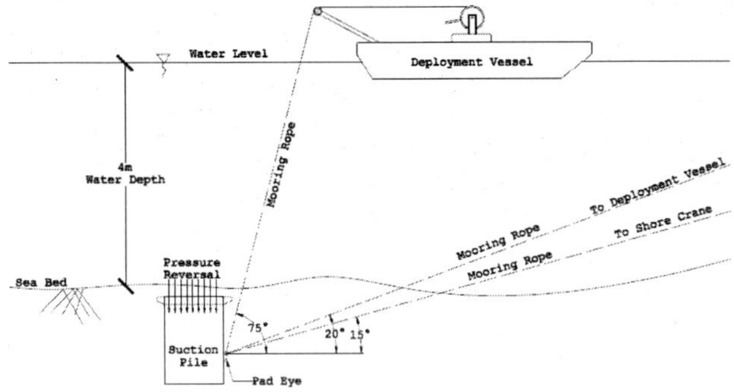

FIGURE 5: Schematic of various loading points during pullout tests

2.6 Results and Discussion

Graphs of ultimate pullout load versus degree of pullout have been plotted for tests carried out on the suction pile models of various diameters. In all cases the L/D remained constant at 1.5. A significant difference in the behavior between the smaller diameter (300 and 500mm) and larger diameter (760 and 1000mm) pile models was observed.

From the model test results it can be observed that as diameter increases the degree of pullout has negligible influence on the pullout capacity normalized value.

Uplift capacity obtained from reverse pumping and lateral pullout at 75° angle pull is observed to be similar. .Maximum load is obtained when the angle of pullout is 20° to horizontal. For lower diameter suction pile models, the difference between static pullout using shore crane and

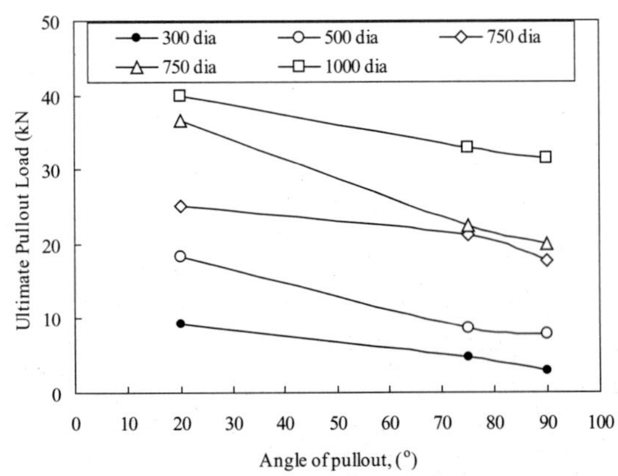

FIGURE 6: Ultimate pullout capacity of model suction pile for various loading conditions

dynamic pull applied from the boat is found to be negligible.

Calibration is obtained for lateral capacity offered by shore crane which is equal to the sum of weights of the wedge of soil offering passive resistance, suction pile model and water column above the model suction pile. In case of uplift by reverse pumping, it is found to be theoretically approximately equal to the sum of external skin friction, soil plug weight and weight to suction pile model.

3. CONCLUSION

The suction pile models have been tested for ultimate pullout capacity under various loading conditions under marine environmental conditions like waves, tides and currents. It is generally observed that for larger diameter models the influence of angle of pullout on the capacity is found to be negligible

It can be observed that as the ultimate pullout increases as the load application changes from vertical to horizontal position upto 15° angle of pullout for suction piles of 750mm and 1000mm diameter. Decrease in pullout capacity can be observed from the results of shore crane pullout and applying pullout load from the -barge which is prominent observed in the case of larger diameter piles (750mm and 1000mm). The reduction in lateral pullout capacity between shore crane and 20° pullout is observed to be about 35% due to effect of environmental conditions like wave action and wind prevailing at the site, which is considered as cyclic and dynamic condition respectively.

ACKNOWLEDGEMENTS

The tests have been carried out by NIOT under the project development of components for offshore structures' funded by the Ministry of Earth Sciences. Significant contributions to the fieldwork provided by Arokia Prabhu, E. Chandrasekaran, G. Chandrasekar, V. Doss Prakash, A. Vadivelu, K. Venkatesan, B. Ravikiran, N.Sureshkumar and P. Jayaprakash is gratefully acknowledged.

REFERENCES

Bang, S., Jones, K., and Cho, Y. (2005) "Transition of Soil Strength during Suction Pile Retrieval in Sand". International Journal of Offshore and Polar Engineering, 15, Issue. 2, pp 155-159.

Bang, S., Jones, K., Kim, Y.S., Kim, K.O., and Cho, Y. (2006) "Vertical Pullout Capacity of Embedded Suction Anchors in Sand", Proceedings of the 16th ISOPE, California.

Goldberg, A. D. (1994), Design and construction of a model suction pile, Offshore Technology Research Centre, Texas A& M University.

Luke, A. M., Rauch, A.F., Olson, R.E., Mecham, E.C. (2005) "Components of suction caisson capacity measured in axial pullout tests". Ocean Engineering, 32, Issue 7, pp 878 – 891.

Narasimha, R. S., Hema, L. K., Pallavi, B., Surendran, S. (2006) "Studies on pullout capacity of anchors in marine clays for mooring systems". Applied Ocean Research, 28, Issue 2, pp 103-111.

Tjelta, T. I. (2001) "Suction Piles: Their Position and Application Today", Proceedings of the 11th ISOPE, Norway, pp 1-6.

Designing of Landfills for Early Post Closure Development

P.P. Savoikar

Department of Civil Engineering, Government Polytechnic, Mayem, Bicholim – Goa

E-mail: psavoikar@gmail.com

ABSTRACT

Municipal solid waste (MSW) landfills may be as high as 100m or more and cover huge surface areas of invaluable land. Sky high prices of land in urban areas has thrown challenge before geoenvironmental engineers to consider option of converting every landfill into potential grounds for developing commercial, industrial and recreational facilities, after allowing 15 years waiting period for consolidation of the waste, which may be too long. Settlement, corrosion problems, methane gas generation and leakage are the major problems faced in the post closure development. This period can be possibly shortened by reinforcing the landfill with geosynthetics or by densifying the closed landfill by using ground improvement techniques like tamping and dynamic compaction for improving bearing capacity of fills, so that less waiting period can be allowed before construction of facilities over closed landfills. However, centrifuge modeling studies need to be taken up to understand the settlement behavior and consolidation behavior of closed landfills.

1. INTRODUCTION

Municipal solid waste (MSW) is normally composed of demolition and construction waste, residential waste, commercial waste, industrial waste, miscellaneous waste and non-hazardous liquid waste in different proportions, which may vary from landfill to landfill. These landfills cover large urban areas of invaluable land, which cannot be used otherwise. Soaring prices of land in urban areas had resulted into converting the closed landfills into potential grounds for commercial, industrial and recreational facilities, after allowing of certain period for consolidation of the waste. However, this development work over landfills has to be undertaken with due care and thorough study of landfill properties as it may lead into several environmental and engineering problems.

Post closure developments of MSW landfills after the closure and waiting period include construction of roadways, recreational facilities like parks, golf courses, etc. which are termed as soft uses and commercial facilities like offices, service stations, shopping centers and residential buildings, etc. which are termed as hard uses, are very common. Figure 1 shows construction of golf course over closed landfill in California and Figure 2 shows construction of service station over closed landfills in San Fransisco (Bouazza and Kavazanjian, 2001).

FIGURE 1: Construction of golf course over closed landfill in California (Bouazza and Kavazanjian, 2001)

The two basic issues associated with post closure work are categorized into the environmental and geotechnical issues. These problems get still worsened when structures on these landfills are subjected to earthquake forces under saturated landfill conditions. However, for post closure development, permissions for development are given subject to completion of long waiting period after closure of landfill, which varies from 10 to 15 years depending of guidelines issued by each country. In India, Central Pollution Control Board (2000) guidelines permits use of closed landfill sites after fifteen years of post-closure monitoring, for human settlement or otherwise only after ensuring that gaseous and leachate analysis comply with the specified standards.

FIGURE 2: Construction of service station over closed landfill inSan Fransisco (Bouazza and Kavazanjian, 2001)

2. POST CLOSURE DEVELOPMENT ISSUES

2.1 Environmental Issues

The environmental issues faced primarily after construction of structures are twofold. These are related to explosions caused due to accumulation of landfill gas and contamination of subsurface or atmosphere with other waste products generated as a result of development activity. Migration and accumulation of methane and carbon dioxide in confined places below foundation or in buildings through the cracks developed in the floors due to differential settlement may result into fire or explosions. Other non-methanogenic organic compounds developed along with landfill gases may contain carcinogenic traces which may cause serious health hazards.

The above environmental issues can be tackled by providing:

- Gas collection systems and engineered covers.
- Additional protective measures like physical and pneumatic barriers, passive venting systems, and active gas collection and treatment systems.
- Proper cover system in the case of parks, golf courses, where more water for vegetation and irrigation is used, which may lead to settlements and generation of gases.
- A gas migration control system at the top, at the perimeter and below the landfill can be installed.

Figure 3 shows a gas detection system installed at the building at Tecnoparc de Montreal, Canada (Rollins and Fournier, 2001).

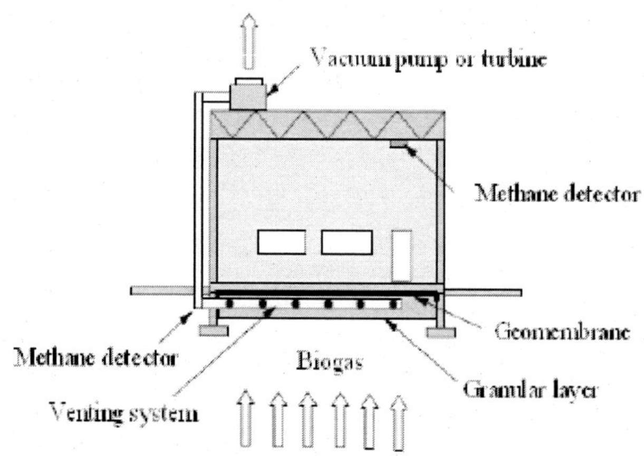

FIGURE 3: A gas detection system installed at the building at Tecnoparc de Montreal, Canada (Rollins and Fournier, 2001)

2.2 Geotechnical Issues

The three basic engineering issues identified are low bearing capacity of landfills, total and differential settlements which dictates use of deep foundations for the multistoried structures or shallow foundations for the low rise structures. Presence of geosynthetic clay liners in lined landfills restricts use of pile foundations as it may damage these liners. Both the total and differential settlements affect the design of shallow foundations directly, and causes additional down drag on piles in case of deep foundations. It affects the design of site utilities, pavements, etc. as well. These settlements result into cracking of structural components like floors, beams, columns and damage to walls, sometimes resulting into total collapse of the structure. For practical purposes, the post-closure long-term settlement is assumed between15% to 20% of the waste mass thickness, whereas the differential settlement is taken as half the total settlement, i.e. 7.5% to 10% of waste thickness as proposed by Bouazza and Kavazanjian (2001). Sharma and De (2007) reported that post closure settlements can be estimated using coefficient of secondary compression due to self-weight and that due to external loads on landfill. Bouazza and Kavazanjian (2001) recommend that if these settlements are too large, ground improvement techniques before undertaking any development works must be carried out. In the case of thin soil cover, Bouazza and Kavazanjian (2001) recommended an engineered fill, composed of geogrids or high-tensile geotextiles, as shown in Figure 4, beneath the foundation, to bridge over non-uniform settlements and to provide higher rigidity to support larger loads.

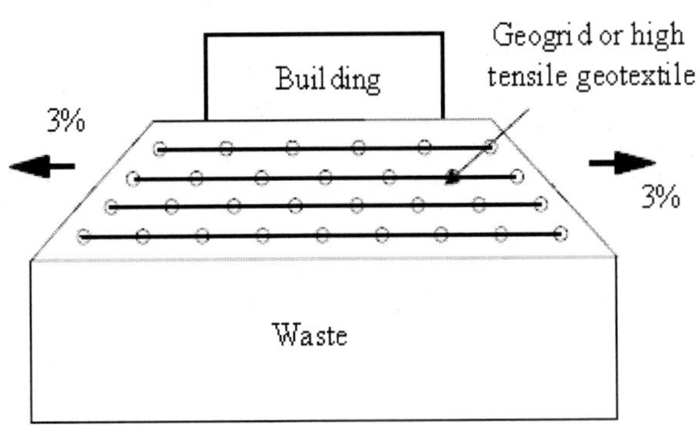

FIGURE 4: Engineered fill with geogrids or geotextiles to support heavier loads (Bouazza and Kavazanjian, 2001)

For deep foundations, driven piles are commonly used to support the high rise structures on landfills without engineered liners, for which prestressed or precast concrete piles are used. Ground improvement techniques like densification of fill using tamping and dynamic compaction may be helpful in improving bearing capacity of fills. For low rise structures, use of conventional spread footings, reinforced concrete mats, and grid foundations are recommended.

In all the above cases, the waiting period of 15 years is allowed so that landfill gets sufficiently consolidated and no appreciable settlement takes thereafter. To cut short this high waiting period, the technologies like dynamic compaction and geosynthetic reinforcement can be resorted to.

3. STATEGIES FOR EARLY POST CLOSURE DEVELOPMENT

Post closure development works are resorted to due to very high cost of land in urban area and land scarce situations. In such cases, first the attempts should be made to save the existing land used for landfills. There can be two basic alternatives to save the cost of land or use of new land for landfills, viz.,

- To save land by designing and constructing the landfill as sustainable landfill using bio-reactor technology such that the waste can be converted into compost which can be taken away, thus creating empty space for reuse for new landfill.
- Early post closure development by
  a. Reinforcing the landfill with geosynthetics, or
  b. Densifying the closed fill by using ground improvement techniques like densification of fill using tamping and dynamic compaction and/or dynamic replacement, or
  c. Construction of piggyback landfill/vertical expansion of landfill

3.1 Sustainable Landfill

Ever growing solid waste problem in both the developed and developing countries has lead to construction of landfills at regular pace and place. This problem is alarming in urban areas also. Landfill constructions take away valuable land with it which is not available to the user again after closure. Hence, in many cases, closed landfills are being looked upon as potential sites for landfill expansions sideways and above the existing closed landfill, popularly known as piggyback landfills. The other possible solution could be redevelopment of closed landfills as suggested above after allowing a suitable waiting period. But this amounts to delay in development works over closed landfills. Presently, the concept of sustainable landfills is gaining wide popularity. Beloti (2006) reported that advanced technologies be used to find out alternate ways and means to meet the requirements of society while preserving the balance of ecosystem.

The concept of sustainability was introduced in the case of landfills by Hettiarachi (2006) and Hsieh et al. (2008). Figure 5 shows the schematic diagram of sustainable landfill technology. This concept involves the stages like anaerobic degradation, aerobic decomposition, and landfill mining in incorporating sustainability in landfills. In this case, part of landfill is converted to compost by adopting bioreactor technology and taken away as manure, so that the empty space is available for new landfill cell.

3.1.1 Anaerobic Degradation

In the anaerobic degradation stage, the biodegradation of waste in the landfills is accelerated by re-circulating the leachate. This stage also involves collection of landfill gas, air injection, and environmental monitoring so as to recover full energy contents of organic waste. In this case, the aerobic reaction takes place until the oxygen present in waste gets depleted and then the controlled anaerobic condition till the majority of the biodegradable organics are exhausted. The waste decomposition takes place in a highly efficient leachate recirculation network which maintains an optimum moisture content of 50- 60% for waste decomposition which also serves as a medium for bacteria and nutrients transfers.

3.1.2 Aerobic Decomposition

In the second stage, aerobic decomposition takes place where in the existing landfill is converted to compost like material which is harmless and usable by-product. In this stage, air is introduced into the landfill and maintained to enhance the aerobic decomposition of waste converting the biodegradable waste into compost like material. The gas extraction system used during the anaerobic stage pumps air into the landfill to create aerobic conditions to accelerate biological degradation of organic waste, with recirculation of leachate. Periodic testing of waste composition from boreholes and the analysis of the leachate needs to be done to ensure the complete biodegradation.

3.1.3 Bio-mining and Recovery

This is the final stage in sustainable landfill concept. It involves practical waste mining after stabilization of organic waste. In this stage the landfill which is converted to compost like material is mined and sold as manure. Waste mining **is** done after the complete biodegradation of waste. The separation of waste is done based on size of organic componenets in the compost. The organic component of size 1-5cm is used as compost for agricultural applications while the other materials like plastic, heavy metals, glass and other non-biodegradable materials are also separated out using the magnetic drum (for ferrous metals) and air stream (for light materials like plastics).

Thus, the existing landfill gets removed from the existing place and the vacant place after mining can be utilized for new landfill, thus saving the costly resource, land. This concept yield multiple benefits like space recovery, elimination of commonly encountered problems such as soil/ water contamination and landfill gas emissions.

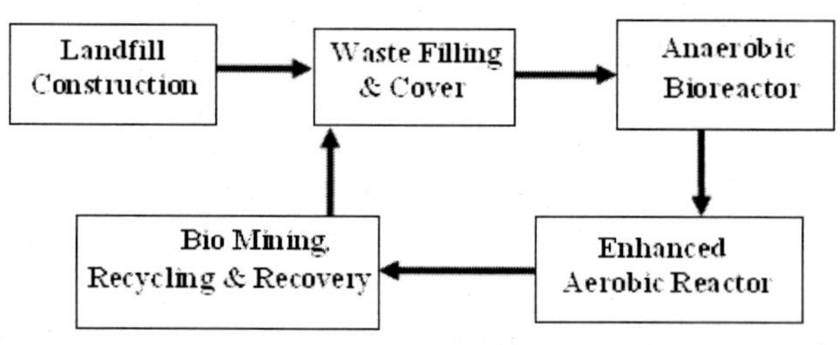

FIGURE 5: Sustainable landfill (after Hsieh et al., 2008)

3.2 Dynamic Compaction for Closed Landfills

Dynamic compaction can be resorted to reduce the settlement of landfill taking place slowly after the closure of landfill. This settlement can be attributed to various factors such as consolidation or settlement of landfill due to its own weight, raveling settlement due to the movement of fine material into larger voids present in the landfill, physical-chemical change occurring in the landfill like corrosion, oxidation and combustion and finally due to bio-degradation of waste (Edil et al, 1990). Compaction process consists of using a heavy tamper (pounder) that is repeatedly raised and dropped from varying heights to impact the ground. The degree of improvement depends upon the energy supplied for compaction which depends upon weight of tamper and height of fall, grid spacing and number of repeated falls at same point in the grid. Normally, a 15 to 20 ton tamper (pounder) is dropped from height of 15-20m delivering energy to the tune of 300 ton-m. Grids of 5m are selected and 8 to 12 repeated blows are applied to each point in the grid. The effectiveness of compaction can be assessed based on results of Standard Penetration Tests and Pressuremeter Tests carried out before and after the compaction process. It has been observed that the compaction reduces the voids and increases the bearing capacity and hence the potential large settlement. However, physical-chemical change and biodegradation cannot be substantially improved as densification can reduce the amount of oxygen in the landfill resulting in slowing down the rate of decay.

Yee et al. (1999) reported that the depth to which improvement in properties occurs, can be calculated as:

$$D = \delta.\alpha.\sqrt{W.H} \tag{1}$$

where D is the depth to which the dynamic consolidation causes some improvement in the fill properties, W is the Pounder weight, H is the drop height and δ is the speed factor to be taken as 0.9 for cable drop and 1.2 for free fall and α is the soil structure factor to be taken as 0.3 to 0.7 for different soil types.

Also, dynamic replacement can be resorted to in addition to dynamic consolidation, wherein dynamic compaction process is used to create large diameter granular columns through the soft soils or waste fill to the underlying dense bearing stratum. This method has found to be effective as it combines the benefits of dynamic consolidation and granular columns. Figure 6 and Figure 7 shows the schematic diagram of dynamic replacement and dynamic compaction and replacement respectively. The important stages involved in the dynamic replacement process are as below:

1. removal of top 2.5m of fill material and replacing it with clean sand
2. dynamic compaction of entire area and dynamic replacement below the column footings/foundation areas,
3. performing the pressuremeter tests to assess the degree of compaction
4. surcharging the entire area with 2m high surcharge and monitoring of settlements, and
5. removal of surcharge and commencement of foundation works.

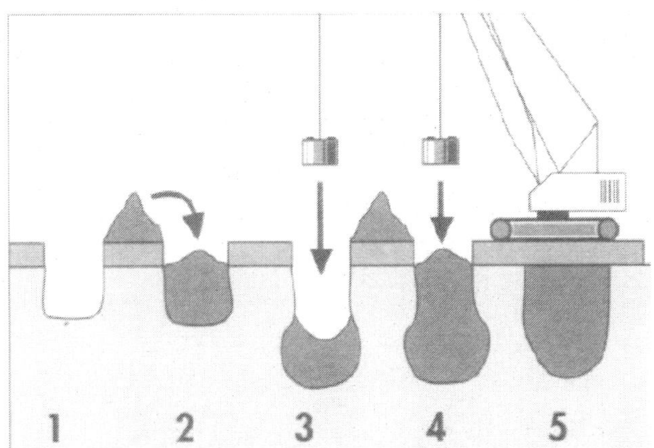

FIGURE 6: Schematic diagram of dynamic replacement (Yee et al., 1999)

3.3 Expansion of Closed Landfills

Due to increasing cost of land in urban areas for construction of new landfills, capacity expansion of existing landfills is one of the most viable solution. This capacity expansion can be done in two ways, viz. vertical and or/lateral expansion by encapsulating the old landfill or by construction of new landfill over the existing landfill by providing geosynthetic liners. The later one is popularly known as piggyback landfill. This vertical expansion serve various advantages apart from saving of huge cost of land like optimum use of landfill area, high waste volume per unit area, low cost, less public opposition and easier permitting (Qian et al., 2001). Also, expansion of landfill capacity may cause settlement of existing as well as new landfill which further leads to liner system and slope stability issues. Leachate collection system, gas collection system and liner system may also get affected due to the settlement of the old and new landfill, which needs to be taken care of.

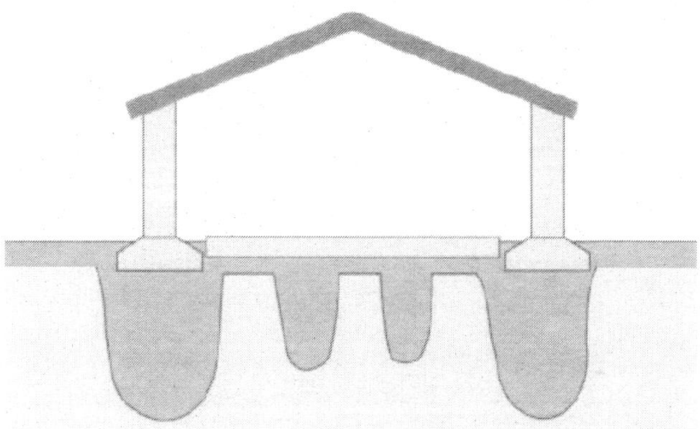

FIGURE 7: Schematic diagram of dynamic replacement and dynamic compaction (Yee et al., 1999)

Qian (1996) suggested the following design considerations for designing of vertical expansion of landfills:

- Selecting suitable composite liner system between the old and the new landfill,
- Estimating total and differential settlement of old landfill due to new landfill
- Reinforcing the liner system to minimize the differential settlement,
- Determining the deformations and stability condition of leachate and gas collection system and
- Stability of soil, waste mass, liner and cover system during excavation, construction, operation and closure of landfills.

4. CENTRIFUGE MODELING OF THE SETTLEMENT BEHAVIOUR OF LANDFILL

Yegian et al. (1995) conducted shaking table tests for determining dynamic interface shear behaviour and strength properties of geomembrane/soil system. Centrifuge model tests have been conducted by Kockel et al. (1997) to determine the waste-structure interaction and to investigate deformation induced stress redistributions within the waste body near a structure. Also, centrifuge model tests were performed by Thusyanthan et al. (2006a & b) to study settlement profile of landfill and to investigate dynamic behaviour of solid waste landfills. Thusyanthan et al. (2006a) developed a model whose physical, chemical and dynamic properties are more or less similar to real MSW using mixture of peat, E-grade kaolin and fraction-E fine sand. Thusyanthan et al. (2006a) tested the above three mixes by varying the composition of above three ingredients and also obtained their unit weight, compressibility and shear strength. Centrifuge tests were performed on model solid waste at 50g by Thusyanthan et al. (2006b) to study the settlement profile of MSW landfills. It was reported that the settlement of MSW landfills depends upon the compaction method and waste constituents. It was also observed that typically a landfill can settle by 5% to 30% of initial height, after closure. The settlement profile obtained can be used in designing the cover liner, leachate collection and gas recovery system.

Viswanadham et al. (2012) examined the performance of geosynthetic clay liners (GCL) subjected to continuous differential settlements in a centrifuge at 40g. Comparison of behavior was made between 0.6m thick compacted clay liner (CCL) cover system and prototype GCL cover system. During the test, both the models were subjected to an overburden equivalent to that of landfill covers and sealing efficiency at the onset of differential settlements was measured. It was reported that the sealing efficiency of GCL cover system was superior as compared to that of CCL cover system and has ability to withstand differential settlements without losing their sealing efficiency. It was also reported that GCL has the ability to maintain the desired hydraulic conductivity and can sustain large differential Settlements.

Grisolia and Napoleoni (1996), Machado et al. (2002) presented the constitutive models for compression behavior of landfills. Wang Wei

(2012) reported that landfill settlement continues for a longer time even after closure of landfill due to physico-chemical action and biodegradation. Considering the post closure hard and soft developments over the landfill, it is essential to predict these settlements correctly since different types of shallow or deep foundations may rest on these landfills. Wang Wei (2012) presented a model for settlement behavior of closed landfill which is a 3-parameter model giving settlement-time relationship based on physico-chemical and biodegradability. According to this model, settlement of fill is controlled by settlement energy of physico-chemical action (*Vp(s)*) and biodegradable action (*Vb(s)*) . This energy is given by the Eq (2) given below:

$$E_t = \frac{ds}{dt} = k_3\left(Vp(0) - \Delta Vp(s)\right).\left(Vb(0) - \Delta Vb(s)\right) \quad (2)$$

or $$E_t = \frac{ds}{dt} = k(a-s).(b-s) \quad (3)$$

where E_t is post closure settlement rate, t is the time and k, a and b are the positive constants which depend upon waste composition and loading conditions. The final three-parameter model is given by

$$S = f_3(t) = \frac{a.b.e^{-k(b-a)t} - a.b}{a.e^{-k(b-a)t} - b} \quad (4)$$

The above equations were validated with the data obtained by Zhang and Chen (2004) for the 128m high Tianziling municipal solid waste landfill in China and by Liu et al. (2004) for Laogang municipal solid waste landfill which is 4m high testing landfill in Shanghai, China and the results obtained from Eq (4) were in close agreement with data presented by above researchers. It was also reported that this model overcomes the deficiency of the traditional settlement models and can be used for design, management and reutilization of landfills.

The above developed model can be used to predict the settlement rate of closed landfills. However, centrifuge model studies need to be undertaken to confirm the results obtained above, to understand settlement behavior of closed landfill with time and possibility of reducing the 15 year long waiting period.

5. CONCLUSIONS

Closed landfills are continuously used as potential sites for post development works. However, geotechnical and environmental issues need to be handled carefully. Construction over closed landfills cannot be commenced before the waiting period of 15 years. Methods such as dynamic compaction have been found to be effective in reduction of differential settlement and increase in bearing capacity. Hence, there can be the possibility of reducing the waiting period of 15 years, for which detailed centrifuge modeling studies need to be taken up. Also, the behavior of landfills provided with geosynthetics reinforcements at various locations in landfill need to be evaluated properly by performing centrifuge model studies to ascertain whether construction can be commenced over closed landfills before the normal waiting period. Other solutions include sustainable landfills wherein a landfill is converted to compost and finally compost is mined so that landfill vanishes and place is available for fresh landfill to be constructed. Alternatives like piggyback landfill have also been tried successfully to save cost on costly urban land so as to maximize the waste storage space. Centrifuge model studies need to be done to understand the settlement behavior of closed landfills and possibility of reducing the fifteen year long waiting period for post-closure development work to begin.

REFERENCES

Beloti, M. (2006) Personal Communication. Santa Clara University.

Bouazza, A., and Gassner, F. (2005) "Geosynthetics reinforcement in landfills design: Australasian perspectives". International Perspectives on Soil Reinforcement Applications, Geotechnical Special Publication, ASCE, 141, pp 1-15.

Bouazza, A., and Kavazanjian, E., (2001) "Construction on former landfills". Proceedings of 2nd ANZ Conference on Environmental Geotechnics, Newcastle, pp 467-482.

Edil, T. B., Ranguette, V. J., and Wuellner, W. W. (1990) "Settlement of municipal refuse". Geotechnics of Waste Fills –Theory and Practice, ASCE STP 1070, pp 225-239.

Gourc, J. P., Reyes-Ramirez, R., and Villard, P. (2005) "Geosynthetic reinforcement in landfill design: European perspectives". International Perspectives on Soil Reinforcement Applications, Geotechnical Special Publication, ASCE, 141, pp 1-12.

Grisolia, M., and Napoleoni, X. (1996) "Geotechnical characterization of municipal solid waste: Choice of design parameters". Proceedings of 2nd International Congress on

Environmental Geotechnics, Osaka, Japan, Balkema, Rotterdam, The Netherlands, 2, pp 641-646.

Hettiaratchi, J. (2006) Bio-Cell Project. www.eng.ucalgary.ca/resrch_civil/bio-cell-project/Hettiaratchi-bio-cellproject.htm

Hsieh, H., Meegoda, J., Hettiarachi, J., Haggar, S., and Stressel, R. (2008) "Towards the development of sustainable landfills". Proc. of Geocongress 2008: Geotechnics of Waste Management and Remediation.

Kockel, R., Konig, D. and Syllwasschy, O. (1997) "Three basic topics: mechanics of waste mechanics". Proceedings of Fourteenth International Conference on Soil Mechanics and Foundation Engineering, 3, pp 1831- 1837.

Liu, J. Y., Xu, D. M and Zhao Y. C. (2004) "Long-term monitoring and prediction for settlement and composition of refuse in Shanghai Laogang municipal landfill". Environmental Management, 34, Issue 3, pp 441-448.

Machado, S. L., Carvalho, M. F., and Vilar, O. M. (2002) "Constitutive model for municipal solid waste". Journal of Geotechnical and Geoenvironmental Engineering, 128, Issue 11, pp 940-951.

Palmeira, E. M. (2005) "Geosynthetic reinforcement for soft foundations: Brazilian Perspectives". Geotechnical Special Publication , ASCE, 141, pp 1-11.

Qian, X. (1996) "Design of vertical landfill expansions". Michigan Department of Environmental Quality, Waste Management Division, Lansing, MI.

Qian, X., Koerner, R.M., and Gray, D.H. (2001) Geotechnical aspects of landfill design and construction, Prentice Hall, New Jersy.

Sharma, H. D., and De, A. (2007) "Municipal solid waste landfill settlement: Postclosure Perspectives". Geotechnical and Geoenvironmental Engineering, 133, Issue 6, pp 619-629.

Thusyanthan, N. I., Madabhushi, S. P. G., and Singh, S. (2006a) "Centrifuge modeling of solid waste landfill systems-Part 1: Centrifuge testing of model waste". Geotechnical Testing Journal, 29, Issue 3, pp 217-222.

Thusyanthan, N. I., Madabhushi, S. P. G. and Singh, S. (2006b) "Centrifuge modeling of solid waste landfill systems-Part 2: Centrifuge testing of model waste". Geotechnical Testing Journal, 29, Issue 3, pp 223-229.

Viswanadham, B. V. S., Rajesh, S., and Bouazza, A. (2012) "Effect of differential settlement on the sealing efficiency of GCLs compared to CCLs: Centrifuge study". Geotechnical Engineering Journal of the SEAGS & AGSSEA, 43, Issue 3, pp 55-61.

Wang, W. (2012) "A new model for settlement process of closed municipal solid waste landfill". Electronic Journal of Geotechnical Engineering, 17G, pp 947-958.

Yee, I. K. (1999). "Upgrading of existing landfills by dynamic consolidation" Master Builders Journal., pp 21-29, 210-214.

Zhang, Z. Y. and Chen Y. M. (2004) "Degradation settlement model of organic substance of municipal solid waste landfill". Rock and Soil Mechanics, 25, Issue 2, pp 238-241.

Zornberg, J. G. (2005). "Geosynthetic reinforcement in landfill design: US Perspectives". International Perspectives on Soil Reinforcement Applications, Geotechnical Special Publication, ASCE,141, pp 1-25.

Performance of Slopes Reinforced by Anchored Geosynthetic System Under Seepage Conditions: Centrifuge Study

A. Rajabian, H. Ghiassian[1] and B.V.S. Viswanadham[2]

Department of Civil Engineering, Science and Research Branch, Islamic Azad University, Fars, Iran
[1] *Department of Civil Engineering, Iran University of Science and Technology, Tehran, Iran*
[2] *Department of Civil Engineering, Indian Institute of Technology Bombay, Mumbai, India*
Email: viswam@civil.iitb.ac.in

ABSTRACT

Anchored geosynthetic system (AGS) can be utilized for stabilization of slopes as well as erosion control in coastal areas. In this study performance of slopes reinforced with AGS under seepage conditions are examined through centrifuge modeling technique. Results of two centrifuge model tests carried-out on 2Vertical:1Horizontal reinforced slopes having nine and twelve anchors. Anchor and geosynthetic layer were modeled according to the scaling considerations. Behavior of AGS slopes were evaluated by keeping anchor inclination and spacing of anchors as constant and varying number of anchors. With the help of linearly variable differential transformers (LVDTs) and pore water pressure transducers (PPTs) surface displacements from crest of the slope and development of phreatic surfaces within the slope were measured. Result of limit equilibrium analysis performed on AGS slope models show that the analysis can satisfactorily predict behavior of AGS slopes.

1. INTRODUCTION

Stability of infrastructures can be influenced by effects of climate change in urban areas. Global warming as an aspect of climate changes may result in sea level rise and increase in ground water level (Yasuhara et al. 2007). This issue may lead to seepage induced failure in infrastructures due to decrease in effective stress. Over the years, conventional armor systems such as riprap, gabions, concrete blocks and so on, have been utilized to protect coastal areas against erosive forces tending to reduce mass stability. Recently, using geosynthetics materials, new methods such as geotextiles wrap around revetment (Recio-Molina and Yasuhara 2005; Yasuhara and Recio-Molina 2007; Faure et al. 2010) or geotextiles containers (Saathoff et al. 2007; Lee and Douglas 2012; Yasuhara et al. 2012) are utilized for coastal protection.

Anchored geosynthetic system (AGS) was introduced by Koerner (1984) and Koerner and Robins (1986) to stabilize slopes and mitigate erosion. AGS includes a geosynthetic fabric such as geotextile, geonet or geogrid, provided with reinforced openings on the given grid pattern and small diameter metal rods as anchors. In practice, the fabric is draped on the ground surface and then anchors are driven to approximately 75 to 90 percent of their designed depth through the openings. Next, the anchors are fastened to the geosynthetic fabric and then driven to their final depth. Consequently, the fabric is tensioned and a curvature is imparted to the soil-geosynthetic interface. Due to tension induced in the fabric and curvature of the interface, compressive stresses develop on the slope surface leading to increase in confining pressure. As a result of installing of AGS, normal stresses on the potential slip surface increase, and shear strength parameters of the soil are improved due to compaction/consolidation of beneath soil. If geosynthetic fabric used in AGS is geogrids or geonets, vegetation covering can be combined with AGS to give more shear resistance and protection of the fabric against ultraviolet radiation.

Further studies on AGS were performed by Hryciw (1990, 1991), Vitton (1991), Hryciw and Haji-Ahmad (1992), and Hryciw and Irsyam (1992). Ghiassian et al. (1997) based on infinite slope theory presented an analytical solution giving factor of safety of AGS slopes under uniform seepage conditions. Experimental studies performed by Ghiassian (1995) showed that failure surface experienced by infinite slopes reinforced by AGS diverts from transitional and approaches to a curvilinear surface. In this respect, Ghiassian et al. (2009) suggested that limit equilibrium approach with adoption of a circular arc or composite failure surface can predict reasonably good results in infinite AGS reinforced slopes.

In this paper performance AGS slope models are evaluated under seepage conditions through centrifuge model tests at 50g. Two centrifuge model tests were carried-out on the AGS reinforced slopes having inclination of 2Vertical:1Horizontal and different number of anchors to examine influence of number of anchors on the stability and deformation behavior of AGS slopes. This paper presents a part of work performed by Rajabian et al. (2012).

2. CENTRIFUGE MODELING AND SCALING CONSIDERATION

The aim of geotechnical centrifuge modeling is providing identical stress field in model and prototype. In this respect, a model scaled down by 1/N prototype is placed in a centrifugal acceleration field of N times the normal gravity. With attention to mechanism of loading in centrifuge modeling, utilization of centrifuge modeling in studying performance of reinforced slopes such as soil nailing, AGS reinforced slopes, geotextile reinforced slopes and etc. is relevant.

The large beam centrifuge facility having 4.5 m radius available at the Indian Institute of Technology Bombay (IITB), India was used in the present study. The centrifuge has a capacity of 2500 g-kN can carry a maximum payload of 25 kN at 100 gravities. Specifications of the centrifuge were given elsewhere by Rajesh and Viswanadham (2009).

In order to relate data achieved from centrifuge based model tests to those of corresponding prototypes, it is necessary to develop scaling relationships linking model behavior to the prototype. In centrifuge modeling of AGS slopes, the scaling relationships pertaining to slope, anchors and geosynthetic layer are necessary to be established. With ignoring bending and shearing resistance of the anchors as conservatively, only pull-out strength and axial stiffness are adopted for scaling considerations of anchors.

The geosynthetic fabric is tensioned through anchors in AGS. Additionally, under seepage conditions, the fabric must act as a filter to prevent washout of fine particles and pore pressure build up. Hence, criteria pertaining to dual function of reinforcement and filtration were adopted to establish scaling relationships between model and prototype geosynthetic.

The geosynthetic fabric model was a geocomposite layer composed of a non-woven geotextile and a geogrid. This geocomposite can provide dual function of filtration and reinforcement. Two criteria including 1) tensile load-strain behavior and 2) frictional bond behavior along soil-geogrid interface were adopted to establish scaling consideration for reinforcement function of the fabric. Scaling considerations presented by Viswanadham and König (2004), Rajesh and Viswanadham (2009), Izawa and Kuwano (2010) and Rajesh and Viswanadham (2011) were used to achieve the modeling considerations for these criteria. In order to establish filtration function, characteristics of opening sizes and cross-plane flow of the nonwoven geotextile were selected as criteria. Therefore, equivalent opening size (EOS) and permittivity (ψ) were adopted as indices to achieve scaling factors for opening sizes and cross plane flow characteristics, respectively. Derivation of scaling factors for filtration function can be found in Rajabian et al. (2012). Table 1 presents a summary of scaling factors for centrifuge modeling of AGS slopes.

Table 1: A Summary of Scaling Factor for Centrifuge Modeling of AGS Slopes

Parameters	*Prototype*	*Model*
Anchor parameters		
Length, l_a(m)	#1	1/N
[a]Axial stiffness AE_a/l_f (kN/m)*	1	1/N
[a]Pull out strength, F_a (kN)	1	1/ N^2
Geosynthetic parameters		
Tensile strength, T_g(m)	1	1/ N
Secant stiffness, J_g (kN/m)	1	1/ N
Percentage open area, *f**(%)	1	1/ N^2
Interface friction angle, δ_{sg}	1	1
[a] Permittivity, ψ	1	1
[a] Equivalent opening size	1	1

* E_a: Elasticity modulus of tensile element of anchor; l_f: Free length of anchor; $f = a_l a_t/[(a_l+b_l)(a_t+b_t)]$, where a_l and a_t are grid opening sizes in longitudinal and transverse directions and b_l and b_t the widths and ribs in longitudinal and transverse directions, respectively; N: scale factor or g-level, $\#(l_a)_p/(l_a)_m = N$, $(l_a)_p$ and $(l_a)_m$ are length in prototype and model dimensions, respectively.

[a] After Rajabian et al. (2012)

3. MODEL MATERIALS

3.1 Model Soil

A silty sand was used for fabricating slope models. The soil was formulated using a blend of locally available fine sand from Goa state of India and commercially available kaolin in 4:1 ratio (by dry unit weight). According to Unified soil classification system, the sand was classified as SP. Hydrometer analysis carried out on kaolin showed that the kaolin has 58 % silt and 42 % clay size particles.

The Koalin was found to have liquid limit and plasticity index of 42.6% and 18.4%, respectively. It is classified as CL according to USCS. Properties of the soil used in modeling are presented in Table 2.

Table 2: Summary of Properties of Model Soil

Property	*Value*
Optimum moisture content (%)*	9
Max. dry density (kN/m^3)*	18.75
USCS	SM
Friction angle (°)#	27
Cohesion (kN/m^2)#	11.6
Coefficient of permeability (m/s)+	1.54×10^{-6}

* resulted from conducting standard Proctor test.

resulted from performing direct shear test at normal stresses: 50 kN/m^2, 100kN/m^2 and 150 kN/m^2

+ resulted from performing falling head test on moist compacted soil.

3.2 Model Anchor

In present study, pre-tensioned anchors were utilized for model anchor in AGS slope models. Mechanism of AGS in enhancing stability of slopes is irrespective of anchorage type. Polyester strands coated with polyvinyl chloride having 3 mm in breath and 1 mm in thickness were adopted to model the tensile element of anchors in free length. By performing tension tests on five polyester strands samples, a single polyester strand was found to have an average axial stiffness of 16.125 kN/m, tensile strength of 0.032 kN at 2 mm elongation (i.e. 2 % strain) and an ultimate tensile load of 0.211 kN at an ultimate strain of 6.8%. Based on observed failure surface of 2V:1H un-reinforced slope, reported by Raisinghani (2011) at IITBombay,

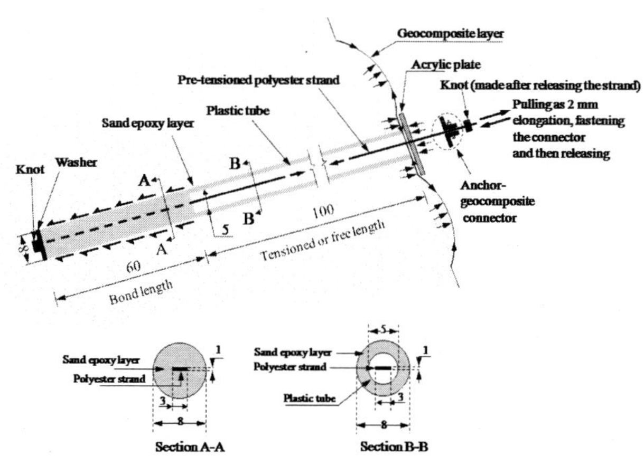

FIGURE 1: Model anchor designed for AGS slope models (all dimensions are in mm)

free length portion of the anchor was designed as 100 mm. With adoption of this length, the anchors exceed the failure surface. A length of 60 mm was designed for bond length portion of the anchor to ensure sufficient pull-out strength. A uniform blend of epoxy resin and coarse quartz sand (having particle sizes ranging from 0.5 mm to 1 mm) was used to grout the strand in the bond portion. A thin plastic tube with diameter of 5 mm coated with grout, was used in free portion to prevent the polyester strand from coming in contact with the around soil. Figure 1 shows details of model anchors designed in the present study.

3.3 Model Fabric

A geocomposite layer including a non-woven geotextile and a scaled-down model geogrid was prepared and used as AGS fabric in slope models. Table 3 present a summary of properties of non-woven geotextile, geogrid and geocomposite utilized in present study.

Table 3: A Summary of Properties of Components of Geocomposite Used in Present Study (after Rajabian et al. 2012)

Property	*Non Woven Geotextile*	*Geogrid*
Mass per unit area (g/m^2)	46	10.34
Thickness t_g (mm)	0.66	[a]-
Percentage open area f (%)	[a]-	93.4
Apparent opening size (mm)	0.18	[a]-
Ultimate tensile load (kN/m)	0.23	0.89
Ultimate strain (%)	5.77	47.25

[a] not relevant

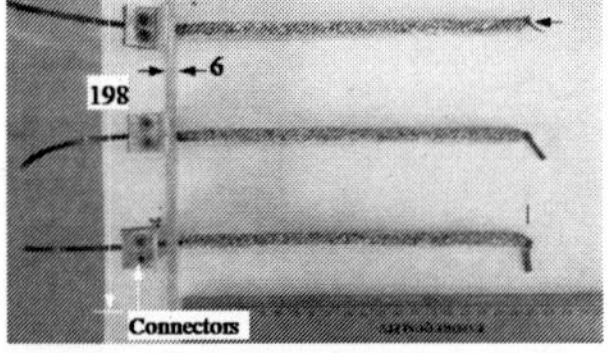

FIGURE 2: A view of acrylic plate, connectors and anchors

Results of two wide width tensile tests (ASTM D-4595, 2005) performed on model geocomposite revealed that the geocomposite has an average ultimate tensile load of 1.22 kN/m at an ultimate strain of 15%. In order to transfer the anchor loads to the geosynthetic fabric, a connection system including custom made L-shaped connectors and a 6 mm thick acrylic plate having 198 mm in length and 30 mm in breadth was employed. Figure 2 shows acrylic plate, connectors along with the anchors fabricated in present study.

4. TEST PACKAGE, MODEL PREPARATION AND TESTING PROGRAM

A model container having internal dimensions of 760 mm in length, 200 mm in width and 410 mm in height was used to house the model in the present study. The container includes a thick transparent Perspex sheet to monitor status of the model during testing through a digital camera. The other walls of the container are made of well-machined mild steel plates. The container was equipped with a seepage simulator system including seepage tank, solenoid valve and water reservoir to simulate rising of ground water table in the model. Water was allowed to move from reservoir tank to seepage tank through the solenoid valve. The seepage tank was provided with 2 mm diameter pre-punched holes in its side contacting with soil. Moreover, this side was covered with a thin non-woven geotextile layer to prevent washing of fine particles.

For monitoring soil movement during the flight through image analysis technique, a rectangular grid of permanent markers with horizontal and vertical spacing of 350 mm and 210 mm, respectively were firmly pasted on the inner side of front wall. The models were instrumented by linear variable differential transformers (LVDT) and pore water pressure transducers (PPT) to measure surface settlements of slope models and developed pore water pressure within the models, respectively. A stand pipe installed at the bottom of seepage tank to maintain head of water as 5 mm less than slope height. Figure 3 depicts cross section of model test set up developed in present study.

In order to reduce friction between container walls and soil, internal sides of front and rear walls were covered with a thin layer of white petroleum grease and thin polythene sheet strips of 100 mm. In present study two AGS slope models having height of 240 mm and inclination of 2V:1H (63.4°) were fabricated on the base layer having thickness of 60 mm. Three LVDTs and five PPTs were placed at specific points as shown in Figure 3.

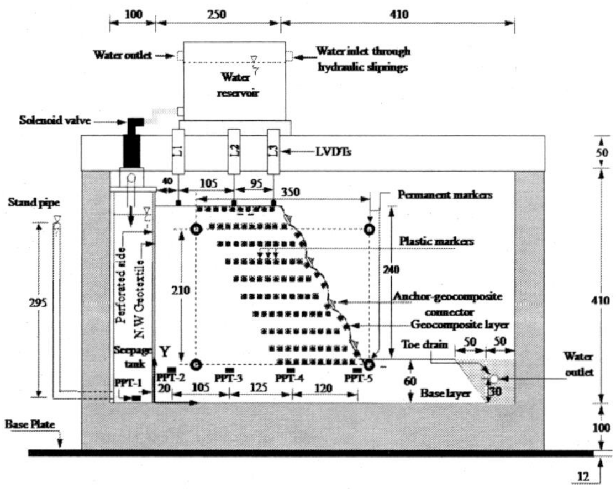

FIGURE 3: Cross section of model test set up developed in present study

All models were moist-compacted in layers having thickness of 30 mm at optimum moisture content and maximum dry unit weight. Inclination of slope models was achieved by employing a temporary wooden support while compacting the layers at normal gravity. In order to trace movement of the soil through image analysis technique, ten L-shaped plastic markers spaced 20 mm centre to centre were placed on the top of soil layers. A red colored food dye was placed at selected locations on the top surface of each layer to trace the movement of water within the slope body.

A custom designed template was used to drill the soil and install the anchors at 15° inclination with horizontal. After inserting the anchors, areas between anchors row were slightly shaped as circular arc. Whit this arrangement the anchor loads is well transferred on slope face through inducing tension in fabric. The geocomposite layer having pre-punched holes was laid on the face of the slope. Two ends of the geocomposite were sufficiently anchored at toe and crest of the slope. Next, the strands were passed through the geocomposite and acrylic plates and were pulled for about 2 mm elongation (corresponds to 32.1 N according to tension test results on polyester strand). While holding the tensioned strand, connectors were fastened and then the strand was released. In order to prevent possible slippage of the strand through connectors, a knot was made at the exposed portion of the strands. Figure 4 shows a view of prepared AGS model placed on swing basket before spinning.

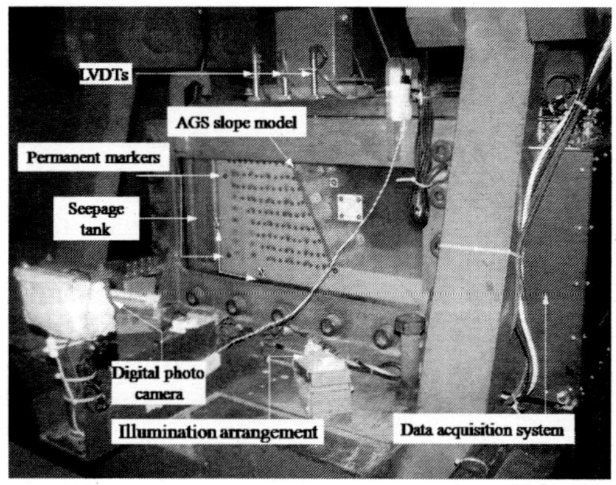

FIGURE 4: A view of prepared AGS slope model before spinning on the swing basket

In this paper, two AGS slope models having nine and twelve number of anchors were tested at 50g acceleration. Table 3 gives details of AGS slope models tested in present study.

Table 3: Details of AGS Slope Models

Parameter	*Test legend*	
	RRS 4	*RRS 5*
No. of anchors	9	12
S_V	60(3000)[a]	60(3000)
S_H	60(3000)	60(3000)
Anchor load (kN)	0.032(80)	0.032(80)
Anchor inclination (°)[b]	15	15
Anchor length (mm) $[L_b+L_f]$	160 [60+100]	160 [60+100]
Slope inclination (°)	63.4	63.4
Slope height (mm)	240	240

S_H: Horizontal spacing of anchors, S_V: vertical spacing of anchors, L_b: bond length of anchor, L_f: free length of anchor

[a] values given in parenthesis are in prototype dimensions at 50g

[b] with horizontal

After reaching to 50g corresponding to 104 revolutions per minute, solenoid valve was opened to allow percolation of water through the model. Using a digital photo camera mounted along with the model status of front elevation of the model was monitored. Each test was lasted up to occurrence of failure or large deformation. LVDTs and PPTs data were recorded at one second intervals through the data acquisition system.

5. RESULTS AND DISCUSION

In this study performance of AGS slope models were evaluated with raising ground water table at 50g acceleration. An unreinforced slope model with inclination of 63.4° was tested by Raisinghani and Viswanadham (2011) at 30g. the unreinforced model slope was stable at 30g, but underwent catastrophic failure with inducing seepage within 5.06 days. Figures 5a-b present variation of pore water pressure measured by PPTs. After 5 to 6 days, steady state seepage conditions were established in AGS models. Figures 6a-b present surface settlements measured by LVDTs with seepage time in days for AGS slope models. In these figures, values of normalized pore water pressure during various time of seepage are also shown. The normalized pore water pressure (u/γH) is defined as the ratio of measured pore water pressure obtained from PPT3 placed at 125 mm (6.25 m) from crest of the slope to the product of bulk unit weight of the soil and height of the slope H. Figures 7a-b present the variation of measured surface settlements with the horizontal distance from crest of the slope for different seepage times in prototype dimensions. In these figures, values of u/γH (for PPT3) for AGS models are also shown. Seepage time is time elapsed after attaining to 50g.

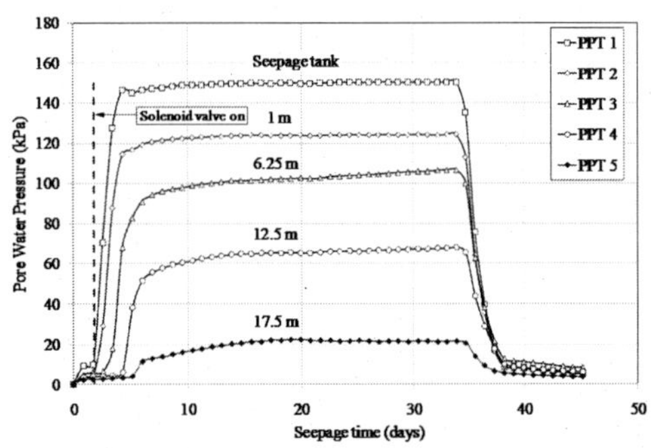

(a) AGS slope model with nine anchors (Model RRS 4)

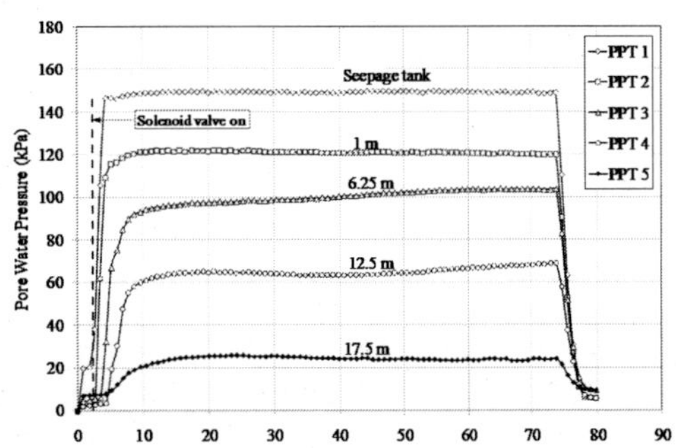

(b) AGS slope model with twelve anchors (Model RRS 5)

FIGURE 5: Variation of measured pore water pressure with seepage time in prototype dimensions at 50g

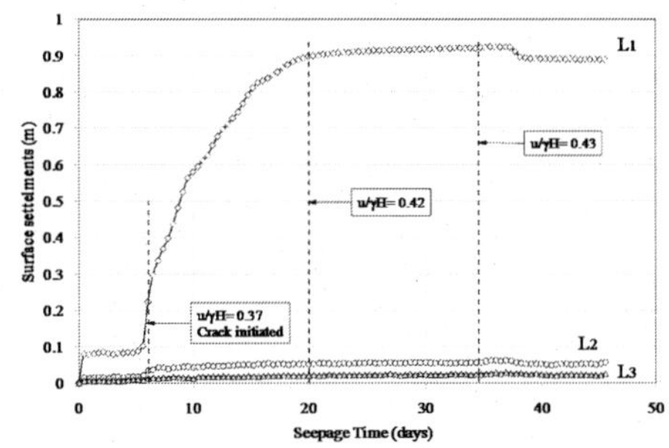

(a) AGS slope model with nine anchors (Model RRS 4)

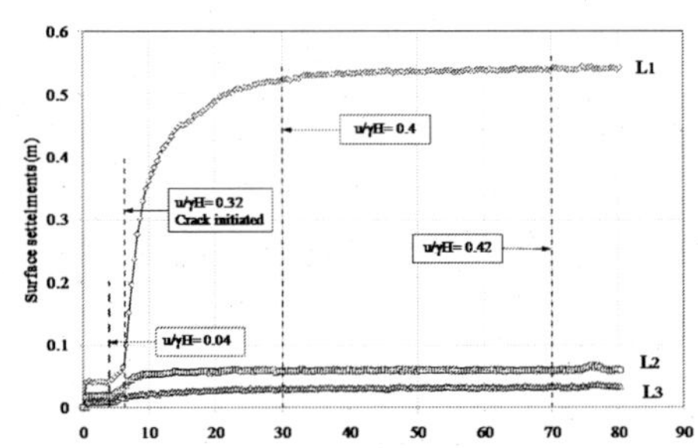

(b) AGS slope model with twelve anchors (Model RRS 5)

FIGURE 6: Measured variation of surface settlements with seepage time in prototype dimensions at 50g

As can be observed from Figures 6 and 7, surface settlements increase with development of pore water pressure within slope models. Unreinforced slope model was found to have crest settlement of 2.5 m in prototype dimensions at 30g (Raisinghani and Viswanadham (2011). In comparison, AGS slope model RRS 4 with nine even at 50g experienced maximum crest settlement of 0.92 m (7.67 % height of the slope) at seepage times of 34.5 days. This indicates effectiveness AGS in crest settlements and deformation behavior. Maximum crest settlement for AGS slope model RRS 5 with twelve anchors was 0.54 m (4.5% height of the slope) at 72.2 days seepage times showing influence of increase in number of anchors in decreasing crest settlements of AGS slope models under seepage conditions.

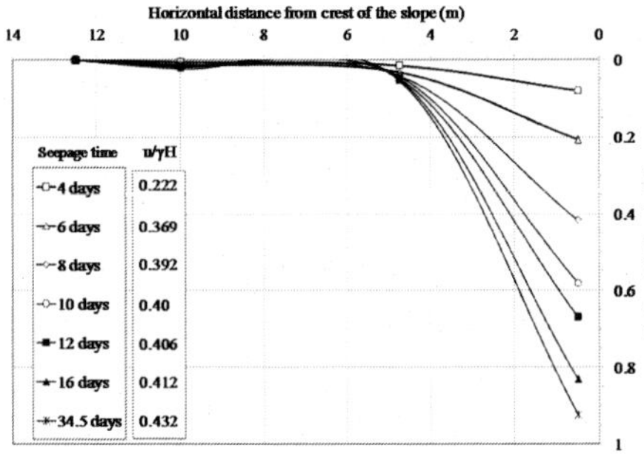

(a) AGS slope model with nine anchors (Model RRS 4)

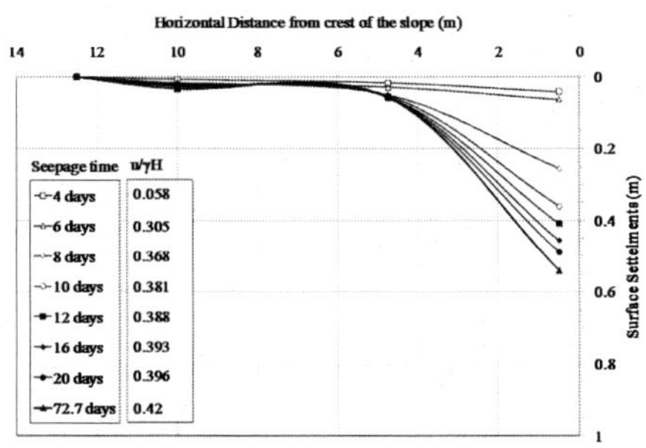

(b) AGS slope model with twelve anchors (Model RRS 5)

FIGURE 7: Variation of surface settlement with horizontal distance from crest of the slope in prototype dimensions at 50g

Both AGS slope models suffered progressive failure at onset of seepage. A circular failure surface was observed for AGS slope models. Figure 8a-b shows the front elevation of the slope models during penultimate stages of centrifuge tests at 50g. Exhumation of anchors embedded within the slope models after the tests revealed that anchors experienced ruptured during the test due to effect of seepage. Status of exhumed anchors is also shown in the Figure 8. Bulging of the geocomposite fabric at the toe of slopes during failure shows that fabric has a significant role in

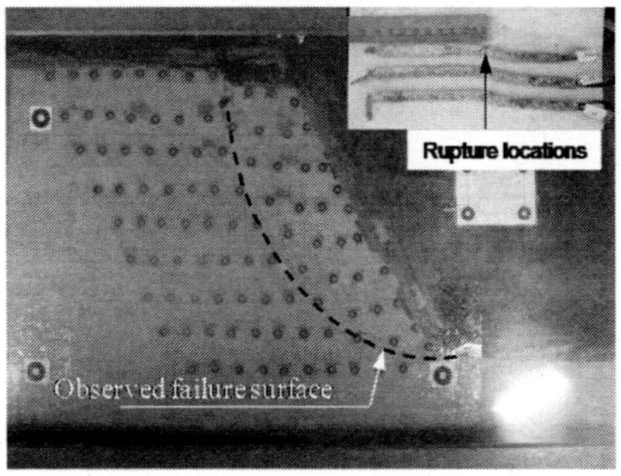

(a) AGS slope model with nine anchors (Model RRS 4)

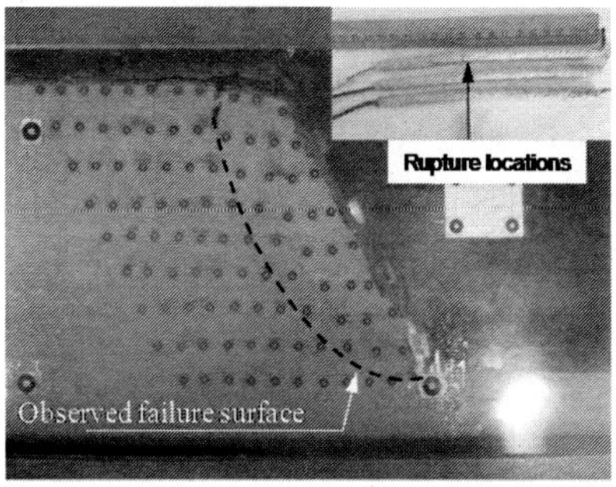

(b) AGS slope model with twelve anchors (Model RRS 5)

FIGURE 8: Front elevation of AGS slope models during penultimate stages of centrifuge tests at 50g

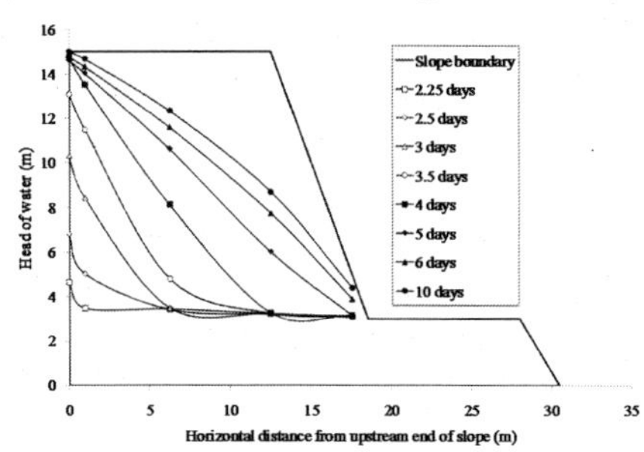

(a) AGS slope model with nine anchors (Model RRS 4)

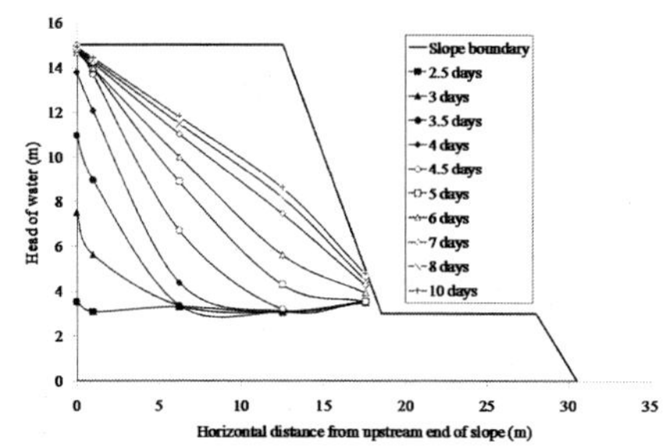

(b) AGS slope model with twelve anchors (Model RRS 5)

FIGURE 9: Developed phreatic surfaces at different seepage times in prototype dimensions at 50g

controlling and mitigating deformation of AGS slopes after rupturing the anchors.

Using pore water pressure data obtained from PPTs, phreatic surfaces developed at different seepage times can be achieved. Figures 9a-b depict developed phreatic surfaces corresponding to different seepage times for AGS slope models in prototype dimensions at 50g.

5.1 Image Analysis

Digital image analysis technique was employed to monitor and measure displacement vectors of the soil particles. With the help of this technique failure surface can be identified in slope models. The image analysis was performed using GRAM++ software (Gram ++ 2004). Captured images during various stages of the tests were scaled and calibrated using permanent markers pasted at specific rectangular pattern on inner side of acrylic sheet at 1g (Fig. 3). Figures 10a-b depict displacement vectors since moment of setting solenoid valve to on and to penultimate stages of the test in prototype dimensions for two AGS slope models. Figure 11a-b gives face movement normalized to the height of the slope (S_f/H) for AGS slope models against normalized height (y/H) at different seepage times in prototype dimensions at 50g. y is referred as height measured from base layer. Values of normalized pore water pressure at the toe are also shown in these figures. As can be noted, normalized face movement for AGS slope model RRS 4 is S_f/H = 0.084 m within 34 days subjecting to seepage. On the contrary, AGS slope model RRS 5 with twelve anchors experienced normalized face movement of 0.070 m after 70 days subjecting to seepage. These findings highlight influence of increase in number of anchors in decreasing face movement of AGS slope models.

5.2 Stability Analysis

The limit equilibrium based stability analysis was

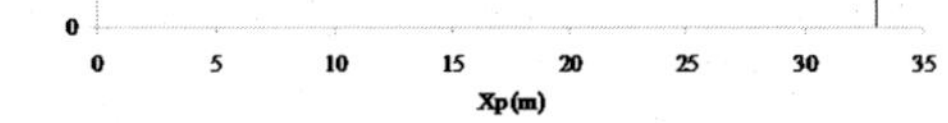

(a) AGS slope model with nine anchors (Model RRS 4)

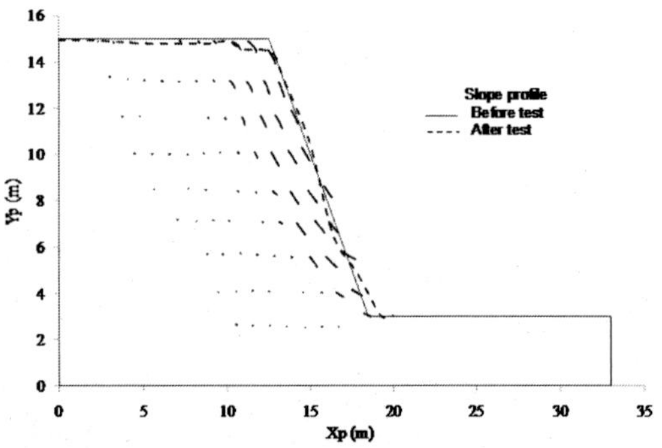

(b) AGS slope model with twelve anchor (Model RRS 5)

FIGURE 10: Displacement vectors resulted from image analysis in prototype dimensions at 50g

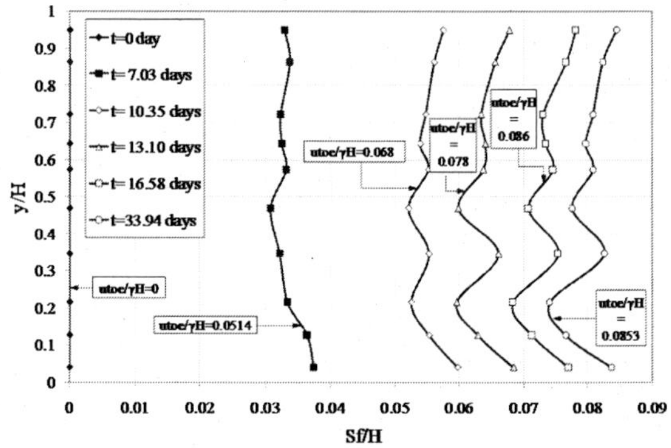

(a) AGS slope model with nine anchors (Model RRS 4)

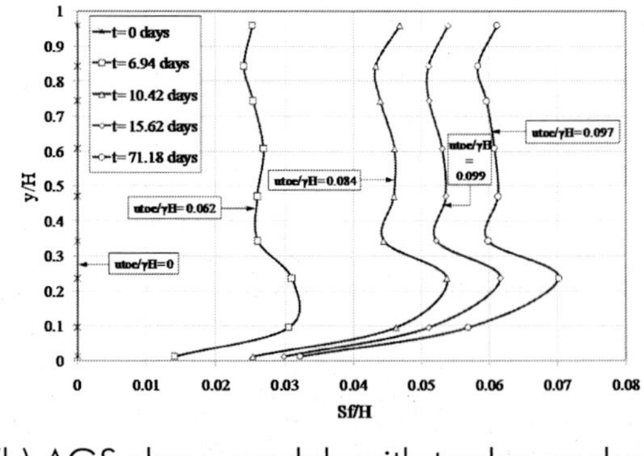

(b) AGS slope models with twelve anchors (Model RRS 5)

FIGURE 11: Face movement of slope models a different seepage time in prototype

carried-out for AGS slope models in prototype dimensions. The analysis was performed with adoption of Bishop's modified method and using Slide software (Rocscience, 2005). With the help of image analysis technique, displacement of end of the anchor were calculated and corresponding forces was obtained from tensile load-elongation curve resulted from tension test. Summation of anchor loads with inclination of 15° with horizontal was uniformly distributed over slope face. Further, phreatic surfaces obtained from PPTs (Fig. 9) were given to the analysis. Figure 12 presents variation of factor of safety with u/γH for AGS slope models. As can be observed, Value of factor of safety decreases with increase in u/γH. The factor of safety calculated for AGS slope models RRS 4 and RRS 5 are 1.043 and 1.062 at verge of failure. Therefore results of stability analysis are in good agreement with those of centrifuge models.

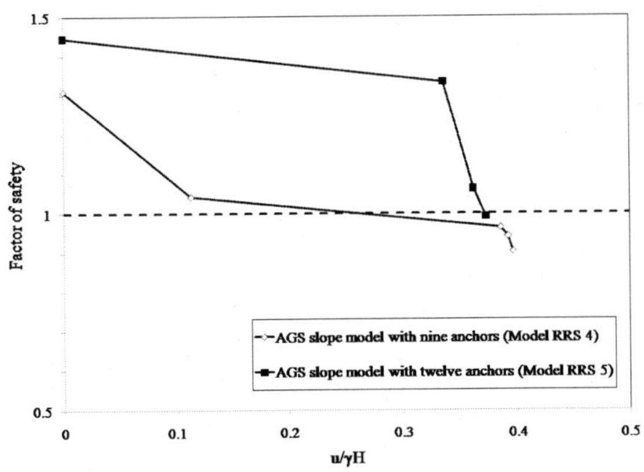

FIGURE 12: Variation of factor of safety with u/γH for AGS slope models

6. CONCLUSIONS

In this paper, performance of slopes reinforced by AGS is evaluated through centrifuge modeling technique. Based on results of centrifuge model tests and analysis carried-out on AGS slope models having different number of anchors, the following conclusions can be drawn:

1. Unreinforced slope with an inclination of 63.4°(2V:1H) was found to be stable before rising of ground water table at 30g. This slope catastrophically failed with rising ground water table. However, AGS slope model having nine anchors with pattern of 3x3 underwent a progressive failure at 50g. This finding confirms effectiveness of AGS in improving stability of slopes.
2. AGS slope model with nine and twelve number of polyester strand anchors were found to have maximum crest settlement of 3.03 m (0.252 H) and 0.05m (0.0042H) during penultimate stages of the test, respectively. Further, normalized face movements for AGS slope models with nine and twelve anchors were found to be 0.084 m and 0.070 m, respectively. Hence, with increase in number of anchors, deformation behavior of AGS slope is improved.
3. Results of stability analysis were found to be in good agreement with those of centrifuge tests.

REFERENCES

American Society for Testing and Material (ASTM). (2005) Standard Test Method for Determining Tensile Properties of geotextile by Wide-Width Strip Method. American Society for Testing and Materials. Philadelphia, USA. D 4595

Faure, Y-H., Ho, C.C., Chen, R-H., Le Lay M., Blaza, J. (2010) "A wave flume experiment for studying erosion mechanism of revetments using geotextiles". Geotextiles and Geomembranes, 28, Issue 4, pp 360-373.

Ghiassian, H., Gray, H. D., Hryciw, R. D. (1997) "Stabilization of coastal slopes by anchored geosynthetic systems". Journal of Geotech. and Geoenv. Engrg, ASCE, 123, Issue 8, pp 736-743.

Ghiassian, H., Jalili, M., Kasemi, D. (2009) "Model study of sandy slopes under uniform seepage and reinforced with anchored geosynthetics". Geosynthetics International, 16, Issue 6, pp 452-467.

GRAM++. (2004). Technical Document. CSRE, Indian Institute of Technology Bombay. http://www.csre.iitb.ac.in/gram++.

Hryciw, R. D. (1990). Load transfer mechanism in anchored geosynthetic systems. Report No. 88-0166, Univ. of Michigan, Ann Arbor, U.S.A.

Hryciw, R. D. (1991). "Anchor design for slope stabilization by surface loading". Journal of

Geotech. Engrg, ASCE, 117, Issue 8, pp 1260-1274.

Hyrciw, R. D. and Haji-Ahmad, K. (1992) "Slope stabilization by anchored geosynthetic systems: anchorage optimization". Proceedings of the ASCE Specialty Conference on slopes and Embankments, Berkeley, CA, ASCE, Reston, VA, USA, Geotechnical Special Publication, Issue 31, pp 1464-1480.

Hyrciw, R. D. and Irsyam, M. (1992) "Pullout stiffness of elastic anchors in slope stabilization subjected to pseudo-static loading". Int. Journal of Physical Modeling in Geotechnics, 10, Issue 1, pp 1-18.

Lee, E.C. and Douglas, R.S. (2012) "Geotextile tubes as submerged dykes for shoreline management in Malaysia". Geotextiles and Geomembranes. 30, pp 8-15.

Koerner, R. M. (1984) "In-situ soil stabilization using anchored nets". In: Fang, H. Y. (Editors.), Low Cost and Energy Saving Constraction Methods, Rio De Janiero, Brazil, Envo Publishing, pp 465-478.

Koerner, R. M. and Robins, J. C. (1986) "In-situ stabilization of slopes using nailed geosynthetics". Proc. of Third Int. Conference on Geotextiles, Vienna, Austria. Industrial Fabric Association Intl. pp 395-400.

Raisinghani , D. V., Viswanadham, B. V. S., (2011) "Centrifuge model study on low permeable slope reinforced by hybrid geosynthetics" Geotextiles and Geomembranes. 29, Issue 6, pp 567-580

Rajabian, A., Viswanadham, B.V.S., Ghiassian, H., and Salehzadeh, H. (2012) "Centrifuge model studies on anchored geosynthetic slopes for coastal shore protecton". Geotextiles and Geomemranes. 34, pp 144-157.

Rajesh, S. and Viswanadham, B. V. S. (2009) "Evaluation of geogrid as a reinforcement layer in clay based engineered barriers". Applied Clay Science, 46, Issue 2, pp 153-165.

Rajesh, S. and Viswanadham, B. V. S. (2011) "Hydro-mechanical behavior of geogrid reinforced soil barriers of landfill cover systems". Geotextiles and Geomembranes, 29, Issue 1, pp 51-64.

Recio-Molina, J., and Yasuhara, K. (2005) "Stability of modified geotextile warp-around revetment (GWR) for coastal protection". Geosynthetic International Journal. 12, Issue 5, pp 260-268.

Rocscience, 2005. Slide Ver. 5.0-2D Limit Equilibrium Slope Stability Analysis. Rocscience, Inc., Toronto, Canada.

Saathoff, F., Oumeraci, H., and Restall, S. (2007) "Australian and German experiences on the use of geotextile containers". Geotextiles and Geomembranes, 25, Issues 4-5, pp 251-263.

Viswanadham, B. V. S. and König, D. (2004) „Studies on scaling and instrumentation of a geogrid". Geotextiles and Geomembranes. 22, Issue 5, pp 307-328.

Vitton, S. J. (1991) "Load transfer mechanism in anchored geosynthetic systems". Ph.D. Thesis, Univ. of Michigan, Ann Arbor, USA.

Yasuhara, K., and Recio-Molina, J. (2007) "Geosynthetic-wraparound revetments for shore protection". Geotextiles and Geomembranes, 25, Issues 4-5, pp 221-232.

Yasuhara, K., Komine, H., Murakami, S., Chen, G., Mitani, Y. and Duc, D.M. (2012) "Effects of climate change on geo-disasters in coastal zones and their adaptation". Geotextiles and Geomembranes. 30, pp 24-34.

Yasuhara, K., Murakami S., Mimura N., Komine H., and Recio J. (2007) "Influence of global warming on coastal infrastructural instability". Journal of sustainability science, 2, Issue 1, pp 13-25.

Seepage Performance of Hybrid Geosynthetic Reinforced Slopes

B.V.S. Viswanadham and D.V. Raisinghani[1]

Department of Civil Engineering, Indian Institute of Technology Bombay; Powai, Mumbai
[1]Textile Manufactures Department, Veermata Jijabai Technological Institute, Mumbai
E-mail: viswam@civil.iitb.ac.in

ABSTRACT

The use of granular backfill is recommended and widely preferred for slopes prone to seepage. However the cost involved in its transportation and construction often leads to the use of locally available low permeable soil for these structures. The present study is aimed at studying the seepage behaviour of low permeable backfill slope subjected to seepage using centrifuge modelling technique. The low permeable backfill slopes were constructed with and without using hybrid geosynthetics and the seepage was induced by raising the ground water table using seepage flow simulator. Centrifuge model tests could demonstrate the effectiveness of hybrid geosynthetics in dissipating excess pore water pressures.

1. INTRODUCTION

The geosynthetic reinforced systems have wide acceptance as a soil reinforcement technique due to the ease of construction and low cost benefits offered by it. The geosynthetic reinforcement layers provide resistance to driving forces developed in the soil mass while the resistance against seepage forces can be achieved with the use of granular backfill recommended for its construction due to its good drainage characteristics. The need of such a select fill within the reinforced zone completely eliminates the economic advantage offered by such structures as the cost of the fill depends on its availability. The absence of such granular fill material forces the need to use locally available low permeable soil as a fill material. The structure thus actually designed for drained condition actually works in undrained condition resulting in excess pore water pressures being generated within the soil mass due to lack of drainage. Cases have been reported about failure of reinforced soil walls constructed with relatively low permeability and investigations have revealed the use of low permeability backfill coupled with insufficient precautions considered for drainage as the prominent reasons (Yoo and Jung, 2006).

Nonwoven geosynthetics possess good drainage capability due to its highly porous structure and could be a possible solution for the use of locally available low permeable soil. Ling et al. (1993), Ghosh and Yasuhara (2004), Hara et al. (2007) and Raisinghani and Viswanadham (2010) have studied the in-plane flow characteristics of nonwoven geosynthetics within fine grained soils. However their low tensile stiffness and high deformability characteristics restricts its use to serve the reinforcement function. The combination of geogrid and a suitable nonwoven geosynthetic, termed as "Hybrid Geosynthetic" can serve the dual function of drainage and reinforcement in geosynthetic reinforced structures constructed with locally available low permeable backfill. Some hybrid geosynthetics already exists in commercial market and are in use in areas of landfill construction, road construction, soft ground improvement, etc. Yasuhara et al. (2002) and Garcia et al. (2006, 2007) have demonstrated the use of geocomposites for some applications.

This paper presents the significance of hybrid geosynthetic layers on the stability and deformation behaviour of a low permeable backfill reinforced slope subjected to seepage. The influence of seepage on the integrity of an unreinforced, geosynthetic reinforced and hybrid geosynthetic reinforced slope has been evaluated by performing centrifuge tests. The seepage phenomenon in model slopes reinforced with and without reinforcement layers has been studied by Resnick and Znidarcic (1990), Hiro-oka et al. (2001) and Viswanadham and Raisinghani (2010). The centrifuge modelling technique was chosen for the present study as it gives identical stress levels as in prototype and lesser time required for studying the seepage behaviour of low permeable backfill slope. Scaling considerations for modelling hybrid geosynthetic layers in a centrifuge are also presented. The centrifuge tests were performed in a 4.5m radius beam centrifuge facility at IIT Bombay, India (Chandrasekaran, 2001). The centrifuge tests were performed at 30 gravities by maintaining slope inclination (2V:1H) and vertical spacing of reinforcement layers as constant. The model preparation, analysis and interpretation of centrifuge test results for slopes with and without hybrid geosynthetic layers are presented followed by stability analysis of the centrifuge models carried out using SLOPE/W software packages (Geo- Slope, 2007).

2. SCALING CONSIDERATIONS FOR MODELING HYBRID GEOSYNTHETICS

The centrifuge modelling technique requires scaling down of various parameters that determine the prototype response. The modelling of hybrid geosynthetic reinforced slopes requires the scaling of parameters related to soil, geosynthetics and reinforced slope. Scaling requirements for the soil fill material (soil density and soil friction angle) can be satisfied by building the model with the same fill material as used in prototype. For modelling of geosynthetic reinforced slopes, dimensions of the slope and reinforcement layers as well as the reinforcement spacing need to be modelled. Table 1 summarizes the scale factors relevant to modelling of hybrid geosynthetic reinforced slopes in a centrifuge. The scaling considerations relevant to the use of geosynthetics in the present study involve modelling of hybrid geosynthetics to serve the dual function of drainage and reinforcement.

2.1 Modelling Reinforcement Geosynthetic

The hybrid geosynthetic used in the present study is a combination of model geogrid and permeable nonwoven geosynthetic. The selection criterion for the model geosynthetic reinforcement layer was based on scaling down the tensile strength–strain behaviour as well as reproduction of identical frictional bond between soil-geogrid-soil by selecting model geogrid with identical percentage open area (f) as prototype geogrids. The percentage open area (f) is the ratio of area formed by grid opening sizes to area formed by grid opening sizes measured up to centre of width of ribs expressed in percentage. The modelling considerations for the reinforcement component have been based on the scaling considerations given by Viswanadham and König (2004), Rajesh and Viswanadham (2009) and Izawa and Kuwano (2010).

2.2 Modelling Drainage Geosynthetic

The modelling considerations for the drainage behaviour have been discussed by Raisinghani and Viswanadham (2011). The scaling consideration for the drainage behavior is based on the in-plane flow characteristics of the geosynthetic material determined by its transmissivity θ, expressed as a product of coefficient of in-plane permeability of geosynthetic k_h and its thickness t_g (Equation 1)

$$\theta = k_h \, . \, t_g \tag{1}$$

For similitude in the in-plane flow behaviour of model and prototype geosynthetic material,

$$\theta_m = \theta_p \tag{2}$$

The physical dimension t_g (thickness) of geosynthetic is scaled down by a scale factor N. Thus, from equation (1) and (2)

$$(k_h)_m = N\,(k_h)_p \tag{3}$$

Thus, the scale factor for the coefficient of in-plane permeability of geosynthetic $((k_h)_m$ should be N times that of prototype $(k_h)_p$.

Several other parameters of geosynthetics also influence the performance of hybrid geosynthetic and hence perfect scaling down of the prototype materials to the desired scale factor may not be feasible. Hence an attempt was made to model hybrid geosynthetic based on the above scaling considerations and representing bandwidth of the characteristics of the existing commercially available prototype hybrid geosynthetic. The scaling relationship for the in-plane permeability of geosynthetic remains to be validated.

Table 1: Scale Factors for Centrifuge Modelling of Hybrid Geosynthetic Reinforced Slope

Quantity	*Ng Model*
Geosynthetic parameters	
Tensile load, (T_g)	1/N
Thickness (t_g)	1/N
Transmissivity (θ)	1
In-plane permeability (k_h)	N
Reinforced Slope parameters	
Height (H)	1/N
Length of reinforcement (L_r)	1/N
Vertical spacing of reinforcement (S_v)	1/N
Seepage parameters	
Coefficient of permeability (k)	N
Seepage velocity (v_s)	N
Time (diffusion) (t)	$1/N^2$
Pore water pressure (u)	1

N - Scale factor or gravity level

3. MODEL PREPARATION AND TEST PROCEDURE

3.1 Model materials

3.1.1 Soil

The model soil used to represent a low permeable backfill for the present study was formulated using a blend of locally available fine sand and commercially available kaolin in proportion of 80:20. The mix properties were selected such that it has percentage fines equal to 20% so as to represent low permeable backfill. The coefficient of permeability (k) for this mix determined by performing falling head test was 1.54 x 10^{-6} m/s. The formulated soil is classified silty sand (SM type soil) according to Unified Soil Classification System (USCS). It had a specific gravity of 2.62, a maximum dry unit weight of 18.75 kN/m^3 and optimum moisture content of 9% (standard Proctor compaction). The cohesion of the soil mix and angle of internal friction determined using direct shear test on moist compacted sample was found to be 11.6 kN/m^2 and 27° respectively.

3.1.2 Model Geosynthetics

Based on the modelling considerations, a model nonwoven geotextile (N3) and a fine model geogrid (G2) with hexagonal shape openings were selected for forming a hybrid geosynthetic, to serve the dual function of drainage and reinforcement. The model geogrid (G2) was purposefully selected with low tensile properties so as to induce failure for geogrid reinforced slopes in centrifuge and develop a hybrid geosynthetic with a nonwoven geotextile component that would have drainage capabilities. The tensile load-strain characteristics of the model geosynthetic materials were determined by performing wide-width tensile tests at a strain rate of 10 mm/min. The model geogrid G2 had ultimate tensile load of 0.31 and 0.23 kN/m with an ultimate strain of 9.39 % and 10.87 % in machine and cross machine direction respectively. The model nonwoven geosynthetic N3 had ultimate tensile load and strain values of 0.23 kN/m and 5.77% in machine direction and tensile load of 0.11 kN/m and strain of 48.14% in the cross machine direction respectively. Based on the requirements of scaling consideration along with compatibility in tensile load and strain values with the nonwoven geotextile counterpart N3, it was chosen to lay the model geogrid G2 in cross-machine direction and model nonwoven N3 in the machine direction for forming the model hybrid geosynthetic G2N3. The properties of the model geosynthetic materials are summarized in Table 2. Studies have shown that the straining of geosynthetic layers in a geosynthetic reinforced slope occurs over a smaller length. This is due to the confinement of the geosynthetic material on either side of the potential failure surface by the soil. Hence, tensile load tests with closer spacing of the clamps (about 5 mm) were conducted for the model geosynthetic samples G2 and G2N3, to represent the effect of confinement. The values are reported in Table 3 and were used for the stability analysis of the reinforced slope models.

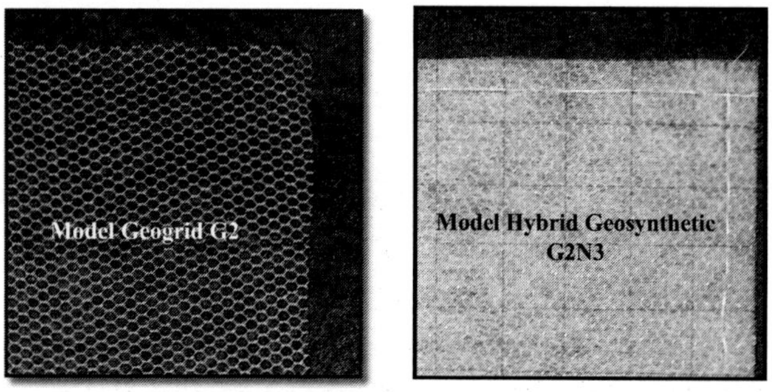

FIGURE 1: Model Geosynthetic materials used for present study

Table 2: Properties of Model Geosynthetic Material

Property	*Geosynthetic Type*		
	G2	*N3*	*G2N3*
Composition and type	Fine geogrid	NPNW	Fine geogrid + NPNW
Mass per unit area (g/m^2)	11.25	46	-[a]
Thickness (mm)	-[a]	0.66	-[a]
# Ultimate Tensile strength (kN/m)			
MD	0.31	0.23	0.56*
XMD	0.23	0.11	-[a]
Ultimate Strain (%)			
MD	9.39	5.77	17.36*
XMD	10.87	48.14	-[a]
Zero span test (kN/m)	0.45	-	1.29*
@ Transmissivity θ (x 10 -6 m^2/s)	-[a]	0.38	1.72

Wide width tensile tests , * Strength in the direction laid in the model slope, -[a] Not relevant/Carried-out, @ - Radial transmissivity; MD- Machine Direction, XMD – Cross Machine Direction, NPNW - Needle Punched Nonwoven

The model hybrid geosynthetic was prepared by loosely stitching the model geosynthetics G2 and N3 at the edges so that they function together as a single material. The ultimate tensile load and strain for the model geocomposite (in the laid direction) are 0.6 kN/m and 17.8% respectively. The thickness of model hybrid geosynthetic is 0.25 mm at 2 kPa and its transmissivity is 0.62×10^{-6} m^2/s at 20 kPa.

Figure 1 shows the photograph of the model geogrid G2 and developed model hybrid geosynthetics G2N3 used for present study.

3.2 Model Preparation and Test Procedure

A model container having internal dimensions 760 mm in length, 200 mm in width and 410 mm in height was used to house the reinforced slope model. The front wall of the model container was made of thick transparent Perspex sheet that enabled to view the model continuously during flight with the help of an on-board digital camera. The other three walls of the model container were made from mild steel plates. A rectangular grid of permanent markers 350 mm × 210 mm in size was pasted firmly on the inner side of the Perspex sheet. These served as a reference for observing the relative displacement of discrete markers glued to the geosynthetic reinforcement layers during various stages of the test with respect to these permanent markers. For unreinforced slope model, discrete plastic markers were embedded within the soil prior to compaction of each layer. A thin layer of white petroleum grease was applied to the inner side of the Perspex glass and the rear wall of the model container and flexible polyethylene strips of 100 mm width were placed above it. This was done to reduce friction between the walls of the container and the model soil and simulate plane strain conditions.

The seepage in the slope models was induced with the help of a seepage tank having internal dimensions 80 mm in length, 360 mm in width and 200 mm in height. The seepage tank was housed at the left end of model container and had perforations on the wall adjacent to model slope to allow seepage of water. A thin nonwoven geotextile layer was used to cover this perforated wall to prevent clogging of perforations by the soil particles during the preparation of the model as well as during centrifuge tests. The water to this seepage tank was supplied from a water reservoir fixed at the top of the model container with the help of a solenoid valve (Figure 2). This water reservoir received water supply from overhead tank through a supply pipe passing through one of the hydraulic slip rings of the centrifuge. A stand pipe connected at the base of seepage tank helped to maintain a constant water head of 5 mm less than the slope height. A pore pressure transducer (PPT1) kept within the seepage tank helped to monitor its water level during centrifuge tests.

The slope height (H) for all the centrifuge models was 240 mm (7.2 m) with a 60 mm (1.8 m) base layer. The slope model was constructed in layers with vertical spacing (S_v) of 40 mm (1.2 m) by compacting them at maximum dry unit weight and optimum moisture content. The slope inclination of 63.43° (2V:1H) was achieved using temporary wooden block placed on the right hand side of the model container, after compaction of the base layer. Four pore water pressure transducers (PPTs) were placed along the base layer at a distance as shown in Figure 2 from the perforated face of the seepage tank, to measure the pore water pressure variation at the onset of seepage during centrifuge test. PPT3 was vertically below the mid-length of the slope top surface and PPT5 was placed at a distance of 20 mm away from the toe.

The slope models reinforced with geosynthetic layers were constructed with a wrap-around technique to represent a flexible facing. An anchorage length equivalent to 0.25H and reinforcement length equal to 0.85H, where H is the slope height, was used for each geosynthetic layer. A grid of 20 mm x 20 mm was drawn on the geosynthetic layers prior to placing them in the soil layer, for observing the straining and ruptures along the length of reinforcement layers. The geosynthetic layers were numbered for detailed observations after the test, with the layer one placed at the base of the slope. For the model test (DVR-4) having geogrid reinforcement layers G2, a thin nonwoven geotextile layer was pasted along the wrap-around portion of the slope face using a rubber based adhesive. This was done to prevent the movement of soil particles through the model geogrid openings during seepage.

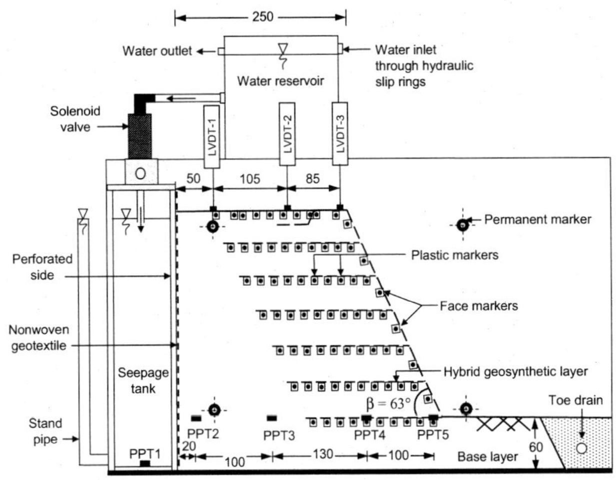

FIGURE 2: Schematic cross-section of a model test package (all dimensions are in mm)

The geosynthetic layers were placed above the compacted soil layers with the nonwoven geotextile component of hybrid geosynthetic layer facing outside towards the slope. The model was constructed in layers of 40 mm thickness using moist compacted soil. After compaction of soil for each layer, a wrinkle free geosynthetic layer was placed above it. L-shaped plastic markers made from thin transparency sheets were glued onto geosynthetic layer using an instant adhesive at a distance of 20 mm from the slope face. These markers were used to trace their movement and to compute strains of geosynthetic layers at the onset of seepage during centrifuge tests. To allow free movement of the markers, white petroleum grease was applied on the sides of the markers in contact with the Perspex sheet. The soil layer was placed above this with geosynthetic reinforcement layer wrapped-back into the slope. Coloured food dye was placed at intervals on the top of each layer to trace the movement of water during seepage. For all the slope models a crest width of 250mm in model dimensions was kept constant. A toe drain was provided at the downstream end of slope on the right end of the container. The surface settlements of the slope model was studied with the help of three Linearly Variable Differential Transformers (LVDTs) placed at the top of the slope at a distance of 50 mm (L1), 155 mm (L2) and 240 mm (L3) from the edge of the seepage tank. Figure 2 shows the cross-section of model test package used in the present study.

(a) Front view (b) Side view

FIGURE 3: Perspective view of hybrid geosynthetic reinforced slope model (DVR-5) before centrifuge test (all dimensions are in mm).

Figure 3 shows a perspective view of a hybrid geosynthetic-reinforced slope model with 6 reinforcement layers and vertical spacing S_v of 40mm, prepared for this study. In all three centrifuge tests were performed, one unreinforced and two reinforced, one with geogrid reinforcement and other with hybrid geosynthetic reinforcement. The slope inclination (2V:1H) and vertical spacing of reinforcement layers (S_v = 40mm) was kept constant for all the tests. The centrifuge model tests were performed at 30g as the critical factor of safety for the unreinforced slope was just above 1 at this gravity level. After the ramping time, the centrifuge was rotated at a constant angular speed of 81 rpm corresponding to 30g. The seepage was induced in the slope models after achieving 30g by operating a solenoid valve provided to regulate the inflow of water and achieve steady state seepage conditions within the slope model. The seeping of water into the slope model caused the spreading of colour dye that enabled to view the development of phreatic surface. The in-flight proceedings of the test were recorded with the help of an on-board digital camera and images captured at every 30 second interval. The behaviour of the unreinforced and reinforced slope models was observed until failure or time duration of 60 minutes (after activating solenoid valve), whichever occurred first. The data from the LVDTs and pore pressure transducers was obtained using a computerized data acquisition system. Table 3 presents the summary of the centrifuge tests performed in the present study.

Table 3: Details of Centrifuge Test

Test Legend	*Geosynthetic Type*	n_{hg}	n_g
DVR-1b	Unreinforced	-[a]	-[a]
DVR-4	G2	-[a]	6
DVR-5	G2N3	6	-[a]

G2 – Fine Geogrid; G2N3 – Fine Geogrid + Needle punched nonwoven; n_{hg} – No. of hybrid geosynthetic layers; n_g – No. of geosynthetic layers; -[a] Not relevant

4. ANALYSIS AND INTERPRETATION OF CENTRIFUGE MODEL TESTS

4.1 General

The data recorded with the help of LVDTs and PPTs and the images captured using digital camera was used to study the deformation behavior and the developed pore water pressures within the model slopes. The results of the three centrifuge tests, one unreinforced and two reinforced slopes models have been presented here. The unreinforced slope model was stable at 30 g without any seepage. On subjecting to seepage, the pore water pressure rose within the unreinforced slope model (Model: DVR-1b) with the phreatic line touching the toe, and the slope model experienced a catastrophic failure. The geogrid reinforced slope (Model: DVR-4) underwent a progressive or gradual failure whereas the hybrid geosynthetic reinforced slope (Model: DVR-5) attained steady state seepage conditions after some time and its behavior was observed for one hour in model dimensions corresponding to 37.5 days of seepage time in prototype dimensions (Table 4). The following sections present the analysis and interpretation of the test results.

4.2 Deformation Behavior of Reinforced Slopes

The data obtained from the LVDTs was used to determine the surface settlements of the top surface of the slope. Figures 4a and 4b present the variation of measured surface settlements with horizontal distance from upstream end of slope for an increasing seepage time (in prototype dimensions) for geogrid reinforced and hybrid geosynthetic reinforced slope models. The surface settlements are observed to increase with an increase in seepage time and are found to be maximum for LVDT L3 placed at the crest. The unreinforced slope (Model: DVR-1b) and model reinforced with model geogrid G2 (Model: DVR-4) attained failure at seepage time of 5.31 and 7.81 days respectively and hence the surface settlements were observed to increase sharply after this time. A surface settlement of 1.8 m at crest (in prototype dimensions) was measured for model DVR-4. In comparison, the slope reinforced with hybrid geosynthetic (DVR-5) did not experience instability and the maximum settlement was found to be 0.07m (in prototype dimensions) till the termination of the test. For hybrid geosynthetic reinforced slopes, surface settlement variation with seepage were found to be uniform till the termination of the test and this was noted to be attained in short seepage time of 1 day.

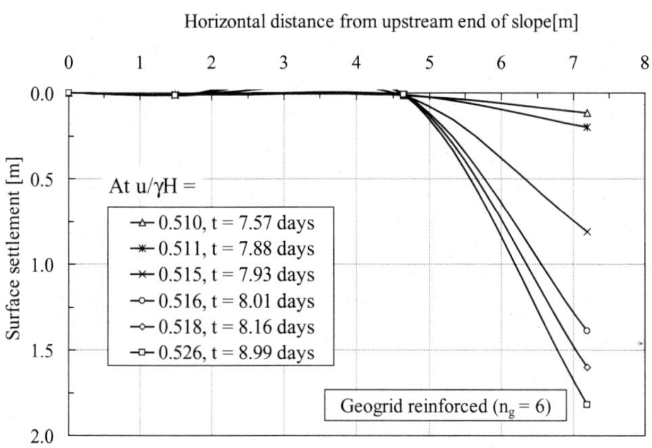

(a) Geogrid reinforced slope model (DVR-4)

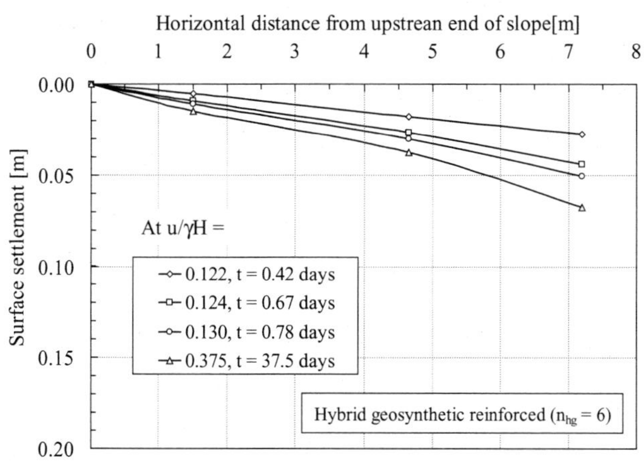

(b) Hybrid geosynthetic reinforced slope model (DVR-5)

FIGURE 4: Variation of surface settlements with horizontal distance from upstream end of slope (in prototype dimensions)

Figure 5 presents the variations of measured vertical settlement with time for LVDT L3 placed at the crest of the slope. also shows a similar behavior. The unreinforced slope model (DVR-1b) is just stable at 30g and is observed to fail on subjecting to seepage at 5.31 days in prototype dimensions. The geogrid reinforced model (DVR-4) depicts a sharp increase in the crest movement on subjecting to seepage at 7.8 days. This indicates an occurrence of catastrophic failure for this model However, the geogrid reinforcement layers helped to delay the time of initiation of the failure compared to the unreinforced slope (DVR-1b). The model reinforced with hybrid geosynthetic layers (DVR-5) was observed to be stable with a constant crest settlement of about 0.07 m on being subjected to seepage for 37.5 days (60 minutes in model dimensions) after which the test was terminated. The stable behavior and extremely low surface settlement values for hybrid geosynthetic reinforced slope model could be attributed to the efficacy of the permeable component of hybrid geosynthetic in draining the seeped water and thus minimize the build up of pore water pressure within the slope model.

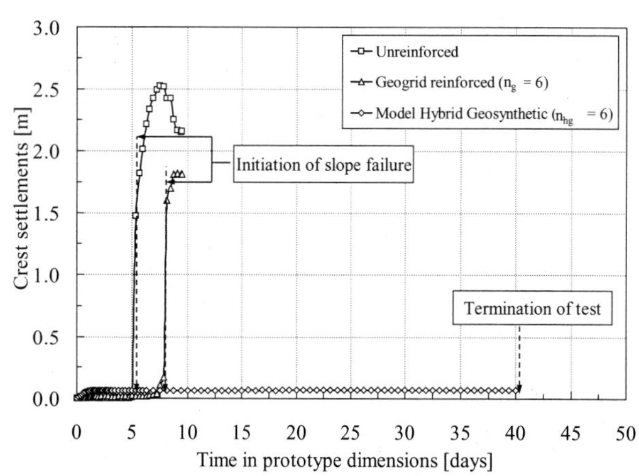

FIGURE 5: Variation of crest settlements with time

4.3 Development of Pore Water Pressures in Reinforced Slopes

The seepage phenomenon in slopes was monitored using PPTs placed in the seepage tank as well as above the base layer of the slope. The pore water pressure readings were used to determine the phreatic surfaces developed within the slope model. Typical plots for the pore water pressure variation with time for the geogrid reinforced (DVR-4) and hybrid geosynthetic reinforced slope (DVR-5) are presented in Fig. 6a-6b. High pore water pressures of about 35 kPa and 26 kPa were measured by PPT4 and PPT5, placed towards the toe for the geogrid reinforced slope (DVR-4). This exhibits the building up of pore water pressure within the toe region of slope resulting in its failure. Contrary to this,

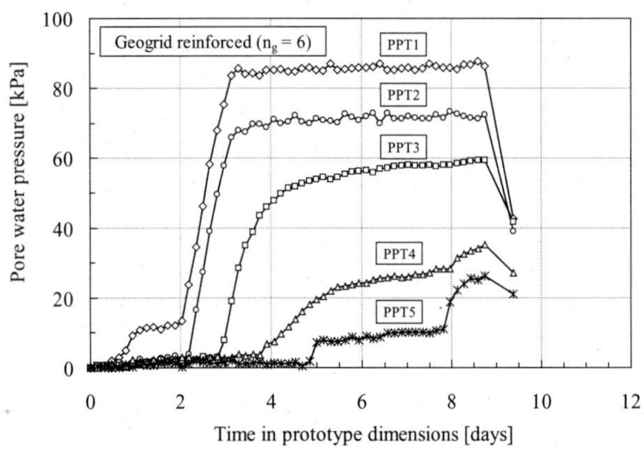

(a) Geogrid reinforced slope model (DVR-4)

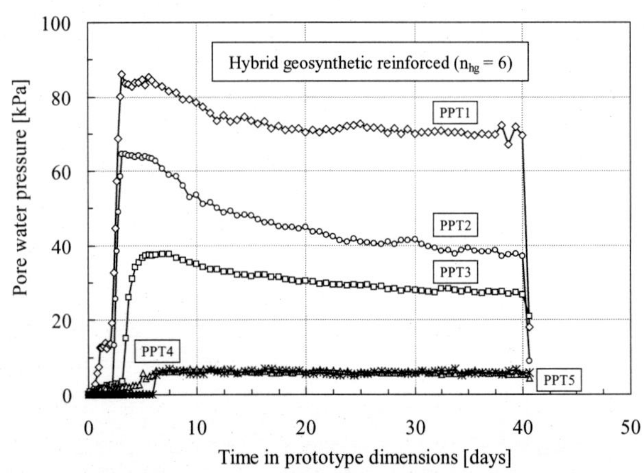

(b) Hybrid geosynthetic reinforced slope model (DVR-5)

FIGURE 6: Variation of pore water pressure with time during centrifuge tests

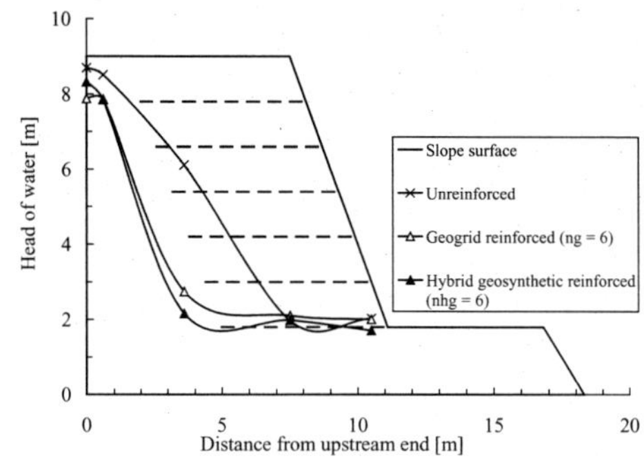

(a) 1 day

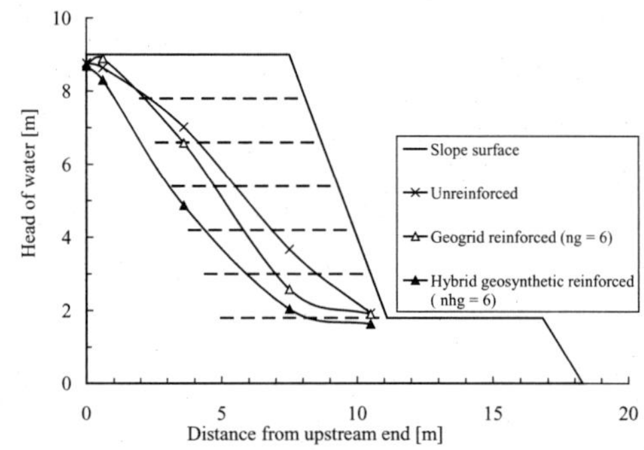

(b) 2 days

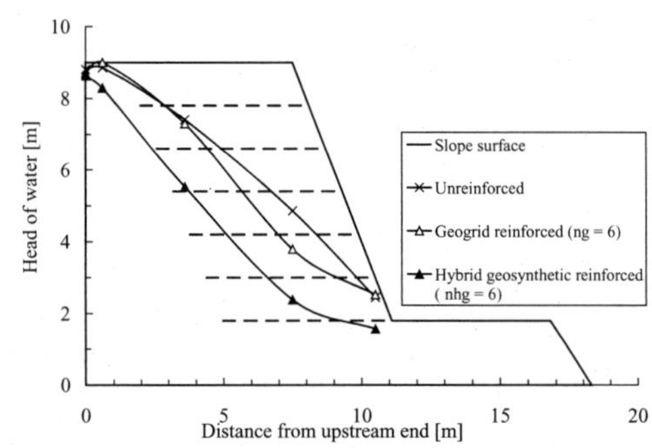

(c) 3 days

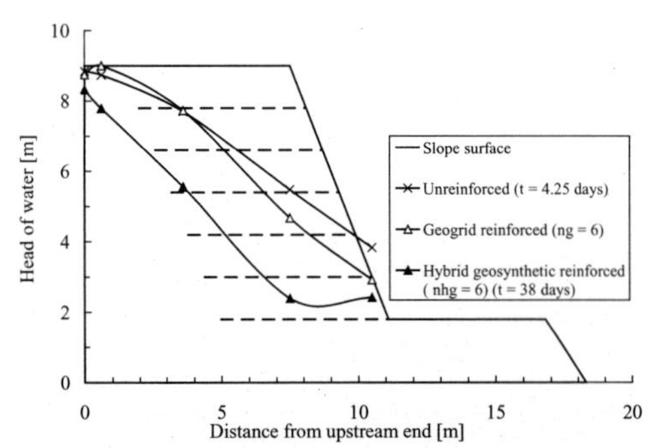

(d) Penultimate stage

FIGURE 7: Development of phreatic surfaces at various seepage times within slope model

substantially lower pore water pressures of about 3-8 kPa are registered by PPT4 and PPT5 placed at the toe region of the hybrid geosynthetic reinforced slopes (DVR-5). The slope reinforced with hybrid geosynthetic layers could attain steady state seepage conditions. The model was tested in centrifuge for 60 minutes (37.5 days) after which the test was terminated. This exhibits the ability of the hybrid geosynthetic layers to function as a drainage layer for a low permeable backfill slope subjected to seepage.

Values of normalized pore water pressure u/γH for uPPT3, defined as a ratio of pore water pressure measured by PPT3 placed at a distance of 120 mm from the left edge of the water tank to the product of soil unit weight and height of the slope, were also determined for seepage time prior to failure and are summarized in Table 4.

Figure 7a-7d present the development of the phreatic surfaces within the model slopes in prototype dimensions for unreinforced (DVR-1b), geogrid (DVR-4) and hybrid geosynthetic reinforced (DVR-5) slope models for seepage time of 1 day, 2 days, 3 days and penultimate stage of the centrifuge test respectively. The horizontal dashed lines indicate the location of geosynthetic reinforcement layers for reinforced slopes. The pore water pressures recorded by the PPTs placed above the base layer were used to determine the water heads. The phreatic surfaces for the unreinforced (DVR-1b) and geogrid reinforced slope (DVR-4) are observed to increase with an increase in the seepage time, as can be seen from Fig. 7b-7d. The phreatic surfaces for the hybrid geosynthetic model reinforced slope (DVR-5) are observed to be lower than geogrid reinforced slope (DVR-4) for the same time duration, confirming the effectiveness of hybrid geosynthetics in dissipation of developed pore water pressures.

4.4 Digital Image Analysis of Centrifuge Models

Images captured during the centrifuge tests were analyzed using digital image analysis software (Image-Pro, 2004) to obtain soil movements and face movements for a particular test. The displacement of discrete markers was traced and analyzed with reference to permanent markers for selected images of model at various stages during the test. The coordinates of the discrete markers obtained with reference to permanent markers was used to determine their displacement and plot the displacement vectors.

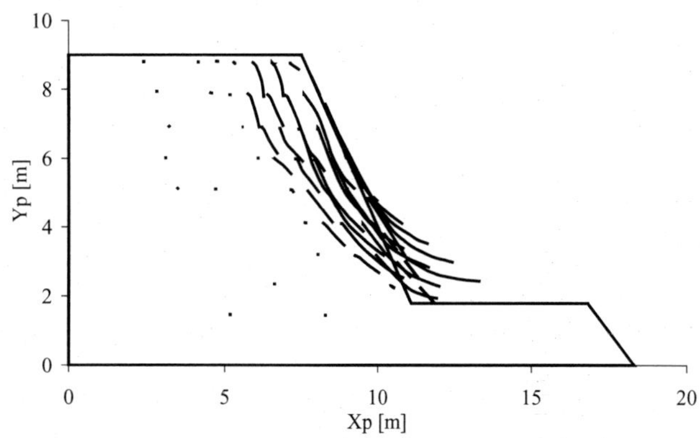

(a) Geogrid reinforced slope model (DVR-4)

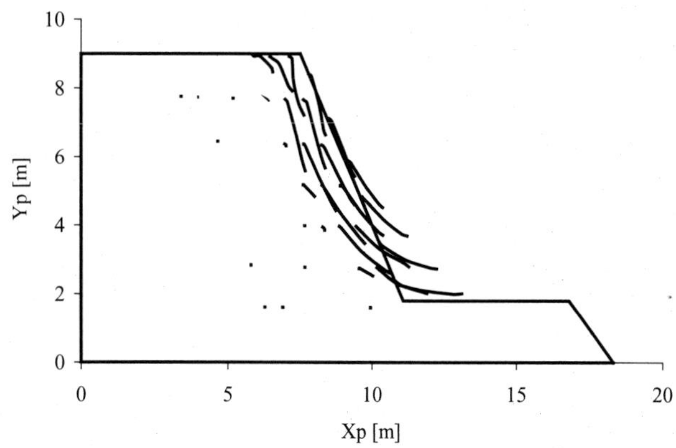

(b) Geogrid reinforced slope model (DVR-4)

FIGURE 8: Displacement vectors obtained using digital image analysis

Figures 8a–8b represent the displacement vectors obtained for the unreinforced and hybrid geosynthetic reinforced slope model from beginning till the onset of failure. The movements experienced by the markers in horizontal and vertical direction are plotted on slope profile in prototype dimensions. The horizontal and vertical distances have been represented as Xp and Yp respectively on the abscissa and ordinate in prototype dimensions. A clear failure surface formation along with the movement of the slope in both, the horizontal as well as vertical direction, can be visualized from the displacement vector plot for unreinforced and slope model reinforced with geogrid layers. As observed from Figure 4b, the hybrid geosynthetic reinforced slope was observed to experience maximum settlement of 0.07m and hence its displacement vectors could not exhibit any movements and hence have not been plotted here.

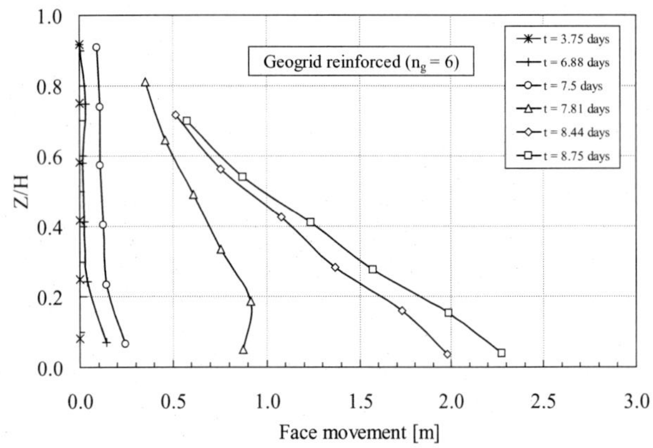

FIGURE 9: Variation of movement of slope face with Z/H at the onset of seepage (a) Model: DVR- 4 and (b) Model: DVR-5

The lateral displacements of the slope face are investigated with the help of coordinates of markers embedded or glued to the slope face. The face movements are plotted considering the slope face to be vertical coinciding with the vertical axis and origin at the toe (as shown in Fig. 2 at the toe). Figure 9 presents the movement of face markers plotted against the normalized height (z/H) for the slope model reinforced with geogrid layers (DVR-4), where z is the height measured from above the base layer. The geogrid reinforced model (DVR-4) exhibits a sudden increase in face movement at the toe from 0.25 m to 2.3 m in prototype dimensions for time corresponding to 7.5 and 8.75 days at the onset of seepage. The snapping of geosynthetic layers can be inferred from the increasing face displacements within this time interval. In comparison, slope reinforced with hybrid geosynthetic layers was found to experience very negligible distress due to dissipation of developed pore water pressures during all stages of the tests and could not exhibit any face movements.

4.5 Stability and Seepage Analysis of Centrifuge Models

Stability analysis was carried out for the unreinforced and geosynthetic reinforced slope models (in prototype dimensions) using the simplified Bishop's method using Slope/W software (Geoslope, 2007). The slopes were analyzed using the phreatic surfaces from the pore pressure data measured using PPTs during the respective centrifuge test. The geosynthetic reinforcement values for the model material were converted to tensile load corresponding to prototype material as per the scaling considerations, for the stability analysis. The bond skin friction value τ_b of 10.07 kPa and 11.47 kPa, determined by conducting modified direct shear tests for the model geogrid and hybrid geosynthetic material G2 and G2N3, was used for the stability analysis. The properties of soil used for centrifuge tests were used for the stability. Both, the unreinforced (DVR-1b) and geogrid reinforced model (DVR-4), become unstable with factor of safety values less than one at u/γH values corresponding to 0.49 and 0.53. The results of the stability analysis are found to be in good agreement with the centrifuge test observations, with the observed failure for these models in centrifuge. In comparison, the hybrid geosynthetic reinforced model (DVR-5) is observed to exhibit nearly constant value of factor of safety with a maximum u/γH value of 0.29 and 0.37. The nearly constant values of factor of safety for hybrid geosynthetic reinforced slopes for seepage time of up to 37.5 days (Table 4) and lower u/γH values exhibits the effectiveness of hybrid geosynthetics as a drainage media.

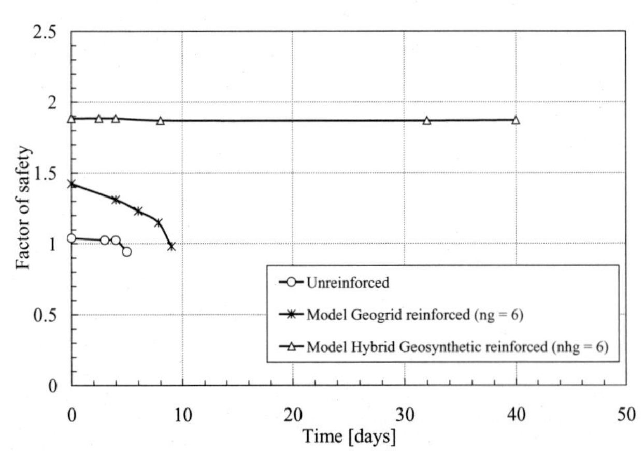

FIGURE 10: Variation of factor of safety with time (in prototype dimension)

Table 4 gives the summary of centrifuge test and stability analysis results. Values of $S_c$max/H, u/γH and corresponding safety factors (measured value of maximum settlement at the crest (with the help of LVDT L3) normalized to height of the slope) are given in columns 3e5. As can be noted, for model DVR-6 reinforced with ng ¼3 and nhg ¼ 3, the value of S_c,max/H is 0.006 even after subjecting to a seepage for about 37.5 days. In comparison, the value of S_c,max/H is 0.35 after subjecting to seepage of about 5.04 days itself. This shows the beneficial use of hybrid geosynthetics in improving the performance of a low permeable slope subjected to seepage.

Table 4: Summary of Centrifuge Test and Centrifuge Analysis results

Test Legend	*Seepage Time t (days)*	*u/γH*	*FOS*
DVR-1b	5.04	0.493	0.94
DVR-4	9	0.525	0.98
DVR-5	37.5	0.30	1.87

H = Height of slope; u/γH = Normalized pore pressure

5. CONCLUSIONS

The results of three centrifuge model tests carried out on unreinforced and hybrid geosynthetic reinforced slopes with low permeable backfill subjected to seepage are presented. The impact of seepage on the stability and deformation behavior of model slopes was simulated by raising the ground water table within the model slope during flight at 30 g. Based on the analysis and interpretation of centrifuge model test data and stability analysis carried out for the model tests, the following conclusions can be drawn:

1. The unreinforced and geogrid reinforced slope model undergoes distress and loses its integrity with an increase in pore water pressure and the phreatic surface touching the toe.
2. The lower phreatic surfaces for hybrid geosynthetic reinforced slope (DVR-5) for a given seepage time and the low values of normalized pore water pressure u/γH confirm the effectiveness of hybrid geosynthetic layers in dissipating the excess pore water pressure developed within the low permeable backfill slope. The constant values of pore water pressure variations for seepage time of 37.5 days (in prototype dimensions) observed for these slope models ensures steady state conditions being achieved.
3. The results of stability analysis were found to be consistent with the observed performance of the reinforced and unreinforced slope models subjected to seepage in centrifuge.

Presence of permeable nonwoven geosynthetic in a hybrid geosynthetic along with a geogrid layer helps in reducing the excess pore water pressures developed and maintain the integrity of low permeable backfill reinforced slope. Further studies need to be focused with different types of hybrid geosynthetics for its effectiveness for their use marginal backfill materials.

REFERENCES

Chandrasekaran, V.S. (2001) "Numerical and centrifuge modelling in soil structure interaction". Indian Geotechnical Journal, 31, Issue 1, pp 1-59.

Garcia, E. F., Gallage, C. P. K., and Uchimura, T. (2006) "Unsaturated infiltration on artificial embankments reinforced with geosynthetics". Proceedings of the 8th International Conference on Geosynthetics (8ICG), Vol. 4, pp. 1417-1420.

Garcia, E.F., Gallage, C. P. K., and Uchimura, T. (2007) "Function of permeable geosynthetics in unsaturated embankments subjected to rainfall infiltration". Geosynthetics International, 14, Issue 2, pp 89-99.

Geo-Slope, 2007. SLOPE/W and SEEP/W Ver. 7.15 User's Guide. Geo-Slope International Ltd, Calgary, Canada.

Ghosh, C., and Yasuhara, K. (2004) "Clogging and flow characteristics of a geosynthetic drain confined in soils undergoing consolidation". Geosynthetics International, 11, Issue 1, pp 19-34.

Hara, K., Mitsui, J., Kawai, K., Shibuya, S., Hongoh, T., and Lohani, T.N. (2007) "Evaluating in-plane hydraulic conductivity of non-woven geotextile and plastic drain by laboratory test". New Horizons in Earth Reinforcement, IS Kyushu '07, pp. 273-279.

Hiro-oka, A., Kobayashi, M., Nagase, H., and Shimizu, K. (2001) "Performance of geotextile reinforced slopes subjected to seepage flow". Landmarks in Earth Reinforcement, pp. 207-212.

Image-Pro Plus (2004), Version 5.1, Media Cybernetics, Inc., USA.

Izawa, J. and Kuwano, J. (2010) "Centrifuge modelling of geogrid reinforced soil walls subjected to pseudo-static loading". International Journal of Physical Modelling in Geotechnics 10, Issue 1, pp 1-18.

Ling, H. I., Tatsuoka, F., Wu, J. T. H. and Nishimura, J. (1993) "Hydraulic Conductivity of Geotextiles under Typical Operational Conditions". Geotextiles and Geomembranes, 2, Issue 6, pp 509-542.

Raisinghani, D.V., and Viswanadham, B.V.S. (2010) "Evaluation of permeability characteristics of a geosynthetic reinforced soil through laboratory tests". Geotextiles and Geomembranes, 28, Issue 6, pp 579-588.

Rajesh, S., and Viswanadham, B.V.S. (2009) "Evaluation of geogrid as a reinforcement layer in clay based engineered barriers". Applied Clay Science, 46, Issue 2, pp 153-165.

Resnick, G.S. and Znidarcic, D. (1990) "Centrifugal modeling of drains for slope stabilization". Journal of Geotechnical Engineering, ASCE 116, Issue 11, pp 1607-1624.

Viswanadham, B.V.S., and König, D. (2004) "Studies on scaling and instrumentation of a geogrid". Geotextiles and Geomembranes, 22, Issue 5, pp 307-328.

Viswanadham, B.V.S., and Raisinghani, D.V. (2010) "Centrifuge model studies on the behavior of Geocomposite reinforced soil slopes". 7th International Conference on Physical Modeling in Geotechnics (ICPMG 2010), Vol. 2. pp. 1157-1162.

Yasuhara, K., Murakami, S., Ghosh, C., and Recio-Molina, J. A. (2002) "Advantageous features of the geocomposite learned from small-scaled model tests". Proceedings of the 7th International Conference on Geosynthetics (7ICG), Vol. 1, pp 435-440.

Yoo, C., and Jung, H.Y. (2006) "Case History of Geosynthetic Reinforced Segmental Retaining Wall Failure". Journal of Geotechnical and Geoenvironmental Engineering, ASCE, 132, Issue 12, pp 1538-1545.

Use of Geocomposite as an Internal Drain in Levees: Centrifuge Study

Rishabh Kumar Saran and B.V.S. Viswanadham

Department of Civil Engineering, Indian Institute of Technology Bombay, Mumbai, India

E-mail: viswam@civil.iitb.ac.in

ABSTRACT

The aim of this paper is to investigate the performance of levee sections at the onset of flooding and subsequent seepage with the help of centrifuge model tests at 30 gravities. Two levee sections were modeled viz. levee section without drainage layer and with horizontal geotextile drainage layer. The flood was induced at 30 gravities with the help of remotely operated in-flight flood simulator. Levee section without any horizontal drain or clogged drain was observed to experience a catastrophic failure. The analysis of centrifuge test results shows that horizontal geotextile drainage layer is effective in ensuring stability of the levee section at the onset of flooding and subsequent seepage because of effective to dissipation of pore water pressure. An attempt has also been made to compare results obtained from centrifuge studies with seepage and stability analyses using SEEP/W and SLOPE/W software. Further, parametric variation was performed with the help of numerical modeling of seepage and stability considering Influence of normal stress on in-flow characteristics of geotextile. Finally, this paper concludes the use of geotextile horizontal drainage layer as an internal drain as a replacement of conventional sand drainage layer as a viable option for imparting economy to construction projects. However, Performance of levee facilitated with geotextile internal drain at the onset of flooding was found to be sensitive to normal stresses on geotextile and their clogging phenomena.

1. INTRODUCTION

Many economically, industrially developed and strategically important cities of the world are situated in the vicinity of rivers and protected by the levees at their periphery at the onset of flooding. The levees is the type of earthen embankments made up of locally available borrow material with low to very low permeability i.e. silty sand to clayey soil. Any breach in levee section as a result of any climatic event such as flooding or tsunami can trigger catastrophic consequences to the environment and society. The breach in levee section can be explained with the help of various mechanisms such as slope failure in downstream at the onset of steady state seepage, erosion in downstream, piping failure in foundation strata, internal erosion because tunneling by animal activities, erosion due to overtopping, rapid drawdown in upstream. An earthen embankment or levee has to accommodate various components such as cut-offs, trenches, drainage layers and reinforcement layers etc. in order to arrest above stated failure mechanism and sustain its stability when subjected to various breaching mechanisms. All these components address several functions such as, drainage, resistance to piping and erosion control in order to maintain stability and facilitate the functionality of levee. Among these, Horizontal drains serve to dissipate pore water pressure in the downstream side. In this study internal drainage system in the form of horizontal drain embedded in downstream side is emphasized. BIS 9429(1990) indicates guidelines for drainage system for earth and rockfill dams where it is discussed that geometry of internal drains are essentially affected by water level at upstream side, height of embankment both in upstream and downstream sides, permeability of embankment and drainage material, gradation of embankment and drainage materials and angle of discharge face. Additionally, several regulations advocate different thicknesses of horizontal drains. US Army Corps of Engineers EM 1110-2-1913 suggests a minimum thickness of horizontal drain layers of about 457.2mm for construction purposes. BIS 9429(1990) proposes a minimum 1.0 m thick horizontal drain in earthen dams. Geometry of horizontal drain is also affected with the availability of well graded material, as per filter requirements. These codes discusses criteria for granular material as a internal drain but due to rapid expansion of infrastructure in developing countries in the world like India, lack of availability of suitable material for drainage imposes economic burden on projects. Hence, in the view of above it is vital to explore avenues for replacing conventional drainage layer with an appropriate material, which can cater the basic requirements of internal drain.

Several researchers adopted different approaches for understanding the hydraulic behavior of the levee section, such as analytical, numerical, large-scale and small-scale studies. Iryo and Rowe (2004) examined the hydraulic response of a geotextile layer embedded in sandy soil subjected to 1-dimensional infiltration using SEEP/W. Giroud(2005) attempted to quantified the behaviour of geosynthetics using theoretical analyses. Jia et al. (2009) examined the stability of the slope model constructed with sandy silt (12% sand, 80% silt, less than 5% clay) subjected to rising and lowering water level by conducting large-scale model tests at normal gravity. Ubilla et al. (2008) investigated the failure and non-failure conditions of levees subjected to Katrina hurricane with help of centrifuge studies on the London Avenue and Orleans Canal South. Das and Viswanadham (2010) proposed an experimental set up for examining the piping behavior of silty sand soil type with and without discrete and randomly distributed geofibers. Raisinghani and Viswanadham (2010) presented the effect of geosynthetic layers in enhancing the in-plane permeability of fine-grained soils in the laboratory by using custom made permeameter tests. It was reported that the use of a permeable geosynthetic layer in the form of a geocomposite (which is combination of woven geotextile and non-woven geotextile or geogrid and non-woven geotextile) helps in lowering phreatic surface and reducing generated pore water pressures by draining water in the plane of geocomposite layers. Successful application of using geocomposite materials for pavement, behind retaining walls, reinforced soil walls and slopes was reported by Raisinghani and Viswanadham (2011).

However, very few attempts have been made to understand the stability and deformation behavior of levee sections without and with geo-composite drainage layer as internal drain with the help of centrifuge modeling technique. Hence this draws the research interest of the present study.

In this present study, hydraulic behavior of horizontal drainage layer of non-woven geotextile at the toe of the embankment constructed with low permeable embankment material was evaluated using centrifuge tests at 30 gravities. For this purpose, two levee model sections with and without horizontal drainage layers subjected to flooding was considered as described in Table 1. Fig. 2 shows cross section of model test package used in this paper. Non-woven geotextile was used as internal drain materials in the horizontal direction located at toe towards downstream side. A set-up was developed to simulate in-flight flooding at 30g. Various parameters such as material properties of levee section, upstream and downstream slopes, levee height, top width and base layer thickness were kept constant for both the tests. The effect of inclusion of horizontal geotextile drainage on the stability and deformation behavior of levee, pore water pressure distribution, phreatic surface development has been evaluated. An attempt was also made to compare observed centrifuge model test results with the help of finite element based SEEP/W and limit equilibrium based SLOPE/W (Geostudio, 2007). Furthermore, influence of in-flow characteristics of geotextile on the seepage and stability behavior of levees was also evaluated using above stated software packages with the help of parametric variations. The parametric variations considered the decrease in transmissivity of prototype geotextile subjected to various normal stresses.

Table 1: Details of Centrifuge Tests

Test Legend	*Drainage Material*	*H (mm)*	*Drainage Length (mm)*	*Thickness of Drainage Layer (mm)*
RL-1	-[a]	200 (6000)	-[a]	-[a]
RL-3	NW Geotextile	200 (6000)	185 (5550)	0.66

-[a] Not relevant; Prototype dimensions are given within parenthesis; NW = Non-woven

2. SCALING CONSIDERATION FOR MODELING NON-WOVEN GEOTEXTILE

In the present study, centrifuge modeling technique was adopted to achieve stress similarity at homologous points by subjecting a model reduced by 1/N times to N time's gravity. The use of centrifuge modeling technique was considered relevant because of the requirement of simulation of stress conditions and pore water pressure variation at the onset of flooding and subsequent seepage through a levee section as that of in the prototype. In addition, climatic events like simulation of flooding at full-scale are difficult to simulate.

In order to scale-down a geocomposite layer, three criteria are required to be satisfied: i) Scaling down of tensile-load strain characteristics, ii) Scaling of in-plane permeability characteristics, and iii) Soil-geocomposite interface friction. Detailed discussion on modeling of geocomposites in a centrifuge was reported by Raisinghani and Viswanadham (2011). Based on the scaling considerations, there is requirement of having a geocomposite layer with reduced thickness, tensile load and secant stiffness reduced by 1/N times that of the commercially available geocomposites and with identical soil-geocomposite interface friction. However, in order to satisfy scaling requirements from hydraulic properties point of view, the geocmposite shall have identical transmissivity as that of in the commercially available geocomposites and shall possess N times more in-plane permeability characteristics in the model. Although geocomposite can have reinforcement and drainage functions, considering the requirement of drainage function, only non-woven geotextile was considered and referred as geocomposite in the present study.

3. CENTRIFUGE MODEL TEST ON LEVEES SUBJECTED TO FLOODING

3.1 Model Materials

Two types of model materials are involved. They are: (i) Model soil for constructing a levee; (ii) Non-woven geotextile as an internal drain and as a replacement for conventional sand drainage layer. Generally, levees are constructed by using low to very low permeable materials so, by blending 80% of locally available fine sand with 20% of commercially available kaolin (by dry weight), silty sand type model soil was formulated. The model soil was found to have maximum dry unit weight $\gamma_{d,\ max}$ of 18.75 kN/m^3 and optimum moisture content (OMC) of 8% according to standard Proctor compaction. The value of cocfficient of permeability of the model soil moist-compacted at $\gamma_{d,\ max}$ and OMC is 1.54 x 10^{-6} m/s. The model soil (moist-compacted at $\gamma_{d,\ max}$ and OMC and saturated) was found to have a cohesion of 11.9 kN/m^2 and angle of internal friction of 27° obtained by conducting drained direct shear box test.

Based on the above set scaling considerations for modeling non-woven geotextile, a non-woven geotextile having thickness of 0.66 mm, a mass per unit area of 46 g/m^2, transmissivity θ of 0.38 x 10^{-6} m^2/s at 20 kPa normal stress. In-plane

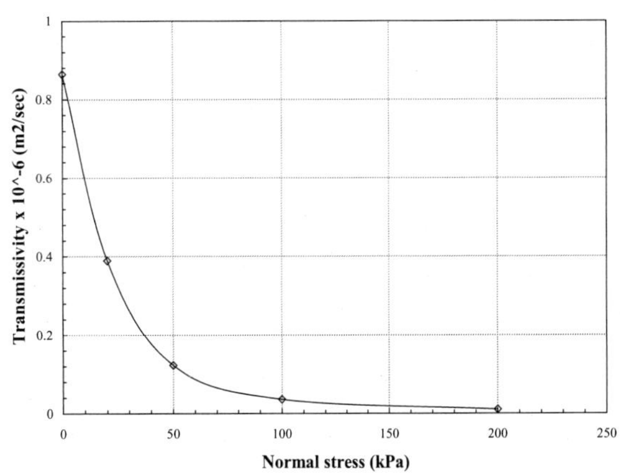

FIGURE 1: Variation of transmissivity with normal stress

characteristics were evaluated for varying normal stresses as shown in Fig. 1. Further, Table 2 summarises values of transmissivity of geotextile (GT) at various normal stresses. It also discusses various geotextile legend details considered for parametric variation in numerical simulation of seepage and stability.

Table 2: Summary of In-flow Characteristics of Model Geotextile

Legends	*Normal Stress (kPa)*	*Transmissivity × 10^{-6} (m^2/sec)*
GC-1	20	0.38
GC-2	50	0.12
GC-3	100	0.04
GC-4	200	0.01

3.2 Model Test Package

The large beam centrifuge facility of 2500 g-kN capacity with 4.5m radius available at Indian Institute of Technology Bombay was used in the present study. The detailed specifications were discussed by Viswanadham and Mahjan (2007). Identical stress-strain behavior can be simulated by subjecting the scaled down model to N-times gravity level in controlled environment. Centrifuge models tests reported in this study were carried out at 30 gravities. Levee model was constructed within the strong box having internal dimensions of strong box were 760 mm in length, 200 mm in breadth and 410 mm in height. Fig. 3 shows perspective view of a model test package. Plane strain conditions were maintained by applying white petroleum grease on Inner surfaces of front and rear walls of strong box. As shown in Fig. 3, a rectangular grid of 430 mm x 210 mm with four permanent markers was glued on the Perspex surface in contact with the soil to serve as reference frame for image analysis. Movable markers were embedded in the soil at different elevations as along the downstream face to traced displacement vectors at the onset of flooding. Relative positions of movable markers were captured and recorded with the help of high resolution digital camera. Four pore water pressure transducers (PPT) were embedded at the levee bottom to measure the variation in pore water pressure with respect to seepage time. As shown in Fig 2, two linear variable differential transformers (LVDT) were housed at both side of crest to obtain crest deformations with respect to seepage time.

To induce in-flight flooding, a remotely operated DC-pump based flood simulator has been developed. As shown in Fig. 2, there are three main component parts of the set-up: a) water reservoir b) Pump assembly and c) voltage controller. Water reservoir of 15Liter capacity having dimensions (650 mm x 160 mm x 140 mm) was housed behind the strong box. Firstly, this water reservoir constantly receives water supply from one of the slip rings provided in centrifuge. Hence, the reservoir serves as constant head reservoir to the inlet of pump. Secondly, pump was housed vertically over the water reservoir. A mild-steel frame was also fabricated to fasten pump firmly to backside of the strong box with help of nuts and bolts. Finally, power supply to this pump was regulated remotely with the help of DC-voltage controller. Thus, for a given g-level, voltage level and levee section, a particular flood rate was induced. One PPT was also housed in upstream side of levee to monitor and evaluate flood level with respect to time.

3.3 Model Test Program and Test Procedure

Base layer was constituted of two different layers viz. 60 mm thick with silty sand and 40 mm thick with kaolin. Kaolin layer was overlaid the base layer of silty sand (80:20 mix). Kaolin having very less permeability was intentionally selected as a model soil to be placed just beneath the levee section in order to provide impermeable foundation and induce potential seepage only through levee section. In Model RL-3, a model needle punched nonwoven geotextile was embedded horizontally at downstream to serve as a horizontal internal drain.

Model Levee sections were constructed having height of 200 mm, top width of 100 mm, upstream slope 1V:1H (45^0) and downstream slope 1.5V:1H (56.3^0) as depicted in Fig. 2. Geometry of levee was selected such that factor of safety at the onset of seepage is close to one. In this paper, results of two centrifuge model tests are presented. They are: i) without drainage layer (RL-1) and ii) with horizontal geotextile drainage layer (RL-3), as summarized in Table 1. Levee models were constructed in layers of 30 mm thickness moist-compacted at $\gamma_{d,\ max}$ and OMC according to standard Proctor compaction. L-shaped movable plastic markers were also placed at the top of each compacted layer horizontally at a centre to centre spacing of 20 mm from downstream slope face. External side of one leg of the movable markers was lubricated with grease in order to reduce friction between Perspex sheet and marker surfaces. Positions of the movable markers were traced with the help of image analysis of front elevation images of levee section captured during the centrifuge test. At selected locations, food dye was placed to trace the seepage of water at the onset of flooding through a levee section during centrifuge test at high gravity.

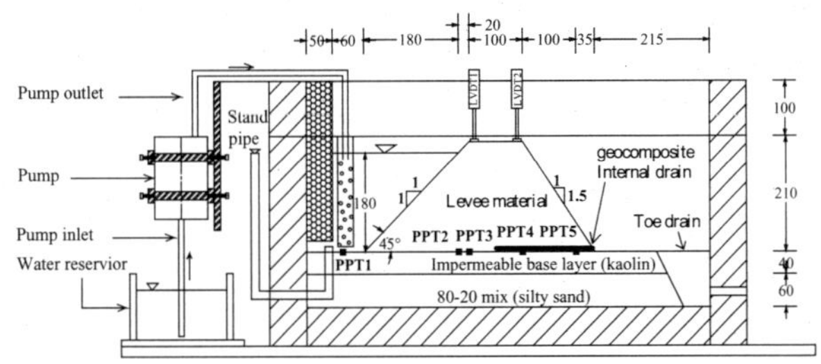

FIGURE 2: Schematic cross-section of model test package (All dimensions are in mm)

In model RL-3, horizontal drainage layer with a model non-woven geotextile layer was placed at toe of levee section towards downstream side positioning from vertical projection of midpoint of top width to the toe in downstream. Length of drainage layer was kept constant throughout the width of levee. It was positioned vertically 10 mm above the base layer, embedded in levee material to ascertain no clogging of geotextile with kaolin particles from the impermeable base layer of kaolin.

All levee models were instrumented with 5 no's PPTs and 2 no's LVDTs to evaluate variations in pore water pressure and settlement behavior of levee crest respectively. As shown in Fig. 1, PPT1 was placed above the base layer towards upstream to monitor the flood level and another four PPTs were embedded along the levee bottom at a distance of 180 mm (PPT2), 200 mm (PPT3), 300 mm (PPT4) and 40 mm (PPT5) from the upstream toe. PPT3 was placed at different position at 250 mm from upstream toe of levee in model RL-3. In model RL-3, PPT3 was placed right below the mid-point of top width of the levee section to capture pore water pressure variations at the junction of levee material and horizontal drainage layer. LVDTs were housed at both left and right side of the crest at a horizontal distance 210 mm and 290 mm from upstream side toe. After completion of the model, in-flight simulator was mounted along with the model and initial checks were performed.

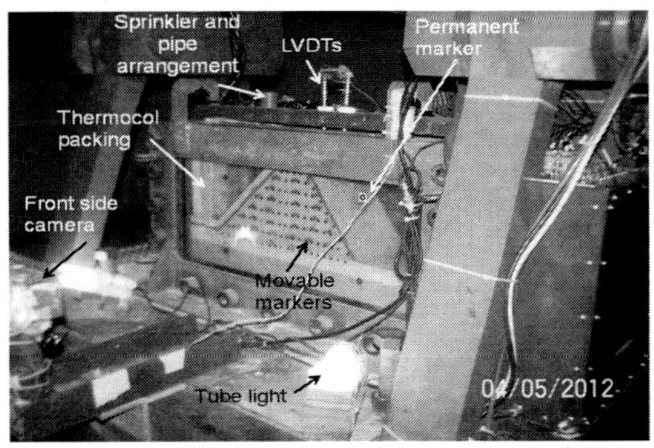

FIGURE 3: Perspective view of model test package before test

For all the models, 30 gravities was induced by rotating a model with 81 revolutions per minute. On reaching 30g, in-flight flood was commenced by activating pump remotely from the control room. Once the pump started, flood level kept on increasing due to constant pumping of water into the enclosure in front of the upstream face, as shown in Fig. 2. In order to prevent early seepage of water into levee section, the upstream side slope surface was coated with a thin layer of bentonite up to desired flood height. Constant High flood level could be maintained with the help of a stand pipe connected to the enclosure in front of the upstream slope face, as shown in Fig. 2. At the onset of flooding and subsequent seepage, seepage of water into levee section could be easily traced with help of diffused colored food dye. The seepage phenomenon through levee section was captured with the help photographs triggered at every 30 seconds from the time of commencement of flooding.

4. ANALYSIS AND INTERPRETATION OF CENTRIFUGE MODEL TESTS

4.1 General

Based on Pore water pressure data obtained from PPTs at the commencement of flood corresponding variation of pore water pressures and development of phreatic surfaces were calculated for both centrifuge models carried-out in the present study. Deformation behavior of the crest of levee sections was evaluated with the help of LVDTs data and profile of levee section in the form of displacement vector assessed with the help of image analysis. The following sections present analyses and interpretation of results of centrifuge tests.

4.2 Simulation of Flood

Figure 4 shows the variation of flood level simulated at 30 gravities with seepage time in prototype dimensions. The variation is plotted by considering the time in prototype dimensions, right after commencement of flooding at 30g. This was calculated by using $t_m/t_p = 1/30^2$ (where t_m = time in model dimensions and t_p = time in prototype dimensions). As shown in Fig. 4, highest flood level could be achieved within four days in prototype time for all the models. It was attempted induce identical flood levels for both the levee section, but due to limitation of pump capacity in high gravity environment water head slightly dipped and fluctuations in the flood level were observed. However, hydraulic response of levee sections in the form of pore water pressure and phreatic surfaces were calculated with respect to function of flood level for individual levee sections.

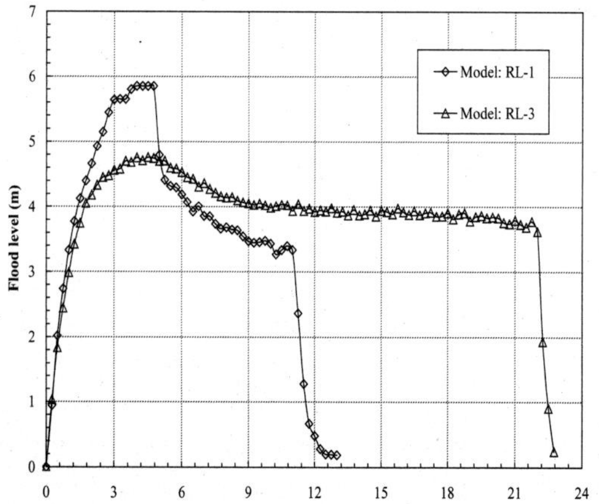

FIGURE 4: Simulated flood function with respect to time in centrifuge models

4.3 Deformation Behaviour of Levees

Data collected with the help of LVDTs housed at top width of levee section were utilized in order to measure crest deformations at both sides of crest viz. landside crest and riverside crest. Measured deformations in prototype dimension of landside crest (i.e. towards downstream side) with respect to time in days for all the centrifuge tests are shown in Fig. 5. It can be noticed from Fig. 5 that levee model RL-1 underwent catastrophic failure right after attaining highest possible flood level. However, levee section RL-3 with horizontal drainage layer was not observed to experience only notable crest settlements. Values of normalized pore water pressures (u/γH) are also labeled in different stages of test in Fig. 5. Normalized pore water pressure can be defined herein as a ratio of the measured pore water pressure with the help PPT3 to the product of bulk unit of the soil (γ) and height of levee section H. At the onset of seepage, normalized pore water pressure within levee section for model RL-1 was observed to increase from 0.613 to 0.665 at 5 days after initiation of flood. On the other hand, steady state seepage conditions were noticed to establish in model RL-3. This emphasizes the importance of un-clogged or proper drainage layer at the toe in the downstream of levee section.

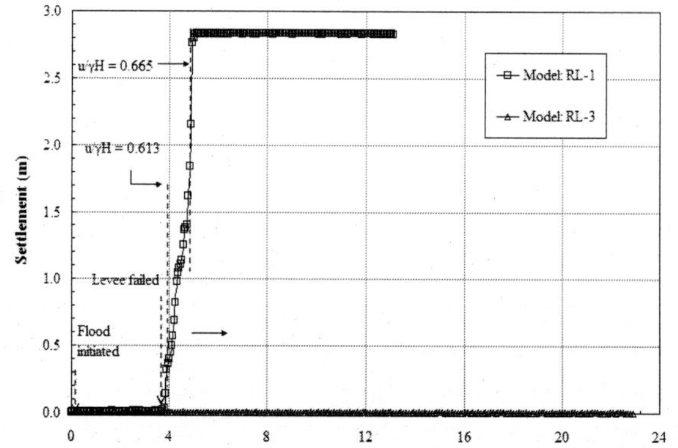

FIGURE 5: Measured variation in surface settlements of land side crest with respect to time

4.4 Development of Pore Water Pressures within Levees

Measured pore water pressure in centrifuge tests were used to plot the phreatic surfaces with respect to time in order to understand the seepage phenomena within the levee section. Development of phreatic surfaces in levee section with prototype time in days for models RL-1 and RL-3 are demonstrated in Figs. 6a and 6b respectively. Two different types of observed phreatic surface development patterns for both the models are presented in Figs. 6a and 6b . Relatively raised elevation of phreatic surfaces was noticed in case of model RL-1. This can be attributed to the absence of horizontal drainage layer at the toe or the presence of clogged drainage layer. Furthermore, in the case of model RL-3, a considerable change in the flow path of the water and significant depletion in phreatic surface within the levee section was observed even after 20days in prototype time. The dissipation of pore water pressure suggests efficiency of geocomposite horizontal drainage layer at the onset of flooding and subsequent seepage in model RL-3.

Normalized pore water pressure variation with respect to prototype time for models RL-1and RL-3 are demonstrated in Fig 7a and 7b respectively. In Fig. 7a, variation of $u/\gamma H$ is depicted which represents normalized pore water pressure corresponding to PPT3. Similarly, in Fig. 7b variation in $u_{toe}/\gamma H$ corresponding to PPT5 placed at the toe is demonstrated. As shown in Fig. 7a, relatively distinct high peaks in normalised pore water pressure at both within the levee section and the toe were recorded for model RL-1. These recorded high peaks of pore-water pressures points toward catastrophic failure immediately at the

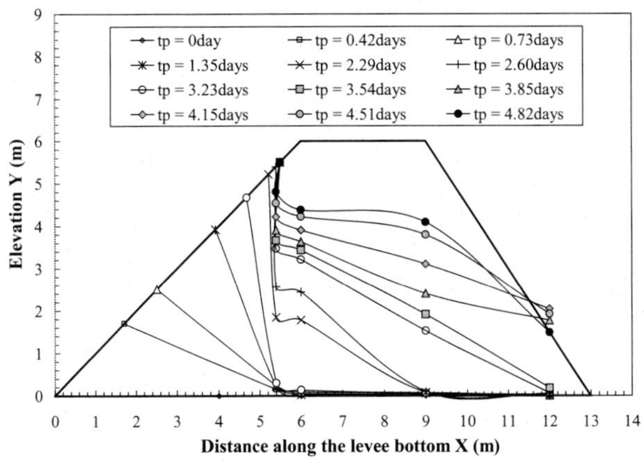

(a) Model: RL-1

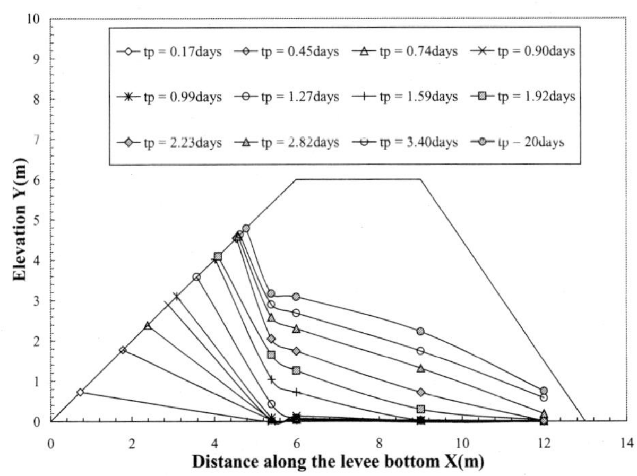

(b) Model: RL-3

FIGURE 6: Development of phreatic surface with seepage time within time (days)

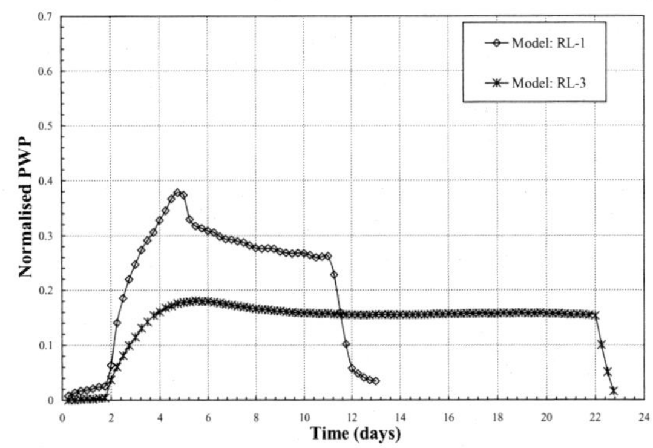

(a) $u/\gamma H$

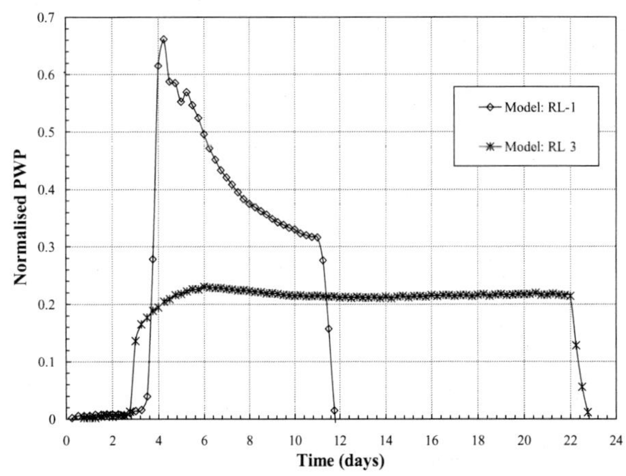

(b) $U_{toe}/\gamma H$

FIGURE 7: Variation of $u/\gamma H$ and $u_{toe}/\gamma H$ with time t days

onset of maximum flood level for model RL-1. In comparison, lower values of $u/\gamma H$ and $u_{toe}/\gamma H$ were recorded for model RL-3 reflecting stability of levee models. Moreover, interestingly, any distinct peaks were not noticed for model RL-3 and this indicates the need of an unclogged internal drain in levee sections. In the case of RL-3, centrifuge tests were terminated after attaining steady state conditions in terms of pore water pressures. Long term hydraulic behavior of drainage measures is beyond the scope of this study.

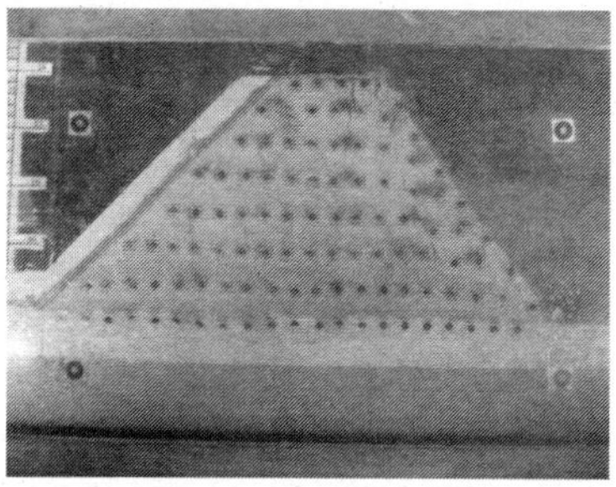

(a) T_p = 3.85 days

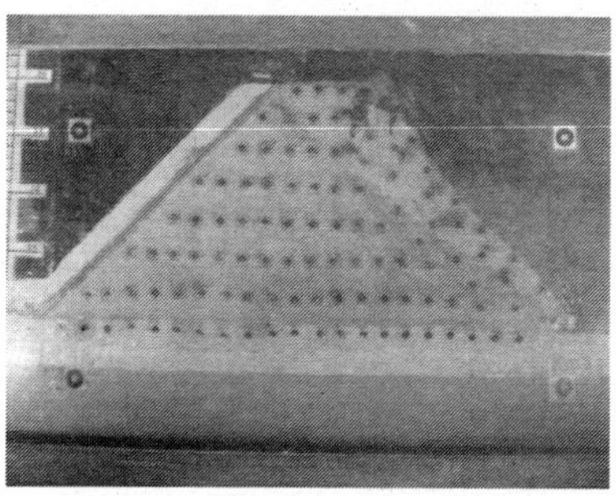

(b) T_p = 4.51 days

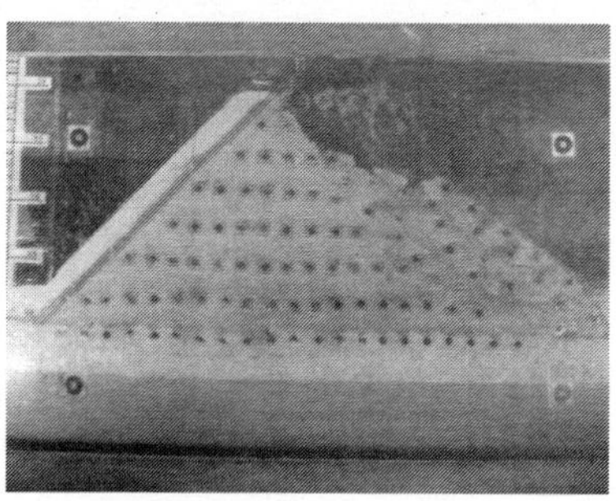

(c) T_p = 5.1 days

FIGURE 8: Various stages of levee failure at the onset of seepage with prototype time in days

4.4 Digital Image Analysis of Centrifuge Models

Images of front elevation of levee models captured with the help of on-board digital cameras at 30g for various stages of seepage were used to arrive at displacement vectors of markers embedded within the levee body and along with the downstream slope face. Image processing software, ImageJ (Version 1.44, 2011) was used for evaluation of relative positions of movable markers. For each image, global co-ordinates were fixed with reference to a grid of permanent markers. This was adopted for analyzing each and every image to prevent any error due to monitor resolution. After fixing global coordinates, positions of all the markers were registered for each image.

Fig. 8 shows typical images captured with the help of onboard camera at 30 gravities. Fig 5a, b and c depict various stages of levee failure at the onset of flooding and subsequent seepage with prototype time. These images were used for the purpose of image analysis where movement of movable markers can be easily observed. Fig. 9 shows displacement vectors from the time of initiation of flooding and up to penultimate stage of the stage. Fig. 9 also depicts levee geometries of model RL-1 at both pre and at failure states. Levee model RL-1 was observed to experience a catastrophic failure at 5.1 days after initiation of flooding. A clear and distinct failure surface extending from the crest towards water side to the toe of the levee section can be observed. However, in case of levee model RL-3, negligible movements were recorded and hence digital image analysis was not performed.

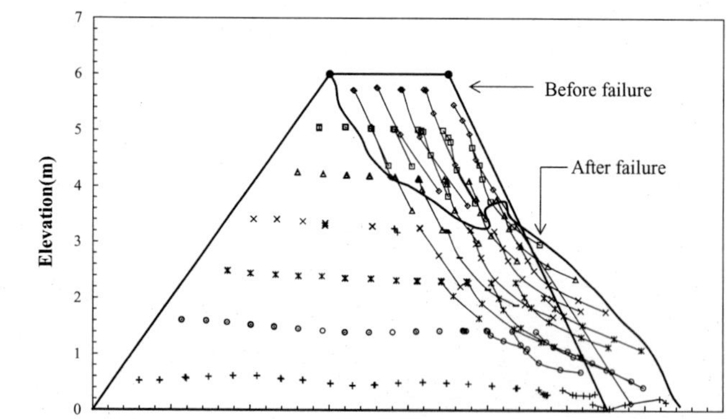

FIGURE 9: Displacement vectors for model RL-1

5. SEEPAGE AND STABILITY ANALYSIS OF CENTRIFUGE MODELS

Finite element based software package SEEP/W and limit equilibrium method based SLOPE/W (GEO-SLOPE, 2007) software packages were used to carry out seepage analyses and stability analyses for both two models. For both models, dimensions of models were maintained corresponding to prototypes and identical soil properties were assigned. As per the scaling considerations, hydraulic properties of geotextiles were scaled-up corresponding to 30 gravities. Geotextile tensile load characteristics were neglected while performing stability analysis with SLOPE/W. Boundary conditions in numerical modeling were kept similar and assigned in the form of total head in both upstream and downstream those adopted in centrifuge model tests. Flood functions were kept identical to simulated flood in a centrifuge test with the help of flood simulator having same transient time span, as shown in Fig. 2. Soil water characteristics curve (SWCC) was obtained using grain size distribution curve in order to consider unsaturated-saturated properties of levee material (Geoslope, 2007). Further, drainage materials sand and geotextile for internal drain were modeled as saturated materials and interface model respectively. Foundation layer was modeled as an impermeable material to ensure potential seepage through levee section only. For seepage analysis, special geometrical points were assigned in numerical models to represent various PPT positions along the levee bottom of centrifuge models. At these geometrical points, pore water pressure variations and phreatic surfaces were obtained from the using SEEP/W. Variations in pore water pressure at levee bottom were plotted for the special geometrical points. For stability analyses with limit equilibrium, slip surfaces were specified to pass through toe at the downstream side and minimum factor of safety is reported using modified Bishop's method of slices.

Figs. 10a and 10b present variations of pore water pressure for both centrifuge model tests. For comparison purposes, measured pore water pressure with seepage time for models RL-1 and RL-3 are also plotted shown in Figs. 8a and 8b. Model RL-1 without any drainage measure registered relatively high pore water pressures along levee bottom in comparison to RL-3. In all the levee sections the delay in pore water pressures built-up at levee bottom was found to be about by 2 days immediately after initiation of flooding. The delay in pore water pressure built-up can be attributed to unsaturated-saturated properties of levee material or thin coating of bentonite in upstream side. PPT2 placed just vertically below the point where water level interacts with upstream levee face was reported to experience highest pore water pressure. Measured pore water pressures by PPT2 in models RL-1 and RL-3 were 48 kPa and 32 kPa respectively. In comparison, computed pore water pressures from the seepage analysis of models RL-1 and RL-3 at geometrical point corresponding to PPT2 location were found

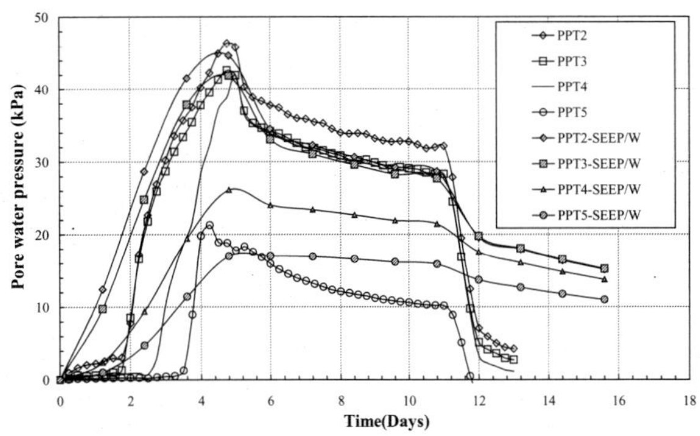

(a) Model: RL-1

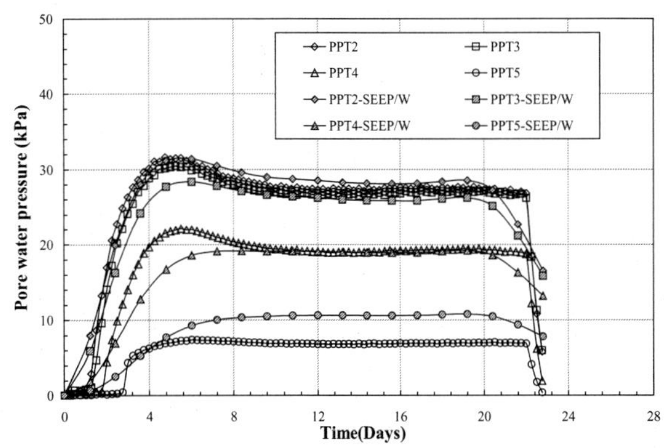

(b) Model: RL-3

FIGURE 10: Comparison of pore water pressure variation in centrifuge tests and seepage analysis

to be 46 kPa and 32 kPa respectively. Measured pore water pressure at PPT5 (near toe) was found to be 22 kPa in case of RL-1, whereas in the case of model RL-3, it was found to be the order of only 7 kPa. Identical trend was observed from the seepage analysis of levee models. Relatively high pressures at the downstream toe in RL-1 caused instability to levee section. Whereas, significant reduction in pore water pressure in model RL-3 at PPT5 was observed.

Figs. 10a and 10b exhibit identical pattern of pore water pressure variation obtained in numerical modeling with measured in centrifuge tests. Pore water pressure in model RL-3 was found 7 kPa from centrifuge tests and 11 kPa through SEEP/W at same geometrical point. The consistencies found in variations of pore water pressures demonstrate a good agreement with observed phenomenon in centrifuge model tests.

Fig. 11a and 11b illustrate seepage analysis carried out with the help of SEEP/W(Geoslope, 2007) based on finite element. Transient seepage analyses were carried out for the same seepage time spans that of centrifuge tests. It was reported that steady state condition of seepage could not established in model RL-1. Various phreatic surfaces within model RL-1are depicted in Fig. 11a. On the other hand, In Model RL-3, steady state seepage could be established without leading to levee failure. Fig. 11b shows various phreatic surfaces within model RL-3.

Figs. 12a and 12b show stability analyses carried out using Bishop's simplified method of slices with the help of SLOPE/W (Geoslope, 2007) based on limit equilibrium analysis. For the purpose of performing stability analysis, phreatic surfaces obtained from SEEP/W were imported in SLOPE/W software.. As depicted in Fig. 12a model RL-1 was found to have a factor of safety 0.933. This is found to represent the observed failure in Centrifuge test for model RL-1 immediately after 5.1 days of seepage time. Whereas, for model RL-3, the computed factor of safety was found to be of the order of 1.203 at high flood level. Marginal increase in factor of safety shows the efficiency of drainage layers to dissipate pore water pressures and enhance stability of levee sections against slope failure in high flood conditions.

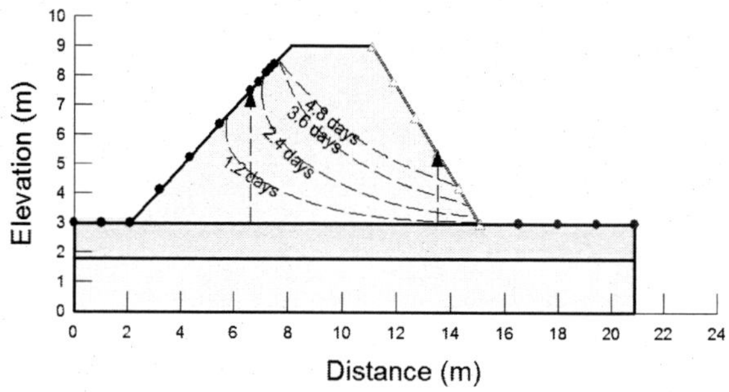

(a) Model: RL-1

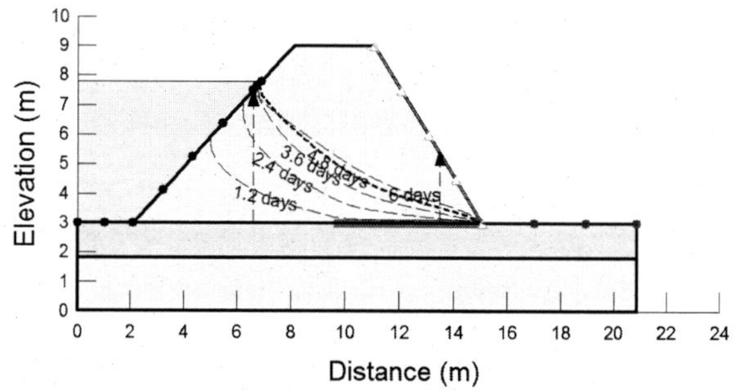

(b) Model: RL-3

FIGURE 11: Seepage analysis of Centrifuge analysis using SEEP/W

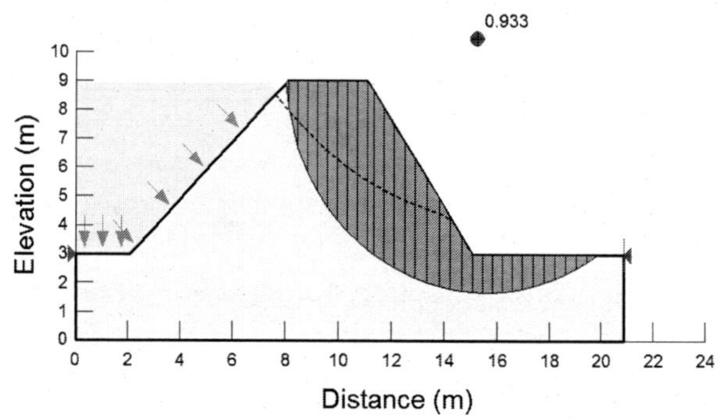

(a) Model: RL-1

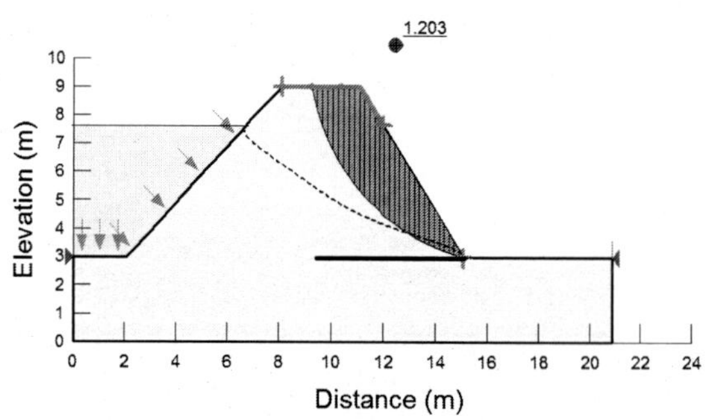

(b) Model: RL-3

FIGURE 12: Stability analysis of centrifuge models using SLOPE/W.

5.1 Parametric Variation

After comparison of centrifuge test results with seepage and stability analysis with numerical modeling, a parametric variation was also performed taking consideration of decrease in transmissivity of geotextile. Four cases were considered for different transmissivity of prototype geotextile as summarized in Table1. Geometry, boundary conditions and levee material properties ware kept constant throughout the parametric variations. In-flow characteristics for model geotextile were scaled-up according to scaling considerations in order to represent prototype drainage material.

Figs. 13 a-d illustrate steady state phreatic surface and equipotential contours for all the four cases. It can be easily noticed that decrease in transmissivity of geotextile leads to rise in phreatic surface at steady state seepage condition. Decrease

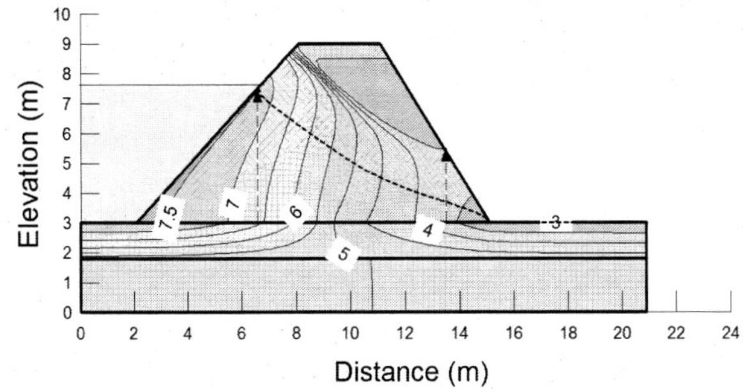

(a) GC-1

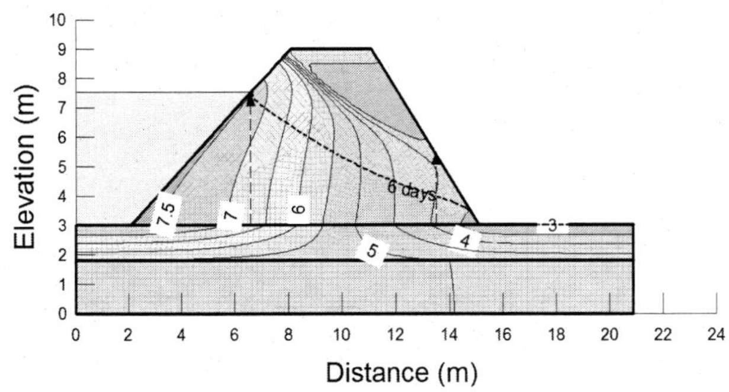

(b) GC-2

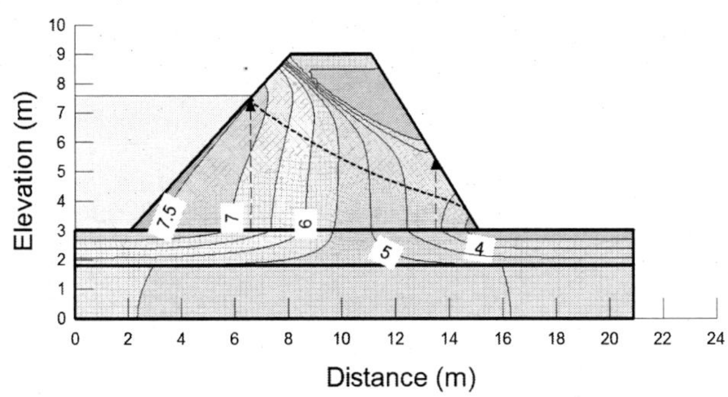

(c) GC-3

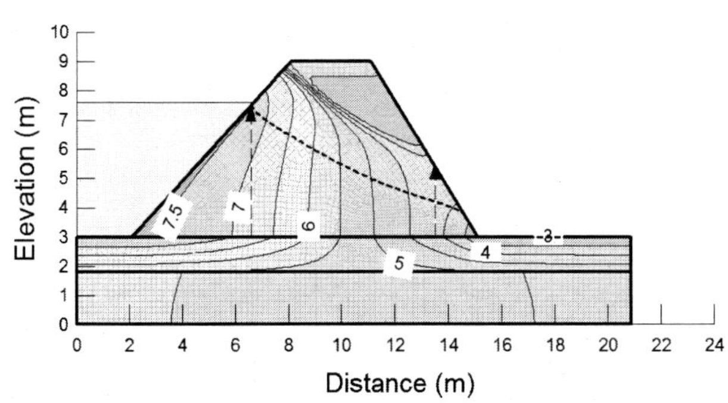

(d) GC-4

FIGURE 13: Pore pressure contours within levee using SEEP/W

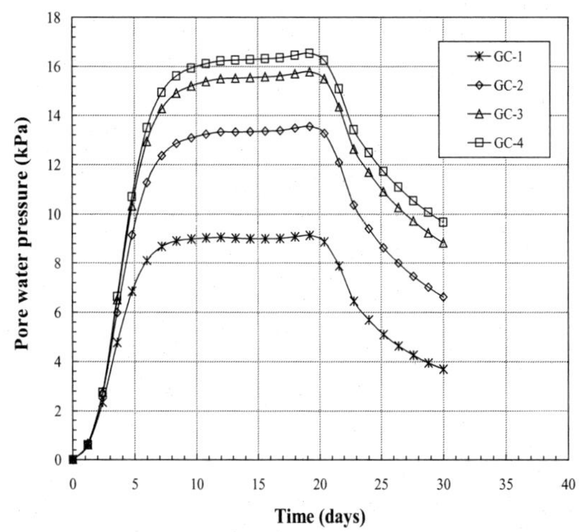

FIGURE 14: Comparison of pore water pressure near toe (PPT5) in SEEP/W

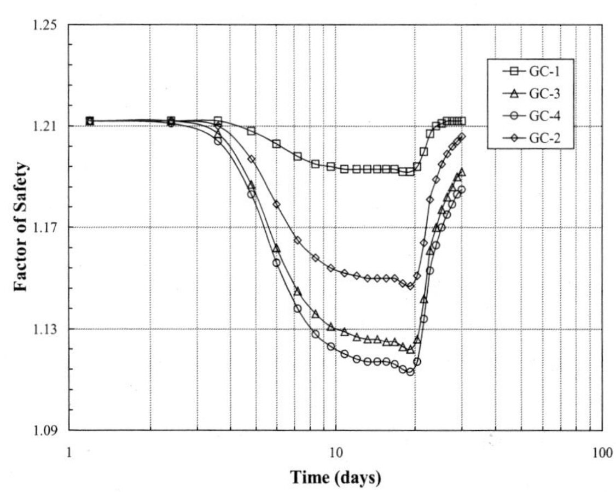

FIGURE 15: Variations of factor of safety against slope failure with prototype time in days

in in-flow characteristics subjected to higher normal stress significantly affects the drainage capacity of geotextile.

Fig. 14 shows variation in pore water pressure variations at the toe in downstream side. The pore water pressure variations were obtained at geometrical point corresponding to the PPT5 in RL-3. It can be noticed that increase in transmissivity leads decrease in the pore water pressure built-up. Pore water pressure built-up were obtained 8kPa and 16.5kPa in GC-1 and GC-4 respectively.

Figure15 shows variations in factor of safety against slope failure for all the cases reported in Table1. It was observed that factor of safety was lower at the high flood conditions. Here prototype time is plotted in Logarithmic scale. Minimum factor of safety was reported to decrease from 1.208 to 1.1 for GC-1 and GC-4 respectively which can be attributed to decreased transmissivity value of geotextile under increased normal stress. Here it is worth noting that decrease in transmissivity was considered only due to increase in normal stresses. Phenomenon of clogging of drains with fines was not accounted in seepage phenomena

6. CONCLUSIONS

This paper evaluates performance of an internal horizontal drainage layer (nonwoven geotextile) positioned at the levee bottom in the downstream side at the onset of flooding and subsequent seepage. For this purpose, two centrifuge model tests were performed. Furthermore, centrifuge model tests were also simulated with numerical experiments using finite element method based SEEP/W and limit equilibrium based SLOPE/W software. Furthermore, influence of varied inflow characteristics on the behavior of seepage and stability of levee was evaluated with help parametric variations using above stated software packages. Finally, based on analyses and interpretations of centrifuge model tests and seepage and slope stability analyses, the following conclusions can be drawn:

1. Levee section (RL-1) without any horizontal drainage layer or clogged drainage layer underwent a catastrophic failure at the onset of flooding and subsequent seepage. The normalized water pressure (u/γH) was registered as high as 0.613 at the initiation of failure and in the case of models with internal drain in the form non-woven geotextile, it is 0.24. Further, reduced values of maximum pore water pressures at the toe of the order of 7 kPa was recorded for model with non-woven geotextile as an internal drain. In comparison, the maximum pore water pressure recorded at the toe for the model without any drainage layer was 22 kPa. This indicates the efficacy of provision of geotextile in the form of internal horizontal drain.
2. Movement of displacement vectors obtained from image analysis of images captured at various seepage stages indicates a catastrophic failure for the levee section without any drainage layer. On the contrary, levee sections with embedded geotextile layer as an internal drain was not experienced any notable movements.
3. Results of transient seepage and stability analyses were found to be very consistent with those in physically centrifuge model test results. From parametric study, it was also concluded that decrease in transmissivity affects efficiency of dissipation of pore water pressure and hence leads to decrease in factor of safety.
4. Finally, hydraulic response of geotextile layer was found to be similar to sand drainage in order to improve the stability of levee. This study indicates that the conventional drainage layer constructed with sand as an internal drain can be replaced with a suitable geosynthetic layer which can impart economy in levee/earth dam construction projects. However, performance of levee section facilitated with geotextile drains were found to be sensitive to reduced drainage capacity of geotextile, which in turn can be attributed to normal stress on the geotextile or their clogging phenomena.

However, further studies are needed to have a better insight into the hydraulic behavior of geotextile embedded in levee sections.

REFERENCES

American Society for Testing and Material (ASTM). Standard Test Method for Determining the (In-plane) Hydraulic Transmissivity of a Geosynthetic by Radial Flow. American Society for Testing and Materials, West Conshohocken, Pennsylvania, USA, D6574.

Das, A. and Viswanadham, B.V.S. (2010) "Experiments on the Piping Behavior of Geofiber-Reinforced Soil". Geosynthetics International, Thomas Telford Ltd, 17, Issue 4, pp 171-182.

EM 1110-2-1913 (2000), Manual for Design and Construction of Levees, Department of the Army, U.S. Army Corps of Engineers Washington, DC 20314-1000.

Geo-Slope (2007). SLOPE/W and SEEP/W Ver. 7.15 User's Guide. Geo-Slope International Ltd, Calgary, Canada.

Giroud, J.P. (2005), "Quantification of geosynthetics behavior". Geosynthetics international, Thomas Telford Ltd", 12, Issue 1, pp 1-26.

Iryo, I., and Rowe, R.K. (2003) "On the Hydraulic Behavior of Unsaturated Non-woven Geotextile". Geotextiles and Geomembranes, Elsevier Ltd, 21, pp 381-404.

Iryo, T., and Rowe, R. K., (2004) "Numerical Study of Infiltration into a Soil-Geotextile Column". Geosynthetics International, Thomas Telford Ltd., 11, Issue 5, pp 377-388.

BIS 9429(1999), Drainage System for Earth and Rockfill Dams-Code of Practice, Bureau of Indian Standards.

Jia, G. W., Zhan Tony L. T., Chen Y. M., and Fredlund D. G. (2009). "Performance of a Large-Scale Slope Model Subjected to Rising and Lowering Water Levels". Engineering Geology, Elsevier B.V., 106, Issue 1, pp 92–103.

Raisinghani, D. V. and Viswanadham, B. V. S. (2011) "Centrifuge model studies on low permeable slope reinforced by hybrid geosynthetics". Geotextiles and Geomembranes, 29, Issue 6, pp 567-580.

Viswanadham, B. V. S., Mahajan, R. R., (2007) "Centrifuge model tests on geotextile reinforced slopes". Geosynthetics International, 14, Issue 6, pp 365-379.

Ubilla, J., Abdoun, T., Sasanakul, I., Sharp, M., Steedman S., Vanadit-Ellis W. and Zimmie T., (2008) "New Orleans Levee System Performance during Hurricane Katrina: London Avenue and Orleans Canal South". Journal of Geotechnical and Geoenvironmental Engineering, ASCE, 134, Issue 5, pp 668-680.

Centrifuge Studies on the Behaviour of Back-to-Back Geogrid Reinforced Soil Walls

B.H. Katkar and B.V.S. Viswanadham

Department of Civil Engineering, Indian Institute of Technology Bombay, Powai, Mumbai

E-mail: bharat.katkar@iitb.ac.in; viswam@civil.iitb.ac.in

ABSTRACT

The main objective of this paper is to bring differences between the deformation behaviour of single vertical and back-to-back geogrid reinforced soil walls constructed using wrap-around technique. Three centrifuge model tests were performed on scaled models of soil walls in the National Geotechnical Centrifuge Facility present at IIT Bombay. Centrifuge test on single wall was carried out to compare its behaviour with back-to-back walls. Whereas two centrifuge tests on back-to-back walls were carried out with an aim to study the effect of unconnected and connected reinforcement layers from either walls. An incremental gravity procedure was adopted for the testing. Gravity level was increased up to 75g or till failure, in increments of 5g from 10 g onwards.. At each gravity level, status of front elevation of the model was captured and subsequently used for digital image analysis. Digital image analysis was performed to analyse the deformation behaviour and strain distribution in the reinforcement layer of the walls. Interpretations based on the observed deformation behaviour are presented. The interaction of failure surfaces occurring in back-to-back walls were also studied and compared to that occurring in single geogrid reinforced soil wall.

1. INTRODUCTION

Reinforced soil walls are nowadays extensively used for achieving the elevation differences required in landforms. Soil can be reinforced in numerous ways either by using geogrid, geocell or soil nails, depending on the applicability of the system, based on the site conditions and economy. Geogrid reinforcement was used for the present study, a scaled down mesh was used with approbation to the scaling laws for tests in geotechnical centrifuge. The load transfer mechanism from soil to the reinforcing geogrid layer is simple and can be found amply explained in the literature (C.J.F.P Jones, 1996). Several investigators conducted centrifuge studies to evaluate the performance of reinforced slopes with steeper inclination and can be found in literature (Viswanadham and Mahajan, 2007). The design procedures for reinforced soil walls are well established and are usually extended for the design of back-to-back walls.

A back-to-back wall is similar to normal vertical soil wall but consists of another wall present at the other end of retained soil. Back-to-back walls are used in construction of bridge abutments, noise barriers, slipways and in highways interchanges. A schematic cross-section of back-to-back wall is shown in Figure 1. The aim of this paper is to showcase the differentiation in behaviour of single wall and back-to-back walls with respect to their deformation behaviour. A wrap-around facing was adopted to represent flexible facing, which also proves to be helpful for construction of scaled centrifuge models. Also additional results were obtained by varying the connection of reinforcement in back-to-back walls. Limited amount of literature is present for experimentally evaluating behaviour of back-to-back walls (Anubhav and Basudhar, 2012). Numerical studies were performed by (Han and Leshchinsky, 2010) and (Hardianto and Turong, 2010) to study the interaction of failure surfaces and to evaluate the seismic stability of back-to-back walls. To the authors knowledge there is an absence of literature consisting the study comparing the behaviour of single wall to back-to-back walls. Though limited numbers of centrifuge tests were performed, the results obtained from these tests can be considered as representative. Currently for the design of back-to-back walls the guidelines issued by FHWA, 2009 are followed. But the deformation behaviour aspect of walls remains uncovered. Also the effect of connecting reinforcements on stability of back-to-back walls is experimentally not known. To address this aspect, the presented scheme of tests was performed. For quantitatively interpreting the data obtained in centrifuge tests, the technique of image analysis was used. Image analysis proves to be useful tool in hands of a Geotechnical engineer for estimation of displacements and strains (Viswanadham and Mahajan, 2007). Hence for the current study aid of digital image analysis software ImageJ was availed. Image analysis was performed on the digital photographs obtained during the in-flight testing of centrifuge model. The strain distribution in geogrid layers and displacement of facing element were captured and quantified using image analysis. Interpretations drawn from the tests are presented at the end of the paper.

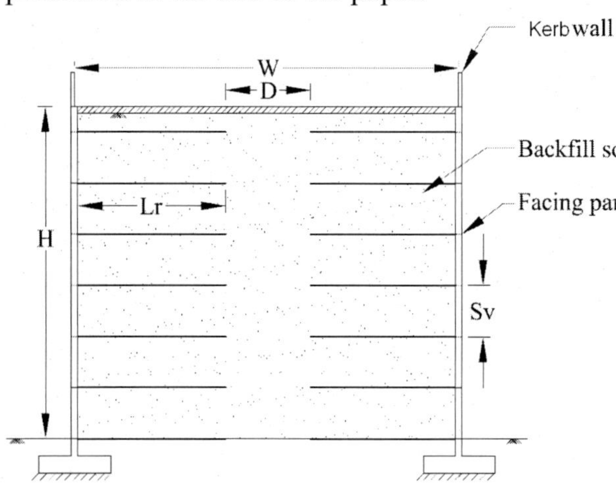

FIGURE 1: Schematic cross-section of back-to-back geogrid reinforced soil wall

2. MODEL MATERIALS

For testing of models in geotechnical centrifuge, it is customary to scale down the materials used in preparation of model. Increase in the gravity level not only increases the dimensions of model, it as well as has a subsidiary effects on strength, stiffness of the material involved. In order to create an exact simulation of conditions prevailing in prototype, it is customary to follow the scaling laws (Viswanadham and König, 2005). Therefore in the present study, the model materials used namely, soil and geogrid, were selected such that the scaling laws were obeyed up to a maximum extent possible. The engineering properties of the materials are described below.

2.1 Sand

Sand used for present study was Goa sand composed of rounded and sub-rounded particles and classified as SP by the Unified Soil Classification System. The details of soil properties are referred and reproduced from (Viswanadham and Mahajan, 2007). The average particle size D_{50} for sand used in this study is 0.21 mm, the coefficient of uniformity is 2.07, and coefficient of curvature is 1.19. The maximum and minimum void ratios of sand are 0.895 and 0.597, respectively. Evaluation of friction angle was carried out by performing direct shear tests at three normal stresses: 50, 100 and 150 kN/m^2. The friction angles were obtained at relative densities of 55% and 85% are 34° and 38°, respectively.

2.2 Geogrid

A black mesh imported from Belgium was used as reinforcement in model soil wall for centrifuge testing. Tensile load-strain characteristics of the model geogrid materials were determined

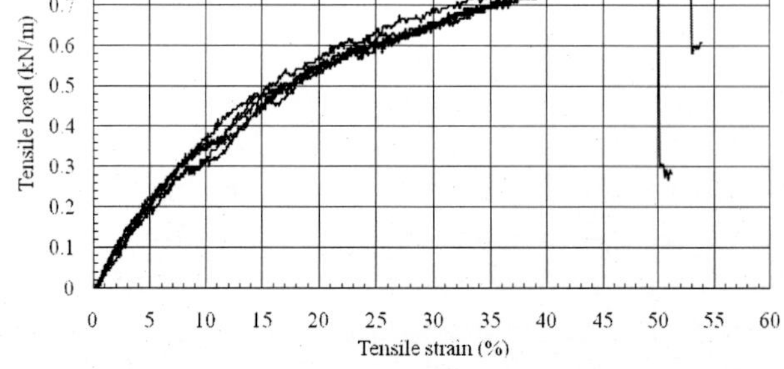

FIGURE 2: Plot of tensile load against percent strain of model geogrid material

by performing wide-width tensile tests (ASTM D4595, 2005) at a strain rate of 10 mm/min. Tensile load testing was carried out in the machine direction of geogrid, the Load Vs. strain graph obtained from the test are displayed in Figure 2. The model geogrid with low tensile properties was purposefully selected to induce failure, geogrid reinforced soil wall during centrifuge testing. Details of model mesh are indicated in Table 1.

Table 1: Properties of Model Geogrid Material

Properties Model Geogrid	*In Model Dimensions*
Color	Black
Aperture shape	Square
Aperture size in machine and cross machine direction	3.5mm × 3.5mm
Material composition	Polyester strands coated with poly-vinyl chloride
Average tensile strength Machine direction	0.796 kN/m
Average ultimate values in machine direction Tensile force Tensile strain	0.159 kN 53.93 %

3. MODEL TEST PACKAGE

The National Geotechnical Centrifuge Facility (NGCF) present at Indian Institute of Technology Bombay was used for the present study. This facility has a balance beam centrifuge of radius 4.5 m with a maximum capacity of 250 g-ton which can be understood as a payload of 2.5 ton at 100g. The maximum capacity of the facility at an acceleration of 200g is a payload 0.625 ton. The model is mounted into a basket hinged to one arm of the beam centrifuge, whereas an adjustable counterweight is permanently mounted at the other end. The counterweight is shifted according to the weight of model test package, to keep the system in equilibrium. An adjustable counterweight plate is present at the top of the counterweight arm of the centrifuge; this plate facilitates the in-flight adjustment of weight to accommodate minor imbalance occurring due to change in geometry of the model. The detailed specifications of this particular centrifuge are discussed by (Chandrasekaran, 2001). Two on-board cameras were installed to watch and record the progress of test on model during in-flight operation. A CCD camera captures images at rate of 25 frames/second, thereby providing a continuous video of the test. Additionally a digital photo-camera was used, which offered an image of higher resolution and was used to acquire images of the model as and when required during the test..The photographic data obtained from these cameras was saved through image-capturing software in the computers installed in instrumentation room, positioned outside the centrifuge chamber. Linear Variable Differential Transformer (LVDT) was used to monitor the surface settlement of the model, data of which was acquired using a data-logger placed onto the hinged basket. LVDT data was acquired in to the instrumentation room computer via data-logger software. The number of LVDT`s were varied according to the test requirement and were mounted on the top of a channel section bolted to the strong box carrying model. Strong box with inner dimension of 760 mm (length), 200 mm (width) and 410 mm (height) was used for test purpose in the present study. Strong box has a three walled honeycomb structure built in order to reduce the dead weight of the box. The inner surface of the box is constructed of mild steel plates with a view of reducing its friction with soil. The box is attached with a Perspex glass sheet on the front side, which is connected by bolts to the steel plates of the box. Model can be viewed through the Perspex glass during in-flight condition using cameras mounted at a fixed distance in front of Perspex. Illumination was provided in in-flight condition by fluorescent lamps placed in front of camera. Perspex and the inner walls of container were smeared with white petroleum grease to reduce friction between wall-soil and Perspex-soil. Plastic strips of very less thickness were posted on the grease smeared surface, which further reduced the friction between soil particles-wall surfaces, thereby idealizing plain strain condition in the model. These plastic strips were cut to 100 mm width and then pasted on the Perspex; this technique helped in reducing friction as well as helped in removing air voids from within the glass-plastic sheet interface, thereby enhancing the image clarity. Also this helped in nullifying the possibility of soil particles from backfill penetrating into front of movable markers, which further helped in improving the clarity of image.

To use digital image analysis technique for extraction of data, it is convenient to obtain scale of image at each gravity level to acquire the coordinates of markers placed into the model. For this purpose permanent markers were installed on the Perspex glass, the distance of which was measured for each photograph and at every gravity increment. This was also important as at higher gravity levels there is a slight change in the resolution of digital image and this makes the image to shrink slightly. Hence four permanent markers were pasted within the area visible by digital camera (540 mm (length) x 320 mm (height)) on Perspex glass as seen in Figure 4. Movable markers were glued firmly to the geogrid layer as well as the wrap around portion of the geogrid, with an aim to trace displacement vectors to aid image analysis.

3.1 Test Program

Three centrifuge model tests were performed with an aim to evaluate the deformation behaviour of single wall, back-to-back wall without and with connection of reinforcement. The dimensional configuration adopted for these test are as given in Table 2. The designation of the parameters can be referred from Figure 3.

Table 2: Test Designation and Dimensions

Test Legend	*W1*	*W2*	*W3*
Width (mm)	-	430	380
W/H	-	1.593	1.407
D/H	-	0.185	0
Sv/H	0.148	0.148	0.148
Lr/H	0.7	0.7	0.7
No. of layers	7nos.	7nos.	7nos.
Remark	Single wall	Back-to-back wall	Back-to-back wall with connected reinforcement

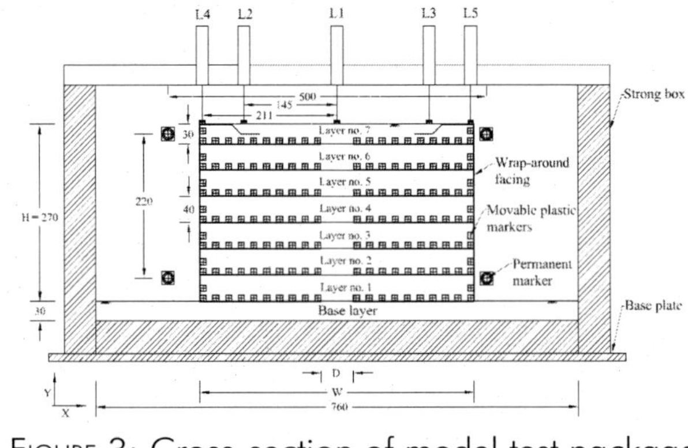

FIGURE 3: Cross-section of model test package

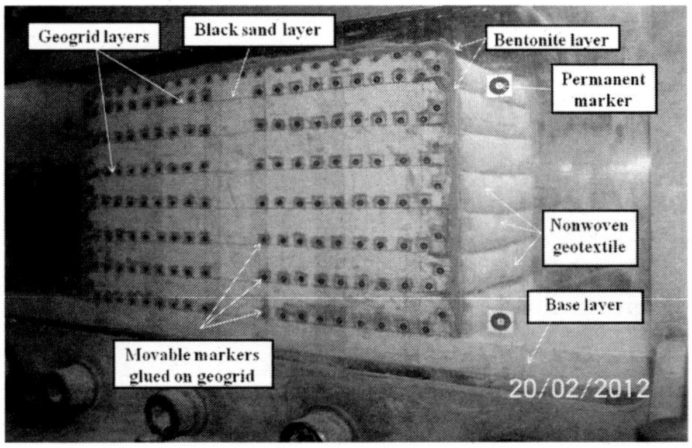

FIGURE 4: Photographic view of model test package with details

3.2 Test Procedure

Geogrid reinforcement layer with length of 0.7 H (189 mm), spacing of 40 mm and anchorage length of 0.25 H (67.5 mm) for all layers and with anchorage length of 0.37 H (100 mm) for top layer were properly cut with a width of 20 mm. Anchorage length was provided to avoid opening up of individual layers. More anchorage was provided at topmost layer to avoid sloughing failure at crest due to lesser confining pressure present above the topmost layer i.e. layer 7 (Figure 4) during centrifuge test. All geogrid layers were numbered by paper whitener for ease of post investigation. Non-woven geotextile material with very less thickness and minimal strength was attached on the exterior of wrap around portion of geogrid (Figure 4), this measure prevented flow of sand through the openings of geogrid along with providing the primary purpose of flexibility to the facing of the wall. It was assumed that geotextile did not participate in any load carrying mechanism. To avoid leakage of soil material from facing of the model the geotextile material was cut with more width, keeping in consideration 20 mm excess into anchorage length. For the sake of image analysis four permanent markers were firmly attached at the visible corners of Perspex glass (Figure 4) and repeatedly checked for accurate distance between them, the distance always was noted before and after centrifuge test. For some model tests the permanent markers were shifted to newer positions in order to avoid their interference with in-flight model displacement. Thin transparency sheets were used to prepare both permanent as well as temporary/movable markers. The movable markers were cut to a size of 10 mm x 20 mm with folded size of 10 mm x 10 mm being marker face and same size was kept for marker leg, which was firmly glued to the geogrid and embedded in the soil. Movement of marker during centrifuge test was aided by the grease applied to the face portion of the marker in contact with the Perspex glass. Movable markers were glued to the geogrid layer at an approximate centre to centre spacing of 20 mm. These movable markers were used to plot the deformation of wall facing, strain distribution in the geogrid and displacement vectors of their position at each gravity level. Same strong box of dimensions of 760 mm (length) x 410 mm (height) x 200 mm (width) was used for all the model tests.

With base layer of 30 mm thickness extending to the full length of the strong box, all models were built with a total height of 300 mm. Base layer was constructed by sand pluviation technique to attain a relative density of 85%, so that firm foundation conditions can be established. Density was verified for each pluviated layer throughout the construction of the model. Wooden supports were placed over the base layer and adjusted according to the width of model to be tested. Supports were smeared with grease to reduce the friction with geotextile material during removal of support. First layer of geogrid was laid and movable markers were glued throughout the length of geogrid at a spacing of 20 mm, facing markers were also attached at the same time. First marker on the geogrid was glued at some distance from the facing edge to avoid overlapping of this marker with the facing marker. Next sand layer with a relative density of 55% was then pluviated and levelled properly. This procedure wall continued until the seventh layer was constructed as shown in Figure 4. Wooden supports were then removed and model was checked for any leakage of sand particles. As a precautionary measure a thin layer of Bentonite clay (as seen in Figure 4) was applied to vertical edges of retaining wall to prevent in-flight constructional failure of model. The interference occurring due to bentonite layer was assumed to be negligible during in-flight deformation of the wall. The model was then carefully hoisted and mounted into the centrifuge basket with least disturbance to the strong box. Strong box was then bolted to the centrifuge basket and LVDTs (three or five) were placed over the top surface of model centrally, depending upon model width, the distances of which are indicated in Figure 3. A bolted unyielding steel channel section was used to support LVDTs.. Two cameras were mounted in front of centrifuge basket to get a front elevation of basket through the transparent Perspex glass. Counterweight was then shifted based on the centre of gravity of the mounted model.

In every test the self-weight of the model was increased in the increments of 5g after initially reaching a level of 20g, with a stabilization time of two minutes at each gravity level. Digital photographs were taken at each gravity level along with a continuous video recording from CCD camera. The model name, gravity level and time of snap were noted for each image. An additional picture was taken even after the failure of the wall during in-flight condition to obtain profile of the model after failure.

The surface movements of the model walls were measured by number of LVDTs placed on the top of the wall. The LVDTs were positioned as shown in Figure 3. And as described previously, the centrifuge models were subjected to an incremental increase in gravity with stabilization

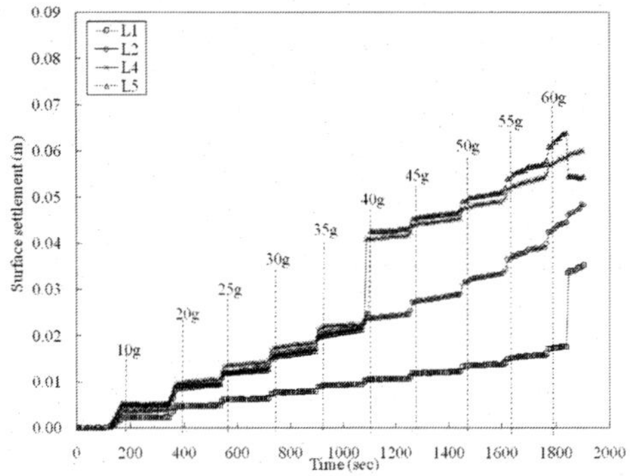

FIGURE 5: Variation of surface settlement with repect to time (Model: W3)

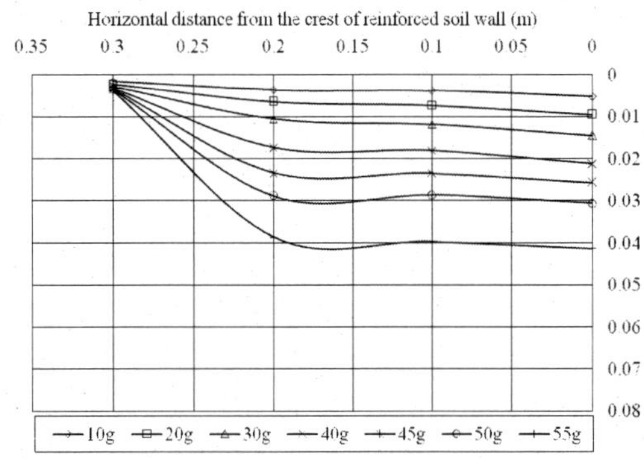

(a) Model W1

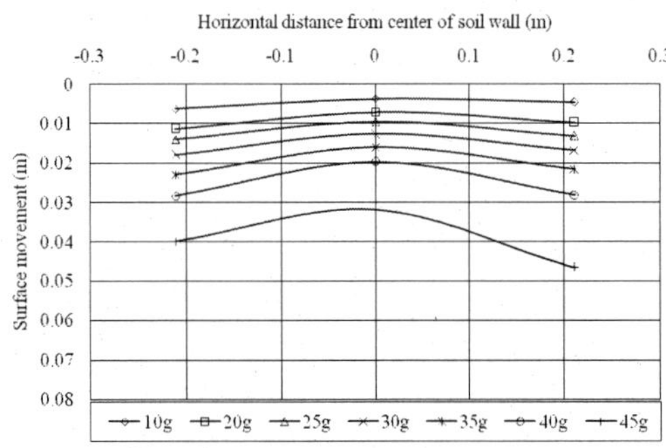

(b) Model W2

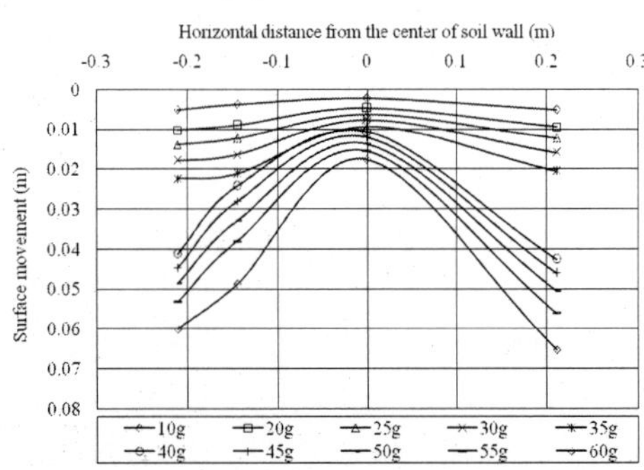

(c) Model W3

FIGURE 6: Variation in surface movements of geogrid reinforced soil wall models

time of two minutes. This increase in gravity can be visualised by observing the surface settlements of the model, as represented in Figure 5. The data from LVDT showing looping values were discarded as it was considered as instrumentation failure. Sudden increase in the settlement of the crest LVDT's indicate the localised failure at 40g.

LVDTs were positioned at centre and crest of the wall, in order to obtain the surface profile with increase in gravity level. The plot of surface settlement with distance are indicated in Figure 6a-c.

4. IMAGE ANALYSIS

Digital image analysis can be used to record the static and dynamic deformations of soil, to identify the failure surface in retaining structures, and also to estimate the straining of geotextile material (Zornberg and Arriaga, 2003). Images captured during centrifuge tests were analysed using digital image analysis software (ImageJ, Version 1.44, 2011). The details about the process of images analysis was followed as mentioned in (Viswanadham and Mahajan, 2007). Nine markers with approximated spacing of 20 mm were glued to each geogrid layer for all models; these movable markers with the help of digital image analysis were used to evaluate the deformations and strains in geogrid layers. For this purpose pictures taken from the digital camera were used for better resolution and better clarity. Image analysis was performed on the series of pictures captured from a particular model test. To calibrate each image properly the distances between the four permanent markers attached to inner side of the Perspex sheet were used to scale the selected image. Scale was adjusted for each picture based on the pixelated distance between the permanent markers. The co-ordinates of permanent markers were fixed relative to picture resolution and size. Care was exercised to select images of same resolution and same size for digital image analysis to have equivalent global coordinates for each image. Movable markers glued onto geogrid layers were digitized for each photographic image, and their coordinates were obtained by selecting locations from each picture at particular gravity level with

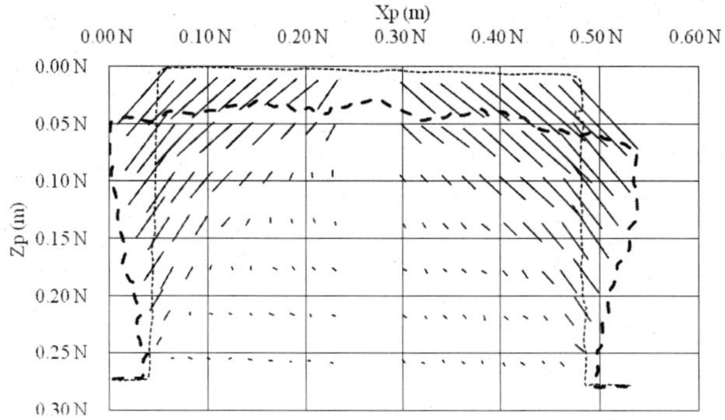

(a) Model W2

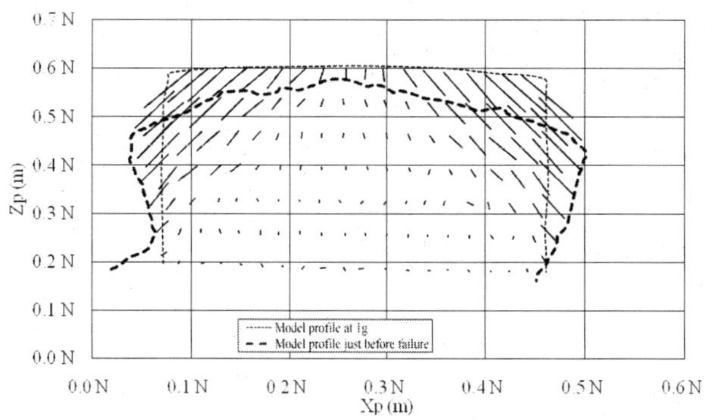

(b) Model W3

FIGURE 7: Displacement vector diagram of back-to-back walls

reference to the global coordinate. Co-ordinates of movable markers from each series of pictures for a particular model were used to arrive at the displacements of wall and strains in geogrid layers with respect to their initial position at 1g level.

Displacement vectors were plotted with an accuracy of 0.05 mm (in model dimensions). Figure 7a-b display the displacement vectors obtained for the back-to-back wall model along-with the profile captured just before failure. The horizontal and vertical distances have been represented as Xp (horizontal distance) and Zp (vertical distance) respectively on the abscissa and ordinate in multiples of N, where N is the scale factor or gravity level (Figure 7a-b). Formation of failure surface can be visualised along-with the resultant movement of walls, from displacement vector diagram. It can be observed from Figure 7a-b that movement of makers with larger magnitude was concentrated towards the wall facing, indicating outward movement of wall before failure.

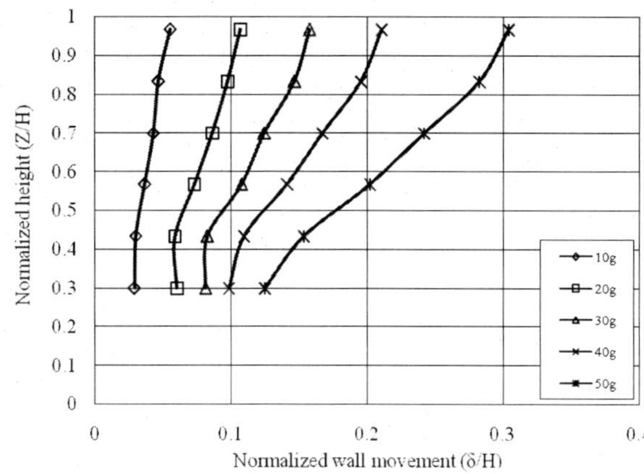

(a) Model W1

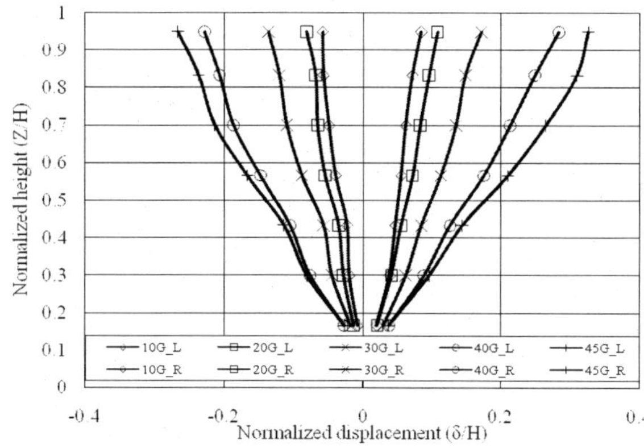

(b) Model W2

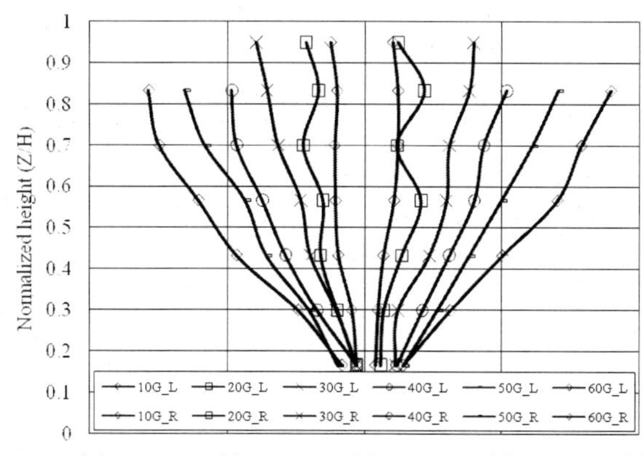

(c) Model W3

FIGURE 8: Displacement diagram of facing element expressed in normalized scale

Co-ordinates of movable markers obtained from image analysis and were used to calculate the facing displacement of wall at each step of gravity level. Displacements were calculated using simple distance formula given as,

$$\delta = \sqrt{(X_2 - X_1)^2 + (Y_2 - Y_1)^2} \qquad (1)$$

where, (X_1, Y_1) are coordinates of the movable markers located at face of the wall at 1g, and (X_2, Y_2) are the co-ordinates of same marker at a particular higher gravity level. Normalized face displacement diagrams were plotted as seen in Figure 8a-c. In some centrifuge tests the top facing markers (in soil layer 6 and 7, Figure 4) were lost during the test because of the opening of the topmost reinforcement layer at intermediate gravity levels, in such a case the displaced position of marker were not plotted. The equivalence in construction of both sides of back-to-back walls can be verified by observing the uniformity of deformation of both walls with increase in gravity-level, as can be seen in Figure 8b-c.

The wall movement indicates an increase in the displacement of facing with increase in height of wall (i.e. increase in gravity level). For models W1, W2 and W3 the maximum normalized displacement at 40g on the right face of wall at normalized height 0.833 (Figure 8a-c) wall was noted to be greater than 0.2. The maximum normalized displacement of the same layer at pre-failure condition (Figure 8a-c) of each model was 0.310, 0.309 and 0.316 respectively. The face displacements were minimal for the model with single wall as compared to other models.

By using image analysis software, the incremental displacements of the markers attached to the geogrid layers were measured. This displacement data was then used to calculate the distance between adjacent markers present in same geogrid layer at a particular gravity level. The distances were then used to compare the length between same set of two markers at 1g level and Ng level, in order to arrive at strain value occurring between the markers at that N^{th} gravity level. Careful consideration was given to select the markers on appropriate geogrid layer. Strain values obtained from analysis were used to plot the strain distribution diagram (Figure 9a-c) along the length of individual geogrid layer, inclusive of an approximate failure surface drawn by connecting the peak strains in geogrid layers. For back-to-back model wall with connected

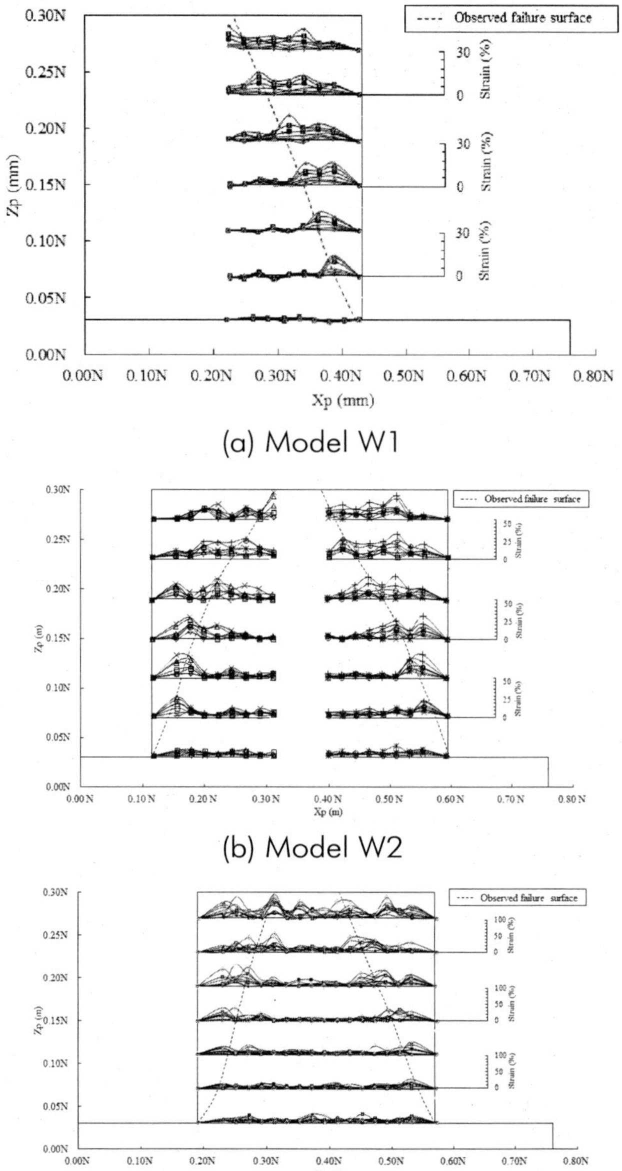

(a) Model W1

(b) Model W2

(c) Model W3

FIGURE 9: Strain distribution along length of reinforcement layer with increment in gravitational field

reinforcement, the mobilisation of strain occurred throughout its length, straining the reinforcement up to its maximum limit. The peak strain variations with respect to the normalised height of the wall for each increment in gravity level are plotted in (Figure 10a-c). The strain was calculated using the equation,

$$\varepsilon = \left[\sqrt{\frac{\{(X'_2-X'_1)^2+(Y'_2-Y'_1)^2\}-\{(X_2-X_1)^2+(Y_2-Y_1)^2\}}{\{(X_2-X_1)^2+(Y_2-Y_1)^2\}}} \right] \times 100 \qquad (2)$$

(Viswanadham and Mahajan, 2007)

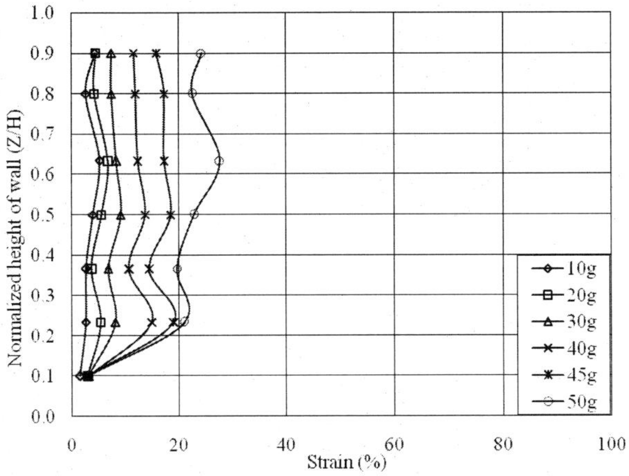

(a) Model W1

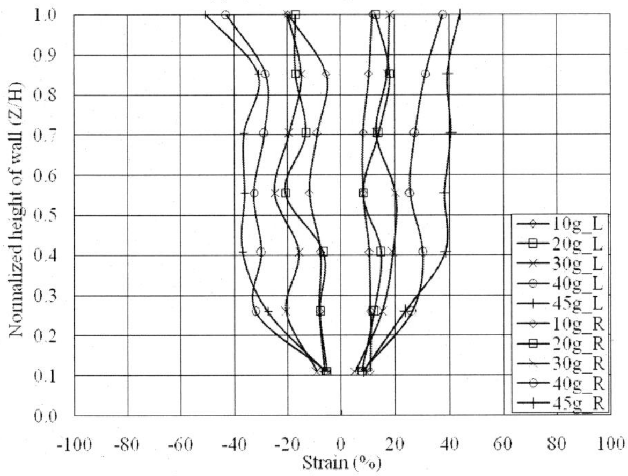

(b) Model W2

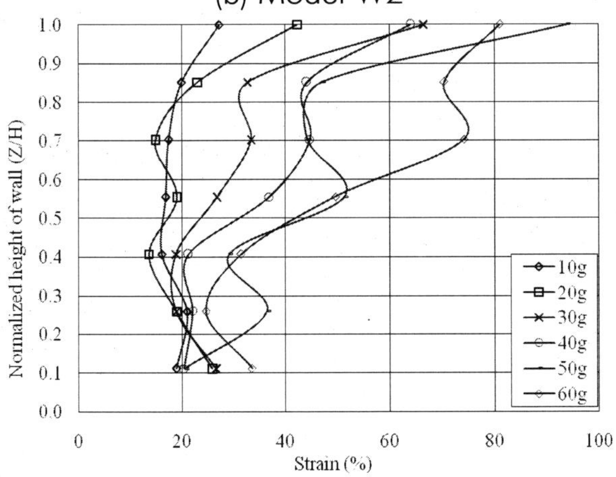

(c) Model W3

FIGURE 10: Variation of peak strain distribution in geogrid reinforcement with increment in gravitational field

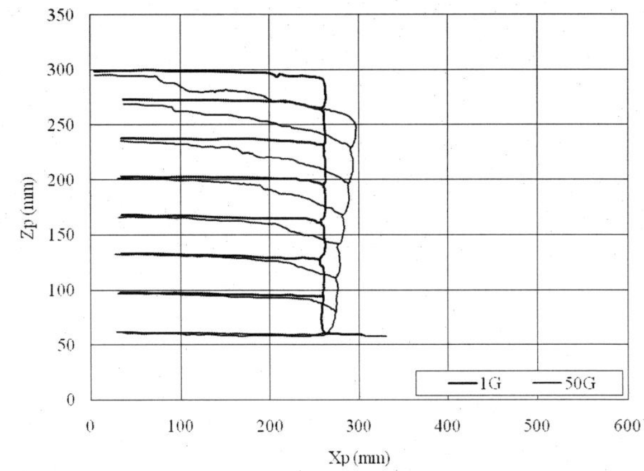

(a) Model W1

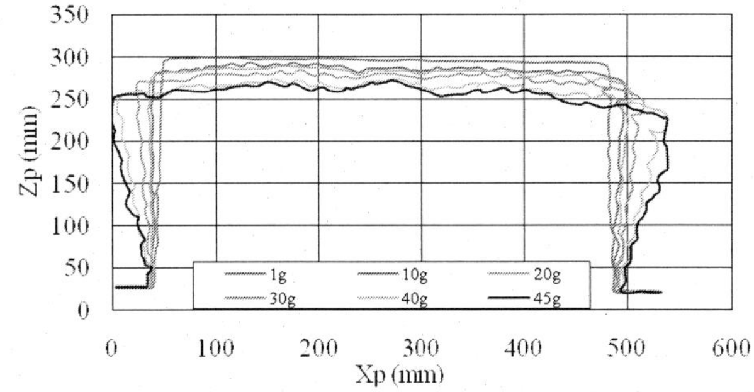

(b) Model W2

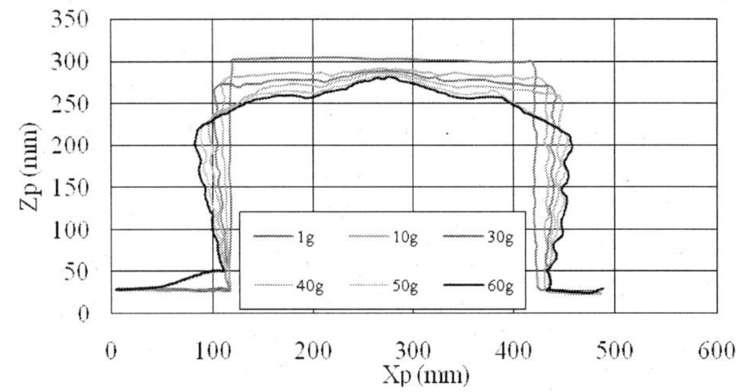

(c) Model W3

FIGURE 11: Displacement profiles of soil walls with increment in gravity level

where, (X_1, Y_1) and (X_2, Y_2) are the coordinates of adjacent markers M_1 and M_2 at 1g level. ($X`_1$, $Y`_1$) and ($X`_2$, $Y`_2$) are the coordinates of same markers at higher gravity (Ng) level. Strain of the reinforcement at the facing edge was assumed to be zero. Though reinforcement layer 7 displayed maximum strain, opening-up of this particular layer in some test provided some difficulty for comparison of it between all the models. For model with connected reinforcement, the strain distribution diagram is drawn continuous for each reinforcement layer from both walls. Figure 10a-c indicates the gradual increase of strain in geogrid with increase in the gravity level. For model W3 with connected reinforcement almost uniform straining occurred along the length of the geogrid. The peak strain in geogrid layer 6 for model W1, W2 and W3 were 22.70%, 39.42% and 70.35% respectively. Model W3 with connected reinforcement developed maximum strain as the possibility of reinforcement failure by pull-out from either end was not possible.

To obtain the deforming profile of the models, in-flight images were carefully marked digitally along the visible outer edges of the model. A minimum of 200 points were selected on digital markings in order to obtain the exact profile in in-flight condition at each gravity level. By this process a qualitative diagram (Figure11a-c) were obtained which were then used to understand the deformation mechanism of back-to-back soil wall. The deformation of the wall with respect to increment in gravity level can be observed through the figure. An approximate visualisation on the mechanism leading to the failure of wall can easily interpreted through Figure 11a-c.

5. RESULTS OF CENTRIFUGE TESTS

For comparison between models tested in centrifuge, graphs were plotted between the normalizeddisplacements (Figure 12a-b), peak strain distribution against normalised height of the wall in (Figure 13a-b), at 40g/45g and at termination of test respectively.

It can be observed Figure 12a that the displacement of the wall were almost similar for single wall and back-to-back wall W2, but at the same time it can be seen that the facing displacement of wall w3 with connected reinforcement was higher compared to other models. Despite of excessive deformations in wall W3 it sustained its stability up to level of 60g (Figure 12b), whereas wall W1 and W2 failed at 50g and 45g respectively.

Strains developed in the reinforcement layers were synonymous to the deformations observed in the wall facing. It can be observed from Figure 13a that at 45g level the strain developed in the top layers of model W3 was far more than that developed for model W1 and W2. Almost uniform strain in all layers except the bottom layer of reinforcement was observed in wall W1. A similar trend was also observed in back-to-back wall W2. The magnitude of strain observed in back-to-back

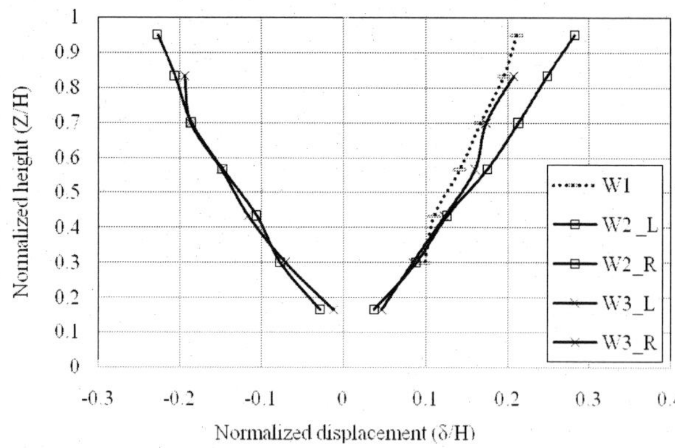

(a) Displacement of walls at 40g

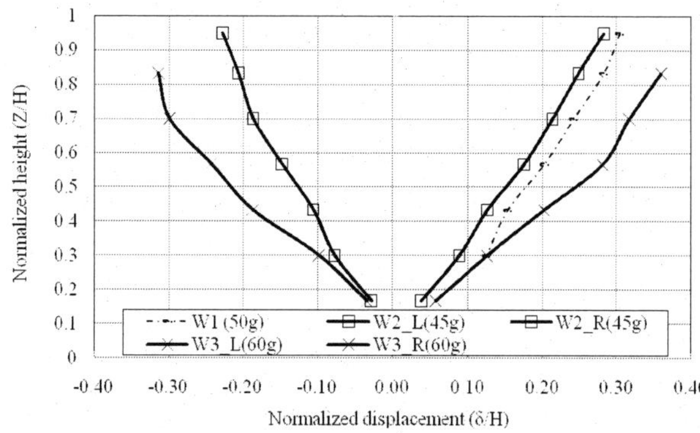

(a) Displacement of walls at termination of test

FIGURE 12: Comparison face displacements for all model cases

wall W3 with connected reinforcement was non-uniform in nature.

Single wall model test W1 displayed lesser surface settlements as well as lower face displacements (Figure 9a and 10a). Though the stability of the wall was found to be lesser than that of back-to-back walls, the deformability was observed to be much lesser. A uniform increase in reinforcement strain with increase in the gravity level was observed. The peak of the geogrid reinforcement strain moved towards inner side of the wall in the upper layers, thereby indicating the movement of failure surface away from facing. An outward face movement away from initial position of the walls was observed immediately after the first increment in gravity level. The horizontal movement of wall facing was noted to increase progressively with increase in gravity level during centrifuge test. The movement at the crest of wall was found to be 2.5 times more than the movement occurring at the bottom of wall at penultimate stage of failure. Lesser movement at bottom of

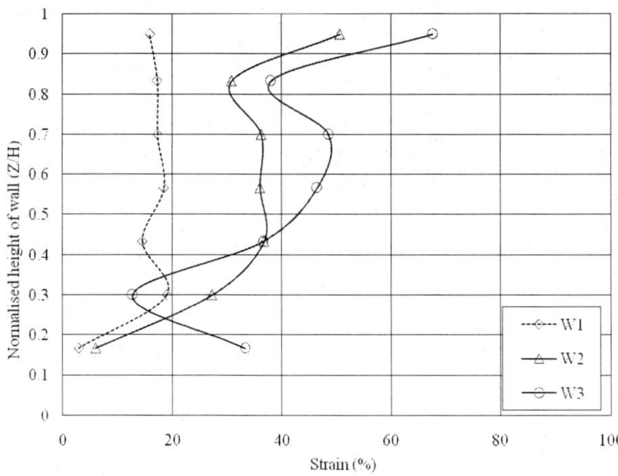

(a) Reinforcement strain developed at 45g

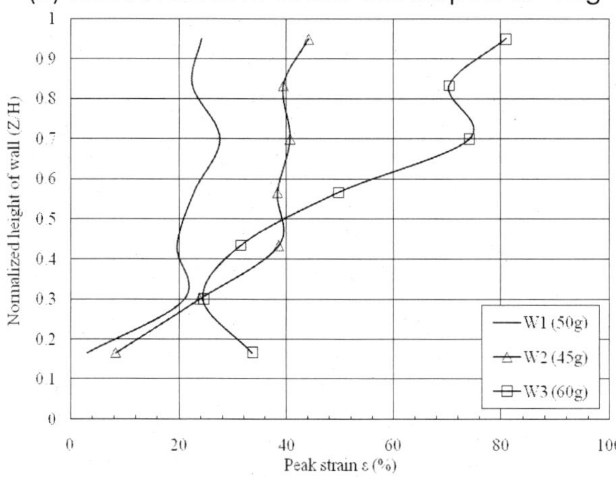

(b) Reinforcement strain developed at termination of test

FIGURE 13: Comparison of peak strain distribution data in reinforcement layers

wall can be attributed to effect of overturning caused due to presence of full active thrust acting from backfill soil on the reinforced zone, thereby rotating the wall about the base. Also at the same time due to relatively smaller lever arm between the downward force exerted by soil in reinforced zone and toe of wall, as compared to that of present in the case of soil walls with inclined facing, the stabilizing moment effective from weight of soil in reinforced zone was also reduced. The maximum mobilized strain of 27.5% was observed to occur in fifth layer of reinforcement, whereas the bottommost reinforcement layer was able to strain up to 3% of initial length. Strains were observed to increase gradually with each increment in gravity level within all layers of reinforcement except the bottommost layer; which can be interpreted as, the failure of wall started in upper layers of the wall and progressed into the lower layers until failure occurred. The failure surface was observed to terminate within the reinforced zone only.

Centrifuge tests on back-to-back soil wall model with a distance of 50 mm between the reinforced zones was observed to undergo excessive facing deformations at 45g, which had an equivalent prototype dimensions of width 19.35 m and height 12.15m. A crown shaped settlement at top surface was observed with lesser settlements occurring at centre of wall as compared to settlements at crests; this occurrence can be attributed to the minimal interaction occurring between the failures surfaces from either walls. The magnitude of crest settlements were almost equivalent to the settlements observed in model with single wall without support. A similar trend of outward movement of facing as that of single wall was observed for both faces of back-to-back walls. The horizontal movement at the crest was observed to be nine times more than the movement occurring at the base of wall at the penultimate stage of the test; it can be interpreted as an imminent failure of the wall was because of overturning of wall about the base due to presence of full backfill thrust acting on reinforced zones. Maximum peak strain was observed in topmost reinforcement and was of order 45%, whereas an uniform strain of 37% was observed in the mid-height of wall and a minimal strain of 8% was observed to occur in bottommost reinforcement layer; which implied that resistance was offered by the upper layers of reinforcement in order to restrict the overturning movement of walls. Failure surfaces from either wall, as signified from kinks in black sand marker layer were observed to leave the reinforced zones of either walls beyond sixth layer of reinforcement, thereby highlighting the effect of presence of wall on the other end affecting the development of failure surfaces.

Model W3 was tested in centrifuge to study the effect of connection in reinforcement layers on the deformation behaviour and strain distribution reinforcement layers. Displacement of the wall was highest for connected reinforcement model as compared to other model walls (Figure 12a-b). The model was observed to fail at 60g which is equivalent to prototype width and height of 22.8 m and 16.2 m respectively. Surface settlement at the crests were observed to be 3.5 times more than the settlements occurring at centre of wall, this can be attributed to formation of structural arc-like portion at the central zone of wall caused by connection of reinforcement. Horizontal movement of wall facing was observed to be maximum at the crest thereby following the same trend as that of single wall. But the magnitude of face displacement was highest as compared to all other back-to-back models tested in centrifuge. Strain developed along the geogrid layer was uniform, with peak strains occurring at two places in all layer and were positioned near the wall facing (Figure 9c). The peak strain distribution was highest for all reinforcement layers except layer 2 and layer 3 as compared to other wall models (Figure 13a-b). Connection in reinforcement created an at-rest earth pressure like condition throughout the height of wall thereby generating higher lateral pressures which consequently were dissipated through straining of reinforcement layers. The maximum peak strain was observed in top reinforcement layer with magnitude of 80% at penultimate stage of failure. Though failure of back-to-back wall with connection in reinforcement occurred at higher gravity level which indicates higher stability of the structure, the deformations in the wall were excessive as compared to other back-to-back models tested in centrifuge.

For all model tests in centrifuge it was commonly observed that, the movement of the upper portion of the wall was always greater than that occurring at base of the walls. It can be understood as; overturning mechanism is prominent in vertical soil walls with single or back-to-back facing. By comparing model W1 and W2, it can be inferred that the presence of wall at other end affects the overall stability and deformation behaviour of reinforced soil wall. The failure surfaces leave the reinforced zone and enter the backfill zone, before commencement of complete failure. The strain developed in the geogrid reinforcement is higher and shows similar trend as of single wall. Unlike single wall with uniform increase in strain up to failure, the back-to-back walls showed lower strain at prefailure condition in layer 2 of the reinforcement. Also unlike single wall the back-to-back walls showed two peaks in upper layers of reinforcement indicating formation of secondary failure surface near facing of the wall.

The effectiveness of connection in reinforcement in back-to-back wall proved helpful from stability point of view, but the serviceability (i.e. deformability) of the wall suffered. The reinforcement showed higher a uniform distribution of strain development, utilizing full strength of geogrid.

6. CONCLUSIONS

In this paper, the performance of two models of back-to-back walls and one model of single wall with same backfill and reinforcement material were studied in a geotechnical centrifuge. A variable gravity level approach was adopted to understand the deformation behaviour of walls and to evaluate magnitude of strain generated in geogrid reinforcement layers. The reinforcement type, reinforcement length and spacing were kept as constant. Following conclusions were derived based on the observations and analysis of the experimental data obtained from centrifuge tests on single wall as well as back-to-back reinforced soil walls.

1. Development failure surface got affected due to presence of wall on the backfill end.
2. Back-to-back walls deformed more as compared to wall with single facing.
3. Soil walls with vertical facing were observed to have deformed more in upper layers of reinforcement.
4. Interaction of failure surfaces was observed to occur in back-to-back walls.
5. Overturning failure mechanism was found to be prominent in soil walls with vertical flexible facing.
6. Connection of reinforcement in back-to-back walls, improved stability of the wall, but the deformation were of higher magnitude.
7. Crown shaped surface settlements, with a hump at the central portion were observed to occur in back-to-back reinforced soil walls.

Additional parametric study with variation in length and spacing of reinforcement will aid in complete understanding the behaviour of back-to-back geogrid reinforced soil wall.

REFERENCES

Anubhav, S. and Basudhar, P. (2011) "Numerical modelling of surface strip footings resting on double-faced wrap-around vertical reinforced soil walls". Geosynthetic International, Volume 18, Issue 1, pp 21-34.

Chandrasekaran, V. S. (2001) "Numerical and centrifuge modelling in soil structure interaction". Indian Geotechnical Journal, Volume 31, Issue 1, pp 30-59.

Han, J. and Leshchinsky, D. (2010) "Analysis of back-to-back mechanically stabilized earth walls". Geotextiles and Geomembranes, Volume 28, Issue 3, pp 262-267.

Hardianto, S. and Truong, K. (2010) "Seismic deformations of Back-to-Back mechanically stabilized earth (MSE) walls". Earth Retention Conference 3, Proceedings of the 2010 Earth Retention Conference, Publisher ASCE, Editors Finno, R.J., Hashash, Y.M.A. and Arduino, P., pp 704-711.

Jones, C. J. F. P. (1996) "Earth reinforcement and soil structures". Thomas Telford Publishing, ASCE press, USA.

Viswanadham, B. V. S. and König, D. (2004) "Studies on scaling and instrumentation of a geogrid". Geotextiles and Geomembranes, Volume 22, Issue 5, pp 307-328.

Viswanadham, B. V. S. and Mahajan, R. (2007) "Centrifuge model tests on geotextile-reinforced slopes". Geosynthetics International, Volume 14, Issue 6, pp 365-379.

Zornberg, J. G. and Arriaga, F. (2003) "Strain distribution within geosynthetic reinforced slopes". Journal of Geotechnical and Geoenvironmental Engineering, ASCE, Volume 129, Issue 1, pp 32-45.

1 g Experiments on Batter Piles Subjected to Dynamic Lateral Loading

R.M. Subramanian and A. Boominathan

Department of Civil Engineering, Indian Institute of Technology Madras, Chennai
E-mail: subbu_geotechengg@yahoo.co.in; boomi@iitm.ac.in

ABSTRACT

Dynamic experiments were carried out on model single batter piles with different inclination angles in a simulated elastic half space. A 500 mm diameter pile is modelled using the available similitude laws based on the flexural rigidity of the pile and is adopted in the present study. Details of the physical modelling of piles, experimental setup, preparation of soil bed, equipments used with their instrumentation and test procedures are presented. The displacement response of batter piles are studied and compared with the vertical pile. It is found that with increase in batter angle the displacement amplitude reduces substantially.

1. INTRODUCTION

Pile foundation is often used in bridges and other structures to support the applied axial and lateral loads. They are slender column members having high length to width ratio and are usually designed to resist high axial loads. It is a preferred approach to utilize the high bearing capacity of soils available at deeper depths by means of adopting pile foundations where the top soil is not strong enough to bear the super structure load. Some structures such as high rise buildings, offshore structures (Quay, harbours) and earth retaining walls are subjected to lateral loads caused by wind, wave action, traffic movement, water pressure and earth quake etc in addition to their vertical loads. For instance in the bridge abutments foundations can be subjected to lateral load induced by water waves and ship impact. Hence, the pile in single as a part of deep foundation has to resist both axial and lateral loads. The lateral load induced on the pile foundation causes deflection, rotation, bending moment and translation of the piles depending upon the geometry and boundary condition of the pile. The type of foundation usually recommended for such loading conditions is combination of vertical and batter piles.

1.1 Batter Piles

A pile which is inclined to an imaginary vertical line or to a vertical pile is referred as batter pile. They are traditionally adopted to carry large horizontal loads as their distinct advantage over vertical piles is that they transmit the applied lateral loads partly in axial compression other than shear and bending. However their recent failures in series of earthquakes have paved way for their removal from foundation codes and the consideration of engineers. The main arguments that have been frequently mentioned by engineers as the real or perceived drawbacks of inclined piles include (Gerolymos et al., 2010): "Parasitic" bending stresses due to soil settlement following an earthquake and or soil consolidation before the earthquake; large forces (of alternating Sign) onto the pile cap; reduction in bending moment capacity due to seismically induced tensile forces; undesirable permanent rotation of the cap when the inclination of the piles is not symmetric and increased structural shear due to the stiffening of the system. It is understood from the above phrase that the behaviour of batter piles under dynamic loading is not fully studied.

In the recent past, due to the improved understanding of the source of the observed poor performance, batter piles are re-establishing their traditional role of withstanding large horizontal loads. Some recent works which include experimental studies by Escoffier et al. (2008) and numerical investigations by Gerolymos et al (2008) and Sheikhbahaei et al (2009) provide evidence to support the use of batter piles as they can be beneficial to the structure they support as well as to themselves. Nevertheless, until today very few case-histories or experimental data are available on the response of batter piles under dynamic loading.

A more appropriate way to capture real field response in laboratory to great extent is by doing a small scale centrifuge experimental study. The stress conditions of the soil in field, soil pile contact pressure, self weight of the soil and pile can be replicated in the centrifuge model studies. However those studies are not feasible always and hence a conventional 1g tests can be performed incorporating appropriate physical modelling of the materials. Also in dynamic testing in laboratory the boundary effect of the rigid walls need to be addressed. This paper gives insight on physical modelling of piles, experimental setup, preparation of clay bed, equipments used with their instrumentation and test procedures adopted in the present study. The displacement response of batter piles with different inclination is discussed and compared with the vertical pile.

2. MODELLING OF PILES

The piles are modeled with similitude law proposed by Wood. et al (2002) based on the flexural rigidity using Eq. (1)

$$\frac{E_m I_m}{E_p I_p} = \frac{1}{n^5} \tag{1}$$

where

E_m = modulus of elasticity of model pile;
E_p = modulus of elasticity of prototype pile;
I_m = moment of inertia of model pile;
I_p = moment of inertia of prototype pile;
1/n = Scale factor for length.

The above relationship was obtained based on a dimensionless factor, Φ used for modelling laterally loaded piles, as given in Eq. (2), (Wood, 2002).

$$\frac{E_m I_m}{G_m l_m^4} = \frac{E_p I_p}{G_p l_p^4} = \Phi \tag{2}$$

G_m = shear modulus of the soil in which model pile is installed;
G_p = shear modulus of the soil in which prototype pile is placed;
l_m = length of the model pile;
l_p = length of the prototype pile.

A RCC pile of 500 mm diameter and compressive strength of 45 N/mm^2 is modeled as an aluminium pipe with outer diameter of 25 mm and thickness of 2.5 mm. Scale factor for length (n) is adopted as 10. Other scale factors that are accounted on credibility of n = 10 for the prototype pile are mass density (1), stiffness (1/10), stress (1/10), strain (1/10), force (1/10^3), acceleration (1), pile flexural rigidity (1/10^5).

2.1 Limitations of 1g Model Tests

The main aim of the 1 g modeling is to produce results in terms of non dimensional values. i.e any specific behaviour of the prototype pile can be addressed in terms of normalized dimensionless values. Similarity between the model and the prototype is ensured when the dimensionless group has the same value in the model and prototype. The scaling law and scale factors used in the present study were verified by modeling the full-scale single pile in clay subjected to static lateral load reported in Rollins et al.(1998). Certain aspects of the prototype behaviour like self-weight effects of the soil, realistic soil-pile stress fields, the stress-strain behaviour of the soil, roughness of the prototype pile are difficult to incorporate in the 1g model tests (Chandrasekaran et al. 2012).

3. ELASTIC HALF SPACE SIMULATION (EHSS)

3.1 Need for EHSS

The piles installed in the field when loaded dynamically produce stress waves which propagate through greater distance within the soil in many directions. Therefore any dynamic tests on piles simulated in laboratory in conventional steel or concrete tanks experience severe wave reflection from the rigid boundary of the tank which inturn adversely affect dynamic characteristics of piles being tested. The response curve for the pile embedded in the rigid tank is scattered and show multiple peaks due to the reflection of long-period waves (Boominathan and Ayothiraman, 2005). Hence, the dynamic model tests are to be carried out in the laboratory by ensuring no reflection of stress waves from the rigid boundary of the model test container.

3.2 EHSS Facility

The EHSS facility is similar to the one adopted by Chandrasekaran et al. (2012) as shown in Figure 1. It consists of a test tank of size 2.0 × 2.0 × 1.5 m, an absorbing layer of saw dust and a boundary element between them. The tank wall of 250 mm thickness is made of brick masonry. The boundary element is made of mild steel basket in logarithmic arc spiral shape, which is covered with geomembrane sheet to separate the soil from the absorbing layer as well as to maintain constant moisture content in the clay layer.

FIGURE 1 Elastic Half Space Simulation (After Chandrasekaran et al. 2012)

Though EHSS absorbs most of the stress waves travelling from the test piles, there is a possibility of very minimal wave reflection from the boundary element due to difference in the stiffness of soil and geomembrane. Hence in order to change directions of the minimum reflected waves going towards the central region where the test piles are being monitored during dynamic excitations a curved profile in the form of logarithmic spiral arc shape (Figure 2) is given to the boundary element. Wet saw dust is used as a wave absorbing medium (Gazetas and Stokoe, 1991) in the space between the wall and boundary element.

FIGURE 2 Logarithmic Spiral Arc Boundary Element

4. EXPERIMENTAL STUDY

4.1 Materials Used

The materials used in the present study are discussed in this section.

4.1.1 Soil

The clay sample used in the present study was obtained from Siruseri, located along the coastal region of Chennai suburb. The properties of the soil are listed in Table 1. The soil is classified as high plastic clay (CH) as per IS 1498- (1970) 2002.

Table 1: Properties of the Soil

Description	*Value*
Sand (%)	12
Silt (%)	28
Clay (%)	60
Specific Gravity, G_s	2.60
Liquid Limit, w_l(%)	69
Plastic Limit, w_p(%)	27
Plasticity Index, I_p(%)	42

4.1.2 Pile and Pile Cap

The pile is modelled as an aluminium pipe as described in section 2. Aluminium plate of thickness 20 mm is used to model pile cap. The top end of the pile which was threaded was screwed into a small solid cylindrical piece that was fixed to the pile cap using allen screws inclined at different angles (10°, 15° & 20°) to the vertical based on pile configuration and it provides a fixity condition. In this way piles can be reused for different pile cap configurations. Also a check nut is provided in the threaded pile to ensure tight connection of pile to the solid piece. The length of the pile was taken based on the embedded length derived from the embedded length to diameter ratio (L/D) along with an additional free standing length of 150 mm above the ground surface.

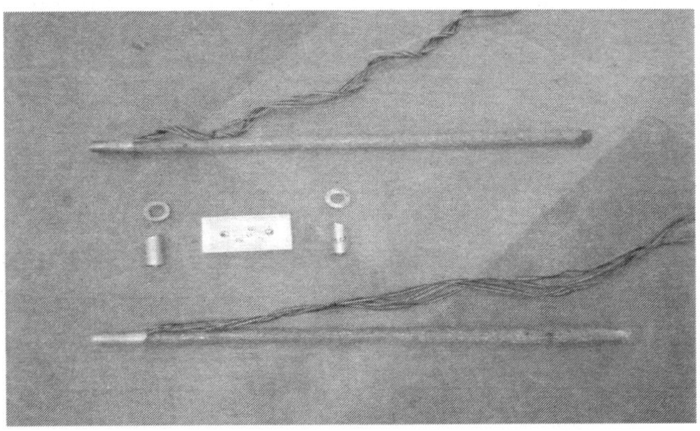

Figure 3 Components of a 1 x 2 Batter Pile Group

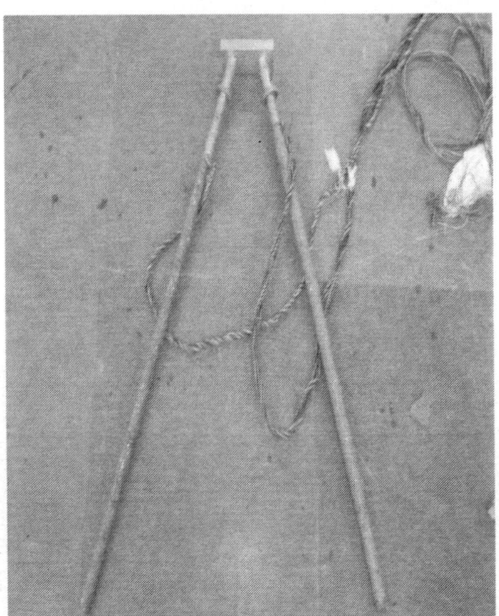

Figure 4 Assembled 1 × 2 Batter Pile Group

Figure 3 shows the individual components used in making 1 x 2 pile group with batter angle 20° and Figure 4 shows the assembly of the same. Though group pile results are not discussed in this paper their photograph is shown for better clarity of the concept.

4.2 Equipments Used

The equipments used for conducting dynamic load tests on piles and acquiring data are presented in this section. A schematic diagram showing complete information of the experimental set up is shown in Figure 5.

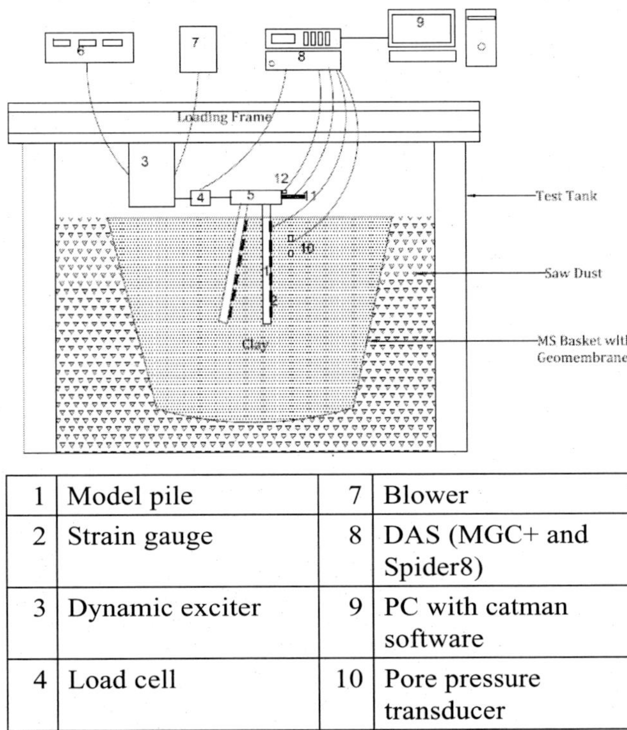

1	Model pile	7	Blower
2	Strain gauge	8	DAS (MGC+ and Spider8)
3	Dynamic exciter	9	PC with catman software
4	Load cell	10	Pore pressure transducer
5	Pile cap	11	LVDT
6	Excitation amplifier	12	Accelerometer

FIGURE 5: Experimental Setup

4.2.1 Electro-Dynamic Shaker

Vibration test system (VTS) make electrodynamic shaker is used in the present study to apply dynamic load. The shaker is capable of producing sinusoidal loading force up to 400 N with operational frequency much greater than 1000 Hz. However for the present work the frequency was varied between 1 to 50 Hz as the resonance is reached well within this range. The excitation is applied to the pile through pile cap by means of steel rod connected between the exciter and pile cap.

4.2.2 *Excitation Amplifier*

Excitation amplifier is used to control the frequency of excitation and force level. It is provided with three knobs to control the amplitude of the voltage and current so as to adjust the force level. The frequency range is controlled by a rotary switch.

4.2.3 *Load Cells*

HBM make load cell with 50 kg capacity is used to measure the applied loads. They are of full bridge strain gauge type and having sensitivity of 0.005 kg. It is connected to the exciter at one end and pile cap at the other end.

4.2.4 *Accelerometer*

HBM make B12 series accelerometer is used in the present study. The measuring range of the accelerometer is +/- 200 m/s^2. Acceleration pick up is placed on one side of the pile cap along the loading direction.

4.2.5 *Data Acquisition System*

A 16 channel data acquisition system is used in the present study to monitor and store the data automatically. Two numbers of carrier frequency amplifier systems (HBM make Spider 8) are used in the present study and are connected to a computer with LPT interface. A CATMAN Professional 5.0 installed on to a PC is used to acquire and analyze data received from various types of transducers through the amplifier system.

4.3 Pile Instrumentation

Piles were instrumented with electrical resistance strain gauges along their length to study their bending behaviour. A HBM make electrical strain gauges having resistance of 120 Ω were used in quarter bridge arrangement along the pile length. A total of nine strain gauges were placed with center to center spacing between successive strain gauges as 75 mm. The desired locations for strain measurement were marked on the cleaned surface of the model pile. The strain gauges were fixed at the marked locations using quick acting adhesive. Thin coating of covering putty (HBM-AK22) was applied over strain gauges to ensure that the strain gauges are water proof.

4.4 Test Procedure

4.4.1 *Preparation of Clay Bed*

The soil is mixed with the required water content in a separate mild steel tank so as to achieve the desired consistency. In the present study the soil bed was prepared to a consistency of soft clay. After fixing the pile group in the centre of the test tank, uniformly mixed clay is placed and hand packed in several layers up to the top of the tank. During placement of the soil layers, care is taken to see that the pile group is installed properly in the correct position. The horizontal surface of the pile cap is ensured by placing spirit level on the surface of pile cap periodically during the installation of piles. Water content measurements were done at different depths and different locations to ensure the uniformity of each soil layer.

4.4.2 *Dynamic Testing Method*

Sinusoidal lateral excitation force is applied to the pile group at different dynamic loads through the dynamic exciter. The time histories of applied

FIGURE 6: Dynamic Load Test on Piles

load, pile head displacement, acceleration, strain along the pile length and excess pore pressure in soil are measured. A typical photographic view of 2 x 2 (10° Batter Vertical) pile group with 3 D spacing subjected to dynamic loading is shown in Figure 6.

5. RESULTS AND DISCUSSIONS

5.1 Time Histories of Load, Displacement, Acceleration and Strain

Typical time histories of measured load, displacement, acceleration and strain are presented in Figure 7 to 10. It can be observed from Figures 7 to 10 that the sampling parameters used for the test are sufficient to capture complete sinusoidal waveforms.

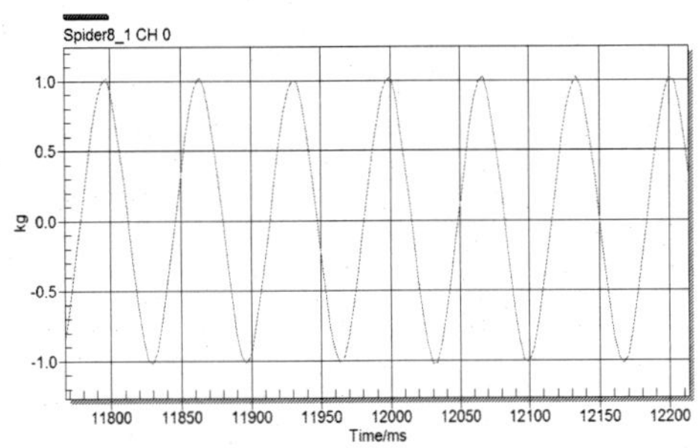

FIGURE 7: Load Time History

5.2 Displacement Frequency Response of Single Vertical Pile

The displacement response of a single vertical pile is shown in Figure 11. It could be observed that resonance frequencies are around 11 Hz and are comparable to the full scale field test on piles subjected to lateral dynamic loads in clay reported in Ayothiraman et.al (2012). The dynamic displacement amplitude increases with increase in load and resonant frequency shifts towards left. The percentage variation of peak displacement with increase in excitation force is presented in Table 2. It can be noted that for an average increase in load of 1.3 N (30 percent) the increase in displacement is about 16 percent.

Table 2: Variation of Peak Displacement Amplitude for Vertical Pile

Force N	*Resonant Frequency Hz*	*Displacement mm*	*Percentage Increase in Displacement*
3.2	11.1	0.138	0
4.6	11.0	0.168	18
6.8	10.9	0.195	14

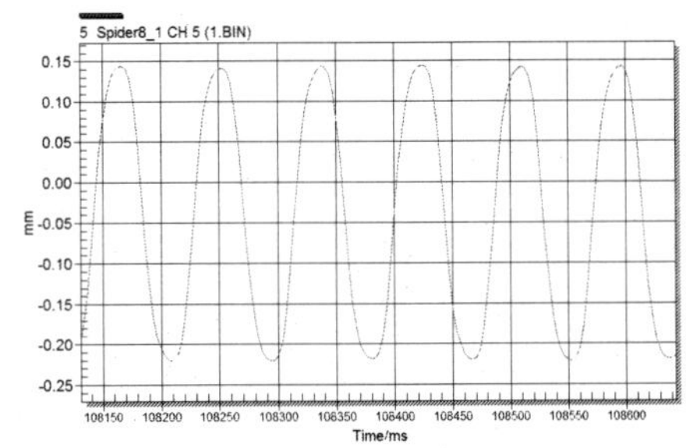

FIGURE 8: Displacement Time History

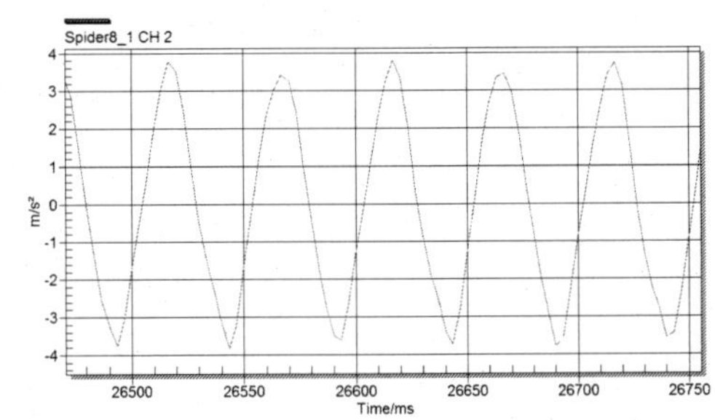

FIGURE 9: Acceleration Time History

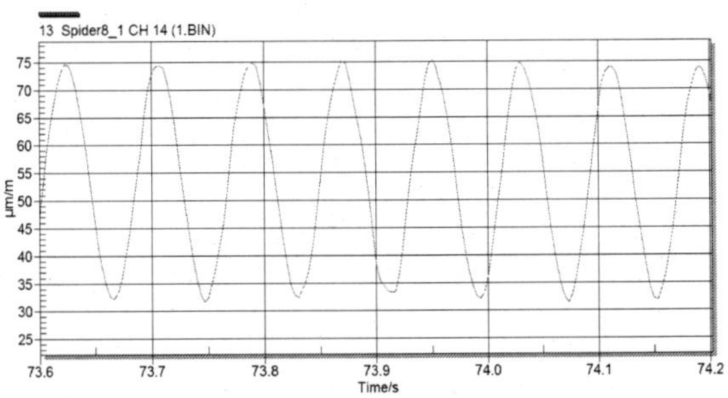

FIGURE 10: Bending Strain Time History

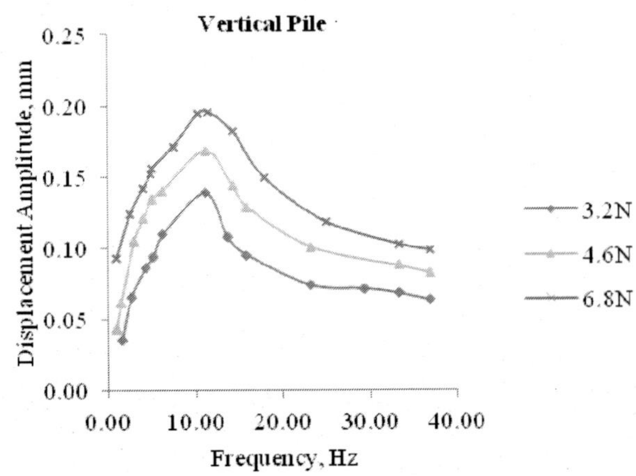

FIGURE 11: Displacement Frequency Response of Vertical Pile

5.3 Displacement Frequency Response of Single Batter Pile

A typical displacement response of a single batter pile (15°) is shown in Figure 12. Like in the case of vertical pile with increase in excitation force the dynamic displacement amplitude increases and resonant frequency decreases. The percentage variation of peak displacement with increase in excitation force is presented in Table 3. It can be noted that for an average increase in load of 0.5 N (10 percent) the increase in displacement is about 7 percent.

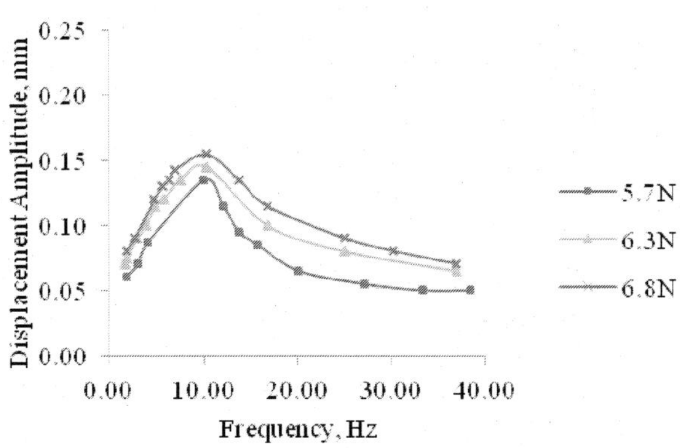

FIGURE 12: Displacement Frequency Response of Batter Pile

Table 3: Variation of Peak Displacement Amplitude for Batter Pile

Force N	*Resonant Frequency Hz*	*Peak Displacement mm*	*Percentage Increase in Displacement*
5.7	10.5	0.135	0
6.3	10.1	0.145	7
6.8	10.3	0.155	6.5

5.4 Effect of Batter Angle on Displacement Response

The displacement response of batter piles with different angle of inclination (10°, 15° & 20°) at an excitation force of 6.8 N is presented in Figure 14. Vertical pile response is also included for comparison. It can be seen that displacement amplitude is considerably reduced with increase of batter angle for the batter piles when compared to the vertical pile. A similar finding is also reported by Gerolymos et al. (2010) based on the finite element analysis carried out on batter piles.

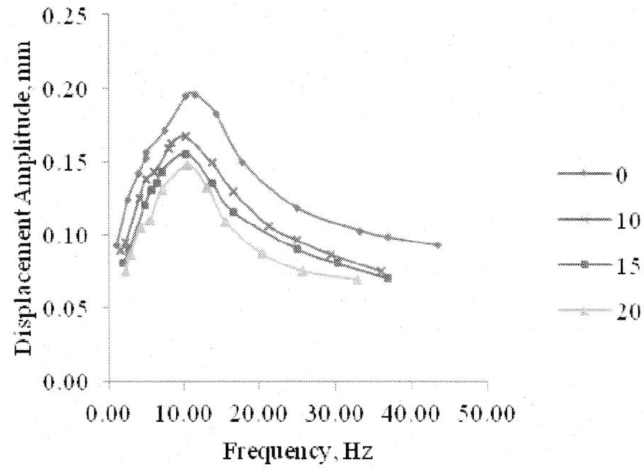

FIGURE 13: Effect of Angle of Inclination on the Response of Batter piles

Table 4: Effect of Batter Angle on Peak Displacement

Pile Inclination (in degree)	*Resonant Frequency Hz*	*Peak Displacement mm*	*Percentage Decrease in Displacement*
0	10.9	0.195	0
10	10.3	0.167	14
15	10.5	0.155	21
20	10.5	0.147	25

The resonant frequency of the batter pile is slightly lower than vertical pile and for both type of piles it occurs within the range of 10 Hz to 11 Hz. The percentage variation of peak displacement with increase in batter angle is presented in Table 4. It can be noted that twenty five percent of reduction in displacement amplitude occurs for batter pile (20°) in comparison to vertical pile. This may be attributed to the ability of the batter piles to carry lateral load in axial direction rather than shear and bending alone as in the case of vertical piles.

6. CONCLUSION

The following conclusions are derived from the present work

- The scaling law adopted here is based on flexural rigidity of the pile and it holds valid for modelling both batter and vertical piles.
- For batter piles like in the case of vertical

piles with increase in excitation force the dynamic displacement amplitude increases and resonant frequency decreases.

- The batter pile shows less displacement amplitude than vertical pile for the same excitation force level.
- For every 10 percent increase in excitation force in case of batter pile (15°) the percentage reduction in displacement is 7 percent.
- The percentage reduction in displacement is about 25 percent for batter pile (20°) when compared to the vertical pile

REFERENCES

Ayothirman, R., Boominathan, A., and KrishnaKumar, S. (2012) "Lateral Dynamic Response of Full Scale Single Piles — Case Studies in India", Full-Scale Testing and Foundation Design, Edited by M.H. Hussein, K.R. Massarsch, G.E. Likins and R.D. Holtz, GSP No: 227, ASCE., pp 710-721.

Boominathan, A., and Ayothiraman, R. (2005) "Dynamic behavior of laterally loaded model piles in clay". Geotechnical Engineering, ICE, GE4, pp 207-215.

Chandrasekaran, S. S., Boominathan, A., and Dodagoudar, G. R. (2012) "Dynamic Response of Laterally Loaded Pile Groups in Clay". Journal of Earthquake Engineering, Taylor and Francis, In press.

Escoffier, S., Chazelas, J. L., and Garnier, J. (2008). "Centrifuge modelling of raked piles". Bulletin of Earthquake Engineering, Vol. 6, pp 689–704.

Gazetas, G. and Stokoe, K. H. (1991). "Free vibration of embedded foundations: Theory versus experiment". Journal of Geotechnical Engineering, ASCE, 117(9), pp 1382-1401.

Gerolymos, N., Giannakou, A., Anastasopoulos, I., and Gazetas, G. (2010). "Seismic Behavior of Batter Piles: Elastic Response", Journal of Geotechnical and Geoenvironmental Engineering , ASCE, Vol. 136 , pp 1187-1199.

IS 1498- (1970- Reaffirmed 2002) Indian Standard classification and identification of soils for general engineering purposes, Bureau of Indian Standards, New Delhi.

Rollins, K. M., Peterson, K. T. and Weaver, T. J. (1998). "Lateral load behaviour of full- scale pile group in clay." Journal of Geotechnical and Geoenvironmental Engineering, ASCE, 124 (6), pp 468-478.

Sheikhbahaei ,A., and Vafaeian, M. (2009). "Dynamic Study of Batter Pile Groups under Seismic Excitations through Finite Element Method", World Academy of Science, Engineering and Technology, Vol. 50, pp 51-57.

Wood, D. M., Crewe, A. and Taylor, C. (2002). "Shaking table testing of geotechnical models." International Journal of Physical Modelling in Geotechnics, Vol. 1, pp 1- 13.

Wood, D. M.(2004) Geotechnical modelling, Spon Press, Taylor and Francis group, First Edition

Comparison of the Performance of Compacted Clay Barriers, Composite Barriers and Geofiber Reinforced Barriers of Landfill Covers

P.V. Divya and B.V.S Viswanadham

Department of Civil Engineering, Indian Institute of Technology Bombay, Mumbai

E-mail: divyaputhumana@rediffmail.com; viswam@iitb.ac.in

ABSTRACT

An attempt was made in the present study to understand and compare the response of impervious barriers of landfill covers subjected to differential settlements. Centrifuge model tests were conducted at 40 gravities using a large beam centrifuge facility available at IIT Bombay to study and compare the performance of compacted clay barriers (CCB), composite barriers (CB) and geofiber reinforced clay barriers (FRSB) of landfill covers at the onset of differential settlements. The thickness of model clay barriers was 25 mm which represents 1.0 m in prototype dimensions. Model barrier materials were selected in such a way to represent a bandwidth of properties of barrier materials used in various parts of the world. Low density polyethylene geomembrane and polyester fibers were used in the present study as model materials. All models were instrumented with Pore pressure transducers (PPTs) to establish water breakthrough through barriers, linear variable differential transformers (LVDTs) to arrive at deformation profiles, digital photo camera to capture the front elevation of the model and CCD video camera to monitor the presence of water above the barrier during centrifuge tests. A 1.0 m thick CCB was found to experience wide and deep cracks penetrating up to full-depth and lose its sealing efficiency at the onset of differential settlements; whereas a CB of identical thickness maintained its sealing efficiency. In the case of FRSB, propagation of cracks was limited to shallow depths with reduced crack widths and water breakthrough was observed to be delayed considerably.

1. INTRODUCTION

Performance of waste containment systems relies on the engineered barriers which can be in the form of compacted clay barriers (CCB), sand bentonite barriers, geosynthetic clay liners (GCL), composite barriers (CB), polymer amended sand-bentonite barriers, geomembranes or a combination of the above. Compacted clay barriers (CCB) can be considered as an economically viable barrier material wherever these soils are abundantly and easily available at sites (Heerten and Koerner, 2008; Gourc et al., 2010). In modern landfills, geosynthetics are used as lining systems. Geomembranes are one of the most commonly used geosynthetics in landfill construction (Rowe and Sangam, 2002; Rowe et al. 2009; Take et al. 2007; Brachman and Gudina, 2008; Bouazza et al. 2008; Rowe, 2011) and are laid above the clay barrier to form a composite barrier (CB) of cover systems. Recently, discrete and randomly reinforced fibers have gained attention as a technique for improving the tensile strength-strain characteristics of soil (Mesbah et al. 2004; Viswanadham et al. 2010; Plé and Lê, 2012; Divya, 2012). The purpose of these impervious barriers is to isolate the contained waste from the environment and thus to prevent the pollution.

Differential settlements in landfill covers are inevitable due to the biodegradation of underlying municipal solid waste (MSW) in MSW landfills (Reddy et al. 2009; Sivakumar Babu et al. 2010) and due to the voids among the waste storage containers in low level radioactive Waste (LLRW) landfills (Keck and Seitz, 2002; Gourc et al. 2010). An effective barrier should be able to accommodate the differential settlements induced and maintain its sealing efficiency. Deformation behaviour of soil barriers was studied for compacted clay barriers (CCB), composite barriers (CB) and geofiber reinforced clay barriers (FRSB) of landfill covers and their response to differential settlements was compared in the present study. Since the differential settlement and resulting cracking is a stress dependant phenomenon, centrifuge model study was adopted in the present study.

2. CENTRIFUGE MODEL TESTS

2.1 Centrifuge Modelling Technique

In a centrifuge test, behaviour of prototype structures can be studied as scaled-down models of 1/N times by providing an artificial gravitational field of N times acceleration due to gravity in a controlled environment. This can be achieved by rotating a small-scale model in a horizontal plane about a vertical axis at a desired angular velocity. A 4.5 m radius large beam centrifuge facility, which is named as the National Geotechnical Centrifuge Facility (NGCF) available at Indian Institute of Technology Bombay was used in the present study. The centrifuge capacity is 2500 g-kN with a maximum payload of 25 kN at 100 g and at higher acceleration of 200 g the allowable load is 6.25 kN. The centrifuge has a swing basket at one end and an adjustable counter weight at the other end. The other specifications are given elsewhere by Chandrasekaran (2001). The centrifuge model tests reported in the present study were conducted at 40 gravities. The scaling considerations for modelling geomembranes and geofibers were deduced with a need to ensure stress similarity between the model and corresponding prototype.

2.1.1 Scaling Considerations for Modelling Geomembranes

Main difficulty associated with the use of geomembranes in a centrifuge study is the scaling down of commercially available geomembranes widely used in the field. Two important criteria to be considered in scaling down of geomembranes are simulation of tensile load-strain behaviour, soil-geomembrane interface characteristics and thickness of model geomembrane material. The scaling considerations for the modelling of geomembranes are explained extensively in Divya et al. (2012a).The linear dimensions like length (l_g) and thickness (t_g) of the geomembrane are reduced by 1/N times that of corresponding prototype dimensions in a centrifuge model.

The tensile load T_g (kN/m), which is the tension developed in the geomembrane per unit width can be obtained using $T_g = \varepsilon_g E_g t_g$. Assuming that tensile strains in the geomembrane with identical material characteristics in model and prototype are identical, i.e. ($\varepsilon_g = \varepsilon_{mg}$ and $E_g = E_{mg}$), the scale factor for the tensile load in model geomembrane in a centrifuge is $T_{mg} / T_g = (\varepsilon_{mg} E_{mg} t_{mg}) / (\varepsilon_g E_g t_g) = 1/N$, where subscript m stands for model dimensions. The tension developed in geomembrane due to subsidence is to be resisted by the interface frictional strength between the soil and geomembrane, in order to avoid the slippage or pullout of the geomembrane. Also, considering the force equilibrium condition that the tension acting in the geomembrane (when it undergoes differential settlement) and the interface frictional force (pullout resistive force of the lateral anchorage) should be same; it can be obtained that the angle of interface friction between the barrier material and geomembrane is same in model and prototype.

2.1.2 Scaling Considerations for Modelling Geofibers

In order to model geofibers in a centrifuge, length, breadth and thickness of geofibers have to be reduced by 1/N times to satisfy complete geometrical similarity between a reduced physical model and prototype. But it is difficult to make and use reduced dimensions for modelling geofibers in small scale physical model tests. The most important properties of fiber reinforced soil are the tensile strength of the fibers and the bond strength along soil-fiber interface. The crack interfaces are being held together by fibers, and thus help to reduce the cracking failure. Thus the length of the fibers (l') should be sufficient enough to enable adequate bond force. The polyester fibers used in the present study were having a special triangular cross section. In order to prevent the failure of fiber, a fiber at the zone of maximum tension should develop an adhesion or bond force (τ_b A') equal to the tensile force (σ_t A) acting on the fiber where σ_t is the tensile strength of the fiber, τ_b is the bond strength, A is the cross sectional area and A' is the circumferential area of the fiber over a length, L. Considering half the length of fiber, maximum adhesive force can be obtained when L = $l'/2$ and maximum tensile force can be obtained when σ_t become equal to ultimate tensile strength of geofiber $(\sigma_t)_{max}$.

When fibers undergo stress within a soil-fiber composite, the force equilibrium is achieved by equating the tensile force acting perpendicular to the cross-section of the fiber and the adhesion force between soil-fiber interfaces along the fiber length, l' over the fiber surface. $(\sigma_t)_{max}$ A = τ_b $(l'/2)$ $(3b')$ where b' is the side width of the fiber. The aspect ratio (l'/b') can be written as $1/2\sqrt{3}$ times (σ_t/τ_b). This is explained in Viswanadham et al. (2011). For identical fiber material properties and fiber strains, the tensile strength and bond strength would also be identical. This implies that fibers can be treated as discrete inclusions and same fibers can be used in centrifuge model as well as in the field.

2.2 Materials used for the Centrifuge Model Tests

The model barrier materials which include soil, geomembrane, geofibers were selected in such a way that it represents the prototype barriers widely used in field.

2.2.1 Soil

According to Benson et al. (1999), in most of the landfill sites, the clay barrier placed in cover systems has liquid limit ranging from 30% to 40% and plasticity index ranging from 10% to 20% and compacted to the wet side of optimum using standard proctor compaction energy. The most important property required for a barrier material is low permeability. A mixture of kaolin and sand in the ratio of 4:1 was selected as the model barrier material which is having the properties in the above mentioned range. The liquid limit of the model soil was 34% and plasticity index was 12%. The maximum dry unit weight was 15.9 kN/m^3 and the corresponding OMC was 22% (according to standard Proctor compaction). The coefficient of permeability of the model soil barrier material compacted at 5% wet of optimum and the corresponding dry unit weight of 14.2 kN/m^3 was found to be 4.4 x 10^{-9} m/s.

2.2.2 Model Geomembrane

A commercially available low density polyethylene (LLDPE) sheet was selected as the model geomembrane in the present study according to the derived scaling considerations. LLDPE geomembranes have greater ability to maintain their integrity under localized differential settlement. The stress-strain behaviour of the model geomembrane is compared with the prototype geomembranes reported by Koerner (2001) and was found to be in good agreement. The thickness (t_{mg}) of model geomembrane was 0.135mm. The ultimate tensile stress of the model geomembrane was 28148 kN/m^2. The ultimate strain was 450% and calculated considering the gauge length segment of the model geomembrane specimen. The initial stiffness J_g was obtained from the tensile load-strain curve by extending the initial linear portion of tensile load-strain curve and the value of J_g was 480 kN/m in prototype dimensions, at 40g.

2.2.2 Model Geofibers

Geofibers (flexible polymeric fibers) are available in different shapes and lengths. They can be made of different materials such as polypropylene, polyester and polyvinyl alcohol, etc. In the present study, polyester fibers (PET) having equivalent diameter 40 microns were used. PET fibers are having a triangular cross-section which may help in interlocking between the soil particles. The breaking load of polyester fiber was 0.326 N at an elongation strain of 19.25% (0.145 N at 5% strain and 0.246 N at 10% strain).

2.3 Test Set-up and Procedure

In the present study three centrifuge tests were described which shows the response of compacted clay barriers (CCB), composite barriers (CB) and geofiber reinforced clay barriers (FRSB) of landfill covers at the onset of differential settlements. A motor based differential settlement simulator (MDSS) developed at IIT Bombay for inducing differential settlements to soil layers was used in the present study for studying the deformation behaviour of soil barriers. The MDSS includes, a PMDC motor, two rigid supports of 180mm length connected to two aluminum plates with a hinge connection. These hinged plates were made to rest on a trapdoor settlement plate. The description of MDSS setup has been explained extensively by Rajesh and Viswanadham (2011). It works on a simple mechanism in which the rotational movement of the motor shaft is converted to translational movement of the central platform through a screw jack and series of gears.

A large box having internal dimensions of 720 mm in length, 450 mm in breadth and 440 mm in depth with back and rear walls made with stainless steel plates and front wall made with transparent Perspex sheet was used for conducting centrifuge tests. White petroleum grease was applied to reduce boundary effects and the friction between walls and soil. Figure 1 shows the model test package used in the present study.

The preparation of the model was carried-out at normal gravity, by maintaining a horizontal platform with the hinge door assembly and central support. Model clay barrier was prepared on top of pre-drained 30mm coarse sand layer followed by 30mm fine sand layer to avoid the stress concentration at the hinges and also to induce smooth continuous differential settlements to the barrier. These layers are separated by thin filter papers. The model barrier was moist-compacted with a moisture content of 5% wet of optimum and at a corresponding dry unit weight for all the centrifuge tests carried-out in the present study. The thicknesses of model clay barriers used in the present study were 25 mm, which represent 1 m in prototype dimensions.

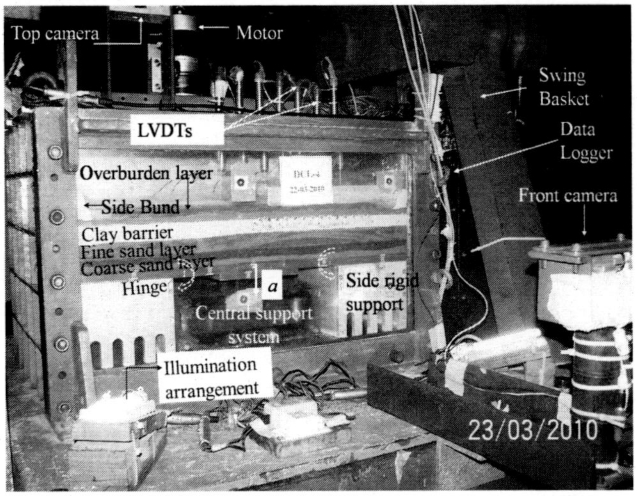

FIGURE 1: Centrifuge model test set-up used in the present study

All models were instrumented with linear variable differential transformers (LVDTs) to arrive at deformation profiles and also with pore pressure transducers (PPTs) on the top to establish breakthrough of water through barriers which was purposely inundated on the top of the barrier. One digital photo camera was used to capture the front elevation of the model and CCD video camera to monitor the proceedings of the tests and presence of water above the barrier during centrifuge tests. After the preparation of model barrier, discrete markers were placed 5 mm below the top surface towards the Perspex side at a spacing of 20 mm centre to centre.

In the case of composite barrier (CB), geomembrane was placed above the clay barrier without any wrinkles and markers were struck to the geomembrane layer also. In the case of fiber reinforced barrier (FRSB), fibers were mixed uniformly in to the soil after mixing thoroughly with half of the desired amount of water. Hand-mixing procedure was adopted to blend the fibers uniformly in to soil. Adequate care was taken to avoid accumulation of fibers. The remaining amount of the water was gradually mixed into the soil-fiber mix. In this way a uniform distribution of fibers throughout the specimen was achieved. Mixing was continued until it was ensured that fibres were evenly distributed throughout the mix by visual examination and the fiber reinforced barrier was prepared as in case of compacted clay barrier.

A water tight seal made-up of a thick bentonite paste was applied all along the sides of the clay barrier and thereafter bunds were constructed along the sides to retain water during centrifuge tests. The overburden pressure of 25 kN/m^2 was created using a sand layer and water, which was placed above the model clay barrier layer. This was achieved by pluviating dry sand of 27 mm thickness with a dry unit weight of 15 kN/m^3 and thereafter it was inundated with water up to 10 mm height above the sand surface. Once the model preparation was completed, the model test package was mounted on the swing basket and all sensors were connected to an onboard data acquisition system. The value of settlement at the centre of the surface of the clay barrier is termed as central settlement, *a*. From here onwards, prototype dimensions are given in the parenthesis. A maximum central settlement of 25 mm (1 m) was induced at a settlement rate of 1 mm/min (36 mm/day).

3. ANALYSIS OF CENTRIFUGE MODEL TESTS

The results were expressed in terms of strain distribution, water breakthrough, cracking pattern and average crack width, w_c and average crack depth, d_c for all the model barriers tested, wherever applicable, which is characterized in terms of settlement ratio and distortion level. The ratio of central settlement, *a* at any stage of deformation to the maximum central settlement, a_{max} is defined here as settlement ratio, a/a_{max}. Distortion level, a/l is defined here as the ratio of central settlement, *a* to the influence length, *l* which is the horizontal distance from the centre of the barrier to a point where induced settlement cease to be negligible. The influence length, *l* is taken in the present study as 200 mm (8 m) uniformly in all centrifuge model tests.

3.1 Deformation Profiles

The deformation profiles of a model clay barrier undergoing settlement can be plotted from the data obtained from the LVDT's provided on the top of the barrier, by performing marker based Digital Image Analysis (DIA) of captured images of front elevation of the barrier at different settlement stages during centrifuge tests. Digital image analysis was carried out using GRAM++ software (GRAM++, 2004).

The coordinates of the markers were obtained from the pictures of the front elevation of the model captured at various stages of the central settlements with reference to the co-ordinate of the permanent markers using map edit module of GRAM++ software. Further explanation of the method is given in Rajesh and Viswanadham, (2011) and Divya et al. (2012a). The deformation profiles obtained for a compacted clay barrier (CCB) is shown in Figure 2. Measured LVDT data was superimposed with deformation profiles obtained from DIA. It was observed that the deformation profiles of composite barriers (CB) and fiber reinforced barriers (FRSB) are quite identical to that of CCB.

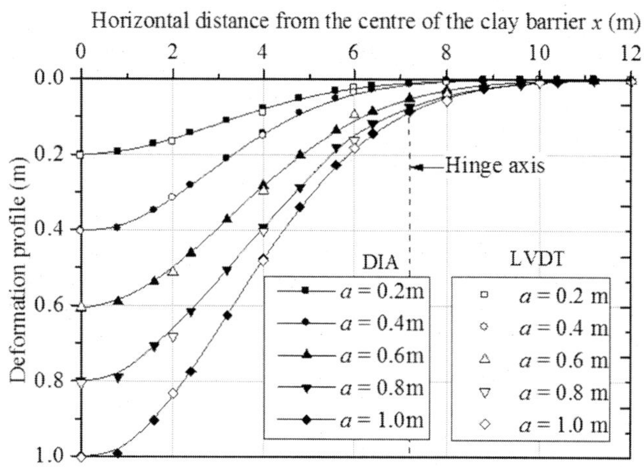

FIGURE 2: Deformation profiles for CCB

3.2 Strain Profiles

The Strain variation along the clay barrier surface can be computed using combined bending and elongation method. The outer fiber strain, $\varepsilon_{of}(x)$ is the strain along the top surface of the clay barrier due to both elongation and bending. The elongation strain $\varepsilon_l(x)$ is the strain due to change in length along the length of the barrier and curvature strain, $\varepsilon_\kappa(x)$ is the strain due to change in curvature of the clay barrier. $\varepsilon_\kappa(x)$ was computed by considering the thickness of the clay barrier, curvature along the length of the clay barrier, and the neutral layer coefficient. In order to compute the geomembrane strain $\varepsilon_g(x)$ along its length at the onset of differential settlement, the geomembrane deformation profiles $w_g(x)$ were obtained from the integral displacement of the markers stuck to geomembrane. The upward/downward thrust distribution from the centre of the clay barrier can be obtained as $q_g(x) = J_g\varepsilon_g(x)\,\kappa_g(x)$. It is explained extensively in Divya et al. (2012a). The strain profiles of a model compacted clay barrier (CCB) is shown in Figure 3.

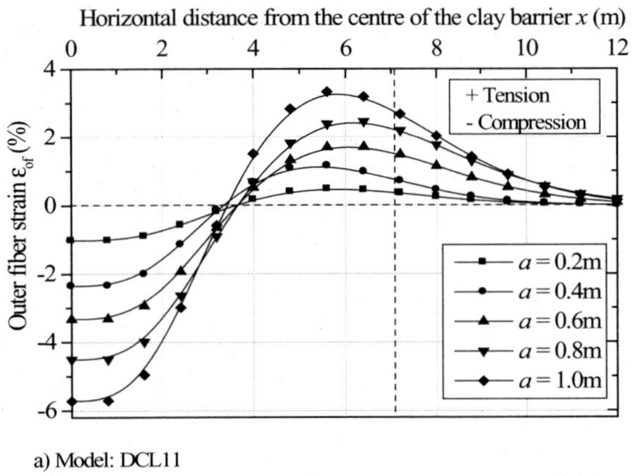

FIGURE 3: Strain profiles for CCB

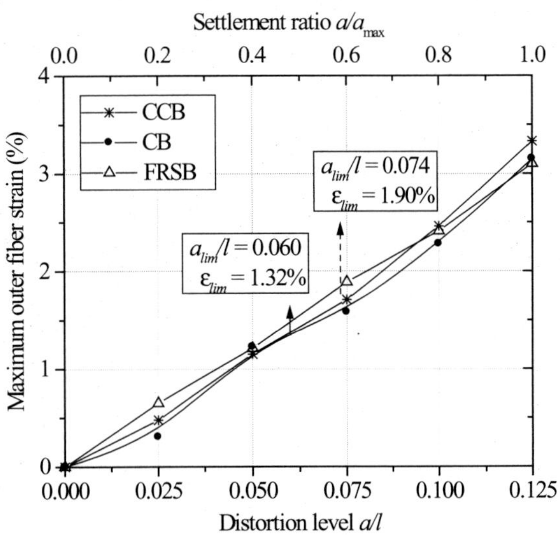

FIGURE 4: Variation of maximum outer fiber strain distribution for CCB, CB and FRSB with a/l and a/a_{max}

The values of maximum outer fiber strain at the zone of maximum curvature were obtained from the variation of $\varepsilon_{of}(x)$ (Figure 3) and plotted against settlement ratios a/a_{max} and distortion levels a/l for all the model barriers, as shown in Figure 4. The maximum outer fiber strain was found to increase for all the model clay barriers, as the distortion levels and settlement ratios. The variation of maximum outer fiber strains with a/l and a/a_{max} is found to be similar for identical thickness of barriers, irrespective of the type of barrier. However there was a considerable delay in strain at water breakthrough ε_{lim} for CB and FRSB compared to CCB as marked in Figure 4. It will be explained in detail in the following section.

3.3 Hydraulic Performance

The hydraulic performance of the clay barrier can be assessed by measuring the infiltration of water through the barrier or through the cracks formed at the onset of differential settlements. This was obtained by measuring the change in height of water using PPT's installed on the top of the barrier surface. The variation of water pressure with time in model dimensions for a CCB is shown in Figure 5.

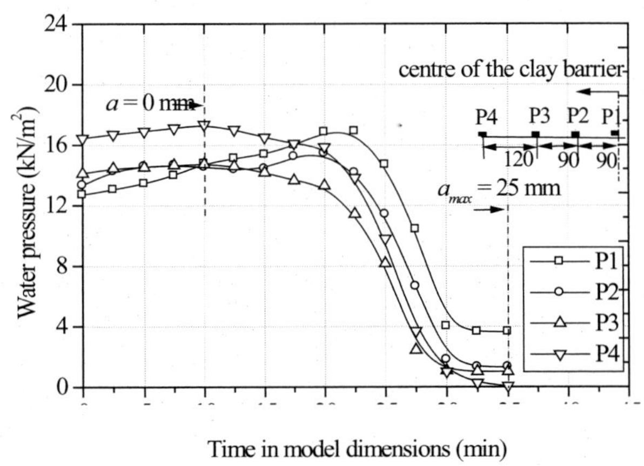

FIGURE 5: Variation of water pressure with time in model dimensions for CCB

Water pressure was almost constant before the commencement of central settlement. After reaching a particular settlement, reduction in water pressure was noticed; which indicates the infiltration of water through the cracks formed in the clay barrier at the onset of differential settlements. But in the case of CB water tend to accumulate at the centre of the deformed barrier without any water breakthrough. In the case of FRSB, reduction in water pressure was observed at higher settlement stages and the water breakthrough was delayed compared to CCB.

Ratio of numerical difference between initial volume of water v_0 and volume of water at any instance v_a to the initial volume of water v_0 is defined as the infiltration ratio IFR. The hydraulic performance was represented in terms of limiting distortion level a_{lim}/l where there is a drastic variation in water pressure. The variation of infiltration ratio, IFR with distortion level a/l and settlement ratio a/a_{max} for CCB, CB and FRSB was shown in Figure 6.

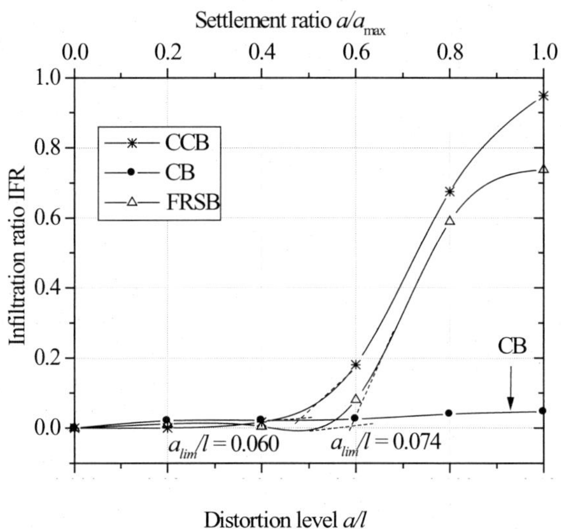

FIGURE 6: Variation of IFR for CCB, CB and FRSB with a/l and a/a_{max}

The a_{lim}/l for CCB and FRSB were observed to be 0.060 and 0.074. The corresponding strain at breakthrough ε_{lim} was marked in Figure 4. It indicates that there is a delay in water breakthrough and hence better sealing efficiency for FRSB compared to CCB. However superior sealing efficiency was observed for CB for which water breakthrough was not observed even after subjected to a maximum of 25 mm (1 m) settlement. This could be attributed to the downward thrust exerted by the geomembrane on the underlying clay barrier. The downward thrust was calculated to be of the order of 22.5 kN/m^2. The method of calculation was explained in Divya et al. (2012a). The observed behaviour of compacted clay barriers and composite barriers were found to be comparable with Plaxis 2D analysis reported by Divya et al. (2012b).

3.4 Cracking Patterns

The cracks were observed at the zone of maximum curvature of the barriers when subjected to differential settlements. Single line cracking was observed for all the barriers. Wide cracks were observed in the case of CCB compared to CB and FRSB. The cracks were extended to the entire width and thickness of the CCB. The maximum average crack width w_c and depth d_c were 2 mm (80 mm) and 25 mm (1 m) respectively for CCB measured at the end of the centrifuge test.

In the case of a CB, cracking was found to be inclined with the horizontal and maximum crack widths w_c were reduced to 0.80 mm (32 mm). However full depth cracks with d_c = 25 mm (1 m) were observed in the case of CB as in case of CCB. It indicates that the superior hydraulic efficiency observed in the case of composite barrier was due to the presence of intact geomembrane over the cracked clay layer. But simply relying solely on the geomembrane alone may be risky due to the chances of defects in geomembranes especially during handling and installation.

In the case of FRSB better resistance to crack propagation was observed compared to CCB and CB. The crack widths and depths were considerably reduced for FRSB. The maximum average crack width w_c and depth d_c were 0.5 mm (20 mm) and 10 mm (400 mm) respectively

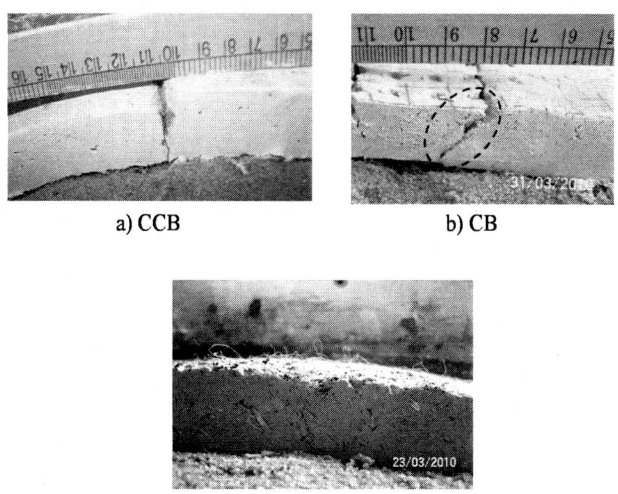

a) CCB b) CB c) FRSB

FIGURE 7: Cross-sections of model barriers at the zone of maximum curvature after centrifuge tests

for FRSB measured at the end of the centrifuge test. When crack appears in a soil-fiber mixture, the fibers in the crack zone impart the entire load. The interfacial friction and bonding between contact area of soil particles and fibers aids in load transfer between them and contributes to better performance and crack resistance of FRSB.

Figure 7 shows the cracking pattern of CCB, CB and FRSB tested in the present study.

Table 1 summarize the results of centrifuge model tests in the present study.

Table 1: Centrifuge Model Test Results

Parameters	*CCB*	*CB*	*FRSB*
d (mm)	25(1000)[a]	25(1000)[a]	25(1000)[a]
a_{max} (mm)	25(1000)[a]	25(1000)[a]	25(1000)[a]
Cracking pattern	Single line	Single line	Single line
w_c (mm)	2 (80)[a]	0.80 (32)[a]	0.50 (20)[a]
d_c (mm)	25(1000)[a]	25(1000)[a]	10 (400)[a]
a_{lim}/l	0.060	-[b]	0.074
$\varepsilon_{of,\ max}$ (%)	3.33	3.15	3.1
ε_{lim} (%)	1.32	-[b]	1.90

d – Thickness of clay barrier ; [a] Prototype values are indicated within the parenthesis; a_{max} - Maximum central settlement; a_{lim}/l - Limiting distortion level; crack width- w_c; crack depth- d_c; max.outer fiber strain- $\varepsilon_{of,\ max}$; strain at breakthrough-ε_{lim};

[b] Not observed.

4. CONCLUSIONS

Centrifuge model tests were conducted in the present study to study and compare the response of compacted clay barriers, composite barriers and geofiber reinforced clay barriers at the onset of differential settlements. From the analysis of centrifuge model tests, following conclusions can be drawn:

Compacted clay barriers found to lose its integrity in terms of sealing efficiency with wide full depth cracks at the onset of differential settlements. The limiting distortion level was 0.060 and the corresponding strain at breakthrough was 1.32%.The maximum average crack width w_c and depth d_c were 2 mm (80 mm) and 25 mm (1 m) respectively for CCB measured at the end of the centrifuge test.

When composite barrier was used, water breakthrough was not observed even after inducing a distortion level of 0.125; though clay barriers found to experience full-depth cracking. Crack widths were reduced to 0.80mm (32 mm). The sealing efficiency was superior compared to compacted clay barriers and fiber reinforced barriers. The performance of composite barriers could be attributed to the downward thrust exerted by the geomembrane on the clay barrier. The maximum downward thrust was found to be 22.5 kN/m^2.

In the case of fiber reinforced barriers, a delay in water breakthrough was observed compared to identical compacted clay barriers. Crack widths and crack depths were found to be reduced compared to identical compacted clay barriers and composite barriers. The limiting distortion level was 0.074 and corresponding strain at breakthrough was found to be 1.90%. The average crack width was 0.50 mm (20 mm) and crack depth was 10 mm (400 mm).

Thus a combination of geomembrane and fiber reinforced barrier could be recommended to landfill cap covers for superior hydraulic performance with a resistance against crack propagation at the onset of differential settlements.

ACKNOWLEDGEMENTS

The authors would like to thank the Centrifuge team at the National Geotechnical Centrifuge Facility of Indian Institute of Technology Bombay, Powai, Mumbai-400076, India for their untiring support throughout the present study.

REFERENCES

Benson, C. H., Daniel, D. E. and Boutwell, G. P. (1999). "Field performance of compacted clay liners." J. of Geotechnical and Geoenvironmental Engineering., ASCE, 125, Issue 5, pp 391- 403.

Bouazza, A., Vangpaisal, T., Abuel-Naga, H. and Kodikara, J. (2008). "Analytical modelling of gas leakage rate through a geosynthetic clay liner-geomembrane composite liner due to a circular defect in the geomembrane". Geotextiles and Geomembranes 26, Issue 2, pp 122–129.

Brachman, R. W. I., Gudina, S., (2008a). "Gravel contacts and geomembrane strains for a GM/CCL composite liner". Geotextiles and Geomembranes 26, Issue 6, pp 448–459.

Chandrasekaran, V. S. (2001). "Numerical and centrifuge modelling in soil structure interaction" Indian Geotechnical journal, 31, Issue 1, pp 30-59.

Divya, P. V. (2012). "Studies on the behaviour of geofiber reinforced clay barriers of landfill covers", Doctoral thesis, Indian Institute of Technology Bombay, India.

Divya, P. V., Viswanadham, B.V.S. and Gourc, J.P. (2012a) "Influence of Geomembrane on the Deformation Behaviour of Clay-based Landfill Covers", Geotextiles and Geomembranes., 34, pp 158-171.

Divya, P. V. and Viswanadham, B.V.S. (2012). "Centrifuge Model Studies on the Behaviour of Composite Covers of Landfills Subjected to Differential Settlements", Geo-congress 2012, Special Publication. 225, ASCE, pp 3467-3476.

Gourc, J. P., Camp, S., Viswanadham, B.V.S. and Rajesh, S. (2010). "Deformation behaviour of clay cap barriers of hazardous waste containment systems: fullscale and centrifuge tests." Geotextiles and Geomembranes. 28, Issue 3, pp 281-291.

GRAM ++ (2004). http://www.csre.iitb.ac.in/gram++/

Heerten, G., Koerner, R. M., 2008. Cover systems for landfills and brownfields. Proceedings of the 24th SKZ-Fachtagung 'Die sichere Deponie', 7–8 February 2008, Wurzburg, pp 1–25

Keck, K. N., Seitz. R. R., (2002). "Potential for Subsidence at the Low-Level Radioactive Waste Disposal Area." INEEL/EXT-02-01154, Idaho National Engineering and Environmental Laboratory, U.S. Department of Energy.

Koerner, R. M., (2001). "Designing with Geosynthetics", Fourth ed. Prentice- Hall, Upper Saddle River, NJ.

Mesbah, A., Morel, J. C., Walker, P. and Kh. Ghavami. (2004). "Development of a direct tensile test for compacted earth blocks reinforced with natural fibers." Journal of Materials in Civil Engineering, ASCE, 16, Issue 1, pp 95-98.

Plé, O., and Lê, T. N. H. (2012). Effect of polypropylene fiber-reinforcement on the mechanical behaviour of silty clay." Geotextiles and Geomebranes, (doi: 10.1016/j.geotexmem.2011,11,004).

Rajesh, S., and Viswanadham, B. V. S. (2011). "Hydro-mechanical behavior of geogrid reinforced soil barriers of landfill cover systems." Geotextiles and Geomembranes, 29, Issue 1, pp 51-64.

Reddy, K. R., Hettiarachchi, H., Parakalla, N. S., Gangathulasi, J. and Bogner, J. E. (2009a). "Geotechnical properties of fresh municipal solid waste at Orchard Hills Landfill, USA." Waste Manage., 29, Issue 2, pp 952–959.

Rowe, R. K. and Sangam, H. P., (2002). "Durability of HDPE geomembranes". Geotextiles and Geomembranes, 20, Issue 2, pp 77-95.

Rowe, R. K., Rimal, S., and Sangam, H. P. (2009). "Ageing of HPDE geomembrane exposed to air, water and leachate at different temperatures". Geotextiles and Geomembranes, 27(2), pp 131–151.

Rowe, R. K. (2011). "Systems engineering: the design and operation of municipal solid waste landfills to minimize contamination of groundwater". Geosynthetics International, 18, Issue 6, pp 391–404.

Take, W. A., Chappel, M. J., Brachman, R. W. I. & Rowe, R. K. (2007). "Quantifying geomembrane wrinkles using aerial photography and digital image processing". Geosynthetics International, 14, Issue 4, pp 219–227.

Viswanadham, B. V. S., Jha, B.K., Pawar, S. N. (2010). "Experimental study on flexural testing of compacted soil beams". Journal of Materials in Civil Engineering, ASCE, 22, Issue 5, pp 460-468.

Viswanadham, B. V. S., Rajesh, S., Divya, P. V. and Gourc, J. P. (2011). "Influence of randomly distributed geofibers on the integrity of clay-based landfill covers: a centrifuge study." Geosynthetics International, 18, Issue 5, pp 255-271.

Sivakumar Babu, G. L., Reddy, K. R. and Chouskey, S. K. (2010). "Constitutive model formunicipal solid waste incorporating mechanical creep and bio degradation induced compression." Waste Management Journal, 30, Issue 1, pp 11–22.

Author Index

Contributors

Askarinejad, A.
Institute for Geotechnical Engineering, ETH Zürich, Zürich, Switzerland
E-mail: amin.askarinejad@igt.baug.ethz.ch

Atmanand, M. A.
National Institute of Ocean Technology, Chennai, India

Babu, G.L.S.
Department of Civil Engineering, Indian Institute of Science, Bangalore 560012, India
E-mail: gls@civil.iisc.ernet.in

Babu, S. M.
National Institute of Ocean Technology, Chennai, India

Boominathan, A.
Department of Civil Engineering, Indian Institute of Technology Madras, Chennai, India
E-mail: boomi@iitm.ac.in

Chazelas, J.L.
IFSTTAR, Nantes, France

Choi, J.H.
Hyundai Engineering and Construction, Co., Ltd., Seoul, Republic of Korea
E-mail: jhchoi@hdec.co.kr

Choo, Y.W.
Department of Civil and Environmental Engineering, KAIST, Daejeon, Republic of Korea
E-mail: ywchoo@kaist.ac.kr

Chouksey, S.K.
Department of Civil Engineering, Indian Institute of Science, Bangalore 560012, India
E-mail: sandeepkc@civil.iisc.ernet.in

Divya, P. V.
Department of Civil Engineering, Indian Institute of Technology, Bombay, India
E-mail: divyaputhumana@rediffmail.com

Escoffier, S.
IFSTTAR, Nantes, France

Gaudin, C.
Centre for Offshore Foundation Systems, The University of Western Australia, Australia
E-mail: christophe.gaudin@uwa.edu.au

Gauffre, C.
Actidyn Systèmes SA, Elancourt, France

Gautray, J.N.F.
Institute for Geotechnical Engineering, ETH Zürich, Zürich, Switzerland
E-mail: jean.gautray@igt.baug.ethz.ch

Ghanekar, R.K.
Institute of Engineering and Ocean Technology (IEOT), Oil and Natural Gas Corporation Ltd., Panvel, Navi Mumbai
E-mail: rkghanekar@gmail.com

Ghiassian, H.
Department of Civil Engineering, Iran University of Science and Technology, Tehran, Iran

Ghosh, C.
Geohazards Division, National Institute of Disaster management, Ministry of Home Affairs, Governement of India
E-mail: cghosh24@gmail.com

Ha, J.G.
Civil and Environmental Engineering, KAIST, Daejeon, Korea
E-mail: jgha87@kaist.ac.kr

Haigh, S. K.
Department of Engineering, University of Cambridge, Cambridge, UK
E-mail: skh20@cam.ac.uk

He, L.Q.
Department of Geotechnical Engineering, Tongji University, Shanghai, China

Hou, Y.J.
Department of Geotechnical Engineering, China Institute of Water Resources and Hydropower Research (IWHR), Beijing, China
E-mail: Houyj@iwhr.com

Iai, S.
Disaster Prevention Research Institute, Kyoto University, Japan

Juneja, A.
Indian Institute of Technology Bombay, Powai, Mumbai, India

Jung, U.S.
Infrastructure Technology Center, K-water Institute, Daejeon, Korea
E-mail: usjung@kwater.or.kr

Kim, D.J.
Hyundai Engineering and Construction, Co., Ltd., Seoul, Republic of Korea
E-mail: djkim@hdec.co.kr

Kim, D. S.
Department of Civil and Environmental Engineering, KAIST, Daejeon, Republic of Korea
E-mail: dskim@kaist.ac.kr

Kim, J. H.
Department of Civil and Environmental Engineering, KAIST, Daejeon, Republic of Korea
E-mail: zlawogus@kaist.ac.kr

Kim, N. R.
Infrastructure Technology Center, K-water Institute, Daejeon, Republic of Korea
E-mail: namryong@kwater.or.kr

Kim, S.
Department of Civil and Environmental Engineering, KAIST, Daejeon, Republic of Korea
E-mail: kimsurin@kaist.ac.kr

Kim, S. H.
Civil and Environmental Engineering, KAIST, Daejeon, Republic of Korea
E-mail: olive2020@kaist.ac.kr

Kim, S.H.
Hyundai Engineering and Construction, Co., Ltd., Seoul, Republic of Korea
E-mail: shjee@hdec.co.kr

Kitazume, M.
Department of Civil and Environment Engineering, Tokyo Institute of Technology, Tokyo, Japan
E-mail: kitazume@cv.titech.ac.jp

König, D.
Institute for Foundation Engineering, Soil and Rock mechanics, Ruhr-Universität Bochum, Bochum, Germany
E-mail: diethard.koenig@rub.de

Krishna, A.M.
Department of Civil Engineering, Indian Institute of Technology Guwahati, Guwahati, India
E-mail: amurali@iitg.ernet.in

Laue, J.
Institute for Geotechnical Engineering, ETH Zürich, Zürich, Switzerland
E-mail: jan.laue@igt.baug.ethz.ch

Leung, C. F.
Department of Civil and Environmental Engineering, National University of Singapore, Singapore
E-mail: ceelcf@nus.edu.sg

Ling, H.
Columbia College, Columbia University, New York, USA

Ling, H. I.
Department of Civil Engineering & Engineering Mechanics, Columbia University, New York, USA
E-mail: Ling@civil.columbia.edu

Ma, X. F.
Department of Geotechnical Engineering, Tongji University, Shanghai, China
E-mail: xf.ma@tongji.edu.cn

Madhav, M. R.
Department of Civil Engineering, Indian Institute of Technology Hyderabad, India
E-mail: madhavmr@gmail.com

Mathur, V. K.
General Manager, Ash Management, NTPC Ltd., A-11,NFL Premises, Sector-24, Noida, India
Email: vkmathur@ntpceoc.co.in

Morales, W. F.
Institute for Geotechnical Engineering, ETH Zürich, Zürich, Switzerland
E-mail: ferney.morales@igt.baug.ethz.ch

Morikawa, Y.
Geotechnical Engineering Division, Port and Airport Research Institute, Japan

Nishimura, S.
Division of Field Engineering for the Environment, Faculty of Engineering, Hokkaido University, Japan

Park, D. S.
Infrastructure Technology Center, K-water Institute, Daejeon, Korea
E-mail: fulgent@kawter.or.kr

Prasad, C. H.
Department of Civil Engineering, Indian Institute of Technology Hyderabad, India
E-mail: hariprasadiith@gmail.com

Raisinghani, D.V.
Textile Manufactures Department, Veermata Jijabai Technological Institute, Mumbai, India

Rajabian, A.
Department of Civil Engineering, Science and Research Branch, Islamic Azad University, Fars, Iran

Rajesh, S.
Department of Civil Engineering, Indian Institute of Technology Kanpur, Kanpur, India
E-mail: hsrajesh@iitk.ac.in

Ramadass, G. A.
National Institute of Ocean Technology, Chennai, India

Rama Krishna, K. S.
Head-Foundation and Geotechnical Engineering Cell, Larsen & Toubro Limited – Construction Division, Chennai, India
E-mail: ksrk@lntecc.com

Ramanamoorthy, M.V.
National Institute of Ocean Technology, Chennai, India

Rames, D.
Actidyn Systèmes SA, Elancourt, France

Ramesh, R.
National Institute of Ocean Technology, Chennai, India

Ravichandran, V.
National Institute of Ocean Technology, Chennai, India
E-mail: vijaya@niot.res.in

Rayabharapu, V.
Department of Civil Engineering, Indian Institute of Technology Hyderabad, India
E-mail: vkraya@gmail.com

Robinson, R. G.
Department of Civil Engineering, Indian Institute of Technology Madras, Chennai, India
E-mail: robinson@iitm.ac.in

Sabard, P.
Actidyn Systèmes SA, Elancourt, France

Saran, R.K.
Department of Civil Engineering, Indian Institute of Technology Bombay, Mumbai, India

Saride, S.
Department of Civil Engineering, Indian Institute of Technology Hyderabad, India
E-mail: sireesh@iith.ac.in

Savoikar, P. P.
Department of Civil Engineering, Government Polytechnic, Bicholim, Goa, India
E-mail: psavoikar@gmail.com

Shin, D. H.
Infrastructure Technology Center, K-water Institute, Daejeon, Republic of Korea
E-mail: shindh@kwater.or.kr

Soundara, B.
Department of Civil Engineering, Indian Institute of Technology Madras, Chennai, India

Subramanian, RM.
Department of Civil Engineering, Indian Institute of Technology Madras, Chennai, India
E-mail: subbu_geotechengg@yahoo.co.in

Sun, Q.L.
Department of Geotechnical Engineering, China Institute of Water Resources and Hydropower Research (IWHR), Beijing, China

Thakur, V.
Faculty for Engineering and Natural Science, Aalesund University College, Norway
Geotechnical Division, Norwegian Directorate of Roads, Norway
E-mail: vikas.thakur@vegvesen.no

Tobita, T.
Disaster Prevention Research Institute, Kyoto University, Japan
E-mail: tobita.tetsuo.8e@kyoto-u.ac.jp

Umashankar, B.
Department of Civil Engineering, Indian Institute of Technology Hyderabad, India
E-mail: buma@iith.ac.in

Vedpathak, S.
Department of Civil Engineering, Indian Institute of Technology Hyderabad, India
E-mail: vedpathaksd@gmail.com

Viswanadham, B.V.S.
Department of Civil Engineering, Indian Institute of Technology Bombay, Powai, Mumbai, India
Email: viswam@civil.iitb.ac.in

Wang, J.
Department of Engineering, University of Cambridge, Cambridge, UK

Wen, Y. F.
Department of Geotechnical Engineering, China Institute of Water Resources and Hydropower Research (IWHR), Beijing, China

Wu, X. Y.
Shanghai Harbour Engineering Research Institute, Shanghai, China

Zhang, C. L.
Department of Geotechnical Engineering, Tongji University, Shanghai, China

Zhang, X.
Shanghai Harbour Engineering Research Institute, Shanghai, China
E-mail: xf.ma@tongji.edu.cn

Zhang, X. D.
Department of Geotechnical Engineering, China Institute of Water Resources and Hydropower Research (IWHR), Beijing, China